TECHNIQUES OF

Climatology

TECHNIQUES OF
Climatology

E. T. STRINGER

UNIVERSITY OF BIRMINGHAM

W. H. FREEMAN AND COMPANY

San Francisco

Printed in the United States of America.

Library of Congress Catalog Card Number: 74-128094

International Standard Book Number: 0-7167-0250-9

1 2 3 4 5 6 7 8 9

To the memory of my parents,

Edward H. Stringer (1888–1957)
Lilian M. Stringer (1891–1954)

without whose encouragement and self-sacrifice this book
would never have been commenced;
and to

Gloria, my wife,

without whose devotion and hard work it
would never have been completed.

Contents

Preface

The science of climatology makes use of three groups of techniques: those of the meteorologist, those of the mathematical statistician, and those of the geographer. My purpose in this book is to present a comprehensive review of these techniques. The reader is assumed to be already familiar with the nature and basic principles of climatology, as set out in my companion work, *Foundations of Climatology*.

This book is not elementary in scope or coverage, but rather constitutes a manual for readers who have mastered the companion volume, and who wish to study further, for professional reasons. Such readers will include: persons taking advanced under-graduate and introductory postgraduate courses in climatology in university depart-ments of geography, meteorology and earth sciences; entrants to state meteorology services taking courses in both climatology and meteorology; and specialists in other fields—physical and biological sciences, engineering, medicine, agriculture, forestry, architecture, planning, public health, etc.—who have, for professional reasons, to turn themselves into climatologists in order to solve a specific problem.

The fundamental notion behind *Techniques of Climatology* is that climate con-stitutes a physical problem. In this, it contrasts strongly with the widely held view of climate as merely "average weather." The important point is that climate as a physical problem involves study of both meteorology and geography; i.e., investigation of both the atmosphere and the Earth's surface is involved.

The meteorologist has long been aware of climate as a physical problem, but he has, until very recently, confined himself to investigations of its atmospheric side. His current incursions into the geographical side of the climatic problem are often geographically naive, and sometimes—especially when "weather control" is involved —positively dangerous. For their part, geographers in general have shown little awareness of the many contributions made to climatology by mathematicians, physical scientists and engineers. The picture of world climates presented by many geography teachers not only lacks intellectual excitement or rigor, but is often scientifically inaccurate. There is a great need for geographers and meteorologists to work together in the study of climate, and I hope that *Techniques of Climatology* will assist this meeting, by showing how each type of specialist has a valid viewpoint *provided* he takes the trouble to understand that of the other.

Although climatology deals with the "normal" behavior of the atmosphere, and as such is clearly the province of the atmospheric scientists (as meteorologists now term themselves), it also deals with local weather and climate. This latter study requires a knowledge of the local landscape and of the basic principles of geomorphology, pedology, and biogeography. Only geographers have formal training both in the rudiments of these sciences and in elementary meteorology. Thus it is to the geography departments of the universities and colleges that one should look for potential climatologists, rather than to the departments of meteorology, physics, or engineering. Although *Techniques of Climatology* is intended primarily as a textbook covering those parts of climatology courses in university geography departments that deal with techniques, I hope it will also guide geography teachers, who often foster in the young an interest in meteorology without at the same time instilling an appreciation of just how scientific this study has now become. As emphasized by Gordon Manley some years ago, properly prepared geographers have much to offer meteorology.* The value of their contribution is proved not only by the work of Manley himself on British weather and climate, but also by the efforts of C. Warren Thornthwaite and F. Kenneth Hare, and their colleagues, in the vastly different fields of micrometeorology and the circulation of the Arctic stratosphere, respectively. It cannot be too strongly emphasized that, although climatology is a geophysical science, whose initial problems and ultimate aims are essentially geographical, its methods and techniques are those of the physicist and mathematician rather than the geographer. If this book helps to provide geography teachers and students with an appreciation of the basic mathematic notation and a few of the physical techniques, both theoretical and practical, that are necessary to follow modern climatology, it will have served a good purpose.

A former President of the Royal Geographical Society once stated that, since geographical exploration in the classic sense is now almost at an end, geographers must extend their interests to new fields if geography is to survive as an active science.† One solution is to extend investigation of the Earth's surface back in time; an alternative is the primary exploration of outer space, and, ultimately, of other planets. Today's pioneer geographical explorers are blazing new trails in the stratosphere and above, rather than on the ground. Climatologists realized the importance of this new development, and the world's first conference on aerospace climatology was held at Washington D.C., in December 1961.‡ Geography teachers have a duty to acquaint

---

* G. Manley, " The Geographer's Contribution to Meteorology," *QJRMS*, 73 (1947), 1.

† Sir R. Priestley, *Geog. J.*, 128 (1962), 257.

‡ See *BAMS*, 43 (1962) 275.

their pupils with these possibilities. I hope that *Techniques of Climatology* may be a suitable primer for arousing the interest of future geographical explorers of our atmosphere and beyond, whether they will be called geographers, climatologists, meteorologists, aerologists, or aerospace scientists.

ACKNOWLEDGMENTS

As I did for *Foundations of Climatology*, I have to thank my former teachers in the Meteorological Office, almost two decades ago, who kindled my interest in the rigorous approach to climatology. In particular, I thank Professor A. F. Jenkinson for considerable guidance in the field of mathematical statistics, and Mr. P. J. Meade and Mr. K. H. Smith for stimulating thoughts in the fields of dynamic meteorology and synoptic meteorology, respectively. Special thanks are due to Professor David L. Linton, Head of the Department of Geography in the University of Birmingham, for his interest in the book at all stages in its preparation. For many years, Dr. Gordon T. Warwick, Reader in the Department of Geography, has been a most valued and constructive critic. The late Professor R. H. Kinvig, and Mrs. B. Eckstein, also helped me in the initial phases of my climatological career.

My climatological ideas have profited from discussions held with American and Canadian climatologists during a period I spent at McGill University in 1964 as Visiting Professor of Climatology. For providing the opportunities for these discussions, I have to thank Dr. F. Kenneth Hare, Dr. Trevor Lloyd, Professor Theo. L. Hills, Dr. H. E. Landsberg, Mr. Morley K. Thomas, and Dr. John R. Mather.

No little word of thanks is due my publishers, for their patience amid the innumerable delays that have attended the writing of the books, and for their care in preparing it for publication; the efforts of Mr. Aidan A. Kelly, of the Editorial Department, are worthy of special mention. I must also acknowledge with gratitude the care with which Professor Gordon Manley read and commented on the book in its early stages.

The labors of typing the book have been shared between a number of ladies: special thanks are due to Mrs. R. Priestman, Mrs. M. L. Stones, Mrs. J. Needham, Miss D. Morgan, and my wife. Indeed, I must record my sincere appreciation of my wife's forbearance of the last six years, during which the demands of two small sons and a book-finishing husband have been, to say the least, somewhat trying.

Birmingham, England,   *E. T. Stringer*
July 1971

TECHNIQUES OF
Climatology

PART I

The Basic Techniques

CHAPTER 1

Observing the Weather

Although both meteorology and climatology are becoming increasingly theoretical, study of weather and climate must always be based on observed facts. The amateur meteorologist is usually interested only in facts, and the academic climatologist, however abstract his theories, has always to test them by means of actual data.

Weather facts may be current or historical. The main problem of current weather information is to ensure speedy dissemination, that of historical information, to ensure as wide a geographical coverage as possible, extending as far back in time as prudent, with homogeneity of data.

Historical Weather Data

The primary sources of historical ("climatological") data are the serial publications of state meteorological services. For Britain, these include: (a) The *Daily Weather Report*, containing data for the preceding day in the form of coded reports at six-hour intervals for 55 land stations and eight weather ships, plus six-hour weather charts; (b) the *Daily Aerological Record*, published several days behind, containing coded upper-air reports at 12-hour intervals for nine or ten land stations and two weather ships, plus charts for midnight covering most of the northern hemisphere; (c) the *Monthly Weather Report*, published several months behind, containing monthly means and frequencies for more than 500 climatological stations in Great Britain and Northern Ireland, plus charts of mean pressure, temperature and sunshine, total rainfall, and depression movements throughout the month; (d) *British Rainfall*, an annual volume published several years behind, containing monthly rainfall totals for the 5,000-odd rainfall stations in Britain, plus maps showing such features as percentage differences from average for specific months or years.

Some publications contain data for the whole world. The most useful of these is *World Weather Records*, which, since 1927, has published data for each decade.[1] These volumes contain year-by-year records of monthly values of temperature, pressure, and precipitation. The 1959 volume includes data for 700 stations, including weather ships in the North Atlantic and North Pacific, Antarctic stations, and remote stations in the Southern Ocean. Such publications are expensive and time-consuming to produce, because much statistical checking is necessary before the data can be regarded as homogeneous.

Since daily weather reports must be prepared very rapidly for prompt publication, they contain occasional errors in analysis. Maps prepared by more leisurely analysis are therefore valuable for research, and several series of such maps have been published, notably, the *Northern Hemisphere Synoptic Weather Map Series* of the United States Weather Bureau. This series covers the period 1899 to the present, and consists of monthly volumes of sea-level charts for 1:00 P.M. G.M.T. each day, plus 500-mb charts for 3:00 P.M. G.M.T. (3:00 A.M. at first) from 1945 onward. The time involved in assembling and screening the data for the entire hemisphere means that the current series is usually published a year behind.[2]

Current Weather Data

Much information can of course be obtained from historical weather data, but real knowledge of the weather can only come from trying to understand its day-to-day vicissitudes, for which one must have current data.

Apart from plain-language reports in newspapers and radio or television broadcasts, which are rarely detailed or recent enough for the serious student, there are five main sources of current data: specialized radio broadcasts (for shipping, in particular), wireless-telegraphy (W/T) and radio-telephony (R/T) broadcasts, and radioteleprinter (RTTY) and facsimile transmissions. All these weather messages are designed to international specifications and cater to a wide variety of interests and types of receiving apparatus, but nearly all of them require some familiarity with the international meteorological codes.

The first code was introduced before the First World War by the Permanent Committee of the International Meteorological Organization; the committee met every five or six years, between the congresses of the full organization. The code was for the exchange of weather reports by wireless telegraphy, and consisted of groups of five figures, because each such group was charged as one word.

The code had two primary groups, represented symbolically by BBBDD FFWTT in the British form. BBB represented barometric pressure, DD wind direction, FF wind speed, W weather, and TT temperature. Two supplementary groups, T'T'RRR MMmmS, represented wet-bulb temperature, rainfall, maximum (MM) and minimum (mm) temperatures, and state of the sea, respectively. The continental form had three figures for temperature, two for rainfall, and only one for wind force.[3]

During the First World War, this code was found to be very inadequate, and, largely on the basis of British proposals, new codes were drawn up. During the Second World War, inadequacies again became apparent, and at the Paris meeting of the IMO in 1946, new codes were discussed. At Washington in 1953, the *Commission for Synoptic Meteorology* of the World Meteorological Organization (WMO) standardized the new code for surface observations, and new international codes for upper-air reports and reports from transport aircraft and meteorological reconaissance aircraft were drawn up. Major changes in the code for surface reports came into force on January 1, 1955, and modifications in the upper-air, analysis, and forecast codes on January 1, 1960.

The present codes still largely adhere to five-figure groups, each identified by a code word. Thus SYNOP indicates surface data for a land station, SHIP denotes surface data from a moving or stationary ship, TEMP indicates upper-air reports from land stations, and TEMP SHIP upper-air data from a ship. MESRAN data are collective summaries of upper-air reports, and CLIMAT and CLIMAT TEMP are summaries of monthly mean surface and upper-air data, respectively.

Codes are also available to describe weather charts, both actual and predicted (prognostic): these are the *International Analysis Codes.* The term BARATIC means the actual surface-pressure chart, expressed in code form. PREBARATIC is the code form of the prognostic surface-pressure chart, UCONAL the code for the actual upper-air chart, and PRONTOUR the coded prognostic upper-air chart. The respective codes are ANAL (S), used for baratic and prebaratic charts, and ANAL (U) for uconals and prontours.

Familiarity with the SYNOP and SHIP codes is essential for maximum use of the *Daily Weather Report,* and the TEMP, TEMP SHIP, and PILOT and PILOT SHIP (for upper winds only) codes must be understood before using the *Daily Aerological Record.* (For further details of these codes, see pp. 6–8.)

Reception of Weather Broadcasts

In accordance with WMO technical regulations, each meteorological region of the world must provide three types of broadcast: (a) *continental,* transmitting data for the region, and capable of world-wide reception; (b) *subcontinental,* transmitting data for part of a region, for reception anywhere in that region; and (c) *territorial,* transmitting data for the territory occupied by one member country of the region, for reception at one or more subcontinental centers in that region. For Region IV (Europe) of the WMO, there are four subcontinental centers: London (or the Meteorological Office transmitting station), Paris, Rome, and Moscow. The continental broadcast (call sign GFL) is from London, as is the subcontinental broadcast (GFA), the latter also serving as the United Kingdom territorial broadcast.

Until recently, most meteorological broadcasts were by W/T, using the Morse code. Wavelengths were secured by IMO delegates to International Telecommunications Conferences at Washington (1927), Madrid (1932), and Cairo (1938); since the Second World War, these have been extensively revised.

Transmissions to outstations within a Weather Service are usually by land-line teleprinter, with links to other countries. For example, the International Meteorological Teleprinter Network in Europe (IMTNE) was developed after the Second World War on a quadrilateral pattern, based on London (Dunstable), Paris, and the headquarters

*Sections of the International Synoptic Weather Message Codes
Normally Published Daily by State Meteorological Services
for Climatological Purposes*[a]

SURFACE OBSERVATIONS

Land stations (SYNOP code)

Mandatory: IIiii Nddff VVwwW PPPTT $NhC_LhC_MC_H$ TdTdapp RRTxTxE
sss (day; RRTnTnE TgTg at night)

Optional: 8NsChshs (repeated if necessary)

Ocean stations (SHIP code)

Mandatory: 99LaLaLa QcLoLoLoLo YYGGiw Nddff VVwwW PPPTT
$NhC_LhC_MC_H$ Dsvsapp   0TsTsTdTd   1TwTwTwtT   3PwPw
HwHw

Optional: 8 NsChshs (repeated if necessary).

AEROLOGICAL OBSERVATIONS

Full upper-air reports

Land stations (TEMP code)
Standard isobaric surfaces: YYGGgg IIiii hhh *TTTdTd*[b]
Significant levels: IIiii *PPPTTTdTd*[b] PtPtPt HtHt TpTp

Ocean stations (SHIP TEMP code): as for land stations, but preceded by
YQLaLaLa LoLoLoGG

Upper wind reports only

Land stations (PILOT code)
Standard levels: YYGGgg *ddff*[c]
Additional levels: *PPP ddff* [c]dmdmfmfm HmHmHmHm

Ocean stations (PILOT SHIP code): as for land stations, but preceded by
YQLaLaLa LoLoLoGG

---

[a] These are the standard forms as agreed on internationally. Individual state publications
may depart from these in minor details.

[b] The group in italics is repeated for each level:
surface, 850 mb, 700 mb, 500 mb, 400 mb, 300 mb, and 200 mb; frequently, also for 150
mb and 100 mb; occasionally, also for 70 mb, 50 mb, 30 mb, 20 mb, and 10 mb.

[c] The group in italics is repeated for each level:
surface, 1,500 meters, 2,100 m, 3,000 m, 4,200 m, 5,400 m, 7,200 m, 9,000 m, 10,500 m, and
12,000 m; frequently, also for 13,500 m and 15,900 m; occasionally, also for 18,300 m,
30,700 m, 23,700 m, 26,400 m, and 30,900 m.

Key to Code Symbols

a	characteristic of barometric tendency[d]
C	type of cloud[d]
C_L	form of low cloud[d]
C_M	form of medium cloud[d]
C_H	form of high cloud[d]
dd	wind direction (approximately, in degrees)
dmdm	wind direction at level of maximum wind (in hundreds and tens of degrees)
Ds	direction of movement of ship[d]
E	state of ground[d]
ff	windspeed (in knots)
fmfm	windspeed at level of maximum wind (in knots)
GGgg	time of observation in hours and minutes G.M.T.
h	height above ground of base of lowest cloud[d]
hhh	height of standard isobaric surface above mean sea level, in geopotential meters below 500 mb, in decameters at 500 mb and above
hshs	height of base of significant cloud (figures 01–50, times 30, gives height in meters; figures 56–80, times 3 gives height in hundreds of meters)
HtHt	height of tropopause (in geopotential hectometers)
HwHw	height of waves on ocean surface (in units of 0.5 meter)
HmHmHmHm	height of level of maximum wind (in decameters)
iii	station at which observation was made[d]
iw	wind indicator (ship observation)[d]
II	country in which observation was made[d]
LaLaLa	latitude of ship in degrees and tenths
LoLoLo	longitude of ship in degrees and tenths (for aerological reports)
LoLoLoLo	longitude of ship in degrees and tenths (for surface reports)
N	total amount of cloud in eighths
Nh	fraction (eighths) of celestial dome covered by C_L (or by C_M if no C_L)
Ns	amount (eighths) of individual (significant) cloud layer or mass
pp	barometric tendency (in tenths of millibars)
PPP	mean sea-level barometric pressure (last three figures, millibars and tenths)
PwPw	period of waves on ocean surface (seconds)
PtPtPt	pressure at the tropopause (millibars)
Q	octant of the globe[d]
Qc	quadrant of the globe[d]
RR	rainfall (figures 01–55, rainfall in millimeters)[d]

---

[d] See M.O. Publication no. 510, *Handbook of Weather Messages,* parts I, II, and III (London, 1963), for full specifications of the codes for these items, plus instructions for plotting the data on synoptic charts.

Key to Code Symbols (Continued)

sss	duration of sunshine (hours and tenths)
tT	tenths figure of air temperature
TT	air temperature (in whole degrees Celsius)
TdTd	dewpoint temperature (in whole degrees Celsius)
TgTg	night minimum temperature on grass (in whole degrees Celsius)
TnTn	night minimum air temperature (in whole degrees Celsius)
TpTp	temperature at the tropopause (in whole degrees Celsius)
TsTs	difference between air temperature and sea temperature (in whole degrees Celsius)
TxTx	day maximum temperature (in whole degrees Celsius)
TwTwTw	sea surface temperature in tenths of a degree Celsius
ww	weather at time of observation[d]
W	weather during the six hours immediately preceding the observation[d]
vs	speed of ship (knots)
VV	visibility (figures 01–50, horizontal visual range in tenths of kilometers)[d]
YY	day of the month on which the observation was made[d]
0	indicator figure for TsTsTdTd group for SHIP data
1	indicator figure for TwTwTwtT group for SHIP data
3	indicator figure for PwPwHwHw group for SHIP data
8	indicator figure for NsChshs group for SYNOP and SHIP data.

---

[d] See M.O. Publication no. 510, *Handbook of Weather Messages*, parts I, II, and III (London, 1963), for full specifications of the codes for these items, plus instructions for plotting the data on synoptic charts.

of the British and American forces in the occupied zones of Germany at Bad Eilsen and Frankfurt a.M. In 1952, it was changed to a triangular pattern, based on Dunstable, Paris, and Frankfurt; and in 1957 the network was extended to link with the land-line teleprinter network in east and central Europe (based on Prague, Potsdam, Budapest, Warsaw, and Moscow) via duplex circuits between Frankfurt and Prague, and between Frankfurt and Potsdam.[4]

Although land-line teleprinter transmissions are private, and intended solely for the Weather Services concerned, the radio broadcasts made under WMO specifications obviously cannot be. In Britain until 1961, it was possible to receive the GFA sub-continental broadcast with a very simple short-wave receiver. Although it was not possible to take the message down directly, owing to the high speed of the automatic transmission (15 five-figure groups per minute), a variable-speed tape recorder was of great use, or a simple automatic-recording device could be made very inexpensively.[5]

In 1956, when many W/T subcontinental broadcasts were nearing saturation, a new WMO requirement stipulated that the 150-mb and 100-mb levels should be included in upper-air messages. Starting January 1, 1956, the continental broadcast from New York for Region IV was by radio-teleprinter, and Region V also adopted this method for its continental broadcast from Australia. RTTY transmission is almost three times faster than W/T Morse, and is easily adapted to automatic tape-relay operation, so that rebroadcast is simple. Therefore the WMO congress at New Delhi (1958) recommended that all subcontinental and continental broadcasts should change from W/T to RTTY not later than the end of 1962. In Britain, GFA changed over by the middle of 1961, after over 30 years of Morse broadcasts.

For private persons or organizations who wish to receive all available current data, RTTY equipment is essential. New equipment is expensive, but government-surplus radios and teleprinters can sometimes be bought relatively cheaply, and details of several circuits have been published that enable workable installations to be constructed.[6] For those not wishing to go to this trouble or expense, however, there are numerous other broadcasts that can be received with ordinary radio receivers.

A WMO regulation specifies that regions must transmit weather messages for shipping that include: (a) in plain language, storm warnings, synopses of weather conditions in specified forecast areas, including positions of fronts and pressure patterns and their expected movements, existing winds and visibilities, and 12- or 24-hour forecasts; (b) reports for land stations and ships using the standard codes in W/T Morse transmissions, together with analyses that will enable the recipient to construct both prebaratics and baratics.

In conjunction with the International Civil Aviation Organization (ICAO), the WMO stipulated that an extensive system of ground-to-air meteorological broadcasts should be set up after the Second World War, and these are very useful for the study of current weather. There are two types of broadcast: (a) short-wave (AIR RADIO) broadcasts from high-power transmitters covering a large area, and (b) VHF (VOLMET) broadcasts covering much smaller areas and giving details of local conditions.

In addition, W/T VOLMET broadcasts on short-wave, carrying for much greater distances, provide weather reports and aerodrome forecasts every hour, using the AERO and TAF codes. The AERO code is a shortened version of the SYNOP code, while the TAF code includes forecasts of type, amount, and height of base of clouds, windspeed and wind direction, visibility, weather, and "supplementary phenomena," including fronts and rapid changes.

If current charts are desired, but not the complete data in code form, *facsimile* broadcasts are the ideal. These include transmissions of complete charts, actual and forecast, both surface and upper-air, tephigrams, and upper winds in tabular form. Apart from the expense of the receiver, a disadvantage with facsimiles is the time required for both transmitting and receiving a given chart, in comparison with tele-printer transmission. For example, the 12:00 noon G.M.T. surface chart for Britain will not be received complete by facsimile until about 5:30 P.M. G.M.T., but would be available by 1:00 P.M. G.M.T. if it were drawn up by hand from data in the 12:15 P.M. RTTY broadcast.[7]

For details of radio, radioteleprinter, and facsimile broadcasts, see the official handbooks.[8] Specially designed charts are available that enable weather maps for specific areas to be constructed very rapidly from the weather broadcasts for shipping that may be received on ordinary radio sets.[9]

The International Meteorological Observing Network

The first step toward an international network was made in 1853, when representatives (mainly naval officers) of ten nations met in Brussels.[10] This meeting followed British proposals that an international conference should be convened to consider a uniform system of meteorological observation on land, and more comprehensive American proposals that the navies of all the maritime nations should cooperate in making ship observations conform to a single system. No definite agreement was reached in Brussels concerning standard units or instruments, but a general preference for the mercury barometer instead of the aneroid was expressed, and a standard form of ship's log was introduced.

Informal discussions were held at Leipzig in 1872, but a large-scale conference of heads of observatories and institutes, held in Vienna in 1873, marked the real founding of International Meteorology. It was decided to hold congresses every five or six years, and a permanent committee of seven members was established to act between congresses. In addition, a number of technical recommendations were made. First, time units were established: the mean solar day at the place of observation, ending at midnight; the calendar month; the civil year; and Dove's 73 five-day periods for temperature means. Second, uniform combinations of observing hours were recommended: either 7:00 A.M., 1:00 P.M., and 9:00 P.M., or 9:00 A.M., 3:00 P.M., and 9:00 P.M. Third, international weather symbols for hydrometeors were introduced. Fourth, the English designations for wind direction (N, S, E, and W) were agreed on for general use. Fifth, five types of meteorological station were recognized: central office, central station, and stations of the first, second, and third order.

A *central office* or *central institute* is the chief office in each country, entrusted by its government with the management, collection, and publication of observations for that country. A *central station* is a subordinate center for the management and collection of data for part of a country. A *first-order station* is an observatory making hourly or self-recording observations, but not collecting data from other stations. A *second-order station* makes regular and complete observations; a *third-order station* makes only some of all the standard observations.

At the Utrecht Conference of Directors of Institutes and Observatories in 1878, the International Meteorological Committee (IMC) was set up. This consisted of 17 directors of the meteorological services of the various member countries, but it did not have the backing of their governments. The IMC met every three years, and did much valuable work, for example, the *Réseau Mondial*. In 1907, Teisserenc de Bort proposed the appointment of a special IMO committee to consider the selection of stations to participate in exchange of data by telegraph, for the purpose of studying the general circulation. Early attempts at such a system had been made by Fitzroy in London in 1861, in connection with the first British weather maps, and by Le Verrier in 1863, in the daily bulletin of the Paris Observatory. Teisserenc de Bort's scheme proved to be too expensive, and instead the Réseau Mondial scheme was introduced. This scheme involved the publication of data for representative land stations, averaging two stations to each 10° square of latitude and longitude. The data were published in annual volumes by the British Meteorological Office, commencing with the publication of data for 1911 in 1915. Data were subsequently published for 1910, and for 1912 to 1918 inclusive.[11] No stations were included from polar regions, except for Norwegian stations in Spitzbergen, Bear Island, Jan Mayen, and northeastern Greenland.

During the interwar period, the IMO continued to standardize techniques. At its Washington conference in 1947, a convention for a new organization, the WMO, was drawn up. The last conference of the IMO and the first congress of the WMO were held in Paris in March 1951, after the convention came into force on March 23, 1950.

The declared purpose of the WMO was threefold: to facilitate worldwide cooperation in the establishment of meteorological networks, to promote the development of centers for meteorological services, and to promote the rapid exchange of weather information and the standardization and publication of observations.

The WMO Secretariat was set up at Geneva, and four constituent bodies were established: the Congress, the Executive Committee, the Regional Associations, and the Technical Commissions. The Regional Associations are: I, Africa; II, Asia, including Arabia; III, South America; IV, North and Central America; V, Southwest Pacific, including Australia, New Zealand, New Guinea, Borneo, Sumatra, and Malaya; VI, Europe, including Great Britain, Ireland, Turkey, Iceland, Greenland, and Spitzbergen. The Technical Commissions are Aerology, Aeronautic Meteorology, Agricultural Meteorology, Bibliography and Publications, Climatology, Instruments and Methods of Observation, Maritime Meteorology, and Synoptic Meteorology. When it was formed in 1952, the WMO had 52 states and territories as members; by its second Congress in 1955, it had 88. The work of the WMO during the first four years of its existence included: introduction of new codes, international ice nomenclature, and barometer conventions; specification of requirements for radiation networks; redefinition of the codes for horizontal visibility; standardization of radiosondes and sferic techniques; introduction of a scheme for the transmission of data from whaling ships; and revision of the meteorological section of the Universal Decimal Classification.

Almost all the conventions of present-day meteorology and climatology are based on recommendations by the IMO or the WMO. For example, the wind-velocity equivalents for the Beaufort scale were fixed in terms of the wind velocity in the open at a height of 6 m (Zurich 1926), the millibar was adopted as the standard pressure unit (London 1928, Copenhagen 1929), standard weather specifications were introduced

(Salzburg 1937), the station model was drawn up (Warsaw 1935–1936), and definite values for the density of mercury and the acceleration due to gravity were agreed on (Berlin 1939).

The IMO recommended spacing observing stations at intervals of 100 to 150 km (Washington 1947), but since actual observing networks serve the needs of individual nations, the coverage is far from even.[12] Until the late 1950's, gaps existed in the southern hemisphere, with its large oceanic stretches. Even making use of all islands, and of all reefs covered by less than 5 fathoms of water and on which instruments can be mounted, leaves two large areas in the Pacific in which there is more than 1,000 nautical miles between fixed stations.

Oceans cover two-thirds of the Earth's surface, yet very little is really known of precipitation distribution over them, since the available data are nearly all from islands, not ships. Aerological data, too, are sparsest over the oceans, particularly for the southern hemisphere. Most aerological stations concentrate on the northern hemisphere troposphere north of 20°N, yet this is a much less important part of the atmosphere, dynamically, than the southern circumpolar vortex or the tropical circulation. Various schemes have been introduced to fill the gaps in the station network: floating ice-islands in the Arctic as instrument-carriers, automatic buoy weather stations, reconnaissance aircraft, and so on. Two of the most successful have been the "Selected Ships" scheme, and the Ocean Weather Station system.

The "Selected Ships" scheme was introduced by the IMO in 1947, when merchant ships of all nations had voluntarily been reporting weather information for almost 100 years. Under the IMO scheme, ships of all types—passenger and cargo ships, whalers, large trawlers, and research ships, particularly those visiting unfrequented waters—were recruited by Port Meteorological Officers. "Selected ships" were provided with mercury barometers, barographs, hygrometers, and apparatus for measuring sea-water temperature. "Supplementary ships" were provided with barometers and air-temperature thermometers, "Auxiliary ships" used their own instruments after these were checked by Port Meteorological Officers, and "Other ships" made non-instrumental observations only, usually because their crews had no time to make more. Some ships, whalers, for example, must not reveal their positions to rivals, and so a special cipher code for weather reports had to be introduced. In 1953, there were 2,385 ships enrolled in the "Selected Ships" scheme, although only 1,100 of these would be at sea on a given day, and most of them would be in the North Atlantic. By 1956, 2,800 ships were enrolled. These ships observe at the main synoptic hours (12:00 and 6:00, A.M. and P.M.). In 1952, 118 reports per day were received from ships in the eastern North Atlantic, in comparison with 52 per day in 1939.[13]

The 1954 ICAO conference in Paris stressed the value of permanent ocean weather stations for the provision of accurate wind data for aviation. Ocean Station Vessel Agreements for the North Atlantic were signed in 1949 and in 1954. Under the terms of the 1949 agreement, ten stations were designated in the North Atlantic, labeled A, B, C, D, E, H, I, J, K, and M, and these were manned by 21 vessels. In 1954, station H (near the New England coast) was eliminated, and the others were reaffirmed. Ocean Weather Station (OWS) A is the most difficult to man, and is occupied by eight vessels in turn: four British, two Dutch, and two French. OWS B, C, D, and E are manned by North American vessels, and OWS I, J, K, and M by European vessels. In general, each weather ship spends about 24 days on station, and 16 days in harbor.[14] Ocean Weather Stations have proved invaluable, particularly for upper-air observation, but they are much needed in the Pacific. More weather ships are to be found in the North Atlantic than in other oceans, yet there are fewer areas distant from land there than elsewhere.

New Zealand is responsible meteorologically for the largest oceanic area in the world, embracing the South Pacific between longitudes 120°W and 160°E. Since a dense weather-ship network is impracticable in such a vast area, every New Zealand ship engaged in regular overseas trade is enrolled in a special Observing Fleet, and all ships of other countries are also invited to make regular reports. The number of reports received in this way increased from less than 5,000 in 1946 to more than 20,000 in 1953.[15]

The International Geophysical Year

The IGY, July 1, 1957, to December 31, 1958, was the occasion of the greatest concentration of meteorological observation ever attempted. International Polar Years held in 1882–83 and 1932–33 had led to the foundations of auroral and Arctic aeroLogical research, respectively. In 1950, L. V. Berkner proposed that a third Polar Year be held in 1957–58, when solar activity would be at a maximum. The WMO considered that a worldwide geophysical year would be more useful.[16]

The WMO and the International Association of Meteorology defined the main meteorological task of the IGY to be the investigation of large-scale physical, dynamic, and thermodynamic processes of the general circulation over the whole world. To this end, three types of observational period were introduced.

1. *Regular World Days* (RWD) allowed one to three days per month for the concentration of observations, particularly upper-air. If all three days were scheduled, two were at the time of new moon, and the third as close as possible to quarter moon, but timed to coincide with any exceptional swarms of shooting stars.

2. *Special World Intervals* (SWI) were announced by alert warnings four to six days before certain predictable events.

3. *World Meteorological Intervals* (WMI) were to include periods of ten days representative of the seasons.

During the IGY, all existing aerological stations were asked to make two radiosonde and four radar-wind soundings daily, and four combined TEMP-wind soundings on the ten consecutive days of each WMI. Observers were to follow the soundings to at least 50 mb, and to 10 mb in the tropics and during WMI's. New aerological stations were established in the Antarctic, and radiosondes were launched from whaling ships. There were 36 Aerobee rockets launched into the thermosphere, in addition to many smaller rockets launched from balloons and aircraft, and sea temperatures were regularly taken down to at least 200 m using bathythermographs. Nephoscope observations were taken in areas where balloon observations were impracticable, and albedo observations were made by astronomical measurements of earth radiation reflected from the moon.

Special attention was paid to certain vertical "cross sections" of the atmosphere, including three pole-to-pole sections. The first lay along longitudes 10°E, 140°E and 75°W, with a tolerance of 5° of longitude. The second, along 75°E, sampled continental conditions in the northern hemisphere and maritime conditions in the southern, via the monsoon regions. The third, along 180°, was mainly oceanic. Other meridional sections in the northern hemisphere lay along 110°E and 20°W, the latter sampling the splitting of the jet stream when it reaches Europe; tropical meridional sections lay along 30°E and 110°E. Zonal sections were taken along the equator between 15°N and 30°S, and along 40°N through the Rocky Mountains and the Sierra Nevada in North America, the latter in order to study the effect of large mountain ranges on the general circulation, for which purpose a partial zonal section across the Andes turned out to be not as useful.

As many surface and aerological observations as possible were taken in temperate zones between 45 and 50°N and between 35 and 40°S, in arid zones between 20 and 30°N and 20 and 30°S, and in the equatorial zone between 20°N and 20°S. Australia and New Zealand set up stations on various Pacific islands. Despite these efforts, the tropical network still remained inadequate. Although a total of 851 ships made observations at 6:00 A.M. daily, of these, 745 were in the northern hemisphere and only 106 in the southern.[17]

Tabulated data from the IGY have been published in the form of microcards, and, most useful for the climatologist, daily weather maps, compiled carefully and without the haste typical of routine synoptic analysis, have been published by the meteorological services of the United States (for the northern hemisphere), South Africa (for the southern hemisphere), and the German Federal Republic (for the tropical zone), in the form of 18 monthly booklets for each zone.[18]

Following the success of the IGY, other observational periods of climatological interest have been recognized. Thus 1964 and 1965 were declared International Years of the Quiet Sun (IQSY). Each Wednesday throughout an IQSY is designated a Regular Geophysical Day (RGD), especially designed for meteorological purposes, and 14 consecutive days in each season form World Geophysical Intervals (WGI), during which seasonal variations and the timing of seasonal changes are of paramount interest. A more comprehensive concept, the World Weather Watch (WWW), provides a world weather service, composed of the integrated activities of the national and international meteorological organizations, which employs satellites, rockets, and other new techniques, in addition to the more usual ones. In connection with the WWW, three world weather centers for the collection and dissemination of data were planned. The first of these, at Suitland, Maryland (near Washington, D.C.), commenced operations on January 1, 1965. Others are located at Moscow and Melbourne. Another development, the International Hydrological Decade (IHD) was declared open in 1965. Three important objectives of the IHD are the standardization of observational techniques and terminology for reporting, collecting and compiling data on precipitation and other hydrological elements, the identification of the principal gaps in the world hydrological data, and the establishment of basic networks.[19]

The Problem of Gaps in the Observing Network

Before the climatologist can fairly complain about lack of data for a given area, he needs to understand the difficulties of establishing and maintaining a network of observing stations. A network has two, sometimes conflicting, aims. Its climatological aim is to provide a broad, long-term picture of the meteorological features of an area, in order to bring the controlling or more stable features of the climate into prominence.

Its synoptic aim is to provide an instantaneous picture of the exact meteorological phenomena within as large an area as possible. But since the latter observations must be representative of a large area—i.e., purely local influences must be minimized— their value for local climatic studies is considerably decreased. Representativeness is secured by employing instruments with relatively large lag coefficients, so that purely local and short-period fluctuations are not recorded, and by locating the instruments (for surface observations) in a standard site.

Differences between various countries' units, methods, and observing times present great difficulties. There is little prospect of complete standardization by international agreement, and the compromise is usually adopted of deciding on minimum acceptable standards of accuracy for the observations. For the timing of observations, *standard times* are fixed by WMO, and the *official times* adopted by the country concerned should be as close as possible to the standard times. The *actual times* of observations should not differ by more than one hour from the standard times. Despite international agreement on such minimum requirements, however, there is no guarantee that these requirements will be fulfilled by a country at a given time. Table 1.1 shows the number

TABLE 1.1.
State of Synoptic Reporting on March 1, 1963

WMO Region	Number of stations asked to make observations at each synoptic hour	
	Surface	Upper-air
Africa	694	124
Asia	963	247
South America	543	47
North and Central America	469	144
Southwest Pacific	286	72
Europe	831	141
Total	3,786	775

of stations requested by WMO to carry out observations at each synoptic hour. A check of these on April 1, 1964, showed that European stations fulfilled at least 97 per cent of their quota at all hours, that North and Central America and the southwest Pacific were fairly satisfactory, but that Africa and South America were very unsatisfactory. Both of the latter only fulfilled 80 per cent of their quota at only two synoptic hours: 6:00 A.M. and 12:00 noon for Africa, and 12:00 noon and 6:00 P.M. for South America.[20]

The question of the optimum number of stations in an observing network is not easy to answer. Like the question "what is a representative observation?" it is really a complex mathematical problem. Furthermore, meteorological networks, particularly the synoptic variety, are very expensive. For example, a single radiosonde ascent in the United States in 1963 cost approximately $50, i.e., about 10 per cent of the income of an average American family per month, and the number of ascents per month is something like 10,000.[21]

For climatological networks, the optimum number of observing stations should be

based on the division of a continent into *coherent climatic regions*, each with an area
of between 200,000 and 2,000,000 square miles (see Appendix 1.1). These regions are
then subdivided into zones of latitude one or two degrees wide and zones of altitude
500 to 2,000 feet deep. The density of stations required for representative purposes
may then be laid down.[22] For each zone of 2,000 to 20,000 square miles, stations
should be set up for the different types of location, such as hilltop, south- or north-
facing slope, coast, valley bottom, industrial city, country town. Each zone needs only
one permanent station, at which observational and instrumental errors are kept as
minimal and exposure conditions as constant as possible. Such a station forms a
benchmark, and climatic conditions at other points within the same climatically
coherent region may be determined either (a) by setting up a temporary station where
needed for a year or so, and correlating its observations with those from the benchmark station in order to obtain long-period estimates, or (b) by taking climatic profiles
or traverses between neighboring benchmark stations, in order to determine the rates
of change of the element in question. The latter measurements are usually made at
times of year or day when extreme values are expected. Table 1.2 gives details of the
spacing of stations needed to obtain representative data.

For synoptic networks, the optimum number of stations is more difficult to determine because two types of user must be satisfied. For the local forecaster, the network must provide qualitative data about weather phenomena, such as cloud types
and precipitation forms; for the theoretician, it must provide quantitative data on
continuous distributions, such as the fields of pressure, temperature and humidity,

TABLE 1.2.
Recommended Networks for Climatological Stations.[a]

Element	Areas of uniform topography		Special topographic situations						
	Rural	Urban	Coastal zone	Plain	Mountains	Hills	Middle of a slope	Island	Valley bottom
Insolation:									
direct plus diffuse[b]	3,000,000[c]		20 to 100	10,000	10,000				
total[d]	120,000		20 to 100	10,000	10,000				
duration	10,000		20 to 100	10,000	10,000				
Sunshine	500	5	20–100	10,000	10,000				
Temperature	10,000	1[e], 4[f]	500			500	500	500	500
Winds	10,000	100	100	10,000	10,000				
Visibility	10,000	4[e], 1[f]	100	10,000	10,000				
Rainfall	10,000	100		5,000	100	500			
Snowfall	5,000	10		2,500	50	200			

[a] This table is an extension of work by C. F. Brooks, *TAGU*, 28 (1947), 845.
[b] Direct and diffuse radiation recorded separately, on horizontal and vertical surfaces.
[c] Each figure indicates the maximum area, in square miles, for which a measurement of the element at a
single station may be regarded as representative under ideal conditions.
[d] Total global radiation recorded on a horizontal surface.
[e] Small town.
[f] Large city.

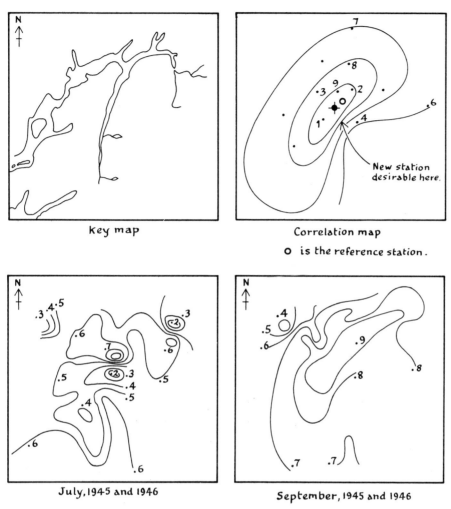

key map   Correlation map

O   is the reference station.

July, 1945 and 1946   September, 1945 and 1946

FIGURE 1.1.
Correlation coefficients for precipitation referred to station Ullensvang in the
Hardangerford region, south of Bergen (after C. L. Godske). The correlation map
represents a hypothetical situation concerning temperature observations.

which are usually smoothed during analysis. An increase in the number of stations in
an area has sometimes resulted in the discovery of new atmospheric phenomena. Thus
Bjerknes increased the density of stations on the southwestern Norwegian coast tenfold
in 1917, and, as a result of the increase in data, was able to develop frontal models
that have been assumed to apply everywhere else in middle and high latitudes without
corroborative evidence from other equally dense networks. Before going to the expense
of intensifying an existing network, however, one should take the logical step of
determining whether it is inadequate or perhaps too elaborate for the purpose it is
intended to serve. By correlation techniques, using a correlation coefficient of at least
+0.80 (and preferably one of +0.90 or more), one can measure the representativeness
of a station.[23] Figure 1.1 gives the correlations between precipitation as measured at
a reference station in the Hardangerfjord region west of Bergen and those at all

neighboring stations. The information content of the data from a station may be measured by its variance: 90 per cent of the information about temperatures at Station 1 is already contained in the observations made at the reference station, and so Station 1 may be unnecessary. By contrast, less than 50 per cent of the information about temperatures at Station 4 can be obtained from data from the reference station, therefore Station 4 is much more necessary than Station 1.

The averaging process to which climatological data are subject soon brings to light apparent errors in observation, which are fairly rare for benchmark stations. The situation is far from satisfactory for synoptic stations, however. The data from the latter are usually made "representative" by checking for personal or accidental observational errors, errors in reduction to sea level, errors due to incorrect timing of observations, and errors during coding, transmission, decoding, and plotting. The necessary assumption is made that if, for example, one observation in a set is apparently different from all the others, then that observation is the incorrect one. A further assumption made in analysis is that a simple pattern of smooth isobars is more likely to be "true" than a more complex pattern of irregular isobars. This latter assumption is necessary if hydrodynamic methods are to be used in forecasting, although the microbarograph in the forecaster's office may often show that it is an incorrect assumption at that station at a given time. Essentially, conventional synoptic practice is to "correct" (or omit from consideration) actual observations because they do not fit in with a preconceived hydrodynamic theory or with the expected course of evolution of the broad-scale synoptic situation. Published synoptic charts obviously obey this convention, and the synoptic climatologist and student of local climate need to bear it in mind when interpreting the charts.

In 1960, the WMO recommended that the density of stations should not be so great as to cause instrumentation (rather than interpolation between stations) to become the dominant source of error. The optimum density of stations depends on the relation between observational error and the gradient of the element being measured: if the gradient is less than the error, then the resulting map analysis will depict the distribution of errors rather than the variations of the meteorological phenomena. A useful measure of optimum spacing is provided by the variance of the observed gradient of the meteorological element that is being measured about the true gradient of that element: the optimum spacing of stations is that for which this variance is minimized. This criterion is readily applicable to upper-air observations. For the United States radiosonde network, the optimum spacing of stations comes to one station every 90 nautical miles, taking a sounding every six hours. The existing spacing averages 220 nautical miles between stations, with soundings every 12 hours.[24]

Assuming that a network has been established and cannot be extended, the problem is how best to make observations that will fill in the gaps. The main techniques involve the use of automatic stations, aircraft, and artificial satellites for synoptic observations, and various types of spot observation and meteorological traverse for local or topo-climatological studies.

Automatic weather stations are meteorological stations taking measurements of the standard elements at appropriate intervals in time, the measurements being transmitted by radio to a central collecting point. They are very useful in sea areas, or in polar or desert regions. An early example was a buoy station developed for use on Lake Ontario: it had three instrumental devices, for measuring water and air temperatures and wind velocity, and the transmitted signals could be picked up on an ordinary communications receiver. Another early type, developed for the United States Navy, transmitted data on pressure, temperature, relative humidity, precipitation, and wind direction, measured eight times every 24 hours, in terms of pulse frequencies generated by a relaxation oscillator that was itself controlled by the measuring instruments. This type could be left unattended for four months (two months in very cold weather). Its power supply was generated by a gasoline-electric plant switched on and off by a clock mechanism set to a predetermined schedule.[25]

An unmanned ocean weather station developed by the U.S. National Bureau of Standards has a transmitting range of up to 1,000 miles and can be left unattended for up to six months. The station is mounted on an aluminum boat, 20 feet long and 10 feet in beam, which can be moored in waters up to 3,600 feet deep. Its instruments measure air and water temperatures, atmospheric pressure, wind velocity, and wind direction; in each case, the readings are converted into variations of electrical resistance that are coded automatically and then transmitted by radio. Power is provided by a combination of 180 dry cells.[26]

Power supply for automatic stations is always a problem, a solution to which may be provided by radioactive isotopes. The first isotope-powered automatic weather station, designed to operate for two years without attention, began transmitting on August 17, 1961, from Axel Heiberg Island. This station is 750 miles from the North Pole in the Canadian Arctic, and fills the gap between orthodox manned stations at Eureka and Resolute. It is powered by a thermoelectric generator that obtains its heat from radioactive decay of strontium-90 contained within three-quarters of a ton of lead shielding mounted in the lower part of an eight-foot long steel cylinder 26 inches in diameter. Electronic equipment and a barometer fill the upper part of the cylinder, and a 37-foot steel tower carries a wind-vane, bimetallic thermometer, and anemometer. The station also includes two 71-foot towers for the radio antennae; a complete data transmission takes only 9 seconds, simultaneous transmissions are made on 3.36 and 4.97 mc, and the maximum range of the transmitter is around 1,500 miles.[27]

Dirigibles or rigid airships offer certain advantages as floating platforms for the location of meteorological stations over such areas as the Arctic Ocean, particularly if they are nuclear-propelled and so may remain airborne for months at a time. Gliders are very simple and relatively inexpensive means of investigating atmospheric phenomena such as thermal currents, and tracking a glider by means of two ground-based theodolites makes it possible to detect variations of air pressure from that given by the hydrostatic equation, so that (provided the glider pilot observes his height and temperature at the same instant as ground observers fix his position) data concerning winds may be computed. However, the powered aircraft is much more widely used than either airship or sailplane for purposes of meteorological observation.[28]

Powered aircraft may be used in several ways as meteorological tools. If navigational aids are available, either radio or radar, then winds may be measured by the aircraft: the required information to compute wind data is the compass heading of the aircraft, its groundspeed, angle of drift, and true airspeed (see Appendix 1.2). If a Doppler navigational system is being used, the computed winds should then be less than 5 knots in error, whereas without Doppler the error may exceed 10 knots. Winds reported by jet aircraft in this manner may be used to construct climatological cross sections (see Figure 1.2) that reveal the dominant seasonal rhythms, despite considerable variation in both time and place of the reports. Photographs taken by aircrews along air routes add life to synoptic cross sections and may reveal unsuspected cloud developments.

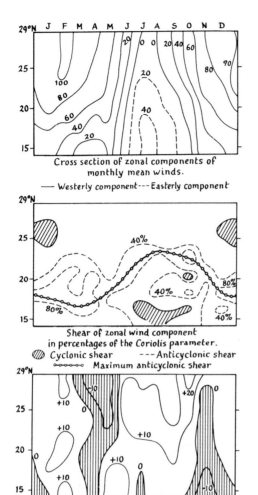

Cross section of zonal components of
monthly mean winds.

—— Westerly component --- Easterly component

Shear of zonal wind component
in percentages of the Coriolis parameter.
▨ Cyclonic shear   --- Anticyclonic shear
∘∘∘∘∘ Maximum anticyclonic shear

Cross section of meridional components of
monthly mean winds in knots.
▨ N-S components.   S-N components left unshaded

FIGURE 1.2.
The 200-mb mean winds on the air route
from El Adem to Aden, derived from
jet-aircraft reports (after P. G. Wickham).

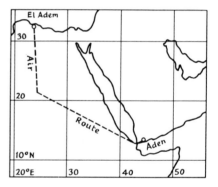

The introduction of compact radiation thermometers has enabled continuous measurements of water and cloud temperatures to be made with an accuracy of $\pm 0.75°F$ over very large areas. The equipment is mounted in high-speed, high-altitude aircraft, such as the Lockheed U-2, and is capable of detecting quite small temperature differences.[29]

High-speed jet aircraft may be employed to obtain quantitative data on mountain waves. In one investigation over southern Norway, a Sabre made only one run, at 25,000 feet. The pilot trimmed the aircraft for straight and level flight, and engaged the automatic pilot. Speed was kept constant, and the vertical airspeed indicator was used to determine the crests and troughs of the orographic waves. The method has several complications (see Appendix 1.3), but very useful streamline cross sections may be obtained in this way (see Figure 1.3). Low-level temperature and moisture cross sections may be obtained by equipping a light aircraft with an aneroid barometer and an electrical-resistance psychrometer. Observations taken along a straight-line route in southern England by a Chipmunk aircraft revealed some features of sea-

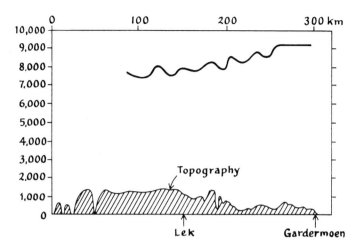

FIGURE 1.3.
Path (solid line) of air particles moving above the mountains of
southern Norway on July 26, 1956, as determined by observations
made from a Sabre jet aircraft (after Y. Gotaas).

breeze structure: the aircraft flew along the route at various heights at a speed of 85 to
90 knots, and psychrometer readings were taken every two minutes.[30]

Other measurements from aircraft are of buoyancy and vertical air velocity, convection, and turbulence, and to identify large airborne sea-salt particles. Many
measurements of radiation, for example, atmospheric long-wave radiation balance,
total and visible solar radiation, cloud infrared emissivity, and albedo, may conveniently be made from aircraft. Aircraft have also been used for icing observations
and for general high-altitude research. All such measurements involve problems for
which special instrumentation is necessary.[31]

Fixed-wing aircraft, particularly jet-powered ones, are in effect mobile meteorological
stations capable of covering great distances in all three dimensions, as is, very locally,
the helicopter. Instruments suspended a certain distance below the front rotor of a
helicopter are in free, undisturbed air at all flying speeds above 5 knots (for many
helicopters this distance is 60 feet, but it depends on the vertical component of the
wake velocity of the aircraft), and low-level measurements may be made in great
detail.[32]

At the other end of the scale, artificial Earth satellites allow routine, global observations of some of the conventional elements. Many meteorological satellite observations
have to be inferred from remote radiometric data; i.e., what the satellite actually
measures, is essentially some aspect of radiation. By means of meteorological theory,
data concerning lapse-rates, atmospheric heating rates, the mixing-ratios of the radiating gases, and so on, may be computed from the radiation measurements.[33]

TIROS I, the first of the meteorological satellites transmitting routine data, was
launched on April 1, 1960. This satellite observed clouds by means of slow-scan
television cameras. One camera supplied pictures of an area of the Earth's surface
about 750 nautical miles square, with the camera axis normal to the surface; the
other camera provided a higher resolution for an area 75 nautical miles square.
TIROS II, launched on November 23, 1960, carried two cameras as before, plus two

radiation experiments. The first of these was based on a five-channel scanning radio-meter with a resolution of about 30 miles when the optical axis was normal to the Earth's surface; this instrument measured radiation reflected or emitted from the Earth in the ranges 0.25–0.6 micron, $0.05–0.75\mu$, $6.0–6.5\mu$, $7–30\mu$, and $8–13\mu$. The second consisted of two thermistor bolometers with a field of view forward in the form of a cone 300 miles in diameter, designed in effect to measure the radiation reflected by the atmosphere. TIROS III was launched on July 12, 1961; it was basically the same as TIROS II, but had a Suomi radiometer, and a second wide-angle camera substituted for the narrow-angle one. TIROS IV, launched on February 8, 1962, omitted the $7–30\mu$ channel from the scanning radiometer and had a new camera lens that gave a field of view 450 miles square with the optical axis normal to the Earth's surface. The inclination of the orbital plane to the equator was 48° for these four satellites; for TIROS V, launched on June 19, 1962, the inclination was 58°, permitting greater latitudinal coverage.[34]

TIROS VIII carried an automatic picture-transmission system that could broadcast television pictures of cloud cover every 3.5 minutes; the broadcast could be received by relatively simple radio receivers on the ground anywhere the satellite was above the horizon. This was a decided improvement; previously, the vidicon camera system carried in the satellite stored the pictures until they were commanded to be " read out " by one of the ground control stations, and delays of five to eight hours were common before a complete nephanalysis could be built up. With the new (APT) system, a complete picture of cloud distribution in the orbited zone can be built up very quickly.

Later developments have included the NIMBUS and TOS satellites. NIMBUS satellites carried more advanced instrumentation than the TIROS series. NIMBUS A was provided with high-resolution infrared equipment for night time cloud photography. TOS, the TIROS Operational Satellite System, was launched in late 1965 in a polar orbit. Every 113 minutes, one of the satellites in the TOS system makes a northward crossing of the equator at a point where it is local noon, so that, every 12 hours, every part of the Earth's surface has come within the field of view of the cameras.

In the first excitement of satellite success, it was encouraging to see from the TIROS photographs that global weather systems were, in general, very similar to predictions based on conventional Earth-bound observations. However, the provision of routine information on global cloud distributions was the basic reason for the TIROS operation. The resulting photographs were transformed into nephanalyses and transmitted by facsimile from the United States to many overseas weather centers, and, in addition, copies of the photographs were published as 35-mm roll films or (for later satellites) on $8\frac{1}{2}$ by 11 inch sheet film. A special grid system was developed for identifying the features on the photographs.[35]

Meteorological satellites provide five types of information: (1) photographs of clouds and precipitation distributions; (2) radiation and radiation-balance measurements; (3) data concerning pressure, temperature, and density in the immediate neighborhood of the satellite; (4) data concerning air movement around the satellite; (5) inferred data concerning meteorological parameters at the Earth's surface and in the lower atmosphere. Clouds and radiation are dealt with in other chapters; our interest here is in the inferences, all of which involve much physical theory. Data can be inferred on the mean temperature of air layers 15 km deep, the radiative temperature of the Earth's surface (which may differ from the locally recorded screen temperatures, depending on the physical nature of the surface and the velocity of the surface wind), the main features of the atmospheric flow pattern and the field of large-scale vertical motion, and the global pattern of heat balance.[36]

Density measurements from satellites show that the upper atmosphere is under direct solar control.[37] The normal minimum operating height for satellites is between 100 and 200 km, and by means of physical theory the density observations obtained at these heights have revealed the rough shape of isobaric features at heights up to 700 km and their variations between day and night and during the sunspot cycle. The resulting data show that the density of the upper atmosphere fluctuates strongly, with a period of 28 days, and also fluctuates diurnally, so that density during the day is 10 per cent higher than during the night at 200 km. The diurnal variation rapidly increases as height increases, so that it becomes the dominant feature of the atmosphere between 400 and 750 km. Both the diurnal variation in density and its average annual value should be lower at times of sunspot minima than at times of sunspot maxima. The satellite observations show two quite distinct solar effects in the upper atmosphere: the day-night variation, which is caused by the position of the sun, is quite different from the fluctuation arising from variations in solar activity. Above 400 km, the isobars are in the form of a permanent hump on the sun-facing hemisphere of the Earth; at the same time, the isobars in both day and night hemispheres pulsate up and down in response to solar activity.

Other possible measurements from satellites include the vertical distribution of ozone and the vertical profiles of pressure and temperature below 24 km; the latter involves making star observations from a telescope mounted on a satellite in a polar orbit, and should result in temperature and pressure data accurate to 0.1 per cent at sea level, and rising to a maximum error of approximately 1 per cent at 30 km. An ambitious scheme suggests using satellites to transmit commands to superpressure balloons, which would make global horizontal soundings of the usual meteorological elements at all levels between 500 and 10 mb. The satellite, in itself a relatively poor meteorological sensor, would be employed to locate the balloons and to provide a communications link. Each balloon would carry measuring elements and would transmit readings at a command from the satellite. Six satellites and 2,000 balloons would be adequate to provide world coverage.[38]

The climatologist concerned with filling in the gaps in the network of surface observations in order to undertake very localized studies must obviously use very different techniques from those just discussed. A distinction may be made between *spot observations*, which are really temporary climatological stations set up for a short time, and *climatic traverses or transects*, which are series of observations made by moving the instrument after each measurement. With spot observations, the "short time" may vary from a few days to a few years, depending on the object of the investigation; the essential point is that enough observations must be made for a stable frequency distribution to be obtained, after which long-term estimates may be made by statistically comparing the observed data with figures for the nearest benchmark station. In other words, observations must be continued until their distribution approaches a normal curve, or some other curve that can be described mathematically, so that the parameters of the element being measured may be deduced from the statistics of the observed data. The statistical comparison may take the form of regression equations, contingency tables, etc.

For a traverse, it is essential to return to the starting point to check the measurements (preferably several times), so that any changes with time in the element under observation may be taken into account. Traverses should normally be made at times when the diurnal rate of exchange is least marked, for example, in the early afternoon or during the night, and a number of traverses should be made under different synoptic situations so that long-period estimates may be made. I.e., suppose that a northerly weather type occurs on 24 per cent of the days of an average year in a certain area. A traverse made on a day with a northerly weather type in control could be taken as representative of conditions during 24 per cent of the year in the area in question, assuming that due allowance has been made for seasonal and other temporal trends.

When making spot observations, it is convenient to use an instrumental arrangement whereby different elements, such as dry-bulb, wet-bulb, and soil temperatures, may all be recorded on the same circular chart. Microclimatological installations for such purposes as agriculture and forestry fall into the spot-observation category, and instrumental devices of greatly varying degrees of sophistication and expense have been produced for this type of work.[39]

For traverses one can choose between simple procedures, in which the observer uses a whirling psychrometer and a hand anemometer to make measurements at selected points along a predetermined route he follows on foot, and more complicated programs using motor vehicles. The simple procedure may be used, for example, to determine local temperature and humidity characteristics of a ridge-top site. Vehicle traverses may take the form of a series of journeys across a city to measure the effects of variations in housing density and the distribution of parks and open spaces on urban heating; here electrical-resistance thermometers mounted in front of the vehicle, recording inside the latter via a bridge apparatus, are very convenient. If several vehicles are used simultaneously, making traverses across the city in different directions, together with wiresondes or other means of obtaining temperature and humidity soundings of the lowest layers of the atmosphere, a quite detailed picture of urban synoptic climate may be built up. Alternatively, a more complex instrumental installation may be adopted in which the vehicle becomes, in effect, a mobile meteorological laboratory. Observations made by such a vehicle, moving at 6 to 10 miles per hour along tracks over Gisborne Mountain, Washington, enabled the thermal belt of the latter to be defined, and the movement of moisture from the upper slopes into the valleys, via drainage winds down the main valley, to be followed.[40]

Climatological Instrumentation

The climatologist needs to consider instrumentation from two points of view. First, what are the limitations of the standard meteorological instruments used for producing the "official" data published by national meteorological services, and what is the magnitude of the resulting probable error in this data? Second, what instruments are available for use if a special investigation has to be made of an area of the Earth's surface or of the atmosphere for which no data have been published? Before we deal with these questions, there are certain preliminary problems with meteorological observations and instruments in general that we should consider.[41]

Three types of meteorological observation and measurement are possible: (a) a *raw observation*, made by the observer's unaided senses, for example, observation of wind-speed by means of the Beaufort scale;[42] (b) a *simple measurement*, in which only a few processes intervene between the phenomena to be measured and the observer's senses, for example, estimation of the height of a low cloud base by observing the time it takes a balloon rising at a fixed rate to reach the cloud; (c) a *complex measurement*, in which a large number of processes, usually mechanical or automatic, intervene between the phenomena and the observer, for example, the measurement of atmospheric pressure by means of a barometer, which must be corrected for temperature, the local value of gravitational attraction, and height above sea level. Obviously, the magnitude of probable error in obtaining an accurate observation will differ considerably in these three cases.

A meteorological measurement, other than a raw observation, usually involves the observation by eye of either (a) a coincidence in space or time, or (b) an estimated small distance. For the latter, the vernier principle is often employed (see Appendix 1.4). Many instrumentally assisted processes may intervene between (a) or (b) and the phenomena to be measured. In all cases, the instruments must be accurate enough for the purpose of the measurements, and sensitive enough to make full use of their accuracy. The distinction between accuracy and sensitivity is important. A sensitive instrument is one in which a small change in the quantity being measured produces a large change in the final message which the instrument conveys to the observer's senses. Accuracy is not necessarily related to sensitivity, and a sensitive instrument may not be as accurate as a less sensitive one measuring the same phenomena. The most important requirement of a meteorological instrument is accuracy, rather than sensitivity; other requirements are reliability, simplicity of design, ease of reading and manipulation, robustness and durability, low cost of maintenance, and low initial cost.

A distinction should be made between *indicating* and *recording* instruments; mercury thermometers read by eye are indicating instruments, for example. With recording instruments, friction is serious, since it affects the indicating mechanism in the sense opposite to that of the phenomenon to be recorded. Hence an error is introduced, and must be controlled.[43] Because of friction, recording instruments are less accurate than comparable indicating instruments, and are not normally used as absolute standards. Recording instruments should be checked regularly against the indicating instrument that is regarded as a standard. Several types of recording device are used: pen recorders, in which a pen makes an ink trace on a chart wrapped around a cylindrical drum rotating at a constant speed; thread recorders, in which a series of dots is made at regular intervals by an inked thread pressed against a chart at appropriate times by a bar mechanism; stylus recorders, in which a pointer traces a line on a specially prepared surface, for example, smoked or waxed; and photographic recorders. Modern Polaroid cameras, which automatically process the film within a short time of its exposure, can turn indicating instruments into recording instruments without sacrificing any accuracy.

STANDARD INSTRUMENTS

Certain instruments have been selected by the various meteorological services of the world as the standards for their benchmark data. Details of pressure, radiation, temperature, cloud, and visibility instruments so selected will be found in other chapters; for other instruments, the reader is referred to the literature.[44] A few comments will be useful here for sunshine, humidity, precipitation, and wind instruments, and for site requirements.

In order to obtain representativeness, so that climatological observations from far distant parts of the world may be compared without complications by local influences, certain standard site requirements are specified by the various meteorological services. In the British Isles, the outdoor instruments (including thermometer screen, earth and grass-minimum thermometers, and rain gauge, and sunshine and wind instruments if possible) should be installed on a level plot of ground, approximately 30 feet by 20 feet, covered with short grass, and enclosed by an open fence or palings. There must be no steep slopes nearby, and the plot should not be situated in a hollow. Any obstructions, such as trees, walls, or buildings, should be at least twice as far from the rain gauge as they are tall. The wind vane should be mounted on a 20-foot mast within the enclosure if it is exposed enough; otherwise the mast should be located separately, for example, on the roof of a nearby building, so that the vane is at least ten feet higher than any obstacle in the vicinity. The anemometer should be mounted outside the plot, at an effective height of 33 ft above the general level of the topography; this usually means mounting it on a 40-foot mast on level, open ground with no buildings or trees within 200 or 300 yards, or on a 50-foot mast if there are buildings not more than 20 feet high in the vicinity. The sunshine recorder should be mounted on a rigid support pillar, not more than ten feet high, so that no obstacle prevents the sun's rays from reaching the instrument at any time of the day or year. The rain gauge should be located near the center of the enclosure, not less than ten feet away from the thermometer screen and not less than twice the height of the support pillar of the sunshine recorder away from the latter. Shelter from wind is considered advantageous for the rain gauge, which should be set with its rim horizontal and exactly 12 inches above the ground, the lower part of the gauge being sunk in the ground. Eddies may be set up around the gauge in the open, so a one-foot-high turf wall, five feet away from the instrument, is usually constructed.[45]

Published data for both climatological and synoptic stations are derived from instruments located as described. It follows that any interpolations or maps based on such data only apply to a standard site. But many localities are not standard sites, and to estimate their climatic conditions from the published figures calls for considerable knowledge of theoretical climatology.

The distinction between climatological and synoptic stations depends not so much on instrumentation as on frequency of observation, in terms of which there is an accepted, three-fold classification: (a) *auxiliary climatological stations*, which make only one set of observations per day, usually at 9:00 A.M. G.M.T.; (b) *normal climatological stations*, which make sets of observations at 9:00 A.M. G.M.T. and at at least one other hour, normally at 3:00 or 9:00 P.M. G.M.T. or both; and (c) *full synoptic stations*, which make eight sets of observations a day, i.e., every three hours beginning at midnight G.M.T.[46] There are other variations on this scheme because of local requirements. *Auxiliary synoptic stations* omit some of the three-hourly observations made by a full synoptic station; *crop weather stations* are auxiliary or normal climatological stations, that have additional instruments for soil observations; and *rainfall stations* normally make only one observation (of rainfall during the preceding 24 hours) per day, at 9:00 A.M.

Climatological and synoptic stations also differ in the way they transmit their observations to the central station. Synoptic stations transmit their data (in coded form) immediately after it is obtained, usually via telephone or teleprinter land line, but climatological stations maintain their data in the form of monthly registers, which are forwarded to the collecting center every month. An exception is *health resort stations*, which, in addition to normal monthly routine, must provide coded data for issue to the press every evening.

The importance of considering site and exposure when examining climatic records cannot be over-emphasized, since changes in either can have a profound effect on the data. For example, there has been an apparent change of 2 or 3°F or even more in monthly mean temperatures for many American stations since the early 1960's, because the Weather Bureau has been replacing the standard screen-sheltered thermometers with hygrothermometers situated in a continuously aspirated thermal shield. Exposure changes remote from the instrument site may have important effects, too; in areas of numerous, well-spaced deciduous trees, for example, windspeeds at 30 to 35 feet above the ground are likely to be up to 40 per cent greater during periods of defoliation than at other times.[47]

When comparing standard climatic data for different countries, one should keep in mind the differing instrumental practices in those countries. The measurement of sunshine duration provides a good illustration. Most countries use the Campbell-Stokes recorder, in which a spherical glass bowl is used as a burning glass to produce a trace on a pasteboard card colored with matt blue ink. In the United States, a thermometric type, the Maring-Marvin sunshine-duration recorder, is preferred.[48] This type is based on the differential heating of blackened and clear air-thermometer bulbs enclosed in a vacuum. The bulbs are separated by a mercury and alcohol column that closes an electrical circuit when solar radiation overheats the air in the black bulb and pushes the mercury far enough along the tube to intercept two wires piercing the glass. The thermometric instrument commences to register at a solar radiation intensity of 0.37 cal per cm² per min, compared with 0.33 cal per cm² per min for the Campbell-Stokes recorder. Both instruments underestimate the true duration of sunshine, but the American instrument will obviously underestimate it more. A complication that prevents a simple conversion between the two is that the difference between the measurements is greater in summer than in winter, because the thermometric instrument responds to diffuse radiation whereas the Campbell-Stokes recorder does not. The conversion factor to transform Campbell-Stokes data into Maring-Marvin data therefore varies seasonally, being greatest in summer and least in winter; it also varies geographically, increasing from 11 per cent of annual total sunshine duration in the northern United States to 14 per cent in the south.

Wind data often presents a difficult problem owing to the manner in which the anemometer is supported. Some standard stations mount their instrument on a mast, others on a tower. A wind-shadow effect may influence the records of an anemometer situated at the top of the mast or tower, and for the latter the effect may be serious. Towers distort the wind flow and reduce its speed, so that the air flow on the lee side of the tower may be reduced to nearly one-half its true value in winds up to 15 miles per hour. Variations in wind speed may be produced, and recorded by the anemometer at the top of the tower, by variations in wind direction as the angle made by the wind with the sides of the tower varies.[49]

Humidity data are, of course, usually based on observations of the mercury level in two ordinary thermometers, one of which has its bulb covered with a muslin wick moistened with distilled water to form a "wet bulb." The physical principle behind this psychrometer is very simple: water will evaporate from the muslin wick, thereby cooling the thermometer bulb by removing latent heat from it. The difference between the temperature recorded by the wet bulb and that recorded by the dry bulb forms a measure of the humidity content of the air: the greater the depression of the wet-bulb temperature below the dry-bulb temperature, the drier is the air. The simplicity of the principle masks a quite complex theory, devised by Regnault in 1845. In their interpretation of this theory, and the use they make of this interpretation in the preparation of hygrometric tables, the various national meteorological services have adopted different versions of the hygrometric formula, which may influence comparisons of the published data (see Appendix 1.5). To complicate matters further, different types of hygrometric formulae have been used to compute humidities in the past. The meteorological services of continental Europe, India, and the United States all employed Regnault's formula (or its predecessors, introduced by August and Apjohn in 1825 and 1835, respectively), whereas the tables used in Britain were, until 1926, based on empirical tables compiled by Glaisher from observations taken at Greenwich and Toronto and in India. That the Glaisher tables differ appreciably from the Regnault tables for low relative humidities, must be taken into account when one is comparing sets of humidity figures compiled by the two methods.[50]

An important result of the adoption of the psychrometer principle as the standard for humidity measurement by national meteorological services has been notice of negative depressions of the wet-bulb. On occasion, wet-bulb temperatures are *higher* than the dry-bulb temperatures, particularly during fog when temperatures are near or below freezing. Such a situation is impossible according to standard hygrometric theory, but the psychrometric equation can be rewritten to show that negative depressions of the wet bulb of up to 0.5°F are quite possible; this theory is still inadequate, because much greater negative depressions have been observed. The implication for the climatologist is that apparent impossibilities or extreme values in the published data may require explanation not in terms of unusual synoptic or microclimatic conditions, but rather in terms of theoretical or computational conventions adopted by the meteorological service that published the data.[51]

When comparing humidity data for different countries, one must take a further feature of instrumentation into account. In the United Kingdom and many other countries, the dry- and wet-bulb thermometers are exposed in a Stevenson screen, in which the actual velocity of air past the bulbs represents the movement termed "light air" even when a strong wind is blowing outside. The British hygrometric tables are computed on the assumption that "light air" applies; on calm days, particularly when temperatures are high, the indicated wet-bulb depression will be greater than the true value. In the United States the standard humidity instrument is the sling psychrometer, a system of wet- and dry-bulb thermometers that are rapidly whirled through the air, and the American hygrometric tables are computed for use with this instrument. The question of adequate ventilation for the wet bulb is very critical, and variations in the manner of its solution may be reflected in the resulting humidity figures. Published data from national meteorological services should, of course, always state how the figures have been computed, but they do not.[52]

Precipitation measurement is difficult to standardize. Although published rainfall data are analyzed in every conceivable way on the assumption that the figures are uniformly accurate, it is very doubtful that they are. Unlike most meteorological instruments, the standard rain gauge does not measure the phenomenon it is intended to observe, but merely samples it. Sometimes, depending on the gauge characteristics and the local exposure situation, the sample is random, but at other times it is biased. Standard gauges, with a diameter of 5 inches, are rarely located to average more than one every ten square miles, so that with the best estimates the area from which rain is actually collected and measured is under one thousand millionth part of the area for which the measurement is assumed to be representative.[53]

The standard rain gauge, an upright cylinder of 5 inches diameter (8 inches in the United States), is mounted with its rim exactly horizontal; it is assumed that the amount of rain collecting in the cylinder per unit area of its aperture is the same as the amount of rain falling on a unit area of the surrounding surface. The principle is very old. The first known rain gauge was that used by Castelli in Italy in 1639, a glass cylinder 9 inches long and 5 inches in diameter, exposed on a roof. Although many different types of gauge were devised during the ninteenth century—G. J. Symons in 1859 began to standardize them—the current rain gauge is basically the same as Castelli's. Errors in rain measurement with such a gauge can develop in several ways, for example: (a) errors due to the aperture of the gauge not being truly circular; (b) errors due to wind eddies, set up by the gauge, carrying excess water into it or preventing rain drops from falling into it; (c) errors in the calibration of the graduated glass measure employed to observe the amount of rain water collected in the receiver within the gauge. To minimize error (a), Meteorological Office gauges must be manufactured to have a maximum error of ± 0.01 inch in the mean of any four equally spaced diameters of their aperture, with a tolerance of ± 0.02 inch for any one diameter. To minimize error (c), the graduations on a tapered glass measure of 0.50 inch total capacity must have a tolerance of ± 0.002 inch at each graduation mark up to the 0.09-inch reading, and of ± 0.004 inch above that, assuming that the measure is calibrated every 0.01 in. To minimize error (b), the convention is to locate the rim at a height of 12 inches above the surrounding ground, a compromise height, because placing the rim closer to the ground would reduce wind eddying due to the presence of the gauge but the amount of rain splashing into the aperture would be increased.

Some countries, e.g., the United States, fit shields to their rain gauges to modify the wind flow around them, so that the resultant flow across the aperture is strictly horizontal, the rain amount caught by the gauge should be independent of wind speed, and the gauge can then be mounted at any height above the ground. The problem of the different quantities of rain that apparently fall simultaneously at different heights above the same point was recognized at an early date. Other countries, e.g., Britain, do not normally use shields. The possible effect of the presence or absence of shields is a further point to be considered when comparing rainfall data from different countries.[54]

The question of exactly how representative a single rain gauge is has been tested by maintaining grids of gauges in very small areas. These tests show considerable variations in rainfall catch within a short distance. Just how much the rainfall caught in a gauge may be regarded as an *absolute* measure of the true rainfall is a complex question. One experiment regarded Lake Hefner, Oklahoma, as a gigantic rain gauge, and took the changes in its level as a measure of the precipitation on it. A comparison of this measure with data derived from 22 orthodox rain gauges located around and on the lake showed that the rain catch in the gauges was 10 per cent less than the rise in lake level, but part of the catch was reduced by strong winds.[55]

The probable error of data derived from the standard instruments is an important consideration for the climatologist. For surface temperature and humidity as measured by thermometers in a Stevenson screen, the instruments are usually read to an accuracy of $\pm 0.1°$F, but the published figures should not be regarded as closer to the true temperature than $\pm 0.5°$F, because there are almost universal, very short-period temperature fluctuations of 0.5 to 1°F that make all but meaningless any differences of less than 1°F in observations of representative temperatures. For upper-air soundings, standardization of instruments and assessment of errors are easier than at the surface, in that uniformity of exposure is no problem, but new difficulties appear. One of these, especially frequent when the soundings are made by radiosonde, is not knowing exactly where the instrument is at the precise moment it is making measurements.[56]

The probable errors in observations derived from radiosonde ascents are as follows: pressure, ± 3 mb at 500 mb, and ± 1.5 mb at 100 mb; temperature, $\pm 0.5°$C up to 80,000 feet, and ± 0.5 to 1.5°C above; humidity, ± 2.5 per cent for temperatures down to $-10°$C and a relative humidity range of 15 to 96 per cent, provided that moisture has not condensed on the humidity-measuring element. When the radiosonde balloon is used for wind-finding in addition to its normal role of measuring pressure, temperature, and humidity at different heights, a ground radar station is employed both to receive data transmitted by the radiosonde and also to act as an electronic theodolite measuring the angles of elevation and azimuth of the balloon. Angular measurements are usually made every two minutes up to 40,000 feet, every five minutes from 40 to 70 thousand feet, and every ten minutes from 70 to 100 thousand feet. The probable error in the angular measurements is $\pm 0.05°$. Other sources of inaccuracy in the published data arise from the fact that the height of the radiosonde is estimated indirectly, by substituting the observed values of pressure and temperature provided by the sonde in the hydrostatic equation, and also from the fact that the radio range observations at certain angles of elevation (below 14° for early observations) may be unreliable.[57]

Since different types of radiosonde are used by different meteorological services, the published aerological data may show discrepancies. The Diamond-Hinman radiosonde, the basic type used in the United States, may be up to 150 feet in error in determining the height of temperature inversions, because in its original form it does not transmit continuous measurements of temperature or humidity. The Väisälä radiosonde, used by the Finnish meteorological service, has a very good response at low temperatures. The British Kew radiosonde designed by Dymond uses a strip of gold-beater's skin as a humidity element (rather than the hair hygrometer of the Finnish instrument), which rapidly loses its response characteristics at low temperatures. Different radiosondes may be used by civil and military services in the same country. The mean difference between pressure and temperature measurements made by U.S. Weather Bureau and U.S. Army radiosondes operating on a frequency of 1680 mc is 2.1 mb and 0.15°C, that is, about as much variation as can be expected from instruments of the same type made by different manufacturers or coming in separate batches from the same manufacturer.[58]

Unlike surface instruments, radiosondes provide data at many removes from the observer, and their possible sources of error are consequently more numerous. For example, errors arise from the effect of solar radiation on the sensing elements of the sonde, and from the time constants of the thermistors that are used to measure temperatures. The magnitude of the errors associated with the different effects are, for temperature, as follows: solar radiation, $\pm 0.2°$C; residual lag of the measuring element, $\pm 0.2°$C; transmitter error, $\pm 0.2°$C; recorder error, $\pm 0.2°$C; calibration error, $\pm0.2°C$; evaluator error, $\pm0.2°C$; lock-in error, if automatic tracking is provided, $\pm0.1°C$. Despite all these sources of error, the over-all probable error in temperature measurement is only $\pm0.5°C$ except at extremely high altitudes. These figures are averages of large numbers of errors: comparison of aircraft temperature measurements with those by radiosondes show that sometimes the latter may be as much as 2.5 to 3.5°C too high. In general, radiosondes seem to provide temperature figures that average 1 to 1.5°C above the true representative temperature because of radiation and lag effects.[59]

Radar winds, in which a microwave radar set is employed to automatically track a reflector attached to the radiosonde, are much more accurate then the conventional electronic theodolite method. In addition to the angles of azimuth and elevation of the balloon that the conventional method provides, the radar-wind technique measures the balloon's slant range plus the time-derivatives of all three quantities. The vector error in wind determination, assuming a balloon ascent rate of 5 m per sec and a mean wind of 30 knots, is only 3 knots at a height of 10 km for the radar techniques as against 10 knots for the conventional method. Even the error in computing radar winds that would be caused by neglecting the Earth's curvature may be taken into account, in view of the relative accuracy of radar winds.[60]

INSTRUMENTS FOR SPECIAL INVESTIGATIONS

The following account concentrates on the measurement of (a) atmospheric water in its various phases, and (b) air movement in various planes and at various scales. Unlike the meteorologist, who is mainly interested in instruments that provide him with measurements of the finest degree of resolution possible, the climatologist often finds that instruments now out of meteorological fashion will satisfy his needs. The current trend is for meteorological instrumentation to increase in complexity and expense, and although the climatologist may find himself involved with costly and complicated equipment, particularly if he is concerned with upper-air climatology or with microclimatology, there is much that can be achieved with simple apparatus, provided the investigator makes himself thoroughly familiar with the theory of his instrument.

Atmospheric water frequently requires measurement in locations where the standard climatological networks do not observe it in sufficient detail. It will be convenient to deal with this element in three phases: evaporation, humidity, and precipitation.

The rate of evaporation from a surface is usually expressed as the volume of liquid water, evaporated from a unit area of that surface in a unit time, which is equivalent to a certain depth of liquid water lost by evaporation per unit time from the whole area of the surface. The factors determining the rate of evaporation include: (1) the surface wind; (2) the temperature and humidity of the surface air; (3) the temperature of the evaporating surface; (4) the physical nature of the evaporating surface; (5) the variation with height of wind, temperature, and humidity in the lowest layers of the atmosphere; (6) the amount of liquid water in the surface that is available for evaporation. The actual process of evaporation may be regarded as taking place in two separate stages, involving first the diffusion of water in liquid or vapor form through solid or liquid matter to the interface between Earth's surface and the atmosphere, and second the removal of the resulting vapor from the surface by air movements.

The first stage depends on factors (4) and (6), the second stage on the vertical profiles of windspeed and specific humidity. The Earth's surface at the point for which measurements of evaporation are desired is an important determinant of the actual rate of evaporation.

The most important principle in evaporation measurement is that the observations should come from an area of the evaporating surface that is small and indistinguishable from its environment. Measurements made with any instrument that, however small, is not indistinguishable from its surroundings cannot be regarded as a reliable indication of the rate of evaporation from that surface. The difficulty is that most evaporation-measuring instruments differ from their environment in many important physical properties, impose their own microclimate on the evaporating surface, and consequently influence the evaporation rate. An additional complication is that under certain conditions there is a net *condensation* of water (i.e., dew), which is not taken account of by the instruments.[61]

The "standard" evaporation instruments include tanks, pans, porous porcelain bodies, and the *Piche evaporimeter*. The latter measures the water lost by evaporation from an exposed disc of filter paper, 3 cm in diameter, which is kept saturated. It is normally mounted in the Stevenson screen, since it is very sensitive to windspeed, and tends to behave like a wet-bulb thermometer under certain conditions. The porcelain bodies include *Livingston's sphere*, a porous sphere 5 cm in diameter, which is kept saturated with water, the loss by evaporation being measured, and the *Black Bellani atmometer*, a black porous disc, 7.5 cm in diameter, attached to the end of a porcelain funnel, into which water is fed from a burette. The Livingston sphere is only 3 mm thick and is therefore very fragile and liable to damage by frost. The Black Bellani atmometer is highly sensitive to wind changes, but less sensitive to changes in net radiation, which is an important indirect control of evaporation rates. The standard British *evaporation tank* is square, with sides 6 feet long and 2 feet high. It is sunk into the ground (or made to float on pontoons in a lake) with its top 3 inches above the surrounding surface, so that the water level in the tank is level with the latter. A rain gauge should be exposed 3 to 10 feet from the tank, to allow for rainfall into the tank. To estimate evaporation, the depth of water in the tank is measured accurately each day by means of a hook gauge inside a small still-water pond fixed to one side of the tank. *Evaporation pans* are usually circular, and are much smaller and shallower than evaporation tanks. The variation in the water level within them due to evaporation and precipitation is estimated by various eye readings.

All the above are, of course, indicating instruments. Various types of recording evaporimeter have also been devised. All the standard techniques have serious defects as sources of representative data. In particular, evaporation tanks and pans suffer from the *oasis effect*; i.e., they form wet oases in the midst of drier surfaces, and strictly speaking measure only the rate of evaporation from water bodies contained within pans or tanks, not the evaporation from any natural surfaces. The energy balance of an evaporation pan affects its rate of evaporation, and the evaporation of a moist natural surface may be found by multiplying the indicated pan evaporation by an empirically determined "pan factor." However, different evaporation pans have different factors for the same surface. In any case, the energy-transfer process of a pan is very different from that of a natural surface, such as a lake, a wood, or a crop. Short-wave radiation is reflected from the bottom of the pan, so that more energy is reflected from the water in an evaporation pan than from a natural water body of greater depth. The lateral extent of shallow water bodies also affects their rate of evaporation: rates are higher from isolated bodies of shallow water than from extended waters of similar depth. In general, the analysis of pan measurements is so complicated that pan-to-pan evaporation ratios determined for one area do not necessarily apply to other areas. Nevertheless, it is possible to obtain quite good correlations between rate of evaporation as measured by an evaporation pan and standard climatic parameters for the locality in which the pan is situated.[62]

The combination of evaporation (due to the evaporating power of the air) plus the water vapor given off by green vegetation forms *evapotranspiration* This may be measured by means of a *lysimeter*, in which a block or column of soil, its top level with the surrounding ground, is enclosed on its lower and vertical surfaces in a waterproof container. Water outflow or percolation through the base of the block is measured, usually by weighing the entire lysimeter and its contents. The soil in the lysimeter must have an undisturbed profile and structure, and the vegetation growing on its surface must be identical with the surrounding vegetation. After installation of the lysimeter, some months must elapse, until the soil has settled and the vegetation has established itself, before the instrument may be used. Once established, it should be almost invisible, so that the main principle of evaporation measurement is satisfied. There are complications; for example, similar vertical temperature and moisture profiles must be achieved both inside and outside the lysimeter, and reliable instruments can be quite complex. Daily rates of *transpiration* may be measured by means of a floating lysimeter.[63]

Since the actual evapotranspiration rate at a given time in a given place depends on the amount of moisture available, it follows that the only possible *standard* measure of evapotranspiration is that for an extensive, vegetation-covered land surface provided with adequate water at all times. By definition, this is the *potential evapotranspiration* for that place. Potential evapotranspiration has been defined by Penman as "the amount of water transpired in unit time by a short green crop of uniform height, which completely shades the ground and is never short of water," and by Thornthwaite as "the water loss from a moist tract of soil which is completely covered by vegetation and large enough for oasis effects to be negligible."[64] With the moisture supply regarded as unrestricted, potential evapotranspiration is limited only by the available energy, so that it cannot exceed the water equivalent of the net radiation, provided advection and changes in soil heat storage can be neglected.

An instrument for the direct measurement of potential evapotranspiration is the *Thornthwaite evapotranspirometer* (see Figure 1.4). It is essential, to use such an instrument, that the soil in the root zone be kept fully saturated with water: this saturation is maintained by a carburetor line feed in the original evapotranspirometer. In a simplified version of the instrument, the carburetor line feed is replaced by a watering can for surface irrigation, thus copying nature, where replenishment is by precipitation. The difficulty with this simplified instrument is that oasis effects become very important: there must be vegetation for an extensive area around the instrument identical to that on it, but such a large area cannot be irrigated all at once with a watering can.[65]

The standard portable instrument for measuring atmospheric humidity is the aspirated hygrometer. This is simply a standard hygrometer in which the airflow past the bulbs is controlled in various ways, for example, by whirling the instrument through the air by hand at a constant rate (as in the British *whirling psychrometer*

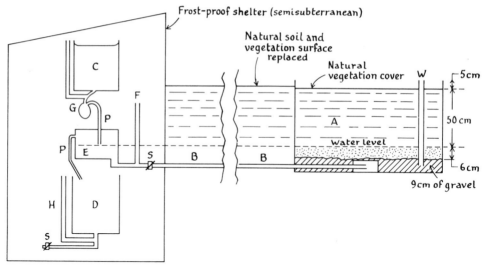

Figure 1.4.
Thornthwaite evapotranspirometer. After C. W. Thornthwaite and J. R. Mather, *AMGB*, 3, series B (1951), 16. A, evapotranspirometer field tank (area, 4 square meters), filled with well-mixed sandy loam; B, underground water-supply pipe; C, water-supply tank (capacity, 40 liters); D, water overflow tank; E, water-level-regulating mechanism; F, glass standpipe, permitting escape of trapped air bubbles; G, sediment bulb; H, glass gauge; P, plastic tubing; S, stopcocks; W, perforated water-level-measuring tube.

and the American *sling psychrometer*), or by using a clockwork or electric fan to drive air past the bulbs (as in the *Assmann psychrometer*). For continuous records of humidity, a *hair hygrograph* is usually employed; it depends on the principle that the change in length of a human hair is a function of the *change* in relative humidity (both above and below freezing), not of the actual amount of water vapor in the air. The response of any given hair to humidity fluctuations is more complex than the response of a thermometer to temperature fluctuations. In steady conditions, the instrument is accurate to $\pm$ 5 per cent for relative humidities above 20 or 30 per cent, but its sensitivity is only about $\pm$ 1 or 2 per cent; given sudden changes in humidity, it may produce serious errors because of the lag.[66]

Various other types of hygrometer are available; they may depend on, for example, chemical absorption, electrical absorption, or diffusion. A most interesting modern instrument is the *infrared absorption hygrometer*, which measures the absolute humidity in a light path 1 meter or more in length. A beam of energy is passed through the atmosphere, and the attenuation of this energy by water-vapor absorption during its passage is measured by means of a lead sulphide photocell and an amplifier. The 1.37μ water vapor absorption band is usually used, and bandpass light filters are used to isolate the measurable quantity. This technique essentially measures humidity by absorption-spectra analysis, and has the advantages of rapid response and high sensitivity at low humidities. Another type of infrared absorption meter may be used to measure relative amounts of precipitable water in the atmosphere.[67]

The lag of the electrolytic type of electrical absorption hygrometer increases rapidly as temperatures decrease. An electric hygrometer using a carbon film as the sensitive element, instead of a thin electrolyte film containing a hygroscopic salt as in the conventional electrical absorption type, has rapid response and a high sensitivity. Measurement of the water content of air at temperatures below freezing is always difficult; it may be carried out, for example, to an accuracy of within 2 per cent by bubbling the air through absolute methanol and then titrating an aliquot of the latter with a reagent.[68]

The direct measurement of dew-point temperature involves an interesting application of thermodynamic principles. The classic method, devised by Regnault, involves cooling a polished surface until atmospheric water vapor just begins to condense on it. More accurately, the temperature at which condensation first begins should be measured, and also the temperature at which the condensate reevaporates. Their arithmetic mean should then give a close approximation to the true equilibrium temperature. The heating and cooling rates to which the polished surface are subjected are made smaller and smaller so that the two temperatures approach each other and should finally reach true dew-point temperature when the heating and cooling rates are zero. Obviously, such a procedure is suitable only for discrete observations and not for providing continuous records. In the *Thornthwaite dew-point apparatus*, an electric circuit switches the heating and cooling component on and off at regular intervals, so that frequent spot measurements are made. In the *Barrett-Herndon dew-point hygrometer*, continuous recording is made possible by a radio-frequency induction-heating system and an electronic proportional heat-control servomechanism.[69]

Dew, a form of precipitation, has been neglected by climatology. Dew is the deposition of water drops by direct condensation of water vapor from the adjacent clear air. It must be distinguished from mist or fog drops clinging to a surface, and also from the water exuded from plants by *guttation*, a process that is fairly frequent on overcast nights when the soil is warm and at or near field capacity. The most common form of dew observation is by means of the *Duvdevani dew gauge*, in which the visual appearance of dew drops accumulating on specially treated blocks is used as a measure of the intensity of the dewfall. Observations made by means of collectors, patterned after the Duvdevani dew block, mounted at ten successive heights between 3 and 72 inches above the ground, show that dew intensities normally decrease with height, as one would expect. However, the height of 24 inches above the ground appears to be critical, at least in part of the Mississippi Delta. Below this height, the vertical variation of dew is a micrometeorological phenomenon, not subject to forecasting by conventional synoptic methods. Above 24 inches, the horizontal (i.e., areal) variation of dew is a macrometeorological phenomenon, subject to synoptic influences. Recording dew gauges, made of apparatus that registers the change in weight of expanded polystyrene (styrofoam) blocks mounted at various heights above a meadow surface in Idaho, showed that the dew deposits ranged from zero on cloudy, windy nights to 0.014 inch on a clear night following a day with rain. The monthly amounts varied from 7 to 29 per cent of the monthly rainfall totals.[70]

Various forms of *dew balance* have been designed for permanent installation, as has an instrument for measuring dewfall. A portable, battery-operated *dew meter* gives direct readings of the total water deposit on a uniform, short grass surface. The formation and persistence of water deposits on plant shoots can be recorded by another type of apparatus.[71]

Fog precipitation may be measured by a fog catcher mounted on a rain gauge. A large amount of water can be precipitated on a surface from fog. At Table Mountain, South Africa, the fog precipitation in one year was 3,294 mm at a rate of 3.75 mm per hour, compared with 1,940 mm for the rainfall at 1.84 mm per hour.[72]

The form, size, and shape of the particles are obvious aspects of precipitation, but are not normally represented in published climatological tables, except in terms of the relative frequency of the different precipitation types. *Hail* may be dealt with fairly easily. The physical characteristics of the hailstones falling in a certain locality may be determined by collecting the stones and examining them in a laboratory. The relative intensity of falling hail may be observed by hail indicators, one-foot-square sheets of styrofoam one inch thick covered with aluminum foil, and mounted one foot above the ground. The different degrees of indentation of the foil surface by the hailstone impact provides a measure of hail intensity.[73]

Raindrops are more difficult to measure. Rain or clouddrops may be collected by a technique in which the droplet makes a physical contact with, for example, a specially coated slide, or the sizes of the drops may be inferred theoretically by various passive techniques that do not disturb the droplets. For each collection technique, there is a minimum measurable drop size. The slide may be coated with magnesium oxide, or alternatively the raindrops may be collected in a flour-filled tray as flour pellets. Once the droplet has been collected the problem is to measure it without disturbing it. A simpler method is to measure the dimensions of the stain the drop produces by impact when it falls on a piece of absorbent filter paper. There are in fact various techniques for measuring raindrops, and it is possible to obtain a continuous record of the changes in raindrop size during a given fall of rain.[74]

The significance of raindrop measurements is that they show different types of cloud and different precipitation mechanisms to be associated with different raindrop-size distributions. In such a study, certain laws governing raindrop size must be taken into account. A rainfall may be completely described by the median diameters of the drops that compose it, and the rainfall intensity. The fractions of the total volume of rain that fall as drops of specific sizes are normally distributed with respect to the fall speed of the drops, the distribution being symmetric about the median drop size. The deformation of the falling raindrops is included in the theory, so that the different terminal velocities of droplets of various eccentricities is taken into account. The raindrop-size distribution discovered by Marshall and Palmer is probably the most used; another was devised by Best. By correlating their distribution—inferred from raindrop-size records on dyed filter paper—with radar echoes, Marshall and Palmer showed that, for rains of a given intensity, there is a definite distribution curve describing the number of raindrops within each range of diameters. For rain reaching the ground from warm-front clouds, the distribution of raindrop sizes is as described by the Marshall-Palmer distribution; for rain originating by the coalescence process, as for example in small shower clouds, the raindrops have rather larger diameters than indicated by the Marshall-Palmer distribution for the same rainfall intensity; and for thunderstorm precipitation, the median drop diameter is very much greater than in the rain falling by the Bergeron mechanism from frontal clouds. Photoelectric spectrometers, which enable raindrop-size distributions to be measured at intervals of one minute, show that, for moderate and high intensity rains, the drop-size patterns are much affected by wind shear and gravity sorting at the time of onset of the rain, and also by impact splashing and aerodynamic breakup of the drops. Despite this, the drop-size distribution at the ground still provides evidence as to which dominant process (coalescence or the Bergeron-Findeisen mechanism) produced the rain (see Appendix 1.6).[75]

Observations of raindrop sizes and their relation to rainfall intensities point to another way of measuring rainfall than by the rain gauge: to measure the rain falling through the air before it actually reaches the ground, by radar. Although radar precipitation studies are expensive to initiate and to maintain, they are the only practical way to obtain a detailed picture of the rain *actually falling* over an extensive area at the time of observation. When instantaneous rather than "historical" measurements of precipitation are required, a rainfall-observing network based on radar will probably prove cheaper in the long run than a network of the required complexity based entirely on rain gauges. Although fully operational commercially produced radar sets are very expensive, government-surplus equipment is available relatively cheaply and can be adapted for rainfall studies by anyone with the necessary electronic skills. Climatologists need some knowledge of radar methods, not only because of this possibility, but also because of the development of the field of *precipitation radar climatology*, which covers the production of maps of average precipitation based on the distribution of radar echoes; the latter are extremely complex, and without an awareness of the factors determining the accuracy of radar-precipitation echoes, one can make very serious misinterpretations.[76]

Early in the Second World War, it was proved that, with the statistical knowledge then available on the relation between rainfall intensity and raindrop size, it should be possible to estimate precipitation rates from the intensity of the echo representing the precipitation on the screen of a radar receiver. This was confirmed in 1947 by actual observation, and radar analyses of precipitation patterns in both the vertical and the horizontal were made soon after. Since then, there have been numerous successful observations of precipitation patterns by radar: five parameters of the pattern can be measured, covering its three space dimensions, its time variations, and the intensity of the precipitation comprising the pattern. If the cathode-ray screen is in the form of a range-height indicator (RHI), then separate maps are required for various bearings to give a complete picture of the precipitation pattern. If the screen takes the form of a plan-position indicator (PPI) display, then separate maps are produced at constant altitudes (CAPPI). In effect, the RHI presents a vertical section through the precipitation pattern, the PPI a horizontal slice through it. The radar echo information may also be presented in the form of *precipitation profiles*, useful for synoptic purposes, and in the form of *echo-rainfall intensity charts*, which enable the intensity of rainfall to be determined from a knowledge of the intensity and range of the radar echo (see Figure 1.5).[77]

The basis of radar precipitation studies is the radar equation, which specifies the average power of the radiation back-scattered to the radar in terms of the precipitation particles that reflect the transmitted pulse, This back-scattered power is then related to the precipitation intensity by means of (a) an assumed drop-size distribution, usually that of Marshall and Palmer, and (b) Mie scattering theory. The attenuation of the radiation by atmospheric gases, clouds, and, especially, precipitation must be allowed for when performing the calculations; attenuation is particularly serious with 3-cm wavelength radar, but much less serious with 10-cm or 5.7-cm radar. Consequently, 3-cm radars have a much lower effective range for precipitation studies than other types. Because of attenuation by the rain itself, the area of the radar echo is

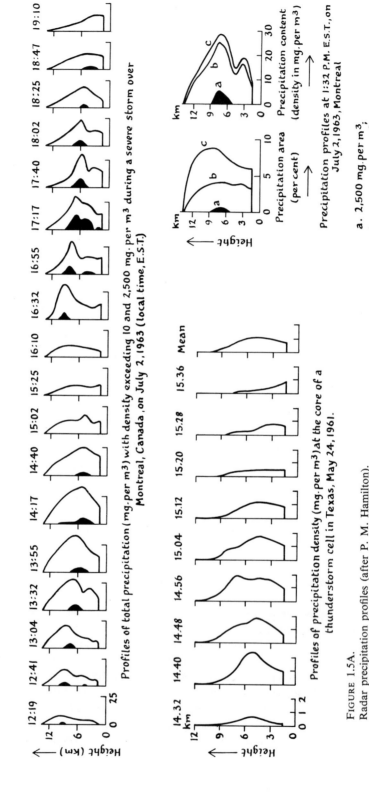

Profiles of total precipitation (mg. per m³) with density exceeding 10 and 2,500 mg. per m³ during a severe storm over Montreal, Canada, on July 2, 1963 (local time, E.S.T.)

Profiles of precipitation density (mg. per m³) at the core of a thunderstorm cell in Texas, May 24, 1961.

Precipitation profiles at 1:32 P.M. E.S.T., on July 2, 1963, Montreal

a. 2,500 mg. per m³;
b.   160 mg. per m³;
c.   10 mg. per m³.

FIGURE 1.5A.
Radar precipitation profiles (after P. M. Hamilton).

somewhat less than the area of precipitation it represents. For a 3-cm radar looking at a point 60 miles away, the total precipitation over a season reported by the radar echo would be only 47 per cent of the actual total precipitation if attenuation were not corrected. From a measurement of the intensity of the radar echo, certain precipitation parameters may be obtained quantitatively, including the rainfall rate, the rain density, and the precipitation volume. In addition, the type of precipitation may be inferred from the echo pattern; convective-type precipitation gives rise to cellular echoes, similar in horizontal and vertical extent, which form linear groups. Stratiform-type precipitation displays a " bright band " at its melting layer, marking the transformation of ice crystals and snow in the upper part of the cloud to rain or drizzle in the lower part. Convective-type precipitation does not display the radar bright-band phenomenon, except in its decaying stage. (On the entire preceding paragraph, see Appendix 1.7.)[78]

Once certain errors from antenna beam effects and those due to attenuation have been resolved,[79] radar is a very valuable tool for filling in gaps in the precipitation-observing network. Comparison with rain-gauge measurements shows that, on the assumption that rain-gauges give the true value, radar measures area-average precipitation better, spot intensity more coarsely. The great advantage of radar is that it provides a quantitative picture of the detailed geographical distribution of precipitation in an extensive area, a picture that may be obtained in no other way (see Figures 1.6 and 1.7).

If a radar precipitation-measuring network is impossible, then obviously various types of rain gauge may be employed to obtain data. Recording rain gauges provide a continuous record of either the total amount of precipitation that has fallen since the record commenced or the rate of rainfall. They are very useful for recording the times of onset and cessation of a rain, and the rate at which it fell between these times. However, they do not form standard methods of measurement, and an ordinary rain gauge must be maintained close by to provide the standard against which the continuous records may be checked. The recording gauges operate on one of several principles: gauges recording the amount of precipitation may, for example, measure the weight of rainwater they have collected; they may have a tipping-bucket arrangement; or they may be float gauges with or without automatic siphoning mechanisms. It is sometimes possible with records from a gauge measuring rainfall rate to estimate total precipitation amount. Home-made rainfall recorders that give reasonable results may be produced very cheaply.[80]

There are several different types of "standard" rain gauge, and it is by no means easy to compare measurements made with the different types. Gauges may be compared by interpreting the data obtained from arrays of rain gauges in terms of the climatic conditions associated with the rainfall. This procedure enables the relative defects of different types of gauge to be determined. Since at least five years' data are required to compute reduction coefficients stable to within 1 per cent, it is impossible to apply the coefficients to measurements obtained during individual storms. Errors due to evaporation losses frequently complicate the comparison between different types of gauge. These losses may be minimized in various ways, for example by the use of oil, after which the error due to evaporation should be negligible for periods of 30 days or more. The standard rain gauges normally measure daily or monthly precipitation totals; gauges measuring weekly totals can be made quite easily.[81]

The gauges so far discussed are mounted with their rims horizontal. By adopting

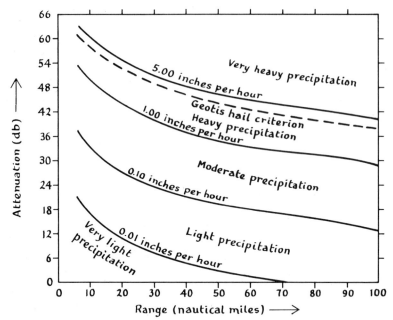

FIGURE 1.5B.
Echo intensity and rainfall intensity chart for WSR-57 radar (after
J. W. Wilson). Given the range and intensity of the radar echo, chart
enables intensity of precipitation producing the echo to be inferred. The
Geotis Hail criterion was derived from radar observation of hailstorms
in New England and does not necessarily apply elsewhere.

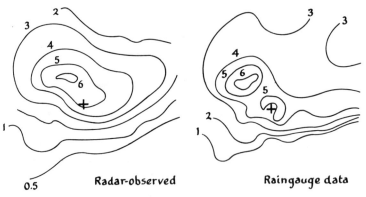

FIGURE 1.6.
The 24-hour rainfall patterns on May 7-8, 1961 (after D. Atlas).

different types of mounting, it is possible to increase considerably the value of rainfall
observations for specific purposes. An octagonal directional rain gauge, for example,
developed to determine how rain impinges on vertical surfaces, shows that walls
facing southwest in the London area receive (with moderate exposure) the most rain
and those facing north receive the least. Southwest-facing walls receive approximately
one quarter of the normal precipitation, and north-facing walls receive about one
sixteenth. By mounting three gauges in an inclined position, symmetrically arranged

A. key map

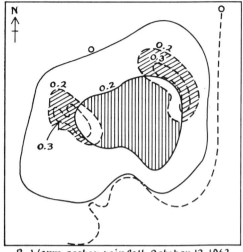

B. Warm sector rainfall, October 12, 1962.
midnight to 1:00 P.M., P.S.T.

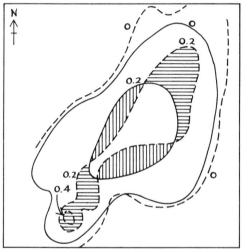

C. Front stationary over Sacramento,
10:00 P.M., P.S.T., October 12, 2:00 P.M.,
P.S.T., October 13, 1962.

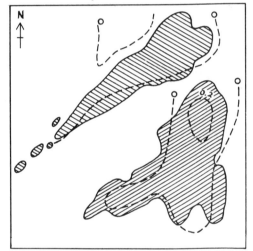

D. Prefrontal rainfall, October 11, 1962.

——— Average hourly precipitation according to Sacramento radar in inches per hour.

– – – – Average hourly precipitation according to rain gauge network in inches per hour.

Overestimate of precipitation by radar relative to rain gauge data.

Underestimate of precipitation by radar relative to rain gauge data.

Weak radar echo and isohyets, 3:00 A.M., P.S.T.

Weak radar echo and isohyets, 6:00 A.M., P.S.T.

FIGURE 1.7.

Radar versus raingauge measurement of precipitation. After R. L. Weaver, *MWR*, 94 (1966), 466.

around a vertical axis, so that their apertures are facing outward at an angle of 40° to the horizontal, the mean direction of travel of a rainstorm may be calculated from the ratio of the catches.[82]

Various types of rainfall-rate recorder have been developed. A very sensitive type, the *rotary rain indicator*, has as its sensing element a rotating grid in an electrical circuit that is closed by the receipt of a drop of water. Small-scale time variations in rainfall rate may be continuously recorded by an instrument that measures the water level in a cylindrical gauge to within 0.001 inch. Another type of recording gauge is sensitive enough to record dewfall. Particularly useful is a recorder that can function for long periods without attention, and can respond to changes in precipitation rate within a few seconds.[83]

Snowfall amount is by no means well-measured at present. It is usually determined by melting the snow that collects in the funnel of a standard rain gauge, and then measuring the depth of the resulting water as if it were rain. With heavy snowfalls, or when weekly or monthly gauges are used, the snow that has collected above the funnel is pressed into it. *Depth of snow lying* is found by pushing a graduated scale vertically into the snow where it lies evenly distributed without drifts. A mean value of several such observations is taken and regarded as the average total depth of snow on the ground surrounding the station. Comparison of figures obtained in this manner is very difficult, because the physical nature of the underlying ground surface and the general topoclimatological situation are also determinants of the depth of snow accumulating at any one place.[84]

Air movement is a three-dimensional vector quantity. The convention is to concentrate on measuring the horizontal component of this quantity for synoptic and macroclimatological purposes, and to leave observation of the vertical component to the microclimatologist. But in mountainous areas or wherever the ground has pronounced changes in slope, the true wind will be considerably inclined to the horizontal; a defect of standard wind data in climatological tables becomes apparent.

The direction from which the wind is blowing is normally specified in degrees clockwise from true north. The horizontal component of wind direction is usually measured by means of a wind vane, which should have the following properties. It should be able to turn about its pivot with the minimum of frictional retardation; it must be properly balanced, so that it does not show a bias toward one wind direction; it should be mounted with its axis exactly vertical and its direction indicators correctly orientated to true north; there should be sufficient damping of the movement of the vane to prevent resonance between the vane and natural wind fluctuations; and it should produce maximum torque, in relation to its moment of inertia, for a given change in wind direction. If a vane has the above properties, it should be accurate to $\pm 1°$ in steady winds, at all except very low windspeeds. (On variable winds, see Appendix 1.8.)

Continuous recording of wind direction suffers from the handicap that the vane may rotate through 360° several times during one period, and this possibility must be taken into account in the design of the mechanism. Because of this difficulty, the standard wind vane is usually an indicating instrument. Simple wind-direction recorders have been constructed for special investigations, as have recorders for obtaining the mean wind direction in turbulent situations where the vane fluctuates considerably.[85]

There are three types of conventional instrument for the measurement of horizontal windspeed.[86] The oldest type of instrument, the *pressure-plate anemometer*, was developed in Italy about 1570 and introduced into England by Hooke in the mid-seventeenth century. It depends on the pressure exerted by the wind on a flat plate exposed at right angles to the wind direction. The pressure is balanced by a restoring force brought about by the deflection of the plate, the deflection providing the required measure of windspeed (see Appendix 1.9).

The *pressure-tube anemometer* is based on Bernouilli's equation connecting the hydrostatic pressure at a point on a horizontal streamline within the airflow to the velocity of the wind at that point (see Appendix 1.10). This instrument is very reliable, but it requires permanent installation on a very securely guyed mast or tower, so that it is unsuited to spot or traverse observations.

The *rotation-type anemometer*, particularly the cup type invented by Robinson, is the usual instrument for measuring windspeeds in special investigations. The cup anemometer comprises three or four (normally three) cups mounted symmetrically about a vertical axis, so that the diametral plane of each cup is in the vertical. The wheel supporting the cups will rotate horizontally because the wind force on the concave side of a given cup will be greater than that on the convex side in a similar position. Because of friction on the bearings of the wheel, a certain minimum windspeed must be attained before the cups begin to rotate, but once the wheel is in motion, its speed of rotation (for a given anemometer) depends only on the windspeed, provided the wind is steady. The rate of rotation of the wheel is measured by various electrical or mechanical counting devices, which provide either the mean windspeed over periods of one minute or more, or the almost instantaneous windspeed. The error in the speed indicated by a cup anemometer in steady winds does not exceed 1 knot for winds of 5 to 100 knots. The sensitivity of the instrument is expressed by the anemometer factor, which depends to some extent on the dimensions of the instrument (see Appendix 1.11).

The advantages of the cup anemometer are its portability and ease of operation, but it has serious disadvantages. In particular, it responds unequally to increasing and decreasing windspeeds, giving too a high speed in gusty winds. The mean windspeed indicated by the instrument in unsteady winds will be greater than the true mean windspeed, because the cup wheel accelerates more rapidly during increasing windspeeds than it decelerates during falling windspeeds (see Appendix 1.12). An instrument operating on an entirely different principle, the *drag-sphere anemometer*, responds to very rapid wind fluctuations and is very useful in unsteady winds.[87]

Anemographs are recording anemometers. The cup anemometer may be coupled to an electrical contact device that provides continuous records, but a much more detailed record (including gusts) is obtained if a pressure-tube anemograph is used. Other types of anemograph have been developed, including an instrument for synoptic use that provides records of the 10-minute average wind vector (i.e., its north-south and east-west components) in digital form.[88]

Sonic anemometers provide much more accurate measures of windspeed than the conventional instruments. One series of observations with such an instrument showed that "standard" surface wind measurements made by a conventional anemometer mounted on an airfield control tower were constantly unrepresentative of the actual wind fluctuations of the runway, the windspeeds being as much as 20 per cent in error. By contrast, the sonic anemometer provided windspeeds accurate to within 1 knot over a range from 0 to 60 knots. The measurement of wind velocity by sonic means is absolute, and the calibration of the instrument may be predicted on theoretical grounds (see Appendix 1.13). An additional advantage is that the method enables both windspeed and wind direction to be measured remotely, with an error of not more than ± 1.0 miles per hour and $\pm 5.0°$, respectively.[89]

Some wind-measuring systems record both windspeed and wind direction. Published climatological data for Britain were for many years provided by the *Dines pressure-tube anemograph*, a very permanent and elegant instrumental system, which records both speed and direction on a single chart by means of two pens. In comparison with the Robinson cup anemometer, however, the Dines anemograph overestimates mean windspeeds (at speeds exceeding 5 m per sec) by as much as 10 per cent. This discrepancy arises partly from the lag of the Dines instrument, and partly from subjective errors that are unavoidable when evaluating the anemogram trace.[90]

A remote-recording electrical anemograph has been devised, and also a relatively inexpensive semiportable recorder for both windspeed and wind direction that can provide long-period records in remote areas.[91]

The climatologist engaged in local studies frequently needs observations of the wind structure of the atmosphere through layers extending from a few thousand feet in thickness to almost the entire troposphere. The most usual, although laborious, technique in use for such studies is *pilot-balloon sounding*. The principle of the pilot-balloon technique is very simple. A balloon with a known (or assumed) rate of ascent is released from a point on the Earth's surface and its progress is tracked by means of a theodolite. Observations of the angle of elevation and azimuth of the balloon are taken every minute. In Figure 1.8, A represents the position of the theodolite and B the position of the balloon at a given instant; C is vertically below B in the same horizontal plane as A. AC may be represented by two displacements, $\overline{AX}$ and $\overline{XC}$ toward the east and north, respectively:

$$\overline{AX} = h \cot E° \cos A°$$

$$\overline{XC} = h \cot E° \sin A°$$

where h is the height of the balloon above the horizontal plane through the observer (i.e., $\overline{BC}$), and $E°$ is the angle of elevation and $A°$ the azimuth of the balloon. Since there are four possible quadrants in the 360° field of vision of the theodolite into which the azimuth $\angle NAC$ may fall, the displacements may have different signs. From the various displacements associated with each successive position of the balloon, the direction and speed of the wind may be computed, on the assumption that the movements of the balloon accurately follow the air motion (see Appendix 1.14).

For normal pibal soundings, rubber balloons are employed that weigh 10, 20, or 30 grams, corresponding to circumferences of 48, 70, and 90 inches, respectively, when the balloon is fully inflated with hydrogen. Balloons of this size ascend at 400, 500, and 500 feet per minute, respectively. The 10-gram balloons are used to determine the heights of low clouds (*ceiling balloons*) as well as for wind observations. The 20-gram balloons are used for soundings to altitudes of not more than 10,000 feet, and the 30-gram balloons for ascents on windy days when the cloud base is more than 10,000 feet above ground level. Some 80-gram and 100-gram balloons are used for special purposes, the former being 150 inches in circumference and rising at 700 feet per min, the latter rising at 1,000 feet per min. For nighttime ascents, a small paper lantern containing a lighted candle is usually suspended from a 30-gram balloon.

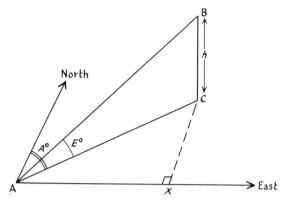

FIGURE 1.8.
Technique of pilot-balloon soundings.

Several methods of determining the height of the balloon are in use. If a single theodolite is used to track the balloon, there are two possibilities. First, it may be assumed that the balloon rises at a constant rate, in which case the height, *h*, equals the rate of ascent of the balloon multiplied by the time elapsed since release of the balloon. Second, in the *tail method*, a length of thread or silk with an attached foolscap-size sheet of paper with rigid edges is suspended from the balloon, and its apparent length is observed by means of a graticule inscribed on the eyepiece of the theodolite. The tail method is much more accurate than assuming a uniform rate of ascent, because convection currents, which increase the actual rate of ascent, and gas escapes, which decrease it, may develop without the observer's knowledge.[92]

The vector error of winds observed by pilot balloons may be computed from theory (see Appendix 1.15). The error for relatively short-range observations may be reduced slightly if two theodolites are used to track the balloon instead of one, but the errors with two theodolites are larger than those with one when the balloon is farther away than the length of the baseline joining the two theodolites. During observations by the single-theodolite method, the balloon's position is usually tracked every minute, and winds are then reported for every 1,000 feet of height above sea level. For relatively low observations, say, to heights not more than 10,000 feet above the ground, observations by double-theodolite made at 20-second intervals give the best results. Triple-theodolite observations are useful on some occasions.[93]

The computation of winds from a pibal ascent involves considerable trigonometric calculation, which may be facilitated by a special slide rule. Although the linear coordinates of the balloon's position change considerably during its ascent, its angular coordinates (i.e., its azimuth and elevation) change little from one observation to the next; the trigonometric scales of the rule may be fixed whilst the linear scales are movable. Double-theodolite observations in particular may be evaluated by means of a computer; such evaluation enables the balloon's position to be determined without any assumptions being made about its rate of ascent or about the absence of vertical currents. Strictly speaking, the line of sight from theodolite to balloon is not a single ray but a cone, whose axis coincides with the ray and whose vertex is at the theodolite. Owing to errors in observational timing and theodolite accuracy, the balloon may be anywhere within a cone whose semivertical angle is 0.05°. If two theodolites are used, the region of intersection of the resulting two cones may be determined by a vector method. Other improvements on accuracy may be made by adapting the theodolite to semiautomatic operation, or, more expensively, by employing a pulsed-light theodolite. The latter electronically measures the slant range of the balloon and automatically computes elevation, windspeed, and wind direction from the slant range, azimuth, and elevation angles.[94]

Many techniques other than pilot balloons may be used to make upper-air soundings of much greater accuracy; these are usually expensive, but the expense may be justified for some investigations, such as wind observations intended to determine the climatology of a rocket- or missile-launching site.

Humidity data provided by standard radiosondes often show anomalies, and efforts have been made to provide better humidity sensors for special investigations. One of these is the *hygristor*, a miniature dewcell; another is a type of modulator that uses a hypsometer. Humidity has been measured indirectly by means of radiation data obtained from a radiosonde, and a lightweight radiosonde (the *transpondesonde*) has been developed. The standard radiosonde may also be adapted to record variations in radio refractive index.[95]

Because water is present in the stratosphere only in very small amounts, special instrumentation is necessary if the variations in stratospheric humidity are to be studied climatologically. Dew-point and frost-point hygrometers have been developed for mounting on aircraft, the error in dewpoint measurement being not much more than 0.7°C, that in frost-point measurement being 1.0°C at -40°C, increasing to 1.7°C at -75°C. An automatic, continuously operating, electronic dew-point hygrometer was first mounted on a radiosonde in the late 1940's; since then many measurements of humidity distribution have been made in the stratosphere. Dew-point hygrometers launched over North America on April 8, 1960, revealed a large and unexpected mass of dry stratospheric air extending from near the North Pole to the latitude of Washington, D.C. Automatic frost-point hygrometers are also available for very low humidity conditions and low (-70°C) temperatures. Water-vapor content of the stratosphere has also been measured indirectly, by means of the absorption of solar radiation by water vapor, observations being made with a high-resolution spectrometer installed in an aircraft.[96]

One disadvantage of radiosondes is that, because of the horizontal drift of the balloon, an actual sounding may be very far from vertical, though it is assumed to be vertical for climatological purposes. *Meteorological rockets* not only ascend vertically, but also gather data from a very much neglected part of the atmosphere, i.e., those layers above balloon level but below satellite level, say, from 100,000 to 300,000 feet. The development of small and relatively inexpensive rockets allows synoptic rocket soundings, and there is now a meteorological rocket network in North America. Another type of rocket enables low wind profiles to be obtained comparatively cheaply. It is fueled by a cold propellant instead of the usual fire system, and may be reused several times. This propellant involves the hydrolysis by water vapor in the air of titanium tetrachloride, which is placed in the rocket in liquid form. The resulting smoke column is photographed simultaneously, every ten seconds, by two cameras approximately 2,000 feet away from the launch site and at right angles to each other. A true vertical wind profile over the launch site may then be obtained by photogrammetry. The only drawback is that the technique can be used only during the day, and only if there is neither rain nor low clouds.[97]

An extension of the rocket technique is provided by the *rocket grenade*, which allows winds between 30 and 80 km to be measured. Grenades are ejected by the rocket during its upward path, and the average horizontal windspeed at each altitude may be calculated from the effect of the wind on the soundwave from the explosion traveling downward.[98]

Other special-purpose methods for measuring upper winds include the observation of smoke drift from rocket flashes, for very high-level winds. The tracking of guided missiles provides almost instantaneous measurements of wind profiles right into the higher stratosphere, because the missile requires less than 2.5 minutes to reach a height of 60 km. The theory for computing the windspeed assumes that the air motion is horizontal throughout the ascent. Upper winds have also been measured by photographing meteor-trains. Simultaneously photographing a meteor from two stations enables atmospheric densities as well as windspeeds to be determined. An easy and inexpensive (but not very accurate) way of estimating upper winds is by nephoscope observations.[99]

The measurement of winds by radar is a particularly useful technique. Microwave radar is used to track automatically a reflector attached to the radiosonde balloon. The radar not only measures azimuth and elevation angles and slant range of the balloon, but also the time derivatives of these quantities. Radar wind finding is much more accurate than other standard methods; assuming a balloon rate of ascent of 5 m per sec and a mean windspeed of 30 knots, the vector error of the radar-determined wind is only 3 knots at an altitude of 10 km compared with an error of 10 knots for the pibal and other conventional methods. Tracking a single balloon by means of two radar sets shows that the radar technique can detect significant wind variations within distances and periods of time that would be too short for normal radiosondes. If steady precipitation is falling in the sector of observation, *Doppler radar* provides much wind information, including instantaneous measurements of wind direction looking crosswind, measurements of windspeed and direction at a fixed point at short intervals (say, every 5 seconds) for an extended period, and measurements of mean wind direction and mean windspeed over the sector at repeatable heights every few minutes. In addition, since Doppler radar essentially measures the radial component of wind, information about convergence is provided too. The standard error of windspeed measurements by Doppler is about 0.25 m per sec compared with 1 m per sec for conventional radar. The limitation of the Doppler technique is that observations are limited to moderate heights (3 to 5 km, for example) and to occasions when precipitation is steady.[100]

Pulse Doppler radar is a further refinement, and provides much more detailed wind profiles than conventional wind-finding radar. The conventional rawinsonde provides wind data as averages within altitude layers approximately 600 m thick. It does not provide data on small-scale wind variations, because of tracking difficulties and also because the balloon responds very poorly to a rapidly changing wind field. By using high-precision pulse Doppler radar to track metalized spherical balloons (the "Rising Observational Sounding Equipment" or ROSE system), wind averages for atmospheric layers 50 m thick may be determined with an error of less than 2 m per sec. The ROSE technique demonstrates that there are mesoscale disturbances in the wind field with a magnitude of 5 to 10 m per sec in windspeed and 5 to 20° in wind direction, and that can reach 2 km in vertical extent. Self-induced oscillations of the balloon set a limit to the accuracy possible with this technique.[101]

The "window" method enables the fine structure of air motions in the free atmosphere to be studied by means of conventional radar. Freely falling strips of light metal foil (*window* or *chaff*) are dropped from aircraft or balloons, and their movement is followed with radar. Each strip is about half the wavelength of the radar signal (usually a 10-cm radar is employed). The strips are dropped in bundles, weighing between 10 and 100 grams, and containing tens of thousands of strips. Since the radar echo from a single bundle approximates that from a sphere 1 m in diameter, the bundle is "visible" to the radar at over 100 km, and its position can be accurately followed up to 50 km. The metal foil has a natural rate of fall of 100 cm per sec in still air, so that wind velocities may be measured to 74,000 feet. Direct measurements of vertical air velocities and divergence are possible, and small-scale turbulence may be observed by measuring the increase in size of the radar echo with time.[102]

The atmospheric sounding techniques so far discussed are concerned with vertical or oblique probing. Horizontal soundings are possible with the *transosonde* system. Constant-pressure balloons that float (usually at 300 mb) provide accurate measurements of pressure and wind for synoptic purposes, but ignore turbulence disturbances insignificant in studying global-scale patterns. The transosonde system is truly a worldwide probing technique: flights every 12 hours from two stations far enough apart in latitude is adequate for upper-air coverage (at 300 mb) for the entire temperate zone. Measures of horizontal and vertical wind velocities are possible, and momentum-flux data may be readily obtained. The relative accuracy of transosonde observations, after allowing for inertial instability effects, means that the technique can be used to estimate the errors in ordinary synoptic analysis. Such a study for the northeast Pacific shows that the ratio of vector-geostrophic-wind error to geostrophic wind varies from 0.15 near the west coast of North America to 0.40 in the central part of the North Pacific. A modification of the transosonde technique would allow it to provide detailed information for the tropopause.[103]

Instruments for synoptic or macroclimatological observations are constructed with large enough lag coefficients to eliminate very short-period fluctuations in atmospheric quantities. Micrometeorological fieldwork uses very sensitive and delicate instruments, and is much more akin to a precise laboratory experiment in physics than to conventional weather observation. The microclimatologist employs micrometeorological instruments primarily to determine mean vertical profiles of air temperature and water content, air movement, and evaporation (radiation and surface properties are dealt with elsewhere). The differences in, say, temperature and air movement within the lowest six feet of the atmosphere are large enough to be measured by many types of conventional instrument at certain times of the day or night, but not precisely enough.[104]

Microclimatological measurements of windspeed may be made with a very sensitive version of the cup anemometer: the "ping-pong-ball anemometer," which has a low starting speed and a rapid response because the three conventional metal cups are replaced by table-tennis balls cut in half. A more usual instrument works on a different principle. This is the *hot-wire anemometer*, which depends for its operation on the thermal conductivity of the air. The rate of heat loss from a body depends partly on the velocity of the airflow moving past it; consequently, if the body is continuously heated its equilibrium temperature (or rather the magnitude of the electric current required to maintain this equilibrium temperature) will be a measure of the air velocity (see Appendix 1.16). The sensitive element in a hot-wire anemometer is usually a platinum wire; one of 0.0025 mm in diameter and only a few cm in length will respond accurately to wind fluctuations of about a hundredth of a second. The advantage of the hot-wire instrument is that its greatest sensitivity is at low wind-speeds, just where the cup anemometer is least sensitive; it can be regarded as a precise indicator of windspeeds in the range from a few cm per sec to 10 or 20 m per sec. It can be made very suitable for use in confined spaces, but it cannot be used routinely in the open air unless a rain shield is fitted to it, which of course affects the wind flow past the sensitive element. A variant of this approach is to use a thermistor as an anemometer: operated at constant resistance, it lags 0.085 second in responding to fluctuations in air speed, and produces an error of 0.5 per cent in the measurement of wind velocity and of 20 per cent in the measurement of wind direction changes.[105]

A sonic anemometer operating with continuous-wave signals enables fluctuations in the vertical components of both wind and temperature to be measured simultaneously with a maximum error of 5 per cent. The eddy fluxes of both heat and temperature can be obtained from this instrument to within about 10 per cent. The pressure-tube principle in anemometry can also be used for microclimatological measurements; accuracy falls off considerably in light winds, but the method provides records of both horizontal and vertical wind components. Resolving anemometers, which automatically divide the actual wind into its standard components, are always useful in microclimatology, because they save considerable time during data analysis. Various other types of anemometers with a fast response are specially suitable for wind-profile determination, for example, a thermocouple-type instrument that is sensitive to windspeeds of less than 1.5 m per sec.[106]

An interesting technique that first brought into prominence the low-level jetstream phenomena involves the determination of wind profiles by the observation of smoke puffs during the day and swarms of pilot balloons during the night. This technique is applicable to the lowest 5,000 feet of the atmosphere and provides windspeeds accurate to within 0.5 m per sec. Another interesting method, for measuring net vertical air movement, is the Thornthwaite windmill-type anemometer.[107]

The damping necessary in the conventional windvane renders it unsuitable for microclimatological work, where a very fast response to wind-direction fluctuations is required. An electronic windvane developed at the Argonne National Laboratory is very useful: unlike the conventional vane, it does not employ the force of the wind itself to drive the vane to the indicating position, but uses a microphone-type sensor to detect any deviation from the null (i.e., indicating) position. This instrument responds not only much faster than the conventional vane, but also over a wider range of windspeeds.[108]

Atmospheric soundings for microclimatological purposes may be made with a *wiresonde*, which consists of a system of temperature and humidity sensors maintained at heights of up to 1,000 feet above the ground by a captive balloon or *kytoon*. The latter is a streamlined balloon, 78 inches long and 39 inches in diameter, which lifts like a kite during strong winds instead of diving downward like a spherical balloon; thus the wiresonde may be kept at a constant height despite fluctuations in wind. Wiresondes enable temperatures and relative humidities to be measured with probable errors of $\pm 0.1°C$ and ± 2 per cent, respectively.[109]

Low-level air movements on a mesoclimatological scale may be measured by means of *tetroons*, which are small, constant-volume, tetrahedron-shaped balloons. These may be arranged to float within 1,700 feet or so of the Earth's surface, and their movements may be followed visually by observers in motor vehicles. They may also be tracked by radar; if ground-clutter is eliminated and transponders are used, the positions of tetroons floating only a few feet above the ground can be accurately located at ranges of 50 miles or more. Mesoscale and microscale low-level air motions may also be followed by means of fluorescent tracers.[110]

Since microclimatological work involves much computation before the data may be interpreted, it is a natural development to incorporate computing devices in the instrumental system. For example, one device is fed with data on the vertical profile of dry- and wet-bulb temperatures and windspeed differences between various heights, immediately after these figures have been recorded, and automatically solves the psychrometric equation to provide vapor-pressure differences. The computer then determines latent-heat and sensible-heat fluxes from the aerodynamic equation. The instrumental system also records data on net radiation and soil-heat flux. The complete system determines the fluxes of net radiation, soil heat, latent heat, and sensible heat once per minute, then automatically integrates these values in terms of time and displays them on dial-counters, which may be photographed at regular intervals of, e.g., one hour. Another approach is to bypass the dial-counter stage by developing a system that transforms the electrical signals from the measuring instruments directly into a digital form suitable for calculations by machine.[111]

The increasing tendency toward complexity in instrumentation, particularly in micrometeorology, means that the climatologist must familiarize himself with fields that are rarely touched on in classical textbooks of instrumental techniques. Quite often, these fields are valuable in themselves for the new light they shed on standard observational procedures. The application of *communication theory* to meteorological measurements is a good example. The performance of a measuring system may be expressed in terms of its frequency response, phase distortion, and linearity, which can be analyzed by means of Fourier and power-spectrum techniques. A linear measuring system is defined as one whose gain or amplification does not change as the amplitude of the input signal varies, a nonlinear system as one whose gain does change. Nonlinearity generates new harmonic oscillations within the measuring system, which will be superimposed on the true signal (i.e., the true "reading" of the instrument) and will give rise to misleading measurements. Among temperature-measuring devices, a thermocouple is a linear system, a thermistor a nonlinear one.

Communication theory shows that it is important to decide whether we wish to measure steady values only, or all the paths by which the observed variable reaches a steady value. The theory indicates how the response of recording systems may be improved; the record produced by the system should be an exact reproduction of the actual variation of the measured quantity, but any measuring system is limited by its power of resolution (i.e., the time it requires to respond to a change in input) and its phase distortion (i.e., the manner in which its response takes place). To improve response, a compensating network whose characteristics are the inverse of those of the sensing element should be inserted in the system. This can only be done with linear systems. With nonlinear systems, a linearizing network must first be included. The response time or response distance of meteorological sensors can be expressed by quantities that can be determined experimentally; then, by assuming a sinusoidal input, one can correct statistical data, derived from the sensors, for sensor characteristics.[112]

There are numerous ways in which a knowledge of *electronics* can help in climatological observation. These vary from ways of providing for the regular generation at field stations of test or standardizing signals that confirm unattended equipment is working properly, to the development of reliable and cheap telemeasuring devices, by which the indications of a sensor are transferred to the indicating or recording part of the instrumental system. Early telemeasuring devices were purely mechanical, but modern devices rely more and more on electronics.[113]

Future prospects for meteorological instrumentation are exciting, and current achievements indicate that the unfortunate distinction between "meteorological" and "scientific" instruments in manufacturers' catalogues may soon be a thing of the past. The climatologist must keep abreast of current developments, particularly of ideas, but it also pays him to keep track of older instruments. Equipment discarded, or superseded by new instruments, may prove to be very useful if expense is an important factor in an investigation. With a knowledge of the principle behind a measurement, suitable instruments can often be made in a modest workshop very cheaply.[114]

Some Unusual Climatic Elements

In this section we shall be concerned with some elements that are unusual in the sense that they do not normally appear in published collections of climatic data, although many of them have been studied for years, and there is a considerable literature dealing with them. We shall not be concerned with how these elements are observed, but with the nature of the elements themselves and their climatological importance.

CHEMICAL CLIMATE

Chemical climatology is concerned with such things as atmospheric gases and aerosols, the chemical nature of precipitation, condensation nuclei, and the various types of matter present in the atmosphere. It is a study with many practical applications, and is now making very rapid progress.[115]

Of the atmospheric gases, *carbon dioxide* and its variations are very significant for climate. Unlike oxygen, nitrogen, and the rarer gases in the atmosphere, carbon dioxide is present in amounts that fluctuate very considerably with season and latitude. Scandinavian observations show that the CO_2 concentration is at a minimum in summer and a maximum in winter in all latitudes, the amplitude of variation decreasing with increasing height above the ground. Above an altitude of 1,000 m, the range of variation in concentration is from 308 to 320 parts per meter, with an average of 314; both the range and the average increase toward the ground. Annual mean values of CO_2 concentration also vary considerably from year to year. For the northern hemisphere as a whole, the annual variation shows a maximum in May and June, and a minimum between August and October, because of emission and absorption of CO_2 by the biosphere, CO_2 being taken from the atmosphere during the growing season and released to the air by decaying organic matter during the winter. The southern hemisphere shows almost no annual variation in CO_2 content.[116]

The exchange of CO_2 between atmosphere and ocean is an important process, and monthly and annual climatological maps have been constructed to show it, on the assumption that atmospheric CO_2 maintains a uniform volume percentage of 0.03. Tropical maritime and polar continental air masses have a high CO_2 content, whereas polar maritime air has a low CO_2 content. The tropical oceans are likely to be the principal sources of atmospheric CO_2, and during stable conditions the net vertical movement of CO_2 should be upward over the tropical oceans and downward over the polar oceans. The average annual exchange rate of CO_2 between ocean and atmosphere is about 2×10^{-3} moles per cm^2 of sea surface. CO_2 molecules remain in the deep sea for about 500 years, and in the atmosphere for four to ten years. Since the average lifetime of a CO_2 molecule in the atmosphere is only ten years before it becomes dissolved in the sea, most of the CO_2 released by artificial fuel combustion during and since the Industrial Revolution must have been absorbed by the oceans. This has a bearing on the problem of climatic change, since industrialization has been assumed to be gradually increasing mean temperatures via a slowly increasing CO_2 amount in the atmosphere. The carbon-dioxide theory of climatic change shows that the average surface temperature of the Earth should increase 3.6°C if the content of atmospheric CO_2 is doubled, and that it should decrease 3.8°C if the CO_2 amount is halved, provided other factors in the radiation balance remain constant. The rate of industrialization would result in average temperature rises of 1.1°C per century due to the addition of CO_2 to the atmosphere, provided all the CO_2 remains in the atmosphere. The theory proves that when the total CO_2 content of the atmosphere is reduced below a critical value (a little less than 132×10^{18} grams, the total CO_2 amount in the atmosphere in the late 1950's), world climate must continuously oscillate between glacial and interglacial periods with lengths of tens of thousands of years, and that no stable situation is possible. That the CO_2 content of continental polar anticyclones during the polar night probably reduces the rate of outgoing longwave atmospheric radiation (because these anticyclones have a minimum water-vapor content) may limit the intensity of cooling during an ice age, but the process would be ineffective during interglacials.[117]

Ozone is another atmospheric gas of great climatological importance. Its connection with the stratospheric general circulation is described in *Foundations*, but the ground ozone pattern is also of considerable interest.[118] Although ozone only constitutes 3 parts in 10 million of the atmosphere, and the total weight of all the ozone in the atmosphere is only about 3×10^9 metric tons, life as we know it on Earth would be impossible without it, because ozone is the chief absorber of ultraviolet solar radiation, which is very harmful in too large a quantity. Ozone is important in pollution: although it has a useful sterilizing action, it irritates the eyes and severely damages such crops as grapes and tobacco. Ozone pollution in cities can be unpleasant during smog, and is partly a function of motor vehicle density. In Los Angeles, road traffic increases result in an increase in ozone concentration, but in Paris, an increase in road traffic coincides with a decrease in ozone concentration, because of the different amounts of nitrogen peroxide in the air in the two cities. Ozone is produced by the photochemical oxidation of unsaturated hydrocarbons (found in motor-vehicle exhaust gases) by hydrogen peroxide, provided the latter is present in a concentration between 10^{-4} and 10^{-7} grams per cm^3 of air. The hydrogen peroxide concentration in Paris air was too low at the time of the observations.

The ground-level concentration of ozone is highly variable; the usual volume-concentration of ozone at sea level is 10^{-8} grams per cm^3 of air, but it may vary from 10^{-10} (almost zero concentration) to 10^{-7} grams per cm^3, and may increase or decrease fifteenfold in less than an hour. The ozone concentration increases with altitude and with isolation; for example, it is about 5×10^{-8} grams per cm^3 in the Kerguelen Islands. Concentration is maximum during the summer and during the day, minimum during winter and at night. Very large irregular variations are superimposed on the annual and diurnal cycles. The ground-level concentration increases three to ten times during storms and as windspeed increases; when winds decrease, ozone content decreases much more slowly. Quiet, foggy air has little ozone.

The main source of ozone is in the stratosphere, but some is generated at the Earth's surface by silent electrical discharges early during storm-cell development (so that ozone concentration increases as a cold front approaches) and by the photochemical effect already mentioned. Ozone is destroyed by animals, plants, and industrial products. The stratospheric ozone provides a very convenient tracer for the study of air movements.

The total amount of ozone above a point on the ground, the *ozone thickness*, is measured by the depth of ozone in a vertical cylindrical air column of unit cross section above that point expressed in microns or in millicentimeters. The usual unit is the *milli-atmosphere-centimeter* (m-atm-cm), which is the depth in millicentimeters reduced to a standard temperature and pressure. The geographical distribution of total ozone content is not predominantly zonal; there are marked changes with longitude. For example, a strip of low ozone values runs across Siberia from India to the North Pole. Maximum ozone content is found at latitudes 75°N and 50°S, with minimum values near 3 or 4°N. The latter form an "ozone equator," which shifts northward during the southern summer. During some seasons, the polar regions show a relatively low ozone content. The Polar Front jetstream over Japan and the U.S.S.R. forms a barrier preventing equatorward transfer of ozone, so that the ozone content of air north of this jet is greater than that south of the jet, particularly in winter.

The ozone content of the troposphere is fairly uniformly distributed within the northern and southern hemispheres, but there is a gap between the hemispheres. Tropospheric ozone has a uniform seasonal variation, as has stratospheric ozone; between them there is a phase lag of two months, during which the ozone is injected into the troposphere from the stratosphere. A 24-hour cycle in ozone concentration is apparent at some Australian stations, which in general show a smaller interdiurnal variation and a more regular seasonal variation in ozone than is found in the northern hemisphere. The 24-hour cycle seems to result from changes in the subsidence regime of ozone-rich stratospheric air rather than from advection between regions of different ozone content, and the ozone cycle may be obscured north of the equator by the greater variability of ozone concentration with the northern hemisphere.[119]

Atmospheric ozone concentration varies greatly, both horizontally and vertically (see Figure 1.9), and correlates with both stratospheric and tropospheric phenomena. Total ozone amount correlates positively with temperature variations in the lower stratosphere, and negatively with variations in the height and density of the tropopause. Since at least one-third of the daily change in ozone content is caused by vertical motions, and the rest by horizontal advection, there must be a connection between ozone fluctuations and surface pressure systems; regions with excess ozone occur to the rear of surface low-pressure centers, and regions of ozone deficit to the rear of surface anticyclonic centers. Regions of increasing ozone surplus are found directly above occluded surface lows, and surface cold or occluded fronts generally divide regions of surplus and deficient ozone content. Daily variations in ozone are generally related to perturbations in the upper troposphere, but are largely controlled by stratospheric perturbations of the polar-night westerlies in winter.[120]

Diurnal temperature changes in the upper atmosphere are brought about by ozone

54

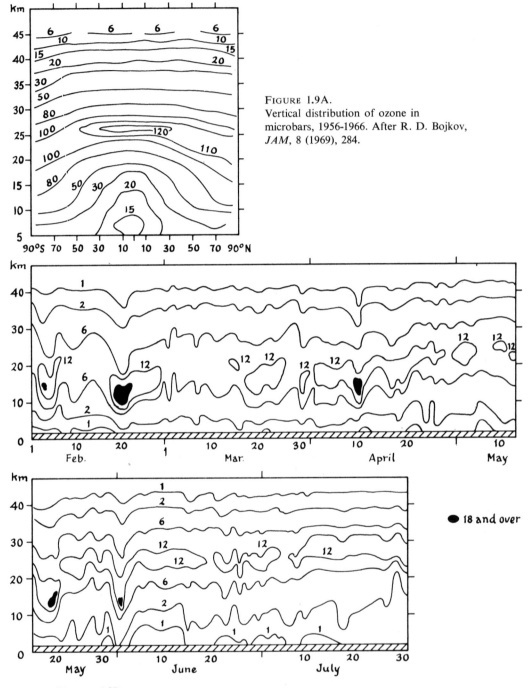

FIGURE 1.9A.
Vertical distribution of ozone in microbars, 1956-1966. After R. D. Bojkov, *JAM*, 8 (1969), 284.

● 18 and over

FIGURE 1.9B.
Ozone densities over Flagstaff, Arizona, 1955, in units of 10^{-3} cm per km (after E. S. Epstein, C. Osterberg, and A. Adel).

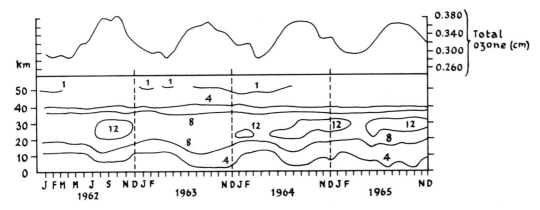

FIGURE 1.9C.
Mean monthly distribution of ozone with height at Aspendale, Australia, in units of 10^{-3} cm per km. After R. N. Kulkarni, *T*, 20 (1968), 305.

absorption. An early rocket flight over New Mexico in 1949 showed that absorption of solar energy by ozone led to the creation of a warm region near 50 km. The diurnal change due to ozone was found to have a maximum value of 5.3°C at 48 km, decreasing to 1°C at 31 km and 70 km.[121]

Precipitation chemistry has received impetus from several directions, notably from agriculture, since the discovery that the quantities of nitrogen, calcium, and (to a lesser extent) potassium supplied to the soil by rainwater are of agricultural significance. Maps based on two years' data from Sweden show the annual amounts of sodium, chloride, calcium, potassium, and nitrogen brought to the Earth's surface by precipitation. Calcium and potassium are both fairly widespread in the atmosphere. The soil is the main source of atmospheric calcium, and its distribution in Sweden is fairly uniform, except for a slight decrease to the north. By contrast, some atmospheric potassium comes from the sea, and some is released in combustion, which causes very small condensation nuclei to form. Ammonia is mainly released from the ground as a gas, particularly in agricultural areas, but some atmospheric nitrogen probably comes from the sea too. Sodium and chloride are derived mainly from the sea, so that their concentrations are highest on the coast, and decrease logarithmically with distance from the coast. The chloride/sodium ratio on the coast is 1.2, compared with the 1.8 characteristic of sea water, and decreases still further inland.[122]

Numerous studies of the composition of atmospheric precipitation have been made, many of which have shown significant geographical differences, for example, in chloride and sulphur. Chloride particles in the atmosphere originate mainly from the sea surface. Particles with densities that range from 10^{-13} to 7×10^{-8} grams per liter can occur in concentrations of up to at least 1,800 particles per liter; the highest concentrations over the sea are found in the lowest air layers, but occasionally high concentrations can be found several thousand feet above the sea surface. Above the cloud tops, the concentration is rarely more than 50 per liter. These concentration variations are important for the mechanism of precipitation. The coalescence theory of precipitation assumes that relatively few water droplets larger than the remainder are present in a cloud, and that such droplets are formed by water-vapor condensation onto particles of atmospheric chloride. The significance of the distribution of sulphur is that it is present in one form of air pollution; one must know the concentration of natural sulphur in the atmosphere before studying the "artificial" variety.[123]

The nitrogen content of atmospheric precipitation depends on how much precipitation has fallen previously, and also on air-mass type; tropical air contains 10 to 30 per cent more nitrogen than polar air, and 50 per cent more than arctic air. The amount of fixed nitrogen that is transferred to the soil by precipitation is between 1 and 5 kg per hectare, which is comparable with the amount given to plants artificially in cultivated areas. The distribution of this fixed nitrogen has a distinct geographical variation in relation to annual totals; with short-period amounts, considerably more geographical variation can be expected. During one storm in Hawaii, nitrogen was brought down by rainfall at a nearly constant rate, although precipitation intensity fluctuated considerably.[124]

The distribution of solid matter in the atmosphere is of considerable climatological interest, particularly when the matter consists of hygroscopic substances. Variations in the concentrations within the atmosphere of solid nuclei onto which water may condense should obviously have some significance for the distribution of clouds and precipitation. Early indications were that condensation nuclei are comparatively scarce over the oceans; at cloud levels, the number of nuclei per cubic centimeter is sometimes less than one, and minimum values found over the North Atlantic and the tropical Pacific were 2 per cm^3 and 130 per cm^3, respectively. Close to the sea surface, much higher concentrations were found, of 100 to 800 per cm^3; some of these concentrations extended up to cloud level, and among them were a few large sea-salt particles. The obvious objective was to find a relationship between condensation nucleus concentration and rainfall drop-size distribution, but the sea-salt particles proved to be interesting in themselves. For example, in addition to the conclusion from the Marshall-Palmer distribution, that for a given intensity of rain there is a definite distribution of raindrops according to their size, the chloridity of rains was found to vary with rain intensity. The weight of chloride present in a certain number of condensation nuclei in a cubic meter of clear air was found to tend toward the weight of chloride dissolved in an equal number of raindrops in a cubic meter of rainy air. This led to the *salt-nuclei hypothesis of rain formation*, which states that raindrops grow on salt nuclei in a manner that prevents any marked change in the distribution of these nuclei during the growth process. The hypothesis indicates that the amount of precipitation falling on oceanic islands should be a function of the number of sea-salt particles in the air, as has been proved in Hawaii. The Hawaiian observations do not indicate that more salt nuclei necessarily cause more precipitation, however.[125]

That sea-salt nuclei in the atmosphere should originate over the sea seems fairly obvious (see Figure 1.10), and the bursting of air bubbles in the white caps of waves in the open sea has been suggested as the major source of these nuclei. The distribution of sea-salt nuclei over the land masses can be inferred from the geographical pattern of chloride concentration in rainwater. In the United States, the chloride concentration drops below its open-sea value along the coast, and then is approximately constant inland because of large-scale vertical mixing within the troposphere rather than because of washout by precipitation. Large-scale precipitation washout seems to be very inefficient, since large salt particles originating as sea spray close to the shoreline can be carried across the entire continent without being washed out of the atmosphere by rainfall. In the westerly belt of the atmosphere, between 6 and 25 days of rainfall are required before the aerosol content of the air can be reduced by washout by a factor of two.[126]

Sea-salt particles bear an important relation to hurricanes. A mature hurricane at sea is an enormous plant for the generation and distribution of salt aerosols. If there is a moist lower atmospheric layer and a relatively dry intercloud layer, latent heat induced by the presence of the salt is likely to be released and may be critical in deciding the stability of the storm. The weight of the salt required to bring about the release of latent heat is very small: only 10 mg per m^3 produces a density increase in the atmosphere equivalent to that produced by a temperature decrease of 0.01°C at ordinary pressures and temperatures. The way the salt mechanism works is that the condensation of salt, falling from relatively dry, clear air at a high level to lower moist levels, releases latent heat, which changes the distribution of heat and water vapor in the cloud layer, and explains most of the increase in potential temperature in the lower layer where air converges toward the center of the hurricane.[127]

Particles of meteoric dust in the upper atmosphere may be detected by various methods, for example, by detecting optically any variations in the rate of change of illumination received from a small part of the twilight sky. The peaks in the time-graph of precipitation received in many parts of the world appear to occur 30 or so days after prominent meteor showers. This suggested the *meteoric dust hypothesis of precipitation*, which states that the nucleating effect of meteoric dust falling into cloud layers in the lower atmosphere is responsible for increased precipitation, 30 days being required for the dust to fall through the upper atmosphere. In an examination of 300 rainfall stations, in both the northern and the southern hemisphere, six precipitation singularities were discovered, five of which occur 30 days after prominent meteor showers. For example, one group, the Giacobinids, has an orbital period of 6.6 years, which results in a meteor shower reaching the atmosphere on October 9 of the appropriate year. The precipitation records show a singularity, involving maximum precipitation on November 8, that is not evident in New Zealand and Australia; since the Giacobinids are incident on the Earth from a declination of 54°N, quite probably dust from them would not reach the southern hemisphere.[128]

The hypothesis that days of extraordinary high mean precipitation are due to the Earth's passing through showers of meteoric dust obviously relies on two types of evidence. The statistical evidence of correlation between precipitation and meteor showers cannot in itself be conclusive; a connection between meteor showers and upper atmospheric dust must also be demonstrated. Twilight observations indicate that significant quantities of dust are found at every level of the atmosphere up to 65 km, with a maximum concentration near 20 km, but there is no correlation definite enough to suggest that meteor showers are the main source of atmospheric dust. Neither is there any definite correlation between hourly rates of sighting of visual meteors for each night of the year and average daily precipitation for 300 stations over a 50-year period. Despite this, the meteoric dust hypothesis is attractive.[129]

The distribution of radioactive matter in the atmosphere is of obvious significance, particularly since the advent of nuclear technology. Quite apart from the effects of nuclear explosions, the atmosphere is somewhat radioactive because of the decay of naturally radioactive isotopes in the land surfaces of the Earth. The natural radio-activity of a given site depends on: (a) the area of land surrounding the site; (b) its distance from the sea; (c) the direction and speed of the wind and the intensity of turbulence at the site. There are large variations in the natural concentrations of the radon and thoron series of isotopes, both seasonally and with change in location and in synoptic type. Radon and thoron behave independently, have different half-lives,

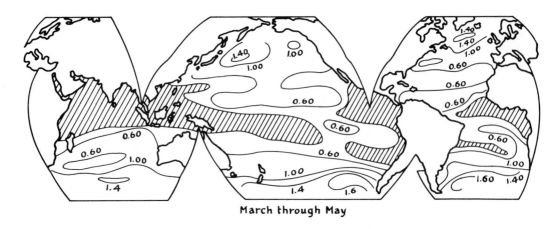

March through May

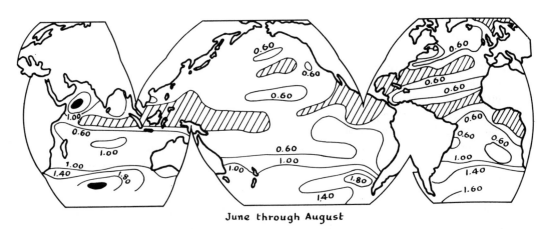

June through August

FIGURE 1.10.
Giant sea-salt particles in the atmosphere. Numbers are average number of salt particles of mass 10^{-10} to 10^{-8} grams at 1 km in units of 10^{-2} per cm³.
After Y. Toba, *T*, 18 (1966), 132.

and are valuable tools for studying the mixing and movement of air masses. For example, since Thorium B has a short half-life (10.6 hours), its concentration in the air depends on local synoptic and topoclimatological site factors but not on air-mass history. Radon, by contrast, is a good index of the history of a prevailing air mass, because it represents the combined effects of accumulation and depletion during several weeks.[130]

Artificial radioactivity in the atmosphere consists largely of the debris of nuclear explosions. Numerous meteorological studies of radioactive fallout have been made. Layers of radioactive debris maintain their identity for long distances and through extensive trajectories. By indicating the persistence in space and time of thin, stable, atmospheric layers, the radioactive debris enables the details of atmospheric meso-structure to be elucidated. These details indicate a gap in the spectrum of mixing processes, between large-scale synoptic phenomena and small-scale turbulence

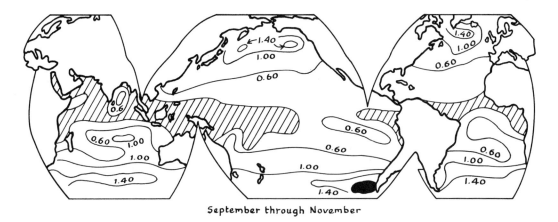

September through November

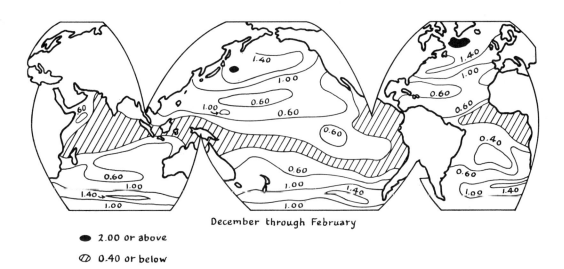

December through February

● 2.00 or above

⬭ 0.40 or below

phenomena. The studies of radioactive climate show that atmospheric mixing processes and the phenomena they give rise to are complex; for example, a simple
surface anticyclone may consist both of air masses from areas uncontaminated by
radioactivity and of air masses infested with nuclear debris from very distant
areas. Nevertheless, some simple relationships emerge. For example, the radioactive
fallout patterns indicate that a small collapsing cold dome travels ahead of a jetstream.[131]

In general, long-lived radioactive material remains more concentrated in the
northern hemisphere than in the southern. Precipitation is an important influence on
fallout, there being often an inverse relation between radioactivity concentration and
precipitation. In the lower atmosphere over Panama, radioactivity does not appear to
cross the "equatorial barrier" that is provided by the belt of continuous precipitation
during the rainy season. The degree of artificial radioactivity at any location generally
seems to depend on the rainfall rate; radioactivity increases when rain intensity
decreases and vice versa. The difference in trajectory of air particles before and behind

a front is of critical importance in determining radioactivity at a given site. Radioactivity may increase considerably during the passage of a cold front; in one situation the "age" of the fission products in the air above a station decreased from 45 to 17 days during the passage of such a front. Radioactive debris in the stratosphere enables mass movements of air above the tropopause to be followed, and also mass changes between stratosphere and troposphere to be inferred. The movement of fission products in the stratosphere shows an annual cycle, with a spring maximum and an autumn minimum, and a semiannual oscillation. The movement of air from stratosphere to troposphere shows an important correlation with baroclinic disturbances: upper stratospheric air is injected deeply into the troposphere in the rear of these disturbances, and then moves equatorward to lower latitudes, whereas tropospheric air flows upward and poleward in advance of the disturbance.[132]

One climatic element whose importance has long been recognized is *turbidity*, which represents the total effect of atmospheric matter in scattering solar radiation within the atmosphere. The variation of the turbidity coefficient throughout the world is important in deciding the radiation and heat balance at the surface of the Earth (see Appendix 1.17).[133]

RADIO CLIMATE

Radioclimatology is concerned with the distribution of atmospheric properties that influence radio propagation and reception. The propagation of radio waves through the atmosphere depends on the physical properties of the latter, particularly on the distribution of temperature and moisture within it; the study of radiometeorology developed considerably as a result of radio and radar progress. The atmospheric physical properties that affect radio propagation are summarized in the concept of *radio refractive index* (see Appendix 1.18), and the field of radioclimatology arose out of the need to predict refractive-index conditions from the synoptic situation.[134]

Refractive-index variations may be studied by means of various parameters. The B parameter (see Appendix 1.19) normally has a range of values from 250 to 400 in the atmosphere. In a standard atmosphere, B does not vary with height, but it does in the actual atmosphere. If the B curve slopes to the right with increasing height, the atmosphere is *subrefractive*; if the B curve slopes to the left, it is *superrefractive*. Superrefraction, a very important phenomenon, involves extra downward refraction of radio waves. Under normal conditions, there is some downward bending of radio waves because of the decrease of air density with increasing height above the Earth's surface. Under superrefractive conditions, radar sets can "see" much further than the geometric horizon, beyond which they cannot normally detect objects. The shorter the wavelength on which the radar operates, the more frequent and the more intense is the superrefraction.

During the Second World War, certain parts of the world were found to be more subject to superrefraction than others. Areas with especially marked superrefractive properties include the Mediterranean region in summer, the Gulf of Aden and the Persian Gulf in summer, the West African coast in the vicinity of the Canary and Cape Verde Islands, the northern part of the Arabian Sea during the Indian hot season, the Bay of Bengal, the west coast of India (except during the southwest monsoon), the coastlands of Australia (except in the northeast) at various seasons, North Island

(New Zealand) in summer, and South Island during föhn winds from the Southern Alps.

The situation in which a low-sited radar can see an object at a much greater distance than usual over water is termed *radar ducting*. Ducting results mainly from a large negative gradient of atmospheric refractive index immediately above the sea surface. In a standard atmosphere, the radius of curvature of a radio ray is approximately four times that of the Earth's surface, so that ducting is unlikely. If the *B* parameter decreases with altitude at the rate of 36 units per 1,000 feet, a ray leaving a transmitter will have the same curvature as the Earth: the ray will therefore remain horizontal and will not leave the Earth, a phenomenon termed *trapping*. Trapping is to some extent related to sea-surface roughness.[135]

Both superrefraction and subrefraction may be present at different heights above a station at the same time. Superrefraction is associated with an increase in temperature and a decrease in humidity with height; subrefraction is associated with increasing humidity and decreasing temperature with height. Ground-based temperature inversions with a decrease in moisture content with height produce a surface superrefractive layer; subsidence inversions (particularly those associated with warm anticyclones and dynamic anticyclogenesis) with moist layers beneath them produce elevated superrefractive layers. Observations at Gibraltar—where superrefraction is the rule rather than the exception—show that there will be superrefraction without a temperature inversion if there is a dew-point lapse-rate.[136]

Maps that have been produced to show the distribution of refractive index are useful in bringing into prominence air-mass contrasts and frontal zones in terms of moisture distribution (see Figure 1.11). These maps are usually based on slightly different versions of the refractive-index equation, and sometimes the distribution of potential refractive index is used, which is referred to 1,000 mb. The refractive-index variations shown on the maps indicate correlations between atmospheric refraction and synoptic parameters such as isobar curvature (see Appendix 1.20). The connection is via the relation between refractive-index and radio-signal strength: for winds of 10 miles per hour or more, refractive index and signal strength are positively correlated, provided no rain, low clouds, fog, or thunderstorms are present. The principle on which the maps are based also enables very small moisture variations in time and space to be measured accurately.[137]

Other types of climate, somewhat connected with radio climate, have been the subject of many investigations: for example, *electrical climate* and *acoustic climate*. Both these topics are important, but a discussion of them would be too lengthy for this book.[138]

SURFACE ELEMENTS AND FACTORS OF CLIMATE

The physical nature of the Earth's surface is a very significant influence on the atmosphere, and, expressed in terms of certain "surface factors," should find a place in all climatic descriptions. The usual factors recognized are the surface stress, the energy balance of the air-land or air-water interface, and the degree of wetness of the Earth's surface. The aerodynamic roughness of the surface and its albedo are properties that are included in these concepts, and the physical properties of soil, vegetation, and water must obviously be embraced by the concepts in some manner. This is a very wide and complex field of study, and only a brief sample of it will be given here.

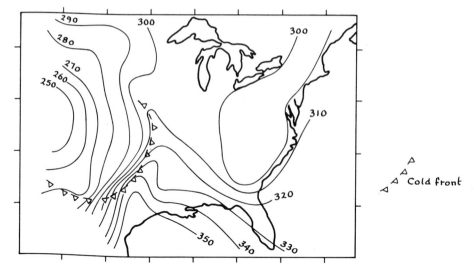

FIGURE 1.11.
Refractive index (*N*-units) at station level, 3:00 P.M. G.M.T., February 19, 1952 (after K. H. Jehn).

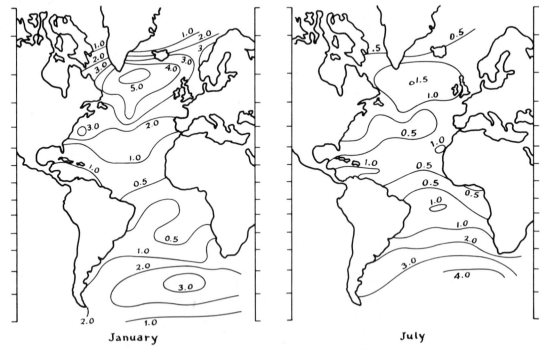

FIGURE 1.12.
Mean wind stress over the Atlantic Ocean, in dynes per cm². After S. Hellerman, *MWR*, 93 (1965), 239.

The *boundary shearing stress* is the horizontal force exerted on a unit area of the Earth's surface by the flow of air across it; it is a function of the texture of the surface, its geometry, and the obstacles to airflow located on it. The surface stress controls wind velocities in the lowest layers of the atmosphere, because it represents the rate at which horizontal momentum is transferred from the atmosphere to the Earth's surface. Mean surface stress (see Figure 1.12) is related to air density and mean wind-speed by the *drag coefficient* (see Appendix 1.21); in turn, the drag coefficient is related to aerodynamic roughness (and hence to the degree of irregularity of the sur-face) by a dimensionless number, von Kármán's constant, and the *roughness length*. The drag coefficient and the roughness length have fairly standard values for land surfaces, but over sea surfaces the drag coefficient increases from 1×10^{-3} at wind-speeds of 3 m per sec or less to 3×10^{-3} at 20 m per sec. Thus maps of the total stress exerted by the ocean on the atmosphere are quite feasible, An alternative, earlier approach was to measure the resistance offered by the Earth's surface to the atmosphere; this was found to have a mean value of 0.006 over the land and 0.0004 over the sea. Maps of estimated aerodynamic roughness may be produced indirectly from land-use information and figures for seasonal changes in the height of vegeta-tion; a complication is that the roughness at a given location varies seasonally, depending on the height of the vegetation. Because of its importance in the control of microclimatic wind profiles, aerodynamic roughness is dealt with in more detail in Chapter 3 of *Foundations*. Aerodynamic drag may be estimated over the sea by means of pilot balloons, and over the land by means of drag plates.[139]

The energy balance of the Earth-atmosphere interface is a complicated problem involving aspects of radiation, heatflow, and dynamic theory. The aspect to be stressed here is the observational one, in particular, the physical factors of site that influence the energy balance at a given point on the Earth's surface. These factors include the thermal capacity, water content, and density of the matter composing the surface at that point—which may take the form of soil, vegetation, rock, water, a man-made substance, or organic life, including man himself—and the rate of flow of heat through the matter. For a soil surface, the study of these physical parameters is well-established: the main difficulty in incorporating these parameters into topo-climatological studies as basic climatic factors is that their numerical value varies depending on whether they are determined *in situ* or from laboratory analysis of soil samples removed from the site. The water and heat capacities of a soil, its density, and the heat flux through it are related by the thermal diffusivity equation (see Appendix 1.22). The measurable quantities needed to solve this equation, and so determine the physical soil parameters for the site, include the vertical temperature profile through the soil, the variation in soil-moisture content, and the degree of compactness of the soil. Since the first 5 cm of the soil contain a large fraction of the soil heat budget, instruments must be buried less than 5 cm below the surface if the entire energy budget of the soil is to be measured. If the soil surface at the locale in question is nonuniform, special measuring apparatus must be used.[140]

The wetness of the Earth's surface will obviously influence the local energy balance and will also provide a moisture source for fog or even for precipitation. This quantity may be measured, but as yet very few figures are available. On occasion, the process opposite to the provision of atmospheric moisture by a wet surface may be observed. This is *sorption* of moisture by the Earth's surface; i.e., the soil serves as a sink of water vapor instead of a source. The process of sorption is defined as " water vapor taken into solid or liquid hygroscopic surfaces whose temperatures are higher than the dew-point temperature of the adjacent air," and it represents the intake of soil water from the atmosphere without the formation of dew. It is known to prevent the formation of fog along the Texas coast of the Gulf of Mexico, and as a climatic factor in the United States is probably restricted to the corn and cotton belts.[141]

In global-scale climatological studies, the orography of the Earth's surface is an important factor. Orography may be taken into account in quite refined dynamic studies of climate, either by means of spot heights determined from intersections of equally spaced grid lines superimposed on a map depicting generalized (i.e., smoothed) land-surface contours, or, more rigorously, by means of Fourier analysis of the height of the surface of the continents. A surface factor of climate that has long been a favorite of climatologists is *continentality*, defined as the degree to which a given climate is controlled by the physical nature of an extensive land surface, rather than by the physical nature of an extensive sea surface. Continentality is usually measured as a function of the surface air temperature of the location. However, the physical validity of the concept of continentality is uncertain.[142]

DERIVED ELEMENTS AND FACTORS OF CLIMATE

The climatic elements so far discussed have been *primitive* or *combined* elements; i.e., they have been either figures obtained by direct observation or measurement, or a mathematical combination of several directly obtained figures. For example, temperature is a primitive element, whereas relative humidity or vapor pressure are elements obtained by combining measurements of dry-bulb and wet-bulb temperature, atmospheric pressure, and windspeed. *Derived* elements, which may be derived either by mathematical physics or by mathematical statistics, cannot be observed directly and cannot be obtained by simple combination of existing elements. Some derived elements could be observed directly, but only with the expenditure of much time or money, so that it is more convenient to estimate them indirectly. For example, microclimatic elements are notoriously difficult to measure routinely, but since all the microclimatic elements must be uniquely determined by the macroclimate and the parameters of the Earth's surface, it should be possible to predict microclimate by knowing the standard macroclimatic observations plus the physical characteristics of the site. This procedure has in fact been followed in the design of a micrometeoro-logical computer.[143]

Many derived elements are concerned with various forms of water, for example, *precipitable water*, which is primarily a concept: the maximum amount of water that can be precipitated out of a column of air between certain pressure levels if the air has a certain moisture content (see Figure 1.13). The term can be misleading, and "total atmospheric water vapor" has been suggested as an alternative, because precipitable water is really a measure of the depth to which water would stand if all the vapor in a column of air were condensed. Mean monthly values of precipitable water at a given station show considerable variation: for example, in the United States, they range from 0.94 cm in February to 2.99 cm, in July, but the annual and overall amounts are much more stable (see Appendix 1.23).[144]

Humidity may be regarded as a derived element by ignoring the various quantities included in the hygrometric equation except relative humidity and temperature, and then determining the statistical relation between the latter. This procedure works

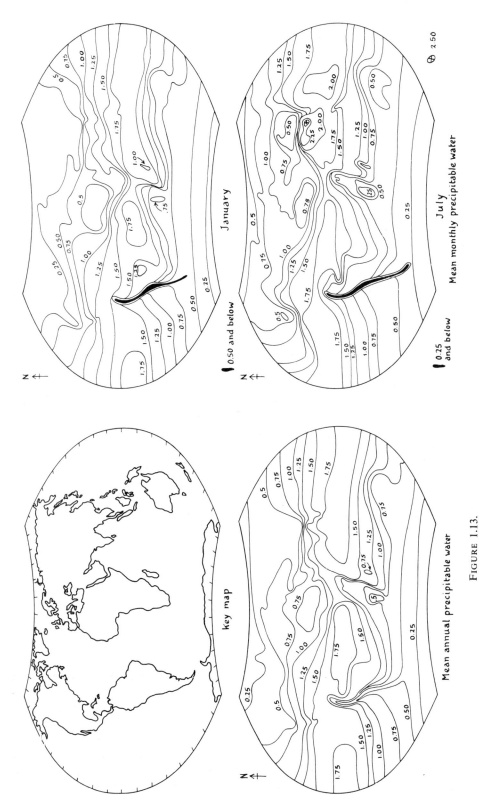

FIGURE 1.13.
Precipitable water, in inches. After S. E. Tuller, *MWR*, 96 (1968), 785.

quite well for monthly and annual mean values.[145] Mean monthly values of dew point may be estimated from mean monthly minimum temperatures and relative humidities in a similar manner (see Appendix 1.24).

Given the complexities in its direct measurement, evaporation clearly lends itself to treatment as a derived element, as which it is particularly interesting, because the treatment makes use of several physical concepts, particularly the energy-balance concept and the aerodynamic concept.

The energy-balance approach to evaporation estimation is based on the fact that since evaporation is essentially the vaporization of liquid water, a process during which latent heat is absorbed at the rate of 591.7 calories per gram for air at a temperature of 10°C and a pressure of 1,013.2 mb, then evaporation can only take place in the presence of a large external supply of energy. The chief source of this energy is the net radiation received by the evaporating surface. Consequently, if all the different elements except evaporation entering into the net-radiation equation are measured, the evaporation can obviously be obtained. There are several instrumental difficulties within this approach if it is to be used to determine evaporation routinely, in particular the measurement of the Bowen ratio, but the principle is sound (see Appendix 1.25).

The aerodynamic concept is concerned with the effects of air movement on evaporation, and is ultimately based on Dalton's formula for evaporation. This formula involves an empirical function that can be represented by measuring the difference between certain meteorological quantities at two heights. For neutrally stable atmospheric stratification, evaporation may be estimated from measurements of windspeed and specific humidity at two levels by means of the Thornthwaite-Holzman equation, which states that evaporation is proportional to the difference in vapor concentration between two reference levels; very precise observations are required if it is to be applied in evaporation estimation. For nonneutral stability, the problem becomes one in atmospheric turbulence, and is dealt with in Chapter 3 of *Foundations* (see Appendix 1.26).

Dalton's formula was adapted to provide evaporation estimates by, for example, Penman, Marciano and Harbeck, and Jacobs. Jacobs was concerned with estimating monthly evaporation values over sea surfaces; Marciano and Harbeck were concerned with lake evaporation; Penman was concerned with obtaining an evaporation estimate from standard land observations. The Penman and Marciano-Harbeck formulae give similar results for the evaporation over areas greater than about 400 acres, but for smaller areas the Penman-computed evaporation is 6 per cent higher than the Marciano-Harbeck value in strong winds, and even higher if winds are light (see Appendix 1.27).

Penman's classic evaporation-estimation method represents a combination of the heat-balance and aerodynamic approaches. The method ignores heat transfer below the Earth's surface, and did not at first take surface roughness into account. An application of the Penman method involves two stages. Firstly, the evaporation from a hypothetical open-water surface at the locality in question is found as a function of net radiation and saturation deficit (i.e., the drying power of the air) at screen level. The evaporation from a surface of short turf is then found by multiplying the hypothetical open-water surface value by a factor that varies from 0.8, in southeast England during the months May to August inclusive, to 0.6 during November-February inclusive, and is 0.7 in the other months. This factor represents the effects of daylight and stomatal opening: the net radiation usually becomes negative at night, during which the stomata of the grass surface close, thus reducing the transpiration component of the evaporation. Later modifications of the Penman equations include one to take into account the effect of surface roughness, and another derived in terms of wet-bulb depression rather than in terms of saturation deficit (see Appendix 1.28).[146]

Certain available aids facilitate use of the Penman method. The equations may be solved graphically, or the water loss recorded by a Piche evaporimeter may be used as an estimate of the aerodynamic term with less error than if a single daily value of vapor pressure is used to calculate the evaporation (see Appendix 1.29). The Penman method is normally intended for estimating evaporation during at least five days. If it is used to obtain daily values, a wind function taking into account surface roughness and a direct measurement of radiation balance are also required. The values for evapotranspiration from crops obtained by the Penman equations are quite likely to exceed the evaporation values obtained from pan measurements in the same locality, particularly if there is considerable advection of heated air.[147]

In estimating potential evapotranspiration, the Penman method assumes that it depends only on atmospheric properties and is independent of soil type and vegetation cover, provided that the soil is completely covered by green vegetation and that the albedo of the latter is approximately constant. An estimation of potential evapotranspiration that is much simpler to carry out may be obtained by Thornthwaite's method, which assumes that mean air temperature at screen level is a good measure of the net radiation, and then computes the evapotranspiration from the statistical relationships between evaporation, as measured by lysimeters in certain well-watered drainage basins, and monthly temperatures. It enables potential evapotranspiration values to be found from screen temperatures, providing corrections are made for length of month and for daylight variations (see Appendix 1.30). These values may be obtained rapidly by means of nomograms or by direct computation. The disadvantage of the method is that screen air temperature is not necessarily a good measure of net radiation at all seasons. At many stations, for example, the potential evapotranspiration as computed by the Thornthwaite method is very close (in terms of energy equivalent) to the net radiation in summer, but is considerably less in spring, owing to the lag of air temperatures behind solar radiation. The application of adjustment for length of day only partly corrects for this lag.[148]

There have been many comparisons of the Penman and Thornthwaite methods, and checks of each against lysimeter observations. In southern Nigeria, the Penman method slightly underestimates the recorded potential evapotranspiration, the Thornthwaite method overestimates it; furthermore, the annual value of potential evapotranspiration according to the latter is 36 per cent greater than that according to Penman. In Scotland, computed values of potential evapotranspiration obtained by both the Penman and the Thornthwaite methods are higher than observed values obtained with simplified Thornthwaite evapotranspirometers at west coast stations, but lower than the observed values at highland stations. In the Thames basin, evapotranspiration observations appear to support the Thornthwaite estimate rather than the Penman one.[149]

A version of the heat-balance method devised by Budyko has been put to considerable use in the U.S.S.R. This version assumes the Bowen ratio is zero, equates the evaporation from a hypothetical open-water surface with the net radiation, and then obtains an expression for evaporability, defined as the moisture loss from a moist surface (see Appendix 1.31).

Evapotranspiration as a derived element provides a convenient method for the estimation of soil moisture content from climatic data. Soil moisture being very difficult to measure by direct observation, an indirect method for estimating it from standard climatic observations has obvious advantages. The essential feature of this indirect method is a budgeting procedure, which begins when the soil is at field capacity and then computes soil moisture content as a balance between income (rainfall) and expenditure (evapotranspiration). There are many complications because of variations in soil factors and root systems, but the method is a useful first indication of the moisture status of a given soil type in a given locality. The moisture status of a soil under permanent grass cannot be found accurately from the Penman, Thornthwaite, or other equations for potential evapotranspiration unless the formulae are modified to take account of the fact that evapotranspiration decreases as soil moisture decreases; i.e., actual evapotranspiration is less than potential evapotranspiration because less and less moisture becomes available for vaporization as the soil dries out. If the decrease is allowed for, then the Thornthwaite estimate generally gives the best result. More accurate results are obtained if the soil moisture is found by a modulated budget, which takes into account changes in ground cover and expanding root systems, as well as the change in soil-moisture stress as the soil dries.[150]

The question of the relation of evaporation to precipitation is interesting. A derived element termed the *evaporation ratio* may be found by dividing the evaporation under natural conditions in a given area by the precipitation falling on that area in the same period of time. The mean value of this ratio for all the land surfaces of the world is about 0.65, and values range from 0.60 for North America to 0.87 for Australia. For a smaller area, the ratio varies from almost 1.0 in desert regions to zero in very well-watered areas, such as Maine and the coastal ranges of the Pacific Northwest. Other derived elements include wind chill, cooling power, and comfort indices.[151]

Notes to Chapter I

1. H. H. Clayton, *World Weather Records*, vol. I (*SMC*, 79, 1927), vol. II (*SMC*, 90, 1934, for 1921–1930 data); H. H. and F. L. Clayton, vol. III (*SMC*, 105, 1947, for 1931–1940 data). United States Weather Bureau and Department of Commerce, Washington, D.C.: *World Weather Records, 1941–1950* (1959); *World Weather Records, 1951–1960*, vol. I, *North America* (1965), vol. 2, *Europe* (1966), vol. 3, *South and Central America, the West Indies, the Caribbean, and Bermuda* (1966), vol. 4, *Asia* (1967), vol. 5, *Africa* (1967), vol. 6, *Antarctica, Australia, the Oceanic Islands, and Ocean Weather Stations* (1968).

2. See also the Weather Bureau's *Daily Synoptic Series of Historical Weather Maps*, prepared during the Second World War and subsequently declassified.

3. E. Gold, *MM*, 84 (1955), 196. On the synoptic interpretation of quantitative observations of hourly rainfall for 63 stations in Britain now included in teleprinter broadcasts, see D. E. Jones, *MM*, 95 (1966), 325. For the effect, on the numbers of rain-days recorded by stations, of the time of day chosen to read rain gauges, see J. Neumann, *JAM*, 8 (1969), 452, using data for Ramle, Israel.

4. *BWMO*, 7 (1958), 135.

5. M. H. O. Hoddinott, *W*, 8 (1953), 3, describes a simple automatic receiver. For a review of meteorological broadcasts available to the nonprofessional, see M. W. Stubbs, *W*, 18 (1963), 147 and 162.

6. For details of amateur RTTY circuits, see A. C. Gee, *Radio Constructor*, 13 (1960), 608, and W. M. Brennan, *Short Wave Magazine*, 23 (1965), 345.

7. For the introduction of facsimile, see C. K. Ockenden, *BWMO*, 3 (1954), 30 and *MM*, 83 (1954), 303.

8. WMO publication no. 9, vols. A-D (Geneva, Switz.), with current amendments, contains full details of transmissions for all countries in the world.

9. For details, see C. E. Wallington, *W*, 19 (1964), 255, 278, and 320.

10. For historical details concerning the development of the international network, see: Napier Shaw, *Manual of Meteorology* (Cambridge, Eng., 1932), I, 160; O. M. Ashford, *W*, 8 (1953), 153 and *BWMO*, 2 (1953), 53; E. Gold, *BWMO*, 9 (1960), 145.

11. See N. Shaw, *op. cit.*

12. A F. Spilhaus, *CM*, p. 705, details the coverage in 1950.

13. For details of the "Selected Ships" Organization, see *BWMO*, 1 (1952). 63; see also C. E. N. Frankcom, *W*, 8 (1953), 140 and *MM*, 82 (1953), 146.

14. For information about the OWS network, see *BWMO*, 3 (1954), 56, and C. E. N. Frankcom, *MM*, 83 (1954), 133.

15. *BWMO*, 3 (1954), 141.

16. For details of the progress of the idea of an IGY, see J. van Mieghem, *BWMO*, 4 (1955), 6 and 52, 5 (1956), 89, and 6 (1957), 96 and 131.

17. C. E. N. Frankcom, *MM*, 90 (1961), 185.

18. *BWMO*, 7 (1958), 2. J. J. Taljaard, *BWMO*, 14 (1965), 90.

19. On the IQSY, see *BWMO*, 12 (1963), 217. During 1964 and 1965, for example, the WGI's were January 13–26, April 13–26, July 13–26, and October 13–26, 1964, and January 11–24, March 8–21, June 14–27, September 13–26, and December 13–26, 1965, all dates being inclusive. On the WWW, see *BWMO*, 12 (1963), 188, and 14 (1965), 86. On the IHD, see M. A. Kohler, *BWMO*, 12 (1963), 193. On IQSY results, see F. G. Finger, H. M. Woolf, and C. E. Anderson, *MWR*, 94 (1966), 651, for synoptic analyses of the 5–, 2–, and 0.4–mb surfaces during the IQSY, based on rocket observations, which confirm the existence of tides and large-scale circulation systems (winter polar cyclone over North America; Aleutian anticyclone) up to at least the stratopause. On the World Weather Watch, see: M. Tepper, *BAMS*, 48 (1967), 94; J. Smagorinski, *loc. cit.*, p. 89; and R. A. Terselic, *WW*, 19 (1966), 112. For an international program of meteorological and oceanographic observations, from ships, satellites, and aircraft, of the air-sea interaction in the Barbados area (project BOMEX), see M. Garstang and N. E. La Seur, *BAMS*, 49 (1968), 627, and B. Davidson, *loc. cit.*, p. 928.

20. *BWMO*: 6 (1957), 146; 2 (1953), 103; 12 (1963), 144; 13 (1964), 146.

21. For the mathematics, see D. P. Petersen and D. Middleton, *T*, 15 (1963), 387. P. O. Thompson, *BWMO*, 12 (1963), 2.

22. C. F. Brooks, *TAGU*, 28 (1947), 845.

23. For details, see C. L. Godske, *BAMS*, 40 (1959), 1.

24. For details, see D. C. House, *MWR*, 88 (1960), 97. See also W. Irwin, *BAMS*, 45 (1964), 388, for a computer evaluation of the ideal radar/rawinsonde network for the United States. For semi-benchmark stations, i.e., stations maintained for a number of years and later abandoned, see D. N. McVean, *W*, 23 (1968), 377, on a year of observations on Mt. Wilhelm, New Guinea, and J. Paton, *W*, (1954), 291, on Ben Nevis Observatory, Scotland.

25. For a review of automatic stations, see M. Ference, *CM*, p. 1207, and *BWMO*, 6 (1957), 70. W. E. K. Middleton and L. E. Coffey, *JM*, 2 (1945), 122. L. E. Wood, *JM*, 3 (1946), 115.

26. B. W. Powell, *W*, 3 (1948), 364.

27. *BWMO*, 11 (1962), 32. For details of automatic stations, see: E. Saunders, *MM*, 97 (1968). 223, on Australia; D. G. Hewitt and P. J. Hardcastle, *W*, 24 (1969), 265, on an automatically recording climatological station; and P. L. Hexter and R. H. Waters, *BAMS*, 49 (1968), 924, on the AMOS III-70 automatic synoptic station of the U.S. Weather Bureau.

28. J. G. Vaeth, *BAMS*, 46 (1965), 50. A. E. Slater, *W*, 10 (1955), 298; D. Pigott, *loc. cit.*, p. 304.

29. P. G. Wickham, *MM*, 90 (1961), 255. T. Hill, *MWR*, 91 (1963), 29. C. F. Konoski, *BAMS*, 45 (1964), 581.

30. Y. Gotaas, *MM*, 87 (1958), 168. J. Findlater, *MM*, 92 (1963), 231.

31. On the measurements mentioned, see: J. W. Telford and J. Warner, *JAS*, 19 (1962), 415; D. R. Grant, *QJRMS*, 91 (1965), 268; E. Kessler, J. T. Lee, and K. E. Wilk, *BAMS*, 46 (1965), 443; A. H. Woodcock and A. T. Spencer, *JM*, 14 (1957), 437; A. W. Brewer and J. T. Houghton, *SPIAM*, p. 571; W. T. Roach, *QJRMS*, 87 (1961), 346; P. M. Kuhn, *MWR*, 91 (1963), 635; M. C. Predoehl and A. F. Spano, *MWR*, 93 (1965), 687; K. J. Hanson and H. J. Viebrock, *MWR*, 92 (1964), 223; D. Mason, *W*, 8 (1953), 243 and 261; N. C. Helliwell, *W*, 13 (1958), 287.

For a general review of the problems and instrumentation, see A. C. Bemis, *CM*, p. 1223, and, for temperature and humidity measurements, *MOHUI*, p. 148.

32. P. L. Randall, *BAMS*, 33 (1952), 416. For examples of aircraft observations, see D. R. Grant, *QJRMS*, 90 (1964), 268, on convection; W. H. Gray, *JAM*, 4 (1965), 463, on cumulus downdrafts in hurricanes; D. N. Axford, *QJRMS*, 95 (1969), 188, on divergence; S. G. Cornford, *QJRMS*, 92 (1966), 105, on precipitation in frontal clouds in southern England; W. E. Bradley and G. E. Martin, *JAM*, 6 (1967), 717, on precipitation drops; and D. H. Lenschow and J. A. Dutton, *JAM*, 3 (1964), 65, on surface temperature variations for different types of surface in southwestern Wisconsin. See also E. J. Frey, *BAMS*, 48 (1967), 658, for an inertial-platform technique enabling the motion through the atmosphere of an aircraft to be measured precisely, and D. N. Axford, *JAM*, 7 (1968), 645, for vertical gust velocities and horizontal wind shear in England as measured by such a technique.

33. For details, see J. I. King, *JAS*, 20 (1963), 245.

34. For general reviews of satellite techniques, see: H. Wexler, *BWMO*, 9 (1960), 2; D. S. Johnson, *BAMS*, 43 (1962), 481; M. A. Alaka, *BWMO*, 11 (1962), 108; D. G. James and I. J. W. Pothecary, *MM*, 94 (1965), 193.

35. W. K. Widger, *W*, 16 (1961), 47. See R. L. Pyle, *BAMS*, 46 (1965), 707, for details of the grid system.

36. For the theory of these inferences, see: R. Frith, *W*, 16 (1961), 364; D. Q. Wark, G. Yamamoto, and J. H. Lienesch, *JAS*, 19 (1962), 369; E. P. McClain, M. A. Ruzecki, and H. J. Brodick, *MWR*, 93 (1965), 445; K. J. Bignell, *QJRMS*, 87 (1961), 231.

37. For the theory, see: D. G. King-Hele, *QJRMS*, 87 (1961), 265.

38. R. Frith, *BMWO*, 11 (1962), 202; F. F. Fischbach, *BAMS*, 46 (1965), 528. V. E. Lally, *BAMS*, 41 (1960), 429.

On satellite techniques, see: R. L. Pyle, *BAMS*, 46 (1965), 707, for the archiving and availability of meteorological data obtained via artificial Earth satellites; D. G. James and I. J. W. Pothecary, *MM*, 94 (1965), 193, for a general review; K. Y. Kondratyev and Y. M. Timofeyev, *GPA*, 71, no. 3 (1968), 164, for problems of satellite meteorology. For descriptions of satellites, see: R. M. Rados, *BAMS*, 48 (1967), 326, for an evolution of the TIROS satellite system; P. J. L. Wildman, *MM*, 96 (1967), 289, on the Anglo-American ARIEL satellites; and F. P. Campbell, *W*, 21 (1966), 422, on UK 3, the first British satellite. For examples of the use of satellites, see: W. Nordberg *et al.*, *BAMS*, 47 (1966), 857, on interpretation of NIMBUS II photographs; J. Bjerknes *et al.*, *BAMS*, 50 (1969), 313, for maps of mean cloud amount in the tropical Pacific; Chin-hua Huang, H. A. Panofsky, and A. Schwalb, *MWR*, 95 (1967), 483, for estimation of ceilings, vertical motion, and surface relative humidity; P. Y. Haraguchi, *JAM*, 6 (1967), 731, on location and maximum windspeed of typhoons in the western Pacific; N. L. Frank, *JAM*, 5 (1966), 541, on hurricane locations; N. A. Streten, *JAM*, 7 (1968), 324, on high-latitude summer circulation in the southern hemisphere; J. S. Kennedy and W. Nordberg, *JAS*, 24 (1967), 711, on progression of stratospheric warmings; J. A. Zak and H. A. Panofsky, *JAM*, 7 (1968), 136, on the average geostrophic winds at 10 mb, inferred from satellite-observed thermal winds; V. L. Gayevsky *et al.*, *GPA*, 69, no. 1 (1968), 332; on surface and cloud temperatures in the U.S.S.R.; D. G. James, *MWR*, 95 (1967), 457, and also M. Wolk, F. van Cleef and G. Yamamoto, *MWR*, 95 (1967), 463, and F. Saiedy and D. G. James, *loc. cit.*, p. 468, on inferred air temperatures; D. Q. Wark and H. E. Fleming, *MWR*, 94 (1966), 351, on inferred atmospheric temperature profiles; D. T. Hilleary *et al.*, *loc. cit.*, p. 367, for experiments on the latter; A. D. Belmont, G. W. Nicholas, and W. C. Shen, *JAM*, 7 (1968), 284, on TIROS inferred temperatures compared with radiosonde measurements; W. L. Smith, *MWR*, 96 (1968), 387, on calculation of tropospheric temperature and moisture from satellite radiometric data, and *MWR*, 95 (1967), 363, on inferred profiles of tropospheric temperature and moisture; A. H. Thompson and P. W. West, *loc. cit.*, p. 791, on inferred average relative humidity below 500 mb in the Gulf of Mexico; E. Raschke, *T*, 19 (1967), 214, for a quasiglobal map of inferred mean relative humidity in the upper troposphere; E. P. McClain, *MWR*, 94 (1966), 509, on the inferred mean saturation deficit of the 1,000- to 500-mb layer; S. Fritz and P. K. Rao, *JAM*, 6 (1967), 1088, on relative humidity estimation; M. Lethbridge, *MWR*, 95 (1967), 487, on inferred precipitation probabilities; E. P. McClain, M. A. Ruzecki, and H. J. Brodrick, *MWR*, 93 (1965), 445, on numerical prediction; and J. R. Hope, *BAMS*, 47 (1966), 371, on GEMINI 4 photography of the path of heavy rainfall in west Texas. On the BOMEX project, see J. P. Kuettner and J. Holland, *BAMS*, 50 (1969), 394. On the archiving of daily mesoscale cloud maps covering the whole Earth, derived from ESSA satellites, see A. L. Booth and V. R. Taylor, *loc. cit.*, p. 431.

39. H. B. Schultz and F. A. Brooks, *BAMS*, 37 (1956), 160, describe a spot-climate recorder. For examples of micrometeorological instrument devices, see I. C. McIlroy, *CTP*, no. 3 (1955), and A. J. Dyer and F. J. Maher, *CTP*, no. 15 (1960).

40. For examples of these procedures, see: P. A. Huxley and M. Beadle, *MM*, 93 (1964), 321; T. J. Chandler, *W*, 17 (1962), 235; F. S. Duckworth and J. S. Sandberg, *BAMS*, 35 (1954), 198; A. Sundborg, *T*, 2 (1950), 221; V. J. Schaefer, *BAMS*, 38 (1957), 124.

For examples of transects and traverses, see: P. A. Huxley and M. Beadle, *MM*, 93 (1964), 312, diurnal and local variation of climate due to dissected topography in southern Uganda; D. G. W. Hurry, *W*, 24 (1969), 214, on a series of spot stations in Glen Nevis, Scotland; K. Smith, *W*, 22 (1967), 363, on a temperature transect in Lima Valley, northern Italy; R. M. Holmes, *W*, 24 (1969), 324, on a wind and temperature transect across the Cypress Hills, in the Canadian southwestern prairies; and T. J. Chandler, *BAMS*, 48 (1967), 394, on humidity traverses across Leicester, Eng.

Climatological expeditions provide examples of both spot observations and traverses. For small-scale expeditions, see: A. K. Kemp and K. L. Durrans, *W*, 23 (1968), 331, on Tibesti in the Sahara; J. M. Havens and D. E. Saarela, *W*, 19 (1964), 342, on the St. Elias Mts. in the Yukon; C. A. M. King, *W*, 10 (1955), 265, on local winds on Svinafellsjökull, Iceland; and J. V. Yorath, *W*, 22 (1967), 291, on Iceland. For medium-scale expeditions, see: B. C. Bishop *et al.*, *JAM*, 5 (1966), 94, on solar radiation in the Mt. Everest region of the Himalayas; and M. M. Miller, *WW*, 17 (1964), 168, on the Khumbu glacier in Nepal. For large-scale expeditions, see: C. S. Ramage, *BAMS*, 43 (1962), 47, on the international Indian Ocean Expedition; E. L. Deacon and J. Stevenson, *CTP*, no. 16 (1968) for observations on the expedition; and R. Suryanarayana and F. R. Miller, *BAMS*, 45 (1964), 644 on the use of the electronic computer. For specialized micrometeorological expeditions, see W. C. Swinbank and A. J. Dyer, *CTP*, no. 17 (1968), and *QJRMS*, 93 (1967), 494, on Kerang and Hay in southeastern Australia.

41. For general discussions of the principles of meteorological instrumentation, see: W. E. K. Middleton and A. F. Spilhaus, *Meteorological instruments* (Toronto, 3d ed., 1953), pp. 3–13; J. C. Bellamy, *AMM*, vol. 3, no. 13 (1957); L. E. Wood, *HM*, pp. 531–72, A. W. Brewer and F. J. Scrase, *QJRMS*, 77 (1951), 3.

42. C. F. Brooks and E. S. Brooks, *TAGU*, 39 (1958), 52, discuss the accuracies of subjective estimates of windspeed at sea, i.e., estimates made in terms of sea disturbance.

43. See Middleton and Spilhaus, *op. cit.*, p. 51, for details of an aneroid barometer control.

44. A. K. Showalter, *BAMS*, 43 (1962), 454.

45. See the current edition of *MOOH* for newer details. These are taken from the 1952 edition, pp. 72 and 151.

46. *Ibid.*, p. 7.

47. B. Ratner, *MWR*, 90 (1962), 89. R. H. Frederick, *MWR*, 89 (1961), 39.

48. For a comparative study of sunshine recorders, see: C. F. Brooks and E. S. Brooks, *JM*, 4 (1947), 105, and *JM*, 5 (1948), 38. See also N. B. Foster and L. W. Foskett, *BAMS*, 34 (1953), 212, for details of a photoelectric sunshine recorder. See: M. Beadle, *W*, 21 (1966), 98, inexpensive modification of the Jordan

photographic recorder; B. G. Collins, *MM*, 97 (1968), 16, on an electrified modification of the Campbell-Stokes recorder, preventing overburn and integrating the total daily sunshine; and C. J. Sumner, *QJRMS*, 92 (1966), 567, on a sensing device for long-period recording.

49. H. Moses and H. G. Daubeck, *BAMS*, 42 (1961), 190. For the effect of towers on wind measurements, see: C. W. Thornthwaite, W. J. Superior, and R. T. Field, *JGR*, 70 (1965), 6047, on the Argus Island tower, near Bermuda; W. F. Dabberdt, *JAM*, 7 (1968), 359, on Brookhaven tower, Long Island, N.Y., and *loc. cit.*, p. 367, on a vertical cylinder 5 meters high at Lake Mendota, Wisc.; and G. C. Gill *et al.*, *BAMS*, 48 (1967), 665, on the accuracy of wind measurements on towers or stacks.

50. See *Phil. Mag.*, 7 (1835), 182, for the Apjohn formula. J. Glaisher, *Hygrometrical Tables Adapted to the Use of the Dry- and Wet-Bulb Thermometers* (London, 1847).

51. E. Gold, *MM*, 76 (1947), 62, 81 (1952), 57, 82 (1953), 54. The theory devised by E. G. Bilham is described with illustrations by S. M. Ross, *MM*, 88 (1959), 313.

52. See R. B. Montgomery, *JM*, 5 (1948), 113, on ventilation. On measurements of humidity, see W. Shackleton, *W*, 23 (1968), 318, for details of an easily constructed humidity slide rule, and R. J. Polavarapu and R. E. Munn, *JAM*, 6 (1967), 699, for an instrument for the direct measurement of vapor-pressure and vapor-pressure gradients to an accuracy of ± 1.2 per cent.

53. *MOHSI*, p. 258. M. J. Brown and E. L. Peck, *JAM*, 1 (1962), 203, point out that, because the precipitation catch of a gauge decreases as wind exposure increases, station histories should give information about changes in exposure.

54. C. C. Warnick, *TAGU*, 34 (1953), 379, described experiments with wind-shields for precipitation gauges. W. Heberden, *Phil. Trans. Roy. Soc. London*, 59 (1769), 359.

55. F. A. Huff, *BAMS*, 36 (1955), 489. G. E. Harbeck, Jr., and E. W. Coffay, *BAMS*, 40 (1959), 348. On rain gauges, see: G. Reynolds, *W*, 20 (1965), 105, on their history; D. J. Holland, *MM*, 96 (1967), 193, on an experiment on the representativeness of rain gauges at Cardington, Eng.; and F. Burns, *MM*, 93 (1964), 308, on their representativeness at Glasgow and Edinburgh, Scotland. Burns found that there is little difference between point rainfall and areal rainfall in circular areas of 100 square miles.

56. For reviews of upper-air sounding systems, see: T. O. Haig and V. E. Lally, *BAMS*, 39 (1958), 401, 40 (1959), 313, and 41 (1960), 31; R. F. Myers, *BAMS*, 43 (1962), 467 (on vertical soundings); and H. J. Mastenbrook, *BAMS*, 43 (1962), 475 (on horizontal soundings).

57. For details, see: *MOHUI*, p. 38; M. Ference,

Jr., *CM*, p. 1207; R. A. Kirkman and J. M. LeBedda, *JM*, 5 (1948), 28.

58. For comparisons of radiosondes, see L. M. Malet, *BWMO*, 3 (1954), 154, 4 (1955), 80, 5 (1956), 157, and 6 (1957), 21. C. A. Moore, Jr., *JM*, 2 (1945), 80. L. Raab, *T*, 6 (1954), 405. *MOHUI*, p. 112. M. W. Hodge and C. Harmantas, *MWR*, 93 (1965), 253.

59. J. W. Reed, *BAMS*, 35 (1954), 253, and 36 (1955), 103. C. J. Brasefield, *JM*, 5 (1948), 147 (on the solar radiation effect). N. K. Wagner, *BAMS*, 42 (1961), 317 (on the time constants), *BWMO*, 4 (1955), 80. F. H. Ludlam and P. M. Saunders, *T*, 10 (1958), 83.

On radiosondes, see: M. W. Hodge and C. Harmantas, *MWR*, 93 (1965), 253, for a comparison of U.S. radiosondes and their errors; F. G. Finger, R. B. Mason, and S. Teweles, *MWR*, 92 (1964), 243, on the U.S. Weather Bureau outrigger radiosonde, providing data on diurnal variations of temperature and pressure in the stratosphere; and D. N. Harrison, *MM*, 98 (1969), 186, on the British "Kew" radiosonde. See also: R. I. Glass, R. D. Reynolds, and R. L. Lamberth, *JAM*, 7 (1968), 141, for a pressure-sensor modification for radiosondes that has recorded unusual pressure variations in the lee of mountains; C. A. Samson, *MWR*, 93 (1965), 327, for a comparison of radiosonde measurements at Denver with data obtained from the same level on Pike's Peak, Colo.; and H. W. Baynton, *MWR*, 93 (1965), 171, for the effect of "motorboating" by radiosondes on the inferred climatology of humidity and refractive index in the West Indies. According to S. K. Cox, J. A. Maynard, and V. E. Suomi, *JAM*, 7 (1968), 691, standard preflight procedures for radiosondes are inadequate at remote tropical locations, and may result in inaccurate data. A record radiosonde altitude of 168,604 feet (51,388 m) was reached on June 9, 1966, in an ascent from Berlin, according to R. Scherhag, *BAMS*, 47 (1966), 982.

60. *MOHUI*, p. 38–81. See also H. Arakawa, *BAMS*, 40 (1959), 343, for the discovery of world-record high windspeeds over Japan by radar-wind soundings. For the mathematics involved, see *MOHUI*, *loc cit.*, and B. M. Singer, *EAFM*, p. 293. A. F. Gustafson, *BAMS*, 35 (1953), 295. On rawinsondes, see: H. W. Baynton, *MWR*, 96 (1968), 47, for stability inferences for Texas; E. F. Danielsen and R. T. Duquet, *JAM*, 6 (1967), 824, on a machine method for computation of winds; and H. M. de Jong, *JAM*, 5 (1966), 436, on the use of theory to adjust observations in order to partially separate instrumental errors from real fluctuations in wind.

61. See *MOHSI*, p. 292, for a review of standard methods of evaporation measurement. See also C. W. Thornthwaite and F. K. Hare, *AMM*, 6 (1965), 163, and "M.O. Discussion," *MM*, 80 (1951), 351, on the important considerations with such measurement.

62. On recording evaporimeters, see C. J. Sumner, *QJRMS*, 89 (1963), 414, and S. Uhlig, *W*, 8 (1953), 9.

L. Wartena and A. J. W. Borghorst, *QJRMS*, 87 (1961), 245. R. A. Dightman, *MWR*, 88 (1960), 101. E. A. Fitzpatrick, *JAM*, 2 (1963), 780, demonstrated how to obtain pan evaporation from measurements of vapor pressure and maximum temperature.

On the measurement of evaporation, see: E. J. Winter and R. A. Moore, *W*, 23 (1968), 82, on a new hook-gauge for evaporation tanks; D. Hook, *MWR*, 95 (1967), 452, on a new type of evaporimeter that measures hourly rates of evaporation, and evaporation during periods of precipitation; L. C. Martinez, *W*, 22 (1967), 470, on a new type of evaporimeter; and R. G. Read, *JAM*, 7 (1968), 417, on white porcelain atmometer measurements of the evaporative power of a tropical forest in the Panama Canal Zone. On evaporation pans, see: W. J. Louw and J. P. Kruger, *NOTOS*, 16 (1967), 29, on a correlation with various climatic factors in South Africa; Shaw-lei Yu and W. Brutsaert, *JAM*, 6 (1967), 265, on pans less than 1 inch deep; W. Brutsaert and Shaw-lei Yu, *JAM*, 7 (1968), 563, on a comparison of shallow pans in terms of mass-transfer equations; J. W. Ellis and A. W. Thomas, *BAMS*, 49 (1968), 940, on a simple, accurate recorder providing a continuous record of evaporation from the water surface in a standard evaporation pan. On pan evaporation versus reservoir evaporation, at Kempton, London, Eng., see D. J. Holland, *MM*, 95 (1966), 22. On the oasis effect, see A. J. Dyer and T. V. Crawford, *QJRMS*, 91 (1965), 345, for micrometeorological profiles of temperature above an irrigated grass field in the midst of dry surroundings at Davis, Calif., that show a leading-edge effect; see also D. C. Davenport and J. P. Hudson, *Agric. Meteorol.*, 4 (1967), 339 and 405, on changes in evaporation rates due to low-level advection across irrigated cotton fields alternating with uncropped, dry, fallow areas at Gezira, Sudan.

63. For the use of lysimeters, see J. C. McIlroy and D. E. Angus, *CTP*, no. 14 (1963). J. Glover and J. Forsgate, *QJRMS*, 90 (1964), 320.

64. H. L. Penman, *Neth. J. Agric. Sci.*, 4 (1956), 8. C. W. Thornthwaite, *TAGU*, 25 (1944), 683.

65. C. W. Thornthwaite, *AMGB*, 3 (1951), 16. B. J. Garnier, *N*, 170 (1952), 286. B. J. Garnier and W. V. Lewis, *W*, 9 (1954), 243.

On evapotranspiration, see: P. C. Ekern, *JAM*, 5 (1966), 431, on actual vs. potential, for bare soil, in Hawaii; L. C. Chapas and A. R. Rees, *QJRMS*, 90 (1964), 313, for a comparison of measurement techniques in southern Nigeria; J. R. Eagleman, *JAM*, 6 (1967), 482, on an empirical estimation of evapotranspiration rate from temperature and relative humidity measurements; J. A. Davies and J. H. M. McCaughey, *AMGB*, 16, series B (1968), 391, on Simcoe, southern Ontario; and R. M. Holmes, *MWR*, 97 (1969), 333, on aircraft observations of the "oasis effect" in southern Alberta. For details of lysimeters, see: J. Glover and J. Forsdyke, *QJRMS*, 90 (1964), 320, on a floating lysimeter, in Muguga, Kenya; D. W. Hand, *Agric. Meteorol.*, 5 (1968), 269, on an electrically weighed lysimeter for daily and diurnal rates of evapotranspiration; and F. J. Lourence and W. B. Goddard, *JAM*, 6 (1967), 489, on a floating lysimeter, giving evapotranspiration rates determined from continuous measurements of sump water level.

66. For details, see *MOHSI*, pp. 171–75, 178.

67. *MOHSI*, p. 157. L. W. Foskett *et al.*, *MWR*, 81 (1953), 267. R. C. Wood, *BAMS*, 40 (1959), 280. W. E. Howell, *BAMS*, 42 (1961), 17.

68. V. B. Morris, Jr., and F. Sobel, *BAMS*, 35 (1954), 226. W. J. Smith and N. J. Hoeflich, *loc. cit.*, p. 60. W. C. Thuman and E. Robinson, *JM*, 11 (1954), 214. On hygrometers, see: R. F. Zobel, *MM*, 94 (1965), 161, on ventilation error in the wet-and-dry-bulb psychrometer; T. A. Bosua, *NOTOS*, 16 (1967), 39, on improvements to the whirling psychrometer; and D. T. Acheson, *JAM*, 4 (1965), 646, on the lithium-bromide dewcell.

69. An alternative method, using a lithium-chloride dewcell, was described by C. B. Tanner and V. E. Suomi, *TAGU*, 39 (1958), 63. C. W. Thornthwaite and H. Hacia, *EAFM*, p. 176. E. W. Barrett and L. R. Herndon, Jr., *JM*, 8 (1951), 40, and 12 (1955), 308. On dew-point hygrometers, see: A. J. Peck, *Agric. Meteorol.*, 5 (1969), 111 and 433, on theory of the Spanner psychrometer, which makes use of the Peltier effect; and H. S. Appleman, *JAM*, 3 (1964), 113, on errors of the dew-point hygrometer, and their effect on relative humidity computations.

70. I. F. Long, *MM*, 87 (1958), 161, and *W*, 10 (1955), 128. S. Duvdevani, *QJRMS*, 73 (1949), 282; see also P. A. Davis, *EAFM*, p. 123. O. H. Newton and J. A. Riley, *MWR*, 92 (1964), 369. M. G. Lloyd, *BAMS*, 42 (1961), 572.

71. E. G. Jennings and J. L. Monteith, *QJRMS*, 80 (1954), 222. J. M. Craddock, *W*, 6 (1951), 300. B. G. Collins, *MM*, 90 (1961), 114. J. M. Hirst, *QJRMS*, 80 (1954), 227. On the measurement of dew, see K. E. Hungerford, *JAM*, 6 (1967), 936, for details of an acetate dew gauge, constructed of matted acetate that is weighed to determine the dewfall. On the formation of dew in semiarid conditions, see W. Baier, *Agric. Meteorol.*, 3 (1966), 103. For experimental observations of the evaporation of dew, see P. E. Waggoner, J. E. Begg, and N. C. Turner, *Agric. Meteorol.*, 6 (1969), 227.

72. J. F. Nagel, *QJRMS*, 82 (1956), 452. On measurement of fog droplets, see B. A. Silverman, B. J. Thompson, and J. H. Ward, *JAM*, 3 (1964), 792, and B. J. Thompson *et al.*, *JAM*, 5 (1966), 342, for details of the laser fog disdrometer, which determines the size distribution of the droplets. For a replication technique for the direct measurement of the size of small fog droplets, see F. K. Odencrantz and P. H. Hildebrand, *JAM*, 8 (1969), 301. On the transition from haze to fog, in terms of water-droplet and aerosol size distributions, see R. G. Eldridge, *BAMS*, 50 (1969), 422.

73. R. List, *BAMS*, 42 (1961), 452. R. A. Schleusener and P. C. Jennings, *BAMS*, 41 (1960), 372. F. W. Decker and L. D. Calvin, *BAMS*, 42 (1961), 475. On the measurement of hail, see S. A. Changnon, Jr., *JAM*, 5 (1966), 899, for the use of standard weighing-bucket rain gauges without evaporation funnels: the hailstones produce a characteristic "spike" on the recording chart trace. On hailstones, see: K. A. Browning, *MM*, 96 (1967), 202, on models of hailstone growth environment, and *QJRMS*, 92 (1966), 1, on the lobe structure of giant hailstones; A. E. Carte and R. E. Kilder, *QJRMS*, 92 (1966), 382, on hailstone characteristics in the Transvaal; O. Vittori *et al.*, *JAS*, 26 (1969), 148, for a chemical analysis; J. Rosinski, *JAM*, 5 (1966), 481, on their solid, water-insoluble content; and A. B. Lowe, *WW*, 18 (1965), 87, on unusually shaped, large hailstones at Keewatin, Ontario. For geographical aspects of hail, see: K. A. Browning *et al.*, *JAM*, 7 (1968), 603, and A. H. Auer, Jr., and J. D. Maritz, *JAM*, 8 (1969), 303, on the local distribution of the maximum size of hailstones in Oklahoma; A. H. Paul, *W*, 23 (1968), 424, on regional variation in the duration of hail fall and in the maximum size of hailstones in North America and South Africa; R. W. Longley and C. E. Thompson, *JAM*, 4 (1965), 68, on the climatology of the causes of hail; S. A. Changnon, Jr., *JAM*, 7 (1968), 518, on the effect of sampling density on the areal extent of damaging hail in Illinois; and F. A. Huff, *JAM*, 3 (1964), 240, and S. A. Changnon, Jr., *MWR*, 95 (1967), 209, on the geographic pattern of hail in Illinois.

74. R. G. Eldridge, *JM*, 14 (1957), 55 and 573. F. Singleton and D. J. Smith, *QJRMS*, 86 (1960), 454. J. P. Wittman and T. C. Goodale, *BAMS*, 36 (1955), 69. R. H. Margarvey, *JM*, 14 (1957), 182. H. R. Byers, H. Moses, and P. J. Harney, *JM*, 6 (1949), 51, measured the temperature of rain at ground level. E. G. Bowen and K. A. Davidson, *QJRMS*, 77 (1951), 445, developed a raindrop spectrograph. D. C. Blanchard, *TAGU*, 34 (1953), 534, produced a simple recorder for raindrop size. A. T. Spenser and D. C. Blanchard, *TAGU*, 39 (1958), 853, introduced a portable raindrop recorder.

On raindrops, see: C. E. Robertson, *JAM*, 4 (1965), 642, for the use of reversal film as an impact collector; J. E. McDonald, *W*, 19 (1964), 177, on the use of spatter-spots on pavements for estimating drop sizes; U.H.W. Lammers, *JAM*, 8 (1969), 330, for an electrostatic recorder for obtaining the distribution of raindrop sizes; and W. P. Winn, *loc. cit.*, p. 335, for an electrical device for the measurement of raindrop radii. On sampling error in raindrop concentration measurements, see: S. G. Cornford, *MM*, 96 (1967), 271. For examples of raindrop-size spectra, see: M. Fujiwara, *T*, 19 (1967), 392, on warm rain in Hawaii; P. S. du Toit, *JAM*, 6 (1967), 1082, on continuous rain; M. Fujiwara, *JAS*, 22 (1965), 585, on individual storms at Miami, Fla.; P. G. F. Caton, *QJRMS*, 92 (1966), 15, on free-atmosphere rain at Pershore, Eng.; A. D. Duncan, *JAM*, 5 (1966), 198, J. W. Telford, *JAM*, 6 (1967), 434, and S. G. Cornford, *JAM*, 7 (1968), 956, on measurements by means of an airborne foil impactor at the base of a shower cloud. On the terminal velocity of raindrops aloft, see G. B. Foote and P. S. du Toit, *JAM*, 8 (1969), 249.

75. B. J. Mason and J. B. Andrews, *QJRMS*, 86 (1960), 346. A. N. Dingle and K. R. Hardy, *QJRMS*, 88 (1962), 301. For hydrometer size distributions, i.e., simultaneous size distributions of snowflakes and raindrops at two altitudes in the melting layer based on mountain-slope observations in Japan and Alaska, see T. Ohtake, *JAS*, 26 (1969), 545.

76. For general reviews of radar meteorology, see: L. Battan, *Radar Meteorology* (Chicago, 1959); J. S. Marshall, W. Hitschfeld, and K. L. S. Gunn, *AGP*, 2 (1955), 1; J. S. Marshall and W. E. Gordon, *AMM*, 3 (1957), 73; D. Atlas, *AGP*, 10 (1964), 317. See also D. Atlas, *BAMS*, 43 (1963), 457, for comparisons of radar- and gauge-estimates of precipitation. J. N. Myers, *JAM*, 3 (1964), 421, gives examples of mean radar-precipitation maps for central Pennsylvania.

77. J. W. Ryde, *The Attentuation of Radar Echoes Produced at Centimeter Wavelengths by Various Meteorological Phenomena: Meteorological Factors in Radio-Wave Propagation* (Physical Society, London, 1947). J. S. Marshall, R. C. Langille, and W. M. Palmer, *JM*, 4 (1947), 186. H. R. Byers and R. D. Coons, *JM*, 3 (1947), 78, and R. C. Langille and K. L. S. Gunn, *JM*, 5 (1948), 301, confirmed the prediction for analyses in the vertical. Predictions in the horizontal had been analyzed by R. H. Maynard, *JM*, 2 (1945), 214. P. M. Hamilton, *QJRMS*, 92 (1966), 346. For details of Echo Intensity-Rainfall Intensity charts, see J. W. Wilson, *JAM*, 3 (1964), 164.

78. On the bright band, see P. M. Austin and A. C. Bemis, *JM*, 7 (1950), 145.

On radar measurement of precipitation, see: T. W. Harrold, *W*, 21 (1966), 247, for a review; J. W. Wilson, *JAM*, 3 (1964), 164, for examples at Atlantic City, N.J.; L. E. Truppi, *MWR*, 92 (1964), 177, on the climatology of precipitation echoes in the United States; G. E. Stout and E. A. Mueller, *JAM*, 7 (1968), 465, for a review of radar reflectivity and rainfall rate relationships; R. Cataneo and G. E. Stout, *loc. cit.*, p. 901, on raindrop-size distributions in humid continental climates in the United States. On the theory of power-law relationships, see J. S. Marshall, *JAM*, 8 (1969), 171. On occurrences of radar bright bands, see T. W. Harrold, K. A. Browning, and J. M. Nicholls, *MM*, 97 (1968), 327, on rapid changes in the height of the melting layer at Pershore, Eng.; see also E. N. Brown, *T*, 16 (1964), 517, on the microstructure of atmosphere and precipitation in a bright band.

For comparisons of radar and rain-gauge precipitation patterns, see F. A. Huff, *JAM*, 6 (1967), 52, on storm mean rainfall in Illinois and Oklahoma, and R. L. Weaver, *MWR*, 94 (1966), 466, on California rainstorms.

On weather radar, see: W. Irvin, *BAMS*, 45 (1964), 388, on actual and optimum networks in the United States; N. L. Frank and D. L. Smith, *JAM*, 7 (1968), 712, on the correlation of radar echoes with various meteorological quantities; J. H. Hand, *JAM*, 3 (1964), 58, on the climatology of cloud-layer detection at Washington, D.C.; and R. J. Donaldson, Jr., *BAMS*, 46 (1965), 174, for a review of methods for identifying severe storms by radar. For examples of the use of weather radar, see: W. G. Harper, *MM*, 93 (1964), 337, on cloud detection with 8.6-mm radar at Pershore, Eng.; R. H. Blackmer, Jr., and S. M. Serebreny, *JAM*, 7 (1968), 122, on marine precipitation on the west coast of North America; P. M. Saunders, *JAS*, 22 (1965), 167, on tropical marine showers on Barbados; R. R. Braham, Jr., *JAS*, 21 (1964), 640, and W. E. Howell, *JAS*, 22 (1965), 465, on summer rain showers in Missouri; R. G. Pappas, *MWR*, 95 (1967), 577, on inversions at San Diego, Calif.; and J. J. Hicks and J. K. Angell, *JAM*, 7 (1968), 114, on breaking gravitational waves in visually clear air at Wallops Island, Va.

On the uses of optical pulsed-light radar (lidar) in meteorology, see: G. G. Goyer and R. D. Watson, *BAMS*, 49 (1968), 890, on remote probing of water vapor, temperature, density, and aerosols above 10 km; R. T. H. Collis, *loc. cit.*, p. 918, on air motions in forested valleys, surface to 300 m, in Idaho; W. Viezee, E. E. Uthe, and R. T. H. Collis, *JAM*, 8 (1969), 274, on low-cloud structure at Hamilton Air Force Base, Calif.; R. T. H. Collis and M. G. H. Ligda, *JAS*, 3 (1966), 255, on particulate matter in the stratosphere; and B. R. Clemesha, G. S. Kent, and R. W. H. Wright, *JAM*, 6 (1967), 386, on atmospheric density variations up to 65 km. For a review, see R. T. H. Collis, *QJRMS*, 92 (1966), 220; for a general account of lasers in meteorology, see E. W. Barrett, *WW*, 20 (1967), 162.

79. R. J. Donaldson, Jr., *JAM*, 3 (1964), 611, and 4 (1965), 727.

80. See *MOHSI*, p. 258, for details of gauge operation. G. Spurr, *QJRMS*, 80 (1954), 237, showed that the Dines rain gauge records 0.95 times the amount of precipitation recorded by the Bibby gauge in a given period. For rainfall rates, the relation is Dines = (1.03 × Bibby) + 2.5 mm per hour. P. G. Hookey, *W*, 20 (1965), 193.

81. E. R. C. Reynolds, *MM*, 92 (1963), 210. E. L. Hamilton and L. A. Andrews, *BAMS*, 34 (1953), 202. F. B. Gomm, *BAMS*, 42 (1961), 311. E. J. Winter, *W*, 10 (1955), 272.

82. There are other types of horizontal-rim gauges, e.g., small-orifice gauges, described by F. A. Huff,

TAGU, 36 (1955), 689, and H. E. Gill, *JGR*, 65 (1960), 2877. For theory and construction details of the octagonal gauge, see R. E. Lacy, *QJRMS*, 77 (1951), 283. For calculating the direction of travel, see C. W. Rose and H. G. Farbrother, *QJRMS*, 86 (1960), 408.

83. G. S. Raynor, *BAMS*, 36 (1955), 27. J. R. Gerhardt, *BAMS*, 39 (1958), 189. E. Northmann, *loc. cit.*, p. 273. C. J. Adkins, *QJRMS*, 85 (1959), 419. On rain-gauges, see: J. Giraytys and A. Heck, *WW*, 21 (1968), 198, on international comparison of rain gauges; E. R. C. Reynolds, *MM*, 93 (1964), 65, on their accuracy, and 213, on errors caused by dew; A. C. Robinson and J. C. Rodda, *MM*, 98 (1969), 113, on aerodynamic characteristics in relation to rain and wind; A. L. Maidens, *MM*, (1965), 142, on new M.O. rain gauges; P. G. Hookey, *W*, 20 (1965), 160, on a homemade rain gauge; G. E. Goodison and L. G. Bird, *MM*, 94 (1965), 144, on telephone interrogation of automatic rain gauges in remote areas; A. L. Maidens, *loc. cit.*, p. 280, on prevention of icing in recording gauges; A. T. Spencer and A. H. Woodcock, *JAM*, 3 (1964), 105, on a photographic recorder for very light showers in marine air. For examples of rain-gauge networks, see: E. M. Shaw, *W*, 21 (1966), 291, on the Devon river authority, England; G. Reynolds, *W*, 23 (1968), 88, on the North Scotland automatic rain gauges; and F. A. Huff, *JAM*, 6 (1967), 435, on recording rain gauges, providing rainfall gradients in storms, in Illinois. For a difficult area, see B. W. Thompson, *W*, 21 (1966), 48, on Mount Kenya. For dense local networks, using amateur observers, see F. P. Ostby, M. A. Atwater, and F. Perry, *WW*, 22 (1969), 60, on central Connecticut, and D. T. Brissett, *loc. cit.*, p. 64, on Colorado Springs. For measurements of the bearing and inclination of rainfall in Southern Rhodesia, see N. W. Hudson, *QJRMS*, 90 (1964), 325.

84. *MOHSI*, p. 258, describes the standard methods of snowfall measurement. See also R. F. Black, *TAGU*, 35 (1954), 203, who reports that a standard American (eight-inch) rain gauge recorded a snowfall amount during winter 1949–50 that was only one-quarter to one-half the "true" amount as based on snow depth and intensity measurements, probably because of wind effects. See K. Itagaki, *JGR*, 64 (1959), 375, for details of snow measurement by means of gamma radiation. See R. T. Beaumont, *JAM*, 4 (1965), 626, for details of the pressure-pillow snow gauge, which measures the water equivalent of snow, on Mt. Hood, Oregon. On the water equivalent of snow, see R. E. Lacy, *W*, 19 (1964), 353, on snow density measurements, and H. C. S. Thom, *MWR*, 94 (1966), 265, on the climatology of maximum annual water equivalent of snow on the ground in the United States. On snowflakes, see: B. A. Power, P. W. Summers, and J. D'Avignon, *JAS*, 21 (1964), 300, on the relation between the density of newly fallen snow and snow-crystal form in Montreal; E. E. Hindman II and

R. L. Rinker, *JAM*, 6 (1967), 126, on the use of replication techniques to provide a continuous record of snowflake form throughout a snowfall in the Colorado Rocky Mts.; and V. J. Schaefer, *WW*, 17 (1964), 278, on the preparation of permanent replicas of snowflake, ice, and frost patterns. On the use of a transmissometer to provide continuous records of falling snow in Montreal, see C. Warner and K.L.S. Gunn, *JAM*, 8 (1969), 110. Snow beds that remain in summer are useful indicators of local climatic conditions—for examples, see: G. Scott, *W*, 19 (1964), 204, on the Cheviot Hills, Britain; D. L. Champion, *W*, 7 (1952), 180, on Ben Nevis; P. C. Spink, *W*, 17 (1962), 408, *W*, 21 (1966), 127, and *W*, 22 (1967), 298, on Scotland; and T. M. Thomas, *W*, 16 (1961), 171, on southeastern Wales.

85. *MOHSI*, p. 187. *MM*, 83 (1954), 11. G. E. W. Hartley, *MM*, 86 (1957), 111. On the measurement of wind direction, see J. Wieringa, *JAM*, 6 (1967), 1114, on evaluation and design of wind vanes, and R. L. Ives, *JAM*, 5 (1966), 544, on indicator-lamp systems. See D. N. Baker and S. G. Williams, *loc. cit.*, p. 33, for a system that measures and integrates wind directions.

86. For details, see *MOHSI*, Chap. 5.

87. R. L. Ives, *JM*, 3 (1946), 122. W. H. Reed III and J. W. Lynch, *JAM*, 2 (1963), 412.

88. L. P. Reiche and F. L. Ludwig, *BAMS*, 42 (1961), 314. B. G. Collins, *MM*, 83 (1954), 232, described an instrument for recording the frequency distribution of mean hourly windspeed; H. H. Crouser, *MWR*, 90 (1962), 23, described a frequency type of windspeed measuring system; and M. T. H. Key, *MM*, 94 (1965), 113, described a battery-operated cup-counter anemometer system designed to record daily total run-of-wind for seven days without attention.

89. R. S. John, *BAMS*, 41 (1960), 618. G. Kelton and P. Bricourt, *BAMS*, 45 (1964), 571.

90. *MOHSI*, pp. 217–24. P. J. Rijkoort, *MM*, 84 (1955), 137.

91. G. E. W. Hartley, *MM*, 84 (1955), 111. C. J. Sumner, *QJRMS*, 91 (1965), 364. On the measurement of surface wind, see A. L. Maidens, *MM*, 96 (1967), 143, on M.O. instruments, and H. C. Shellard, *loc. cit.*, p. 235, on anemograph stations and records in Britain. On anemometers, see: P. B. MacCready, Jr., and H. R. Jex, *JAM*, 3 (1964), 182, on the theory of response characteristics of propeller and vane wind sensors; M. H. Norwood, A. E. Cariffe, and V. E. Olszewski, *JAM*, 5 (1966), 887, on drag-sphere and drag-cylinder anemometers; and G. Kelton and P. Bricout, *BAMS*, 45 (1964), 571, on sonic anemometers. On cup anemometers, see E. F. Bradley, *Agric. Meteorol.*, 6 (1969) 185, on a small, robust instrument; and H. J. Sayer, *W*, 20 (1965), 383, on a remote-reading instrument. For an amateur instrument, see G. E. W. Hartley, *W*, 22 (1967), 416; for a homemade anemometer employing paper cups, see P. G. Hookey, *W*, 20 (1965), 216 and 287. On the theory of cup and

vane anemometers, see S. Ramachandran, *QJRMS*, 95 (1969), 163. For details of a mechanical seven-day recording anemometer, see R. B. Lowe and H. J. Bergen, *Agric Meteorol.*, 4 (1967), 203.

92. For details, see M.O. pamphlet no. 396, *The Measurement of Upper Winds by Means of Pilot Balloons* (London, 3d ed., 1944), p. 29.

93. For the mathematics, see *MOHUI*, p. 33, and B. M. Singer, *EAFM*, p. 296. F. V. Hansen and P. H. Taft, *BAMS*, 40 (1959), 221, describe a plotting system for the evaluation of double-theodolite investigations. K. M. Barnett and O. Clarkson, Jr., *MWR*, 93 (1965), 377. B. M. Singer, *BAMS*, 37 (1956), 207.

94. For an illustration of the slide-rule, see M.O. 396 (Figure 11). W. G. Biggs, *JAM*, 1 (1962), 268. N. Thyer, *loc. cit.*, p. 66. I. Karmin, *BAMS*, 39 (1958), 473. L. A. Jay, *BAMS*, 41 (1960), 633. On pilot balloons, see: H. Rachele and L. D. Duncan, *MWR*, 95 (1967), 198, for computation of wind velocity; and L. J. Rider and M. Armendariz, *JAM*, 5 (1966), 43, for a comparison of simultaneous pibal and tower measurements of wind at Green River, Utah. On double-theodolite ascents, see: H. M. de Jong, *JAM*, 3 (1964), 624, on the theory of adjustment; M. B. Danard, *JAM*, 4 (1965), 394, on wind variability determinations at Shilo, Manitoba; and K. M. Barnett and O. Clarkson, Jr., *MWR*, 93 (1965), 377, on the relation of time interval between sightings to accuracy of wind determinations.

95. A. F. Bunker, *BAMS*, 34 (1953), 406. C. J. Brasefield, *JM*, 11 (1954), 412. W. Conover, *BAMS*, 42 (1961), 249. E. J. Williamson and J. T. Houghton, *QJRMS*, 91 (1965), 330. R. E. Turner, *BAMS*, 46 (1965), 547. A. H. Clinger and A. W. Straiton, *BAMS*, 41 (1960), 250. On radiosondes for special purposes, see: V. E. Suomi, K. J. Hanson, and R. J. Parent, *JAM*, 6 (1967), 195, on the "Chirp" digital radiosonde, which provides very detailed temperature soundings; C. Magono and S. Tazawa, *JAS*, 23 (1966), 618, on the "snow-crystal sonde" to measure the vertical distribution of snow crystals in a snow cloud in Japan; and M. Fujiwara, *T*, 19 (1967), 403, for measurement of size distribution of raindrops aloft, using a filter paper.

96. D. G. Murcray *et al.*, *JGR*, 65 (1960), 3641. D. R. Grant, *QJRMS*, 89 (1963), 546. E. W. Barrett, L. R. Herndon, Jr., and H. J. Carter, *JM*, 6 (1949), 367. T. W. Caless, *BAMS*, 42 (1961), 467. J. T. Houghton and J. S. Seeley, *QJRMS*, 86 (1960), 358. On the measurement of stratospheric water vapor, see: D. G. Murcray, F. H. Murcray, and W. J. Williams, *QJRMS*, 92 (1966), 159, on mixing ratio in the stratosphere above New Mexico; F. J. Brousaides and J. F. Morrissey, *JAM*, 8 (1969), 431, on an inexpensive dew-point hygrometer for balloon soundings up to 200 mb; and D. Chleck, *JAM*, 5 (1966), 878, and J. F. Morrissey and F. J. Brousaides, *JAM*,

6 (1967), 965, on an aluminum-oxide hygrometer for balloon and parachute soundings. For examples of surface-to-stratosphere soundings, see: J. A. Brown, Jr., and E. J. Pybus, *JAS*, 21 (1964), 597, on the climatology of McMurdo Sound, Antarctica; H. J. Mastenbrook, *JAS*, 25 (1968), 299, on the climatology to 94,000 feet of Trinidad, Thule, and Washington, D.C.; and N. Sissenwine, D. D. Grantham, and H. A. Salmela, *JAM*, 25 (1969), 1129, on that of northern California, to 32 km. For inferred soundings, see P. M. Kuhn and S. K. Cox, *JAM*, 6 (1967), 142, on some inferred from radiometric data, and D. A. Chisholm *et al.*, *JAM*, 7 (1968), 613, on some inferred from surface and aerological data, by means of decision trees.

97. For details of the rockets, see: C. L. Armstrong and R. D. Garrett, *MWR*, 88 (1960), 187; T. J. Keegan, *BAMS*, 42 (1961), 715; W. W. Kellogg, *BAMS*, 43 (1962), 129; R. L. Schumaker, *loc. cit.*, p. 131; K. R. Jenkins, *JAM*, 1 (1962), 196. For information on the rocket network, see: W. L. Webb *et al.*, *BAMS*, 42 (1961), 482; J. Giraytys and H. R. Rippy, *BAMS*, 45 (1964), 382; and R. E. Newell, *JAS*, 20 (1963), 213. On the smoke rocket, see G. C. Gill, E. W. Bierly, and J. N. Kerawalla, *JAM*, 2 (1963), 457.

On meteorological rocket techniques and their uses for measuring high-level winds, see: J. Giraytys and H. R. Rippy, *BAMS*, 45 (1964), 382, on the U.S.A.F. rocket-sounding network; R. S. Quiroz, *BAMS*, 48 (1967), 697, on observations in the U.S.S.R.; F. G. Finger and H. M. Woolf, *JAS*, 24 (1967), 387, on shipboard rocketsonde observations of the southern hemisphere stratospheric circulation; M. D. Kays, *JAM*, 5 (1966), 129, on a comparison of rocket winds with estimated geostrophic winds at 10 mb; R. S. Quiroz, *loc. cit.*, p. 308, on the climatology of atmospheric densities at 30 to 50 km, for 80°N, inferred from rocket data; M. S. V. Rao, *JAM*, 6 (1967), 401, on the diffusion pattern in the equatorial stratosphere above Thumba, India; A. J. Miller, H. M. Woolf, and F. G. Finger, *JAM*, 7 (1968), 390, on the small-scale structure of wind and temperature to 46 km at Wallops Island, Va.; F. G. Finger and H. M. Wolf, *JAS*, 24 (1967), 230, on diurnal temperature variation in the upper stratosphere; and A. J. Miller, *JAM*, 8 (1969), 172, on inconsistencies in rocketsonde temperature profiles for the 20- to 55-km layer. According to M. Kays and R. O. Olson, *BAMS*, 48 (1967), 676, rocketsonde measurements are accurate only to within $\pm 2°C$ for temperature, and to within 5 m per sec for windspeeds. On low-level wind profile determinations from rockets, see P. Morgan and U. Radok, *JAM*, 4 (1965), 551, and H. B. Tolefson and R. M. Henry, *JAM*, 5 (1966), 225.

98. A. G. Weisner, *JM*, 13 (1956), 30.

99. D. D. Woodbridge, *BAMS*, 40 (1959), 549. See also E. Manning *et al.*, *JGR*, 64 (1959), 587, and 67

(1962), 3923, on the use of artificially generated sodium clouds to track upper winds. G. H. R. Reisig, *JM*, 13 (1956), 448. F. L. Whipple, *JM*, 10 (1953), 390. L. G. Jacchi, *JM*, 14 (1957), 34. On nephoscopes, see *MM*, 83 (1954), 174.

On sodium-cloud measurements of upper winds, see A. Kochanski, *MWR*, 94 (1966), 199, for atmospheric motions between 80 and 200 km inferred from the drift of sodium clouds in the northern hemisphere. On gun-launched probes of upper winds, see C. H. Murphy and G. V. Bull, *BAMS*, 49 (1968), 640, on the HARP-McGill project in Barbados, involving vertical soundings to above 85 km by means of naval guns.

100. For the mathematics of radar, see *MOHUI*, pp. 16, 39, 73. See also B. M. Singer, *EAFM*, p. 293, and J. E. Laby and J. G. Sparrow, *JAM*, 4 (1965), 585. See P. G. F. Caton, *MM*, 92 (1963), 213, for the basic principles of Doppler radar, and J. R. Probert-Jones and W. G. Harper, *MM*, 91 (1962), 273, for examples. D. Atlas and R. Wexler, *JAM*, 4 (1965), 598, go into the theory in more detail, and T. Fujita, *MWR*, 94 (1966), 19, describes the use of Doppler winds in the computation of mesoscale wind fields. See also R. M. Lhermitte, *BAMS*, 45 (1964), 587, for a review of Doppler and Pulse Doppler radar as severe-storm sensors.

On Doppler radar wind soundings, see G. M. Armstrong and R. J. Donaldson, Jr., *JAM*, 8 (1969), 376, for details of the PSI (Plan Shear Indicator) display, which gives the location of regions of abnormally large wind shear within precipitation, as applied to damaging winds and hail at Sudbury, Mass. For examples of Doppler radar investigations, see: K. A. Browning and R. Wexler, *JAM*, 7 (1968), 105, on determination of kinematic properties of a wind field, 1.5 to 6 km, at Bedford, Mass.; T. W. Harrold, *QJRMS*, 92 (1966), 31, horizontal convergence of the wind field during precipitation at Pershore, Eng.; R. M. Lhermitte, *JAS*, 23 (1966), 575, on a low-level jet at Norman, Oklahoma; R. Wexler, A. C. Chmela, and G. M. Armstrong, *MWR*, 95 (1967), 929, on a wind field in a thunderstorm in New England; C. C. Easterbrook, *JAM*, 6 (1967), 882, on the circulation pattern in convective storms in Buffalo, N.Y.; R. R. Rogers, *T*, 19 (1967), 432, on height-time patterns of rain in Hawaii; K. A. Browning *et al.*, *QJRMS*, 94 (1968), 498, on vertical and horizontal wind velocities within an isolated air-mass shower at Pershore, Eng.; and B. J. Mason, *QJRMS*, 95 (1969), 449, on microphysical and physical processes within frontal clouds and precipitation at Pershore. Doppler radar also makes it possible to determine the "balance level" in a convective storm, i.e., the height at which the precipitation particles are just balanced by the updraft; see D. Atlas, *JAS*, 23 (1966), 635, on Sudbury, Mass., and R. J. Donaldson, Jr., and R. Wexler, *JAS*, 24 (1967), 139.

101. *JAM*, 4 (1965), 130. G. E. McVehil, R. J. Pilié, and G. A. Zigrossi, *loc. cit.*, p. 146. Below 11 km, a smooth spherical balloon 2 m in diameter is aerodynamically unstable and produces spurious high-frequency oscillations. For details of atmospheric mesostructure between 11 and 20 km as observed by ROSE, see A. I. Weinstein, E. R. Reiter, and J. R. Scoggins, *JAM*, 5 (1966), 49.

For pulse Doppler radar examples, see L. J. Battan and J. B. Theiss, *JAS*, 23 (1966), 78, on vertical motion and precipitation particles in a thunderstorm in the Santa Catalina Mts., Arizona, and T. W. Harrold and K. A. Browning, *MM*, 96 (1967), 367, on mesoscale wind fluctuations below 1,500 m at Pershore, Eng. Pulse Doppler radar enables clear-air radar dots ("angels") to be used as air-movement tracers; see K. A. Browning and D. Atlas, *JAS*, 23 (1966), 592.

With the ROSE system of balloon soundings, according to J. H. McCloskey, *JAM*, 8 (1969), 304, a defect of the AN/FPS-16 radar normally used to track ROSE balloons may introduce an error of more than 1 m per sec into the measured wind velocities. ROSE balloons have a smooth surface; Jimsphere balloons are similar to ROSE, but have roughened surfaces. For examples of mesoscale waves between the surface and 20 km at Cape Kennedy, Fla., detected by Jimsphere balloons but undetected by conventional rawinsondes, see R. E. De Mandel and J. R. Scoggins, *JAM*, 6 (1967), 617, and R. F. Stengel, *JAM*, 7 (1968), 513. For details of metalized, spherical, super-pressure, vertically rising balloons in general, see: P. B. Mac-Cready, Jr., *JAM*, 4 (1965), 504, for a comparison of different types; P. Hyson, *QJRMS*, 94 (1968), 592, for a theory of errors in wind data above 50 km; J. R. Scoggins and M. Armendariz, *JAM*, 8 (1969), 449, on errors in wind profiles; R. D. Reynolds, *JAM*, 5 (1966), 537, on the effect of atmospheric lapse-rate variations on balloon ascent-rates to 100,000 feet; J. Luers and N. Engler, *JAM*, 6 (1967), 816, on a theory of optimum methods for obtaining wind data; and A. I. Weinstein, E. R. Reiter, and J. R. Scoggins, *JAM*, 5 (1966), 49, on the mesoscale structure of 11- to 20-km winds at Point Mugu, Calif., and Cape Kennedy, Fla., derived from vertical ascents, on which see also O. Essenwanger, *JAM*, 6, (1967), 591. Most standard balloons reaching over 100,000 feet produce inaccurate data in arctic and tropical regions at night; for details of U.S. Army all-zone balloons designed to correct this, see M. Sharenow, *BAMS*, 48 (1967), 809.

102. J. Warner and E. G. Bowen, *T*, 5 (1953), 36. L. J. Battan, *BAMS*, 39 (1958), 258. A. D. Anderson and W. E. Hoehne, *BAMS*, 37 (1956), 454, and 38 (1957), 302.

103. M. Neiburger and J. K. Angell, *JM*, 13 (1956), 166. K. C. Giles and R. E. Peterson, *JM*, 14 (1957), 569. J. K. Angell, *QJRMS*, 90 (1964), 472, and *MWR*, 90 (1962), 245 and 391. Wan-cheng Chiu, *MWR*, 89 (1961), 297. On transosondes, see: J. K. Angell, *MWR*,

92 (1964), 203, on observations of jetstream winds east of Japan; P. Morel, J. Fourrier, and P. Sitbon, *JAM*, 7 (1968), 626, and V. E. Lally, *loc. cit.*, p. 958, on occurrence of ice on the balloons. On the Global Horizontal Sounding System (GHOST) of transosondes, see: F. Mesinger, *JAS*, 22 (1965), 479, on a theory of behavior of a very large number of constant-volume balloon trajectories; S. B. Solot and J. K. Angell, *JAS*, 26 (1969), 574, on middle-latitude mean monthly zonal winds at 200 mb in the southern hemisphere, derived from GHOST flights; and R. D. Reynolds and R. L. Lamberth, *JAM*, 5 (1966), 304, on the "dangling thermistor" technique, enabling ambient temperatures to be measured to an accuracy of $\pm 1°C$ by radiosondes flown by GHOST balloons.

104. F. A. Brooks and C. F. Kelly, *TAGU*, 32 (1951), 833. For examples of micrometeorological profiles and descriptions of the instruments employed to measure them, see *EAFM*, pp. 126–51 and 229–42 (winds), 152–75 and 199–227 (temperatures).

105. A. R. Kassander, Jr., and R. M. Steward, Jr., *BAMS*, 36 (1955), 384. L. J. Anderson, *BAMS*, 40 (1959), 49. E. R. Sanford, *JM*, 8 (1951), 182.

106. J. C. Kaimal and J. A. Businger, *JAM*, 2 (1963), 156. J. C. Kaimal *et al.*, *QJRMS*, 90 (1964), 467. A. H. Glaser, *EAFM*, p. 248. I. Karmin, *BAMS*, 40 (1959), 473. L. J. Fritschen and R. H. Shaw, *BAMS*, 42 (1961), 42. G. C. Gill, *BAMS*, 35 (1954), 69. On micrometeorological measurements, see: L. J. Fritschen, *JAM*, 6 (1967), 695, for details of a sensitive cup-type anemometer suitable for measuring horizontal windspeed profiles over fairly narrow crop surfaces; W. H. Read III and J. W. Lynch, *JAM*, 2 (1963), 412, on ping-pong ball anemometers; R. E. Stevenson, *JAM*, 3 (1964), 115, on the influence of a ship on temperatures of the surrounding air and water; and H. W. Baynton *et al.*, *JAM*, 4 (1965), 670, for wind profiles in and above a tropical rainforest in northern Colombia, measured by means of a propellor-bivane anemometer. For approximate estimations of microclimatic exposure based on studies of flag-tattering, see N. Rutter, *Agric. Meteorol.*, 5 (1968), 163, for Aberystwyth, Wales, and D. Thomas, *W*, 14 (1959), 375, for west Wales.

107. L. C. Chamberlin *et al.*, *EAFM*, p. 276. C. W. Thornthwaite *et al.*, *PC*, vol. 14, no. 1 (1961). For visual observations of microclimatic effects, see D. E. Pedgley, *W*, 22 (1967), 42, on the shape of snowdrifts as evidence of wind eddies; R. E. Lacy, *W*, 21 (1966), 135, on frost patterns on brickwork; and J. L. Monteith, *W*, 11 (1956), 8, on the effect of grass-length on snow melting. On the measurement of the vertical component of air movement near the ground, see: J. C. Kaimal *et al.*, *QJRMS*, 90 (1964), 467, for a comparison of bivane and sonic anemometer techniques; R. M. Holmes, G. C. Gill, and H. W. Carson, *JAM*, 3 (1964), 802 on a propeller-type anemometer; and P. B. MacCready, Jr., and H. R. Jex, *loc. cit.*, p.

182, for a theory of response of propellor and vane sensors. On the measurement of wind inclination to the horizontal, see J. K. Angell, *QJRMS*, 90 (1964), 307, for spectra and cross-spectra of wind inclination data in the 600- to 2,600-foot layer at Cardington, Eng.

108. H. Moses and G. R. Hilst, *EAFM*, p. 243.

109. M. Ference, *CM*, p. 1216. On wiresondes, see P. Hyson, *JAM*, 7 (1968), 684, for details of a tungsten-wire temperature sensor, which is carried aloft by the British M.O. "Skua" rocket, and then returns to Earth by parachute, functioning as a wiresonde during its descent. See also R. S. Bourke and G. L. Klein, *JAM*, 6 (1967), 419 and 707, for details of a chronometric tethersonde, which is carried aloft by a tethered balloon, and transmits continuous data on temperature-height profiles within the lowest 500 meters of the atmosphere.

110. J. K. Angell, *MWR*, 88 (1960), 277, and 92 (1964), 465. G. C. Holzworth, E. K. Kauper, and T. B. Smith, *MWR*, 91 (1963), 387. D. H. Pack and J. K. Angell, *loc. cit.*, p. 583. On tetroons, see: J. K. Angell *et al.*, *W*, 23 (1968), 184, on flights over New York City; W. A. Hass *et al.*, *QJRMS*, 93 (1967), 483, on the 300-meter level over New York City; K. R. Peterson, *JAM*, 5 (1966), 553, on estimation of trajectories from New York City and Atlantic City, N.J.; and L. M. Druyan, *JAM*, 7 (1968), 583, on New York City. On the use of tetroons as Lagrangian probes at 300 meters in the Los Angeles Basin, providing unique observations of airflow reversals and diurnal recirculations, see J. K. Angell *et al.*, *JAM*, 5 (1966), 565. For examples of tracers, see: N. Thompson, *MM*, 93 (1964), 193, and P. A. Leighton *et al.*, *JAM*, 4 (1965), 334, on fluorescent particles; P. W. Nickola *et al.*, *JAM*, 6 (1967), 430, on a recorder for atmospheric concentrations of zinc-sulfide fluorescent pigment; and C. V. Smith, *MM*, 96 (1967), 150, for a review of airborne tracers.

111. G. J. House, N. E. Rider and C. P. Tugwell, *QJRMS*, 86 (1960), 215. R. M. Brown, *JGR*, 64 (1959), 2369, gives details of an automatic micrometeorological data-collecting system. R. B. Platt, *TAGU*, 38 (1957), 166, describes a field installation for the automatic recording of micrometeorological gradients. L. J. Fritschen and C. H. M. von Bavel, *JAM*, 2 (1963), 151.

112. F. Hall, *JM*, 7 (1950), 121, and 6 (1949), 160. P. B. MacCready, Jr., *BAMS*, 46 (1965), 533.

113. R. L. Ives, *JAM*, 1 (1962), 60. J. F. Nagel, *W*, 8 (1953), 227.

114. V. E. Lally, *BAMS*, 45 (1964), 568. On building inexpensive equipment, see *School Science Review* and the regular "Amateur Scientist" articles in *Scientific American*. On the tendency toward complexity in instrumentation, see R. Perley, *BAMS*, 45 (1964), 740, on the application of systems engineering to weather-data systems, and R. N. Sachdev, S. B. Nadgir, and

K. K. Rajan, *JAM*, 6 (1967), 969, for details of an automatic data-telemetering system for the transmission of data from remote stations to a central office.

115. See J. P. Lodge, Jr., *AGP*, 9 (1962), 97, on the identification of aerosols, and K. Bullrich, *AGP*, 10 (1964), 99, on how they affect the scattering of radiation. See R. A. Cadle *et al.*, *JAS*, 24 (1967), 100, for evidence that the Antarctic atmosphere, near the surface, shows a much higher sulphur concentration than anywhere else in the world.

116. C. D. Keeling, *T*, 12 (1960), 200. W. Bischof, *T*, 14 (1962), 87, and 12 (1960), 216. S. Fonselius, F. Koroleff, and K.-E. Wärme, *T*, 8 (1956), 176. C. E. Junge, *T*, 14 (1962), 242.

117. B. Bolin, *T*, 12 (1960), 274. A. N. Dingle, *T*, 6 (1954), 342, gives maps for CO_2 exchange. H. Craig, *T*, 9 (1957), 1. R. Revelle and H. E. Suess, *T*, 9 (1957), 18. See G. N. Plass, *T*, 8 (1956), 140, and L. D. Kaplan, *T*, 12 (1960), 204, for details of the CO_2 theory, and E. Eriksson and P. Welander, *T*, 8 (1956), 155, for a mathematical model of the CO_2 cycle. See C. E. Junge and G. Czeplak, *T*, 20 (1968), 422, on seasonal variation of CO_2 in the troposphere, and W. Bischof and B. Bolin, *T*, 18 (1966), 155, on space-time variation of CO_2 in the troposphere and lower stratosphere.

118. See R. E. Newell, *JGR*, 49 (1961), 137, for details of the transport of trace substances, including ozone, through the atmosphere in relation to the general circulation. See A. Vassey, *AGP*, 11 (1965), 115, for a review of the topic of atmospheric ozone.

119. C. E. Junge, *T*, 14 (1962), 363. J. P. Funk and G. L. Garnham, *T*, 14 (1962), 378.

120. For atmospheric cross sections showing the variation of ozone density, see E. S. Epstein, C. Osterberg, and A. Adel, *JM*, 13 (1956), 319, for Flagstaff, Arizona. R. J. Reed, *JM*, 7 (1950), 263, described the role of vertical motion in ozone-weather relationships. On control of daily variations, see Sir C. Normand, *QJRMS*, 79 (1953), 39, and K. Allington, B. W. Boville, and F. K. Hare, *T*, 12 (1960), 266.

121. F. S. Johnson, *BAMS*, 34 (1953), 106.

On ozone climatology, see: G. M. B. Dobson, *QJRMS*, 92 (1966), 549, on the annual variation of ozone in Antarctica; R. D. Bojkov, *JAM*, 8 (1969), 284, on mean cross sections, 5 to 50 km, pole to pole; S. Rangarajan, *JAS*, 26 (1969), 613, on a nearly worldwide anomaly in the seasonal variation of ozone, which is that, above 40 km, the maximum ozone concentration is in winter, and the minimum in summer, but below 35 km, the pattern reverses; A. W. Brewer and A. W. Wilson, *QJRMS*, 94 (1968), 249, on regions of formation of atmospheric ozone; A. B. Pittock, *JAM*, 8 (1969), 308, on the synoptic climatology of the vertical distribution of ozone at Aspendale, Australia, and Boulder, Colo.; R. N. Kulkarni, *T*, 20 (1968), 305, on the relation of ozone to the stratospheric circulation over Australia; R. D. Bojkov,

GPA, 70, no. 3 (1969), 165, on the Mediterranean and southeast Europe during an IQSY period); R. D. Bojkov and A. D. Christie, *JAS*, 23 (1966), 791, on seasonal profiles of ozone, in New Zealand; and G. M. Shah, *JAS*, 24 (1967), 396, on a quasibiennial oscillation in ozone, extending from the equator to polar latitudes in the northern hemisphere. For a comparison of ozone concentrations at night and during the day in Australia, see R. N. Kulkarni, *QJRMS*, 94 (1968), 266. According to G. M. Shah, *JAS*, 23 (1966), 535, any apparent increase in ozone during nighttime is illusory. According to H. C. Willett, *JAS*, 25 (1968), 341, the nature of seasonal changes of ozone in polar latitudes (stratospheric warmings are sporadic in the winter Arctic atmosphere, but explosive in the spring Antarctic atmosphere) suggests that variation in the number of solar crepuscular particles (i.e., variability of the solar wind) is an important influence on ozone, but see A. H. Manson, *JAS*, 26 (1969), 587.

See also: R. N. Kulkarni, *QJRMS*, 92 (1966), 363, on the vertical distribution of ozone at Brisbane and Aspendale, Australia; A. B. Pittock, *QJRMS*, 94 (1968), 563, on time cross sections in southeastern Australia; J. G. Breiland, *JAS*, 24 (1967), 569, on daily soundings at Albuquerque, New Mexico, proving that the ozone and thermal stability structure of the lower stratosphere is stratified; W. S. Hering, *T*, 18 (1966), 329, on transport processes, North and Central America; A. B. Pittock, *JAS*, 23 (1966), 538, on a sharp minimum in ozone content over Boulder, Colo., coinciding with occurrence of a layer of volcanic dust; and R. Berggren, *T*, 17 (1965), 180, on use of ozone as a synoptic tracer at Arosa, Switz. For case studies of ozone situations, see R. Berggren and K. Labitzke, *T*, 20 (1968), 88, on distribution of ozone on pressure surfaces from 150 to 300 mb, and *T*, 18 (1966), 761, on detailed distribution of ozone, both vertical and horizontal, in North America.

On low-level ozone, see: W. Warmbt, *T*, 18 (1966), 441, on its synoptic climatology at Dresden-Wahnsdorf; D. R. Davis and C. E. Dean, *MWR*, 94 (1966) 179, on damage to tobacco plants in northern Florida; R. E. Newell, H. W. Brandli, and D. A. Widen, *JAM*, 5 (1966), 740, on the concentration of ozone in surface air at Boston, Mass.; and D. A. Lea, *JAM*, 7 (1968), 252, on the vertical distribution of ozone in the lowest km of air at Point Mugu, Calif., from Los Angeles pollutant, showing strong microclimatic and topographic influences.

122. A. Chojnacki and M. Kac-Kacas, *T*, 15 (1963). 202. A. Emanuelsson, E. Eriksson, and H. Egnér, *T*, 6 (1954), 261, give the maps of Swedish data.

123. For a study of precipitation composition, see E. Eriksson, *T*, 4 (1952), 215 and 280. For maps of the chlorine and sulphur content of precipitation in Europe, Australia, and the United States, see E. Eriksson, *T*, 12 (1960), 63. For measurements of the vertical distribution of atmospheric chloride particles, see

W. G. Durbin and G. D. White, *T*, 13 (1961), 260. See E. de Bary and C. Junge, *T*, 15 (1963), 370, for maps of Europe, and J. P. Lodge, Jr., A. J. Mac-Donald, Jr., and E. Vihman, *T*, 12 (1960). 184, for details of the natural composition of marine atmospheres.

124. A. Ångström and L. Högberg, *T*, 4 (1952), 31. See A. Ångström and L. Högberg, *T*, 4 (1952), 271, for maps. W. A. Mordy, *T*, 5 (1953), 470. On precipitation chemistry, see: D. H. Yaalon, *T*, 16 (1964), 200, on the climatology of ammonia and nitrate in rain water in Israel; W. Dansgaard, *loc. cit.*, p. 436, on the world climatology of stable isotopes in precipitation; L. T. Khemani and B. V. R. Murty, *T*, 20 (1968), 284, on the chemical composition of rain in Delhi; H. Mrose, *T*, 18 (1966), 266, on the *p*H or rain, snow, and fog water at Dresden-Wahnsdorf; O. P. Petrenschuk and V. M. Drozkova, *loc. cit.*, p. 280, on the chemical composition of cloud water in the U.S.S.R.; and A. W. Gambell and I. Friedman, *JAM*, 4 (1965), 533, on an areal discontinuity in the deuterium/hydrogen ratio in rainfall in a single storm in North Carolina and southeastern Virginia.

125. H. R. Byers, *JM*, 6 (1949), 363. V. E. Jakl, *JM*, 7 (1950), 304. A. H. Woodcock, *JM*, 7 (1950), 161, and 9 (1952), 200. R. M. Griffith, *JAS*, 20 (1963), 198. A. H. Woodcock and C. P. Blanchard, *T*, 7 (1955), 437. A. H. Woodcock and W. A. Mordy, *T*, 7 (1955), 291.

126. A. H. Woodcock, *JM*, 10 (1953), 362. C. E. Junge and P. E. Gustafson, *T*, 9 (1957), 164.

127. A. H. Woodcock, *T*, 10 (1958), 355. On atmospheric aerosol, see: R. K. Kapoor and B. V. R. Murty, *JAM*, 5 (1966), 493, on seasonal variations of sulphate and chloride aerosol at Delhi; W. E. Clark and K. T. Whitby, *JAS*, 24 (1967), 677, on the diurnal variation in the atmosphere of Minneapolis; C. E. Junge, E. Robinson, and F. L. Ludwig, *JAM*, 8 (1969), 340, on aerosol size distributions in Pacific air masses at Cape Blanco and Crater Lake, Ore.; S. Twomey and G. T. Severynse, *JAS*, 21 (1964), 558, on size distributions of particles below 0.1μ at Washington, D.C., and Long Island, N.Y.; and J. M. Prospero, *BAMS*, 49 (1968), 645, on continuous low-level observations of marine aerosol at Barbados. On a theory of the falling speed of aerosol particles, see F. Kasten, *JAM*, 7 (1968), 944. For a synoptic study of the giant particles (i.e., diameter 2.5 to 100μ, of the same order as the wavelength of the incident radiation) formed from mineral dusts over south-central New Mexico, see G. B. Hoidale and S. M. Smith, *T*, 20 (1968), 251. On Aitken nuclei (continental aerosol), see A. W. Hogan *et al.*, *JAM*, 6 (1967), 726, on observations over the North Atlantic, indicating that continental aerosols may be present, at concentrations of the order of 500 nuclei per cm^3, at several hundred miles out to sea, and P. A. Allee, *MWR*, 95 (1967), 925, on the climatology of Aitken nuclei concentration

at Washington, D.C. On sea-salt nuclei, see: Y. Toba *T*, 18 (1966), 132, on their world climatology, and *T*, 15 (1965), 131 and 364, on their theory; and N. Rutter and R. S. Edwards, *Agric. Meteorl.*, 5 (1968), 235, on deposition of airborne marine salts at Aberystwyth, Wales.

On hygroscopic nuclei, see: B. V. R. Murty, A. K. Roy, and R. K. Kapoor, *T*, 19 (1967), 136, on the climatology of nuclei originating from the sea during the Indian monsoon; R. S. Sekhon and B. V. R. Murty, *JAS*, 23 (1966), 771, on variations in concentrations of hygroscopic and nonhygroscopic nuclei at Delhi; E. S. Selezneva, *T*, 18 (1966), 525, on the distribution of condensation nuclei over European Russia; L. F. Radke and P. V. Hobbs, *JAS*, 26 (1969), 281, on the variations in concentration of cloud condensation nuclei in the Olympic Mts., Washington; P. Squires and S. Twomey, *JAS*, 23 (1966), 401, on cloud nuclei comparisons between the Caribbean and Colorado; M. P. Paterson and K. T. Spillane, *QJRMS*, 95 (1969), 526, on condensation nuclei produced by the bursting of bubbles in sea water; J. Warner and S. Twomey, *JAS*, 24 (1967), 704, on cloud condensation nuclei produced in great numbers by smoke from sugar-cane fires in Queensland, Australia; and S. A. Changnon, Jr., *BAMS*, 50 (1969), 411, on urban condensation nuclei and their importance for precipitation over built-up areas in the United States.

128. E. K. Bigg, *JM* 13 (1956), 262. E. G. Bowen, *loc. cit.*, p. 142. *T*, 8 (1956), 394.

129. G. W. Brier, *JAS*, 19 (1962), 56. E. J. Hannan, *JAM*, 1 (1962), 426 and 429. F. E. Volz and R. M. Goody, *JAS*, 19 (1962), 385. The meteoric-dust hypothesis depends on the existence, within the atmosphere, of ice nuclei formed by nucleation around a particle of meteoric dust. For observations of ice nuclei in Hawaii, see: C. T. Nagamoto, J. Rosinki, and G. Langer, *JAM*, 6 (1967), 1123; E. K. Bigg, *JAM*, 7 (1968), 950; S. Price and J. C. Pales, *MWR*, 92 (1964), 207; and E. G. Droessler and K. J. Heffernan, *JAM*, 4 (1965), 442, for some at the surface. For ice-nuclei observations over eastern Australia, see E. K. Bigg, *JAS*, 24 (1967), 226; for details of ice-nucleus storms in eastern Australia, see E. G. Droessler, *JAS*, 21 (1964), 701. For automobile exhausts as a source of potential ice-nucleating particles, see G. M. Morgan, Jr., and P. A. Allee, *JAM*, 7 (1968), 241, and V. J. Schaefer, *loc. cit.*, p. 148. On the close correlation between ice nuclei and precipitation, see B. F. Ryan and W. D. Scott, *JAS*, 26 (1969), 611, who found a significant increase in ground-level ice-nuclei concentrations at Seattle and the Olympic Mts., Washington, and at Comp Cortino, Italy, with the onset of rainfall, possibly due to the release of ice nuclei by evaporation of small raindrops falling below cloud base; see also E. K. Bigg and G. T. Miles, *JAS*, 21 (1964), 396, who found a strong lunar influence in the concentrations of ice nuclei in eastern Australia. On atmospheric dusts, see

F. E. Volz, *BAMS*, 50 (1969), 16, on visual observations of striated dust clouds in the stratosphere, associated with the purple-light zone at Bedford, Mass.; see also M. J. Brown, R. K. Kraus, and R. M. Smith, *WW*, 21 (1968), 66, on a network of gauges for measuring dust deposition in the United States east of the Rocky Mts.

130. For examples of natural radioactivity in the atmosphere, see L. B. Lockhart, Jr., *T*, 14 (1962), 350, for Alaska, and T. Thompson and P. A. Wiberg, *T*, 15 (1963), 313. See S. G. Malahov, *T*, 18 (1966), 643, on the diurnal variation of radon and thoron decay products in the atmospheric surface layer at Moscow; J. Fontan *et al.*, *loc. cit.*, p. 623, on the vertical distribution of radon, boron, etc., in the lowest 30 meters at Magny-les-Hameaux, France; J. E. Pearson and H. Moses, *JAM*, 5 (1966), 175, on the variation with height, to 16 m, and time of boron at Argonne, Illinois; and C. R. Hosler, *loc. cit.*, p. 653, on the use of radon-222 as an indicator of the air-pollution potential of the atmosphere at Washington, D.C. On the use of natural radioactivity as an air-mass tracer, see J. Servant, *T*, 18 (1966), 663, on the Sarclay plateau and Morvan hills, France. For measurements of soil-gas emanations, using radon-222 from soil and vegetation as a tracer in vertical diffusion studies, in Champaign Co., Illinois, and Grants, New Mexico, see J. E. Pearson, D. H. Rimbey, and G. E. Jones, *loc. cit.*, p. 655, and *JAM*, 4 (1965), 349.

131. For fallout studies, see: A. J. Dyer and S.-A. Yeo, *T*, 12 (1960), 195, for Australia; P. B. Storebö, *loc. cit.*, p. 293, and S. H. Small, *loc. cit.*, p. 308, for Norway; L. B. Lockhart, Jr., and R. L. Patterson, Jr., *loc. cit.*, p. 298, for the North Pacific; G. Lindblom, *T*, 13 (1961), 106, for Sweden; and J. R. Gat, *et al.*, *T*, 15 (1963), 89, for Israel. On atmospheric structure, see E. R. Reiter, *JAM*, 2 (1963), 691.

132. L. B. Lockhart, Jr., R. A. Baus, and I. H. Blifford, Jr., *T*, 11 (1959), 83. J. F. Bleichrodt, *et al.*, *T* 11 (1959), 404. J. F. Bleichrodt, W. Bleeker, and F. H. Schmidt, *T*, 12 (1960). 188. P. Storebö, *JM*, 17 (1960), 547. D. O. Staley, *JAS*, 20 (1963), 615, and 19 (1962), 450. On the use of artificial radioactivity as an atmospheric tracer, see: H. W. Feely *et al.*, *T*, 18 (1966), 316, for global studies of stratospheric motions; R. J. List, L. P. Salter, and K. Telegadas, *loc. cit.*, p. 345, on stratospheric motions in general; E. A. Martell, *JAS*, 24 (1967), 113, for a climatological study of stratospheric air motions and transport processes; O. Tanaevsky and J. Blanchet, *T*, 18 (1966), 434, on empirical probability trajectories; and P. Kruger and A. Miller, *JGR*, 71 (1966), 4243, on the movement of rain and air across the trade-wind inversion in Hawaii. See also G. V. Dmitrieva *et al.*, *T*, 18 (1966), 407, for radioactive tracer observations of the movement of surface air from middle latitudes into the tropics by shipboard measurements. According to R. D. Cadle *et al.*, *JAM*,

8 (1969), 348, trace observations show that air is transported from troposphere to stratosphere on the anticyclonic side of a jetstream, during tropopause "folding" episodes in the western United States.

133. On turbidity, see: A. Ångström, *T*, 16 (1964), 64, for an analysis of the parameters of atmospheric turbidity; F. E. Volz, *T*, 17 (1965), 513, on global variations in turbidity observed since the eruption of Mt. Agung volcano on Bali, March 17, 1963; and W. H. Fischer, *JAM*, 6 (1967), 958, for turbidity measurements in Antarctica.

134. For a review of the development of radiometeorology, see *CM*, pp. 1265–1300, especially H. G. Booker, p. 1290. See also J. S. Marshall and W. E. Gordon, *AMM*, vol. 3, no. 14 (1957), and *PM*, pp. 275–317. On radioclimatology, see W. A. Arvola, *BAMS*, 38 (1957), 212.

135. E. E. Gossard, *BAMS*, 38 (1957), 274. For the climatology of the Chios-Lemnos (Greece) line-of-sight radio link, see M. Anastassiades and L. Carapiperis, *GPA*, 67, no. 2 (1967), 173. For the influence of atmospheric stability on refractive index fluctuations in the sea area between Norway and Denmark, see D. T. Gjessing, A. G. Kjelaas, and J. Nordø, *JAS*, 26 (1969), 462. For theoretical aspects, see P. Milnarich, Jr., and W. L. Shepherd, *JAM*, 5 (1966), 722, on the prediction of the degree of refractive bending of radio waves in the troposphere, and E. J. Dutton, *JAM*, 6 (1967), 662, on the attentuation of radio rays by convective rainfall. For examples of radar ducting, see P. J. Bacon, *W*, 20 (1965), 82.

136. C. J. Stapley, *MM*, 93 (1964), 294.

137. K. H. Jehn, *JM*, 17 (1960), 264, and *BAMS*, 41 (1960), 304, gives maps. J. R. Bauer and J. H. Meyer, *TAGU*, 39 (1958), 624, give cross sections of refractive index. D. L. Randall, *BAMS*, 35 (1954), 56. C. M. Cain and J. R. Gerhardt, *BAMS*, 31 (1950), 330. J. R. Gerhardt, C. M. Cain, and H. Chapman, *BAMS*, 37 (1956), 251.

138. For reviews of electrical phenomena in the atmosphere, see *CM*, pp. 101–50, and H. J. aufm Kampe, *AMM*, vol. 3, no. 20 (1957). For an example of electrical climate, see W. E. Cobb, *JAS*, 25 (1968), 470, on the atmospheric electric climate at Mauna Loa Observatory, Hawaii, which appears to be a good indicator of the concentration of fine particles in the atmosphere over a large area. The surface electric elements of the atmosphere on Mauna Loa also respond to solar flares; see W. E. Cobb, *MWR*, 95 (1967), 905. For details of worldwide diurnal variations in atmospheric electricity, see R. V. Anderson, *loc. cit.*, p. 899. On the use of an ordinary radio receiver to detect severe local storms, by means of the atmospheric electricity variations that manifest as radio static, see H. Geise, *BAMS*, 45 (1964), 604. For an example of acoustic climate, see G. H. Gilbert, *WW*, 18 (1965), 166, on the pattern of the atmospheric zones of audibility and inaudibility observed as a

consequence of the controlled explosion of 500 tons of TNT at Suffield, Alberta. Two zones of audibility were observed to the west of Suffield, centered at 150 and 300 miles from the town, with a third (ill-defined) zone at 450 miles. To the east, audibility zones extended from 75 to 100 miles, and from 250 to 300 miles, from the town. The intervening zones of inaudibility represented "skip zones," in which the sound waves from the explosion traveled into the upper atmosphere and then were refracted down to the ground again.

139. On roughness length, see Chapter 3 in *Foundations*. See J. R. Malkus, in W. P. Munk, ed., *The Sea: Ideas and Observations*, vol. 1, for sample total-stress maps for the oceans. On resistance, see R. C. Sutcliffe, *QJRMS*, 62 (1936), 3. Roughness and vegetation height were related for Wisconsin by E. C. Kung and H. H. Lettau, *Dept. Met. Ann. Report, Univ. Wisconsin* (Madison, Wisc., 1961), p. 45. On aerodynamic drag, see J. E. Vehrencamp, *EAFM*, pp. 99 and 104, and M. H. Halstead and R. K. Ono, *op. cit.*, p. 114. Classic measurements of aerodynamic drag over concrete were made by P. A. Sheppard, *Proc. Roy. Soc. London*, 188 (series A, 1947), 208, and over grassland by F. Pasquill, *Proc. Roy. Soc. London*, 202 (series A, 1950), 143. On boundary shearing stress at the sea surface, see S. Hellerman, *MWR*, 95 (1967), 607, on seasonal and annual values of wind stress on the world ocean, and *MWR*, 93 (1965), 239, on wind stress on the Atlantic Ocean, and E. B. Kraus, *BAMS*, 49 (1968), 247, on zonal values of surface wind stress on the North Atlantic Ocean. On the measurement of boundary shearing stress over a turf surface, by means of a drag-plate and shearing-stress meter, see E. F. Bradley, *QJRMS*, 94 (1968), 380.

140. For details of measurements, see *EAFM*, pp. 17–80. See also E. L. Deacon, *QJRMS*, 76 (1950), 479, for a method of measuring and recording the heat flux into the soil. For details of special apparatus, see W. D. Sellers and C. N. Hodges, *JAS*, 19 (1962), 482. In the determination of the temperature of a bare soil, contact thermometry is subject to serious errors. For experiments on this topic, see M. Fuchs and C. B. Tanner, *JAM*, 7 (1968), 303. On the components of energy exchange within a field of barley, see P. Ho, P. Schwerdtfeger, and G. Weller, *AMGB*, 16, series B (1968), 262.

141. K. J. K. Buettner, *JM*, 15 (1958), 155. J. M. Hearn, *MM*, 90 (1961), 174, described a surface wetness recorder.

142. For generalized land-surface contours, see L. Berkofsky and A. E. Bertoni, *BAMS*, 36 (1955), 350. The distribution of sea-ice cover is also a surface element of some importance. For observations of sea-ice limits in the North Atlantic, and their synoptic correlations, see H. H. Lamb, *W*, 18 (1963), 67. For meteorological aspects of ice cover on the Great Lakes, see T. L. Richards, *MWR*, 92 (1964), 297, and

for ice-cover charts at ten-day intervals, see C. R. Snider, *MWR*, 95 (1967), 685. For examples of continentality maps, see R. J. Kopec, *BAMS*, 46 (1965), 54, on the area around the Great Lakes, and D. K. MacKay and F. A. Cook, *Geographical Bulletin*, no. 20 (1963) on Canada.

143. M. H. Halstead *et al.*, *JM*, 14 (1957), 308. See also C. S. Benton and W. Covey, *TAGU*, 33 (1952), 673, for a suggested method for computing micrometeorological variables by measuring the temperatures of many indicators suspended at different heights.

144. B. Vonnegut and C. B. Moore, *BAMS*, 42 (1961), 793. For mean monthly maps, see C. H. Reitan, *BAMS*, 41 (1960), 79. On precipitable water, see: S. E. Tuller, *MWR*, 96 (1968), 785, for annual and mean monthly world values; J. Adem, *MWR*, 95 (1967), 83, for maps of total precipitable water in the northern hemisphere troposphere; and V. A. Myers, *MWR*, 93 (1965), 369, on the climatology of precipitable water in relation to easterly atmospheric moisture transport in Trinidad. On the relation between mean monthly precipitable water and mean monthly surface dewpoint, see C. H. Reitan, *JAM*, 2 (1963), 776. On the prediction of total precipitable water from surface dewpoint in the northern hemisphere, see W. L. Smith, *JAM*, 5 (1966), 726, and L. Berkofsky, *JAM*, 6 (1967), 959; and see S. J. Bolsenga, *JAM*, 4 (1965), 430, for mean daily/hourly values.

145. For details, see W. D. Sellers, *MWR*, 88 (1960), 155.

146. H. L. Penman, *Proc. Roy. Soc.*, 193, series A (1948), 120. C. W. Thornthwaite and F. K. Hare, *AMM*, 6, no. 28 (1965), 163. H. L. Penman, D. E. Angus, and C. H. H. van Bavel, *The Irrigation of Agricultural Lands* (Amer. Soc. Agronomy Monograph, 1964), Chap. 27. R. O. Slatyer and I. C. McIlroy, *Practical Microclimatology* (CSIRO, 1961).

147. J. C. Purvis, *MWR*, 89 (1961), 192. C. B. Tanner and W. L. Pelton, *JGR*, 65 (1960), 3391. For examples of evaporation measurements by heat-balance methods, see W. D. Sellers, *JAM*, 3 (1964), 98, on potential evapotranspiration at Yuma, Arizona, and M. J. Fox, *JAM*, 7 (1968), 697, on evaporation from dry stream beds in the deserts of the southwestern United States. On the measurement of the Bowen ratio by means of a simple psychrometric apparatus, see D. H. Sargeant and C. B. Tanner, *JAM*, 6 (1967), 414. For world maps depicting average values of the Bowen ratio, see W. H. Terjung, *AMGB*, 16, series B (1968), 279; for maps of monthly mean Bowen ratio values for inland and coastal waters in Japan, see T. Arai, *TJC*, 2 (1965), 54. For evaporation estimations by means of the Bowen ratio, see D. A. Rijks, *QJRMS*, 95 (1969), 643, on a papyrus swamp in Uganda; and see E. K. Webb, *JGR*, 65 (1960), 3415, and *JGR*, 69 (1964), 2649, for an estimation of evaporation when the Bowen ratio is fluctuating. On evaporation from agricultural land in southern Sudan, measured by

means of a bulk aerodynamic method, see U. Högström, *T*, 20 (1968), 65. For evaporation rates from soil in relation to the prevailing evaporativity of the air, see W. Covey and M. E. Bloodworth, *JAM*, 5 (1966), 364, and J. R. Phillip, *JAM*, 6 (1967), 581. For evaporation from a wet soil surface, inferred from surface temperatures as measured by infrared radiation thermometers, see J. Connaway and C. H. M. van Bavel, *loc. cit.*, p. 650, and M. Fuchs and C. B. Tanner, *loc. cit.*, p. 852.

148. C. W. Thornthwaite and J. R. Mather, *PC*, 8 (1955), 1. See also C. W. Thornthwaite, *GR*, 38 (1948), 55. T. E. A. Van Hylckama, *MWR*, 87 (1959), 107. W. L. Pelton, K. M. King, and C. B. Tanner, *Agron. J.*, 52 (1960), 387.

149. L. C. Chapas and A. R. Rees, *QJRMS*, 90 (1964), 313. F. H. W. Green, *QJRMS*, 85 (1959) 152. N. J. Cochrane, *loc. cit.*, p. 57.

150. On the budgeting procedure, see J. R. Mather, *BAMS*, 35 (1954), 63. For information about the soil factors, see C. W. Thornthwaite, *PC*, vol. 7, no. 3 (1954). For another evapotranspiration equation, see H. F. Blaney and W. D. Criddle, *U.S. Dept. Agric. Tech. Bull.*, no. 1275 (1962). For the modification of the Thornthwaite equation, see G. W. Smith, *JGR*, 64 (1959), 477 and R. M. Holmes and G. W. Robertson, *MWR*, 87 (1959), 101. On soil moisture, see: J. Grindley, *MM*, 96 (1967), 97, on the synoptic climatology of soil moisture deficit in Britain; R. F. Dale, *Agric. Meteorol.*, 5 (1968), 111, on its climatology in Iowa; W. Baier, *Agric. Meteorol.*, 6 (1969), 165, for measurements at Ottawa, Can.; and R. F. Dale and R. H. Shaw, *JAM*, 4 (1965), 661, on the climatology of soil moisture and atmospheric evaporative demand at Ames, Iowa.

151. On evaporation ratio, see J. E. McDonald, *BAMS*, 42 (1961), 185. See B. W. Currie, *BAMS*, 42 (1961), 371, on wind chill in the Canadian Prairies, G. M. Howe, *W*, 17 (1962), 349, on wind chill during the cold spell of Christmas 1961 in Britain, and O. Wilson, *Intern. J. Biometeorol.*, 11 (1967), 29, on wind chill in the Antarctic. On cooling power, see the abstracts in *MAB*, vol. 6 (1955). See: R. L. Hendrick, *BAMS*, 40 (1959), 620, on an index of outdoor comfort in summer for Hartford, Conn.; P. M. Stephenson, *MM*, 92 (1963), 338, on a comfort index for Singapore; P. R. Wycherley, *MM*, 95 (1967), 73, on comfort indices for Malaysia; and E. Jauregni and C. Soto, *Internatl. J. Biometeorol.*, 11 (1967), 21, on discomfort measures in Mexico. On wind chill, see R. Falconer, *WW*, 21 (1968), 227, and A. J. W. Catchpole, *W*, 24 (1969), 320 (Canadian prairies). On comfort indices, see: H. V. Foord, *MM*, 97 (1968), 282, on an index for London; G. A. Watt, *MM*, 96 (1967), 321, and *MM*, 97 (1968), 310, on indices for Bahrain and Sharjah; P. R. Wycherley, *MM*, 96 (1967), 73, on indices for Malaysia; C. N. McLeod, *MM*, 94 (1965), 166, on an index for Gan; and T. Kawamura, *TJC*, 3 (1966), 1, on an index for Japan. See P. M. Stephenson and C. N. McLeod, *MM*, 94 (1965), 171, on errors involved in computing comfort indices from mean values, and J. S. A. Green, *W*, 22 (1967), 128, on weather and outdoor comfort.

CHAPTER 2

Interpreting the Observations

Climatology is largely analysis of past weather data. This analysis may be statistical, cartographic, or a combination of the two. The statistical approach may be sub-divided into two general classes of techniques, the descriptive and the analytical.[1]

Descriptive statistical methods are used to compress many figures, representing observed data, into a few numerical measures or parameters that represent the original data adequately for a specific purpose. These methods, formerly known as "the calculus of observations," involve the computation of means, modes, medians, vari-ances, and correlations between two or more variables. They involve few assumptions, and consider the figures for what they are, not as samples of larger fields.

Analytical statistical methods use descriptive techniques to determine how well the observations agree with a theoretical model that they are assumed to follow. They enable generalizations of a known degree of accuracy to be made from the data. The observations are regarded as a sample of a population or universe, which is assumed to have certain characteristics whose numerical values can be estimated from the statistical description of the sample. The classical theory of analytical statistics is based on the concept of the random sample. Many meteorological observations, a series of daily temperature readings, for example, do not represent random samples, but are stratified samples, i.e., samples drawn from each of several strata of a population, for example, from each of successive hours or days. This stratification limits the number of standard statistical techniques that have direct and immediate climatological application. Further, for many sets of climatic data, it is impossible to define accurately the population from which the observations represent a sample drawn at random.

In attempting to describe the principles of statistical analysis as applied to meteorological data, it is convenient to divide the climatic elements or "variables" into two types: scalars, i.e., variables having magnitude only, such as temperature, pressure, and rainfall; and vectors, i.e., elements, such as wind or air movements, that have both magnitude and direction. An alternative classification is into linear variables, which either have no limiting values or are limited at one or both ends of a linear scale; and circular variables, which may be regarded as varying continuously through all the angles of a circle. Statistical methods for the interpretation of circular variables were developed much later than those for linear ones, although almost all meteorological elements vary continuously with time during a day, a year, or some other period, and hence are circular.[2]

Two other topics are of increasing importance in climatology, and must be described briefly: the design of statistical experiments, especially significance tests, which enable the reliability of the results of the experiment to be assessed; and statistical decisionmaking, which has developed largely out of the mathematical theory of games.

Statistical techniques are very sharp-edged tools, full of dangers to the unwary, and an apparently simple problem may involve concepts that are extremely difficult for the nonmathematician to use. A good example of this is meteorological time-series, in which the observations are considered in the order in which they were actually made, and an attempt is made to determine whether or not there is any underlying pattern or rhythm in the data.

The Analysis of Meteorological Time-Series

A time-series is a series of data arranged chronologically, the data usually being equally spaced in time. To overcome the irregularity of most time curves, some form of smoothing must be adopted (see Appendix 2.1). The simplest form of smoothing is running or moving means, which are the arithmetic averages of successive terms in the series. The smoothing of the curve becomes more pronounced as more terms are included in the means. If the original series is $a, b, c, d, e, f, g, \ldots$, then two new series may be derived thus:

$$\text{Series A:} \quad \frac{a+b+c}{3}, \frac{b+c+d}{3}, \frac{c+d+e}{3}, \frac{d+e+f}{3}, \frac{e+f+g}{3}, \ldots$$

$$\text{Series B:} \quad \frac{a+b+c+d+e}{5}, \frac{b+c+d+e+f}{5}, \frac{c+d+e+f+g}{5}, \ldots$$

Series A consists of three-unit overlapping averages, and series B of five-unit overlapping averages. In general, the application of moving or overlapping means to a time-series eliminates from it the variations or "cycles" whose period is the same as the period of the average. That is, five-day running means will eliminate all variations due to oscillations recurring every five days, and 30-day means will eliminate all trace of any 30-day periodicities.

The choice of the length of the averaging period is critical. Too much smoothing (i.e., too long an averaging period) will completely destroy the significant features of the curve, too little smoothing will not eliminate the irregularities that may mask any underlying pattern, which the smoothing is intended to bring into prominence. Since it is most unusual to hit on the perfect smoothing interval by trial and error, searching for periodicities by means of running averages can be misleading. For example, moving averages made from a purely random time-series can generate an irregular apparent periodicity by means of a simple property termed the Slutzky-Yule effect.[3]

Simple overlapping means may on occasion convert maxima into minima and vice versa, because the averaging procedure gives too little weight to the middle terms. Weighted overlapping means correct this deficiency. Another form of arithmetic smoothing, the method of differences, is used less than running means, although it can be very useful (see Appendix 2.2).

Climatological time-series usually show considerable evidence of persistence or coherence, an effect arising from the tendency of most meteorological observations not to be independent of preceding observations. This dependence decreases with the length of the time interval between successive events. For example, the probability that rain will fall on a given day is much greater if rain fell on the previous day than if it did not; the probability is much less with monthly rainfall totals, while the annual precipitation at a given place has little relation to the precipitation received there during the preceding year.

Different measures of persistence are used for different types of variable. For discontinuous (discrete) variables such as precipitation, the appropriate measure is Besson's coefficient of persistence, R_B, which equals zero when there is no persistence in the data, and infinity if the persistence is complete. Discontinuous variables may be given the value 1 or 0, implying that the element either does or does not occur, and an unbroken succession of occurrences or nonoccurrences constitutes a run. The introduction of persistence within a time-series reduces the number of runs and increases their average length (see Appendix 2.3).

For continuous variables such as temperature or pressure, the normal measure of persistence is the coefficient of persistence, R_a, which equals zero when persistence is nil, and is negative if successive values in the time-series tend to be alternately high and low. Another useful parameter is the persistence factor, s, defined by $s = 1 + R_B$ for discontinuous variables and $s = (1 + R_a)^2$ for continuous variables (see Appendix 2.4). A phenomenon having a persistence factor s tends to occur in groups of average size s, which results in the lengths of runs being s times as long as would be expected without persistence. Runs in a continuous variable are when the element exhibits an unbroken succession of continuously rising (or falling) values.

Persistence within a set of data can seriously affect the calculation of parameters describing the data. For example, a formula different from the normal one must be used to derive the standard deviation if persistence is present (see Appendix 2.5). The normal formula gives a value of 0.9° for the standard deviation of the departure of daily average temperatures at Edinburgh during 1950 from their 50-year average; using the corrected formula, the actual standard deviation (i.e., after allowing for persistence) is 2.0°F.

Persistence often arises because of geographical influences, especially the conservative properties of certain sea or land surfaces. The seas surrounding the British Isles are of this type, so that very cold (or warm) autumns in Britain tend slightly to be followed by very cold (or warm) winters.[4]

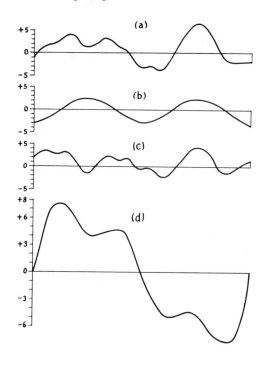

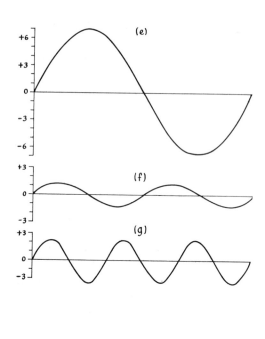

FIGURE 2.1.
Harmonic components of a series.

An alternative approach to smoothing is to separate the curve representing a time-series into regular (periodic) oscillations and irregular (aperiodic) variations, by assuming the curve is made up of the algebraic sum of several regular time-waves plus a random oscillation. One can describe mathematically any curve, however complicated, by a polynomial expression with the same number of constants as there are observations in the original series. Such an expression would be laborious to analyze, however, and it is much more convenient to use Fourier series (see Appendix 2.6). One advantage of the Fourier series is that it can be used for discontinuous as well as for continuous variables, that is, can be applied to elements, such as cloud cover, that may not exist for part of the observational period.

A Fourier series may be expressed as a sum of both sines and cosines, of cosines only, or of sines only, the latter being the one used most frequently in climatology. The technique of studying a time-series by means of the Fourier sine-series is known as *harmonic analysis* (see Appendix 2.7).

Curve *a* in Figure 2.1 consists of a periodic sine curve, *b*, upon which is super-imposed a series of random fluctuations, curve *c*. "Random" in this sense means that there is no apparent reason for the oscillation, and that its effect is not serious for the purpose in mind. Curve *d* is formed by the algebraic sum of three sine waves, curves *e*, *f*, and *g*. Curve *e*, which represents one complete "oscillation" of a sine wave, is termed the first harmonic of the series; curve *f*, representing two complete oscillations, is the second harmonic; curve *g*, with three oscillations, is the third harmonic, and so on.

As the number of the harmonic increases, the amplitude of the sine wave composing it often decreases, and if so, the analysis is continued until the last harmonic defined has a negligible amplitude in terms of whatever is being examined. For example, in the analysis of hourly pressure data, if the total length of the observing period is 24 hours, then the first harmonic gives the diurnal pressure wave (i.e., the pressure values complete a perfect cycle once every 24 hours), and the second harmonic gives the semidiurnal wave (the values repeating themselves twice in 24 hours). The third and higher harmonics usually have negligible amplitudes—tiny fractions of a millibar—and can be neglected for many practical considerations. Since the amplitude of the third harmonic (i.e., the eight-hour pressure wave) is, for most stations, of the same order of magnitude as the accuracy of the standard pressure-recording instruments, there is doubt about whether regular pressure oscillations with periods of less than eight hours actually exist.

Two questions arise in connection with the meteorological application of harmonic analysis. First, do the periodicities brought to light by the analysis really exist? The technique will certainly reveal periodicities, which are all aliquots of the length of the series. For temperature and pressure, simple considerations of astronomic geography indicate that distinct periodicities ought to be present in the data, but for elements such as rainfall and cloud cover, it is uncertain whether or not regular periodicities are feasible, and any Fourier sine waves "discovered" may be purely statistical features that have no physical existence.

Second, if periodicities do in fact exist in the data, are they in the form of sine waves (as is assumed in harmonic analysis) or in another form? Investigation of the time-variations of an element that we have no reason to assume follows regular cycles or periodicities—in mathematical language, a nonperiodic function—can be accomplished by means of Fourier integral analysis, whereas the study of the form of a proved periodicity can be carried out by one or more of the techniques developed by the branch of statistics known as periodography.

Fourier integral analysis is used in many nonmeteorological fields, especially in radio communication theory, where it is used to analyze transient signals (see Appendix 2.8). Its function there is to separate the radio signal, which may or may not have a periodic character, from the "noise," the variations in the received signal that are due to interference sources superimposed on the original signal. The technique is thus appropriate for analyzing climatological time-series, where it can separate the steady-state component, the "climate," (analogous to the radio signal), from the random fluctuations of "weather" (analogous to radio noise).

The basic idea of Fourier integral analysis is to break down a transient signal—i.e., an element which oscillates widely and irregularly with time—into a series of steady-state components. The signal or time-curve in Figure 2.2, A, can, in the interval from $\dfrac{-T_1}{2}$ to $\dfrac{+T_1}{2}$, be expressed by harmonic analysis as the sum of harmonics of lengths $T, \dfrac{T_1}{2}, \dfrac{T_1}{3}, \dfrac{T_1}{4}, \ldots$ Since the origin O can be placed anywhere along the time axis, the interval from $\dfrac{-T_1}{2}$ to $\dfrac{+T_1}{2}$ can cover different parts of the curve. Therefore, to be completely analyzed, the signal must be expressed as the sum of the components of the frequencies of the fundamental period, T, and its harmonics $\dfrac{T_1}{2}, \dfrac{T_1}{3}, \dfrac{T_1}{4}$, etc. For graphic purposes, it is more convenient to use the reciprocals of the harmonics and the fundamental period.

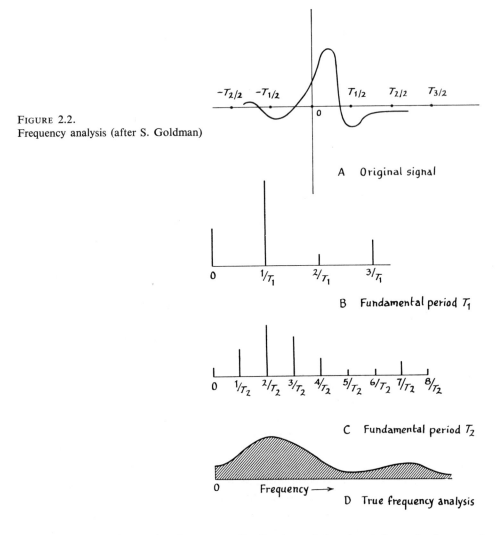

FIGURE 2.2.
Frequency analysis (after S. Goldman)

A   Original signal

B   Fundamental period T_1

C   Fundamental period T_2

D   True frequency analysis

Figure 2.2, B, shows the frequency distribution of the signal for a fundamental period of T_1 and Figure 2.2, C, gives the frequency distribution for a fundamental period of T_2. As T becomes longer and longer, the frequency spacing between the harmonics becomes smaller and smaller, approaching zero as T approaches infinity, when a continuous distribution of frequency components results. When T becomes infinitely long—"infinity" in the climatological sense is relative to the spacing of the original observations in time—the individual harmonics also become infinitely long, so that in effect they are steady-state components. Figure 2.2, D, can then be regarded as giving the true frequency analysis of the original signal, as a graph showing the frequency distribution of its steady-state components.[5]

Whether or not a climatological time-series shows any evidence of periodicities, integral analysis determines all the possible harmonics that the series can contain, and evaluates the frequencies with which the individual harmonics occur. The peak (or peaks) of the resulting frequency curve indicates the periodicity (or periodicities) that is most prominent in the data, even though it (or they) may not be obvious from casual inspection of the empirical graph of the time-series.

Fourier integrals have been comparatively little used in the study of climatological periodicites. Instead, methods more simple mathematically, but often very laborious to compute, the *periodography* techniques, have been widely used. These methods are useful in preliminary analysis, but do not tell the whole story. Only a very brief survey of them will be given here; for full details see the standard texts.[6]

Periodography uses two types of technique, depending on whether the period is known to exist or not. If it is known that a set of data ought to contain a periodicity— a daily, monthly, or annual cycle, for example—then a different method must be applied than where a periodicity may or may not be present.

If the period is known, two different procedures can be followed, depending on whether or not the analysis interval is an integral multiple of the period. For example, if the series is known to contain a regular periodicity of 24 hours, then one technique must be employed if the observed values are spaced at intervals of 3, 4, 6, 8, 12, 48, or 72 hours, and another technique if they are spaced at intervals of 5, 7, 9, 11, 15, 30, or 50 hours.

The Buys-Ballot schedule is used if the analysis interval is an integral multiple of the period; it is essentially a method of "folding" the data. The figures are drawn up into columns, and the value of the average row is found. The average row is then subjected to harmonic analysis, which accurately gives the amplitudes of the periodicities. For example, given ten years' observations of hourly air temperatures for the month of January, we know that these observations must reflect the 24-hour (diurnal) temperature wave. The figures are arranged in 310 rows, since there are 31 days in January, and the average of the 310 rows is found. This average row is then analyzed harmonically. The first harmonic of the resulting Fourier series gives the amplitude of the diurnal temperature variation, the second harmonic yields the amplitude of the 12-hour wave, the third harmonic that of the 8-hour wave, and so on.

If the Buys-Ballot schedule is applied to a time-series that contains a periodicity of which the analysis interval is not an integral multiple, a nonharmonic wave results, which consists of a whole spectrum of Fourier terms or harmonics, in which the true period may or may not be present. A convenient technique to use in this case is Brooks' difference periodogram, which is a special type of graph, plotted as follows. The series is divided into a number of equal sections of length L, and the mean values of the observations in each of these sections are found. Differences between these means are then formed, and then means of the successive differences. The latter means are plotted against a horizontal time-scale, and this graph forms the difference periodogram. If the graph shows an apparent periodicity, we then know that the original data must contain a periodicity of length l_1 or l_2. To determine which of these two lengths is the correct one, another value of L is taken, and the periodogram is recomputed. Once the true length of the periodicity has been found, its amplitude and phase are calculated by harmonic analysis (see Appendix 2.9).

A more rapid method is Carruthers' periodoscope. A value V is chosen such that the length of the periodicity to be examined is an interval falling between $2V$ and $6V$. Sums are formed of the V values, which are then combined (see Appendix 2.10). The combined sums form the ordinates of a graph that represents one curve of the periodoscope. Different values of V are taken, close enough together for each periodicity under investigation to appear in at least two curves of the periodoscope. The completed periodoscope consists of a set of curves, which enable a wide range of possible periodicities to be studied. Only those periodicities that show up in at least two curves of the periodoscope are accepted as genuine.

If the periodicity is unknown, the investigation becomes a search for hidden periodicities. The classic method, introduced by Schuster in 1898, involves constructing a curve that represents the amplitudes (or the squares of the amplitudes) of the periodic components of the time-series as a function of their frequency, period, or wavelength.[7] A number of maxima always appear in the Schuster periodogram curve; some of these indicate true periodicities, others are purely random statistical features. The "real" maxima are separated from the spurious maxima by a special criterion introduced by Schuster. This method has been used very widely for many years, but it is doubtful if the vast labor it entails is now worthwhile in view of newer methods to be described later.

Schuster periodogram analysis is, in fact, an elaborate form of harmonic analysis, since it assumes that the periodic components whose amplitude it describes are perfect sine waves. But there may be periodicities in the data with lengths unequal to those of any of the computed harmonics. Schuster's method will bring into prominence more of these nonharmonic periodicities than simple harmonic analysis can, but it may leave others out.

The technique known as chain analysis is more rapid than the Schuster method, and often gives better results, but is not valid if the periodicities to be isolated are not made up of regular repetitions of sine waves.[8] Since climatological periodicities are often not regular, it will usually pay to adopt one of the nonregular periodicity techniques right from the start. These techniques involve either correlogram or autocorrelation methods.

The standard correlogram technique involves calculating the correlation coefficients between values of the element at different times. The arithmetic mean of the series must first be computed, then the data expressed as differences between the observed values and the mean. The correlation coefficient between this series and series separated from it by one, two, etc., time units is then computed (see Appendix 2.11). The final correlogram is a graph of lag periods as abscissae against correlation coefficients as ordinates. The computation is halted when there are less than 35 pairs of values available for calculating the correlation coefficients. Maxima on the graph indicate periodicities, but not their phase or amplitude. Alter's correlogram, a more rapid method, employing mean differences instead of correlation coefficients as the basis for a graph, is normally used if the smallest number of observations, minus the lag period, exceeds 100 (see Appendix 2.12).

Correlogram techniques are useful in that they make no assumptions about the form or regularity of the hidden periodicity. They may therefore be applied to series in which transient oscillations occur, i.e., periodicities whose phase and amplitude constantly change, or which are not present throughout the whole of the time-series.

A more accurate technique is Fuhrich's autocorrelation method. This provides the various periodicities present in a series in the order of their importance, so that when a periodicity of negligible amplitude is reached, the computation may be halted. Fuhrich's method involves computing the autocorrelation coefficients of the deviations of the original observations from their arithmetic mean, and then computing the autocorrelation coefficients of the new series so provided. If the original series is

$$y_1, y_2, y_3, \ldots, y_{n-1}, y_n,$$

then the "first transformed series" is

$$y'_1, y'_2, y'_3, \ldots, y'_{n-1}, y'_n,$$

where y'_k is the autocorrelation coefficient between the first $n - k$ and the last $n - k$ values (see Appendix 2.13). Repeating the operation results in a second transformed series,

$$y''_1, y''_2, y''_3, \ldots, y''_{n-1}, y''_n.$$

If this latter series is plotted as a time-graph, it will usually be found to take the form of a series of cosine waves. If the waves are irregular, then a third or possibly a fourth transformed series must be computed until a smooth cosine wave-train results. This cosine curve represents the most important periodicity—i.e., that with the greatest amplitude—present in the data. Its amplitude and phase can be taken approximately from the graph, or may be determined precisely by harmonic analysis. The cosine curve is then substracted algebraically from the original curve, and the autocorrelation process repeated on the series provided by the residuals. The process is continued until a smooth cosine curve is obtained; this represents the second most important periodicity, and so on.

The autocorrelation technique can only be used when the original series comprises several hundred values, and breaks down if two periodicities occur of nearly equal amplitudes. It has the advantage that the length of the original series need only be 1.5 to 2 times that of the longest period present in the data to allow all the periodic components to be discovered. The Schuster method, by comparison, needs at least five complete repetitions of the longest period in the data. The Fuhrich technique can also be applied to time-series that may contain the effects of damped oscillations, and its results are not influenced by the analysis interval, so that it may be employed to detect periods that are not aliquots either of the latter or of the total length of the series. It may also be used to isolate the effects of secular trends and aperiodic influences, because these will give rise to residuals when all the cosine waves have been subtracted from the original data. Even with this method, however, periodicities cannot be accepted as real until a test of their statistical validity has been made (see Appendix 2.14).

The application of periodographic analysis to meteorological time-series has revealed many apparent weather cycles. At least 55 periodicities more than one year in length have been discovered, their length varying from 1.03 to 36 years. The most universal periodicities are 2.5, 3.5, 5 to 6, 11 to 12, 19 to 24, and 30 to 35 years, in records of pressure, temperature, precipitation, and extreme weather conditions at many points on the Earth's surface. Some of these are solar-induced, others are dynamic or kinematic; the geographical positions of the subtropical anticylones, for example, vary in accordance with periods of 2,000 to 5,000 and 70,000 to 250,000 days, and are somehow connected with variations in the geographical distribution of vorticity.

A complicating factor in the isolation of fundamental atmospheric periodicities is the existence of forced oscillations and beat frequencies superimposed on the fundamental vibrations. For example, the regular 27-day period of solar rotation and the irregular 11.3-year sunspot cycle combine to form forced vibrations of pressure that are reflected in forced temperature oscillations. The lower harmonics (5 and 7 years) of the primary ($2\frac{1}{3}$ year) Southern Oscillation show interference with the solar cycles, leading to beat frequencies of pressure.

In general, the complexities of time-series analysis have caused a new approach to be developed within the last decade, largely based on the work of electrical-engineering

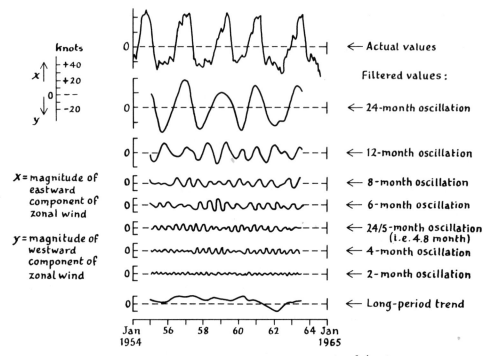

knots

$x \uparrow$
$y \downarrow$

+40
+20
0 --
-20

← Actual values

Filtered values:

← 24-month oscillation

← 12-month oscillation

X = magnitude of eastward component of zonal wind

y = magnitude of westward component of zonal wind

← 8-month oscillation

← 6-month oscillation

← 24/5-month oscillation (i.e. 4.8 month)

← 4-month oscillation

← 2-month oscillation

← Long-period trend

Jan 1954   56   58   60   62   64 Jan 1965

A. Zonal wind components at 50 mb above Canton Island.

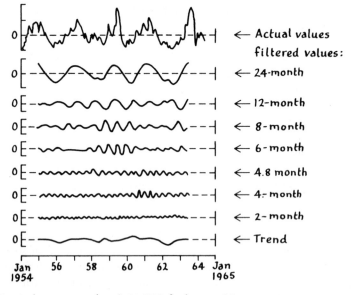

← Actual values

filtered values:

← 24-month

← 12-month

← 8-month

← 6-month

← 4.8 month

← 4-month

← 2-month

← Trend

Jan 1954   56   58   60   62   64 Jan 1965

B. Zonal wind components at 60,000 feet over Singapore.

FIGURE 2.3.
Use of band-pass filters to reveal amplitudes of oscillations with various periods (after G. E. Edmond).

theorists. This approach regards smoothing as a special type of filtering, and uses many standard terms of electrical filtering theory in a meteorological context. These terms are now becoming commonplace in climatological discussions.[9]

The purpose of time-smoothing is to attenuate the amplitudes of the high-frequency periodicities in the time-series, without significantly affecting the low-frequency components. Attenuation is roughly porportional to frequency, becoming complete for a given purpose above some frequency. The assumption is made that the high-frequency oscillations are either random error ("noise") or of no significance for the given purpose.[10]

Running means are only one of many numerical filters, many of which are extremely useful in climatology. Low-pass filters retain the low-frequency variations, but eliminate or at least attenuate the high-frequency oscillations. Smoothing is a form of low-pass filtering. High-pass filters retain the high frequencies but eliminate the low frequencies, and band-pass filters retain the frequencies within a specific intermediate band, but eliminate very low or very high frequencies. The frequency response of a filter is the ratio of the amplitude of a wave of given frequency after filtering to its amplitude before filtering.

In time-series analysis, numerical filters are usually termed *smoothing functions* (see Appendix 2.15). They normally consist of a series of fractional weights that determine in what proportion each observation in a series contributes to the final smoothed value. If the weights are all of equal value, an ordinary running mean results; such a function may reverse the polarity of a wave, as already noted. If the weights are made proportional to the ordinates of the normal probability curve, a normal curve-smoothing function results, which does not have this disadvantage. A third type of function, exponential smoothing, does introduce a phase error; this form of smoothing is performed automatically by many viscous-damped instruments, for example, a simple mercury-in-glass thermometer. Figure 2.3 illustrates the effects of different band-pass smoothing functions.

The lowest frequency at which the response of a filter reaches zero (for all practical purposes) and remains at zero for all higher frequencies is termed its *cutoff frequency*. *Preemphasis* is the amplification of certain frequency bands before the series is analyzed. *Equalization* is the process of restoring the original balance of amplitudes of the component waves in a series that has been altered by filtering or preemphasis; inverse smoothing is the numerical equalization used to accentuate the higher frequencies in order to restore the original balance of frequencies. *Quantizing error* is a random error introduced if smoothed values are rounded off; the high-frequency noise is amplified as well as the true periodicities, and may completely obliterate the latter.

If the curve representing the original time-series is found to possess periodicities of different frequencies or wavelengths in equal amounts, it is said to represent a *white noise*, analogous with white light. *Prewhitening* is a statistical process that removes from the observed curve any known variations (e.g., the annual temperature cycle) that might otherwise swamp the periodicities it is hoped to discover; in effect, it is a high-frequency preemphasis that tends to equalize the amplitudes at all frequencies. *Aliasing* is an important property that results in high-frequency variations appearing as spurious lower-frequency cycles if the original observations were spaced too far apart in time for these shorter periodicities to be portrayed.

Power-spectrum analysis is the field in which these concepts have been developed to the greatest extent in climatology (see Appendix 2.16). This technique, based on

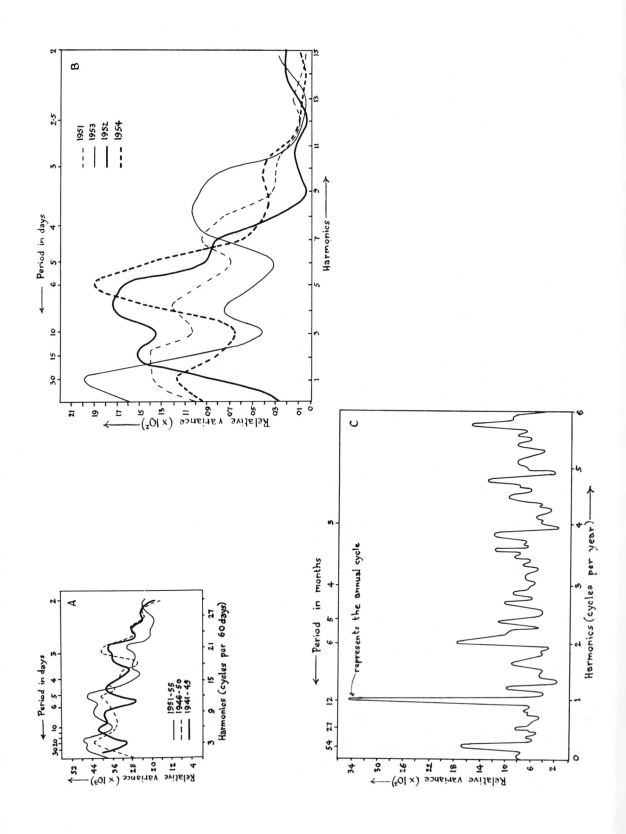

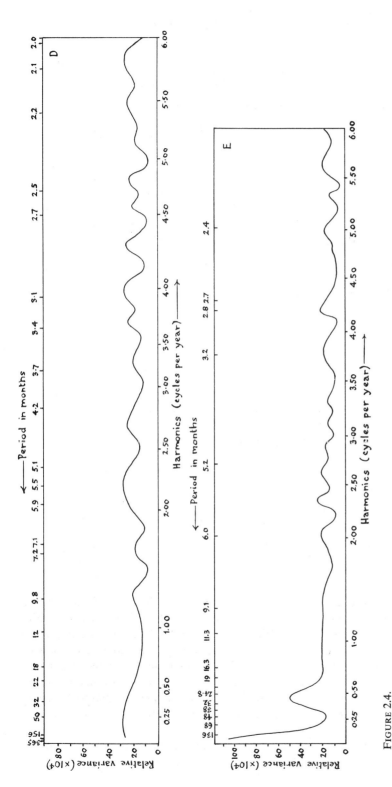

FIGURE 2.4.

Power-spectrum analyses for Woodstock College, Maryland. A, Daily precipitation data; maximum lag 30. B, Daily average temperature, 1951-1954, for high winter (January 1 to February 13); maximum lag 30. C, Monthly total precipitation, 1895-1956, 743 months; maximum lag 132, seasonal march retained. D, Monthly total precipitation, 1870-1956, 1,044 months; maximum lag 546, seasonal march removed. E, Monthly average temperature, 1870-1956, 1,032 months; maximum lag 546, seasonal march removed. Spectra in A are unsmoothed. Spectra in B were smoothed by the Hamming formulae. Spectra in C are unsmoothed. Spectra in D and E were smoothed by averaging the line powers in consecutive groups of ten. (After H. E. Landsberg, J. M. Mitchell, and H. L. Crutcher.)

TABLE 2.1
First 25 harmonics of the 546-month maximum lag spectra of temperature and precipitation at Woodstock College, Maryland.

No. of Harmonic	Inclusive period (years)	Power[a] Tempera-ture	Power[a] Precipita-tion	No. of Harmonic	Inclusive period (years)	Power Tempera-ture	Power Precipita-tion
0	182–∞[b]	182	20	13	6.7–7.3	30	25
1	60.8–182	214	80	14	6.3–6.7	37	21
2	36.4–60.8	99	20	15	5.9–6.3	51	25
3	26.0–36.4[c]	−9	28	16	5.5–5.9[e]	55	17
4	20.2–26.1[d]	14	23	17	5.2–5.5	24	23
5	16.6–20.2	42	22	18	4.0–5.2	4	53
6	14.0–16.6	75	58	19	4.66–4.9	0	42
7	12.1–14.0	75	60	20	4.44–4.66	1	24
8	10.7–12.1	99	31	21	4.24–4.44	4	76
9	9.6–10.7	84	18	22	4.05–4.24	21	79
10	8.7–9.6	16	20	23	3.87–4.05	39	29
11	7.9–8.7	24	16	24	3.71–3.87	16	9
12	7.3–7.9	38	22	25	3.57–3.71	4	12

[a] Units are variance times 10^4.
[b] Secular trend.
[c] Brückner cycle.
[d] Double sunspot cycle.
[e] Second harmonic of single sunspot cycle.

TABLE 2.2
Statistical significance of peaks in a 546-month spectrum, with null hypothesis that the population spectrum is a white noise.
Mean power of white spectrum = $(1/546) \times 10^4 = 18.3$.
(Source of data: H. E. Landsberg, J. M. Mitchell, Jr., and H. L. Crutcher, 1959.)

Number of harmonics contributing to the peak	Degrees of freedom	Significance level (per cent) 95	Significance level (per cent) 99	Significance level (per cent) 99.9
1	3.3	46.7	66.9	95.0
2	6.6	37.0	49.6	65.2
3	9.9	33.6	42.7	54.4
4	13.2	31.6	38.7	49.0

theory published by Wiener in 1930 and G. I. Taylor in 1938, ascribes the observed variation shown by a time-series to an infinite number of small oscillations with a continuous distribution of periods. This contrasts with Fourier analysis, which ascribes the observed variation to a finite number of oscillations with discrete periods. For example, an apparent weekly oscillation in precipitation discovered at a station would be interpreted by Fourier analysis as due to the existence of a definite seven-day cycle. The power-spectrum interpretation would be that periodicities with a length of about one week are more probable than those with lengths of four, five, nine, or ten days.

The power spectrum is a graph depicting the distribution of the probability of oscillations with differing frequencies in a given series. A good example is the analysis of temperature and precipitation data (1870–1956) for Woodstock College, Maryland. The original data comprised daily, weekly, and monthly values. The resulting power spectra, which take the form of graphs with the harmonics of the various periodicities as abscissae, and the variances of the values represented by each harmonic as ordinates, are given in Figure 2.4 (see Appendix 2.17).

The spectra for short-period variations indicate well-marked five- to seven-day periodicities in winter temperatures; these also show up in precipitation, as do oscillations in the range 15 to 25 days and around three days. The last is associated with a fast-moving pressure wave that travels completely around the Earth; the five- to seven-day periodicity is related to the passage of the long waves in the westerlies. Both temperature and precipitation periodicities are not fixed in time, but vary markedly from year to year.

The temperature data had to be prewhitened to remove the annual cycle, which would otherwise have swamped other long-period variations. The long-period spectra show no evidence of either the 22-year (double sunspot) or the 35-year (Brückner) cycles in temperature or precipitation, but the temperature spectra show two highly significant maxima with periods of 1.8 to 2.7 years and above 50 years. The 11-year sunspot cycle is prominent, as also is the second harmonic (5.6 years) of the solar cycle. An unexpected result is that the sunspot cycle contributes 3 per cent to the total variance of annual mean temperature. The precipitation spectra show no significant periodicities at all; in fact, they show only very minor departures from randomness. Tables 2.1 and 2.2 compare the periodicities brought to light by the power spectra with those discovered by earlier methods. The latter isolated very few of the real periodicities, and "discovered" some that are not really present in the data.

Figure 2.4, A, presents a power spectrum of daily total precipitation at Woodstock College (1943–1954) which illustrates a time-series that approximates more closely a red noise than a white noise. A red-noise spectrum has more power (i.e., variance) at its lower frequencies. Many climatological spectra are of this type, with their degree of "redness" directly related to the degree of persistence present in the data. This complicates the isolation of the longer periodicities. A further complication may arise because of the finite length of many time-series. This results in truncated spectra in which end-effects may cause spurious results. Jump discontinuities at the ends of the original series may be smoothed by the process termed *hanning*, but this increases the bandwidth of the filter.[11]

Another example of a power spectrum is given in Figure 2.5, F, of daily mean temperatures at University Park, Pennsylvania, processed by a slightly different procedure from the Woodstock data. The spectrum shows that a very large part of the daily variation of temperature is caused by short-period fluctuations; the maximum at four days, for example, represents the average time between frontal passages. An analysis of the spectra of horizontal windspeeds at Brookhaven, from 76 to 125 m above the ground, discovered a similar peak at four days, of eddy energy because of synoptic disturbances. A second peak was found with a periodicity of about one minute, representing mechanical and convective microturbulence. Between these peaks is a broad spectral gap, centered on oscillations of one to ten cycles per hour. This gap has been proved to exist elsewhere, under different terrain and synoptic conditions; its exact position in the spectrum depends on the severity of the terrain.[12]

A further advantage of power spectra over other methods of time-series analysis is that cross-spectra may be computed. *Cross-spectrum analysis* is useful for comparing

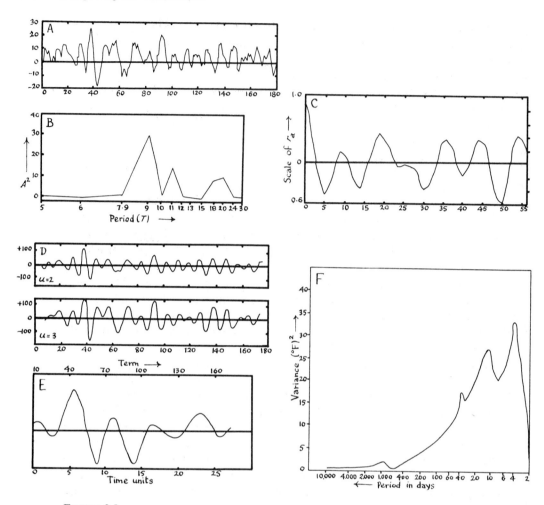

FIGURE 2.5.
A test of climatological time-series techniques (after C. E. P. Brooks and N. Carruthers).
A, The test series (arbitrary units). B, Schuster periodogram of test series. C, Correlogram of test series. D, Carruther's periodoscopes for test series, showing a wave of length 9 or 10 units.
E, Brooks' difference periodogram for test series, showing an oscillation of 5.5 units, equivalent to a periodicity of 8.8 or 18.9. F, Power spectrum of daily mean temperature, University Park, Pa.
(after H. L. Griffiths, H. A. Panofsky, and I. Van der Hoven).

two time-series that are believed to be related but show different components with different lags. It involves computing two spectra, the *cospectrum* and the *quadrature spectrum*, both of which are determined by Fourier analysis of the lag correlation coefficients between the two variables. If the quadrature spectrum vanishes at a certain frequency but the cospectrum does not, the latter being positive, then the two oscillations are in phase at that frequency. If the cospectrum vanishes but the quadrature does not, the oscillations are 90° out of phase. If both cospectrum and quadrature differ from zero, their ratio determines the phase lag between the two series at that frequency; if they are both positive and equal to each other, the phase lag is 45°.[13] Cross-spectra are also useful to test the validity of physical relationships; if the apparent relationship between two variables can only be proved at low frequencies, then it is doubtful if a true relationship exists.

In all method of time-series analysis, the effect of a known secular trend in the data must be allowed for, otherwise an overestimate will be made of the variability resulting from random influences. In general,

$$V_n = T_n + P_n + R_n,$$

where V_n is the deviation of the nth term in the series from the over-all mean, T_n is the contribution to this term by the long-term trend, P_n is the contribution from the periodicities, and R_n is the random contribution (i.e., variations that have no known cause). If a trend is known to exist, then the computed parameters of the data must be corrected. Trends are not easy to isolate, but they have been discovered, for example, by fitting orthogonal polynomials to the data. Maps showing values of the coefficients of the first-degree polynomials then enable regional comparisons of the linear trend to be made.[14]

The standard techniques of time-series analysis all presuppose the existence of periodicities in the data. One can also determine if the variations shown by the series are purely random. If they are not random, then cyclic or periodic effects must be present. Two methods are appropriate here. Kendall's method involves counting the number of turning points (maxima and minima) in the series, and then testing the significance of their deviations from the numbers expected if the series is a random one. Wallis and Moore's method examines the frequency distribution of phase lengths in the series, and tests the significance of the deviations of these frequencies from those calculated for a random series.[15]

Statistical Analysis of Scalar Quantities

We will suppose that we have a series of figures representing measurements of some scalar quantity, e.g., temperature or pressure, at a certain station for some length of time. The observations are instantaneous measurements, not averages over an hour or a day, and are equidistant, i.e., were made at the same time each day. If the series is long, it will be impossible to study it usefully without reducing it to a set of parameters that adequately represent its main characteristics. The first step toward this aim is to construct a *frequency distribution* for the series. The quantities derived from the frequency distribution, i.e., its *statistics*, are then termed its *parameters* if they depend on the form of the distribution. The study of statistics that do not depend on the form of the underlying distribution is *nonparametric statistics*. We shall be concerned with parametric statistics for the remainder of this chapter.[16]

FREQUENCY ANALYSIS

Frequency analysis involves dividing the data into classes: a rough guide is that the number of classes should not exceed five times the logarithm of the number of observations. Thus if there are not more than 50 observations, there should be not more than eight classes, not more than ten classes if there are 100 observations, and so on. The observations are then sorted into their respective classes, and a *frequency table*

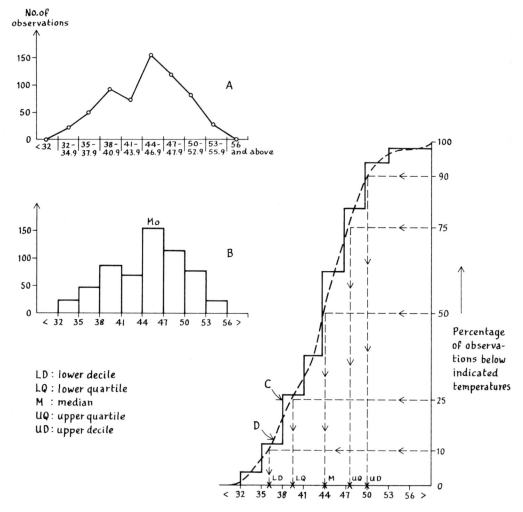

FIGURE 2.6.
Representation of frequency statistics. A, Frequency polygon for a set of temperature data (°F) for a hypothetical station. B, Frequency histogram for the same data (Mo, mode). C, Cumulative frequency histogram for the data. D, Ogive for the data.

results that numerically describes the frequency distribution of the data. The main features of the frequency distribution may be brought out by graphing the figures, either by means of a *frequency polygon* or by means of a *frequency histogram*. (See Figure 2.6). Owing to the different ways in which the class limits may be assigned, several polygons and histograms may be constructed from the same data, and the class limits must be carefully decided with a view to the ultimate purpose for which the investigation is intended. In general, the histogram presentation gives the most useful diagram, particularly for data concerning continuous quantities (see Appendix 2.18). An important feature of this form of presentation is that the width of any bar in any histogram is equal to the difference between two successive possible values of the variable. The height of any bar represents *frequency or probability per unit width*.[17]

The data portrayed in our histogram represent a limited sample drawn from a population or universe that includes all possible values of the element in question. Before deriving parameters from the frequency distribution represented by the histogram, it is obviously important to know if we can modify the distribution so that it forms the best possible estimate of the total population. We do not want our parameters to become redundant when more observations become available. Fortunately, there is a logical basis for deriving a "population" histogram from the "sample" ones. If we had had, say, ten times the number of observations for the construction of a histogram that we originally had, we could have reduced the width of the classes. If the vertical scale was at the same time reduced to one-tenth of its original length, then the resulting histogram would be much more even than the initial one. In the limiting case, in which the class intervals and the vertical scale are made smaller and smaller, the histogram will become a smooth curve. According to the *theory of sampling*, therefore, there is a smooth curve corresponding to any set of observations that encloses the same area as does the histogram of the data, but which represents the population from which the sample of data was drawn more closely than do the original observations. This smooth curve is the *frequency curve* of the data, and to obtain it from the histogram, a curve must be fitted to the latter in such a manner that the total amount added to the frequencies of certain classes equals the total amount subtracted from the frequencies of other classes. The most expedient way to fit this curve is as follows.

1. Superimpose on the histogram by eye a smooth curve that describes its general shape.

2. Adjust this subjective curve so that the probabilities add up to 1.00, by reading the curve at each possible value of the variable, adding the readings, and then increasing or decreasing every point on the curve by the same proportional amount. This second stage is a very important one: it should be realized that the smoothing of a frequency distribution is not just the process of drawing a smooth curve by eye through the histogram, but rather a procedure based on probability theory for which a logical basis exists.[18]

An alternative method of presentation is the *cumulative frequency diagram* and the cumulative frequency curve or *ogive*. The former is obtained simply by plotting cumulative totals for each successive class as one goes from left to right through the frequency table instead of actual class frequencies. The ogive is then obtained by smoothing the cumulative frequency diagram by the two stages described for the determination of the frequency curve (see Figure 2.6, C and D).

Once the ogive of the frequency distribution has been determined, certain parameters of the distribution may be determined directly. The vertical axis of the graph is graduated on a percentage basis, and the value of the variable corresponding to the 50 per cent level then represents the *median* of the distribution (see Figure 2.6, D). The values of the variable corresponding to the 75 and 25 per cent levels represent the *upper quartile* and *lower quartile* of the distribution respectively, and those corresponding to the 90 and 10 per cent levels represent the upper and lower *deciles* respectively. The median, the quartiles, and the deciles are *fractiles* of the frequency distribution. In general, any value of the variable which is both (a) equal to or greater than a percentage f of the values in the set of observations and (b) equal to or less than a percentage of $100 - f$ of the values is the f fractile of the data. Thus the median is the 50 per cent fractile, the lower decile is the 10 per cent fractile, and so on.

Specifying the fractiles of a set of observations makes certain facts about the latter immediately evident. The median is the middle value of the distribution derived from the data: one-half the observations must have values greater than that of the median, and one-half must have values less than that of the median. One-quarter of the observations must have values greater than that of the upper quartile, one-quarter must have values less than that of the lower quartile, etc. One-half of the observations must have values which lie within the *interquartile range*, i.e., the interval between the upper and lower quartiles, and 80 per cent of the data must fall within the *interdecile range*.

Another statistic that may be derived from the original frequency histogram is the *mode*. This is the most frequent or most "fashionable" value; for a histogram, it lies within the range of values covered by the class which includes the largest number of observations. The *average* is a statistic that is not usually found from frequency distributions, but is derived by arithmetical operation on the original data. There are three types of average, which each describe the most likely "central point" about which the observed figures appear to cluster. The *arithmetic mean* is defined as the arithmetic sum of the set of values divided by number of values in the set, and applies to values that appear to increase in an arithmetic progression. The *geometric mean* is defined as the *n*th root of the product of a set of *n* values, and applies to values that appear to increase in a geometric progression. The *harmonic mean* is defined as the reciprocal of the arithmetic mean of the sum of the reciprocals of the original observations, and is appropriate for derived elements that are reciprocals of the raw elements from which they were obtained—for example, geostrophic winds or air densities—or for rates of change or observations that are separated by unequal time intervals. The arithmetic mean is appropriate for observations, such as temperature and pressure, that are equidistant in time; equidistant values that differ considerably from one observation to the next—daily rainfall amounts, for example—are best averaged by the geometric mean. The mean, median, and mode, with frequency distributions that peak in the central part of their range, are related approximately by the expression

$$\text{mode} = \text{mean} - 3\,(\text{mean} - \text{median}).$$

The mode is not as common as the median and the mean in climatology, but it is sometimes useful, for example, in specifying prevailing wind directions in terms of the "modal wind" instead of the usual concept of resultant winds. (See Appendix 2.19.)

Fractiles and means are quite different approaches to the problem of describing the main features of a set of observations. Fractiles constitute *measures of location* in a frequency distribution: they specify the location of a histogram on the horizontal axis of a graph but do not tell us anything about the shape of the histogram. Means represent *expectations*; i.e., they point to the most probable value toward which we can expect the data to tend (see Appendix 2.20). There is a difference, too, in the starting point of the computations or graphic manipulations involved. Means are calculated directly from the original data, whereas fractiles are obtained from the data grouped into classes; means may also be obtained from grouped data.

The shape of a frequency distribution is controlled by certain constants that are termed its parameters, and that in turn are based on the *moments* of the distribution. The *first moment* of the distribution is the algebraic mean of the deviations of the individual observations about some arbitrary value. If this arbitrary value coincides with the mean, the first moment becomes zero. The *second moment* of the distribution

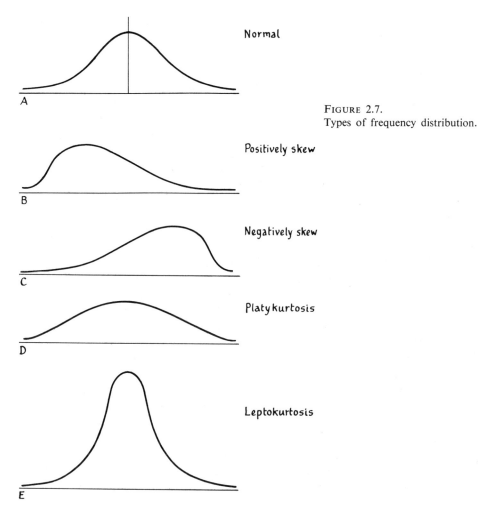

FIGURE 2.7.
Types of frequency distribution.

is the mean of the squares of the deviations about the arbitrary value. If the first moment is zero, the second moment is then defined as the *variance* of the distribution. The *third moment* of the distribution is the algebraic sum of the cubes of the deviations divided by the number of observations; this moment describes the *skewness* of the distribution, i.e., its departure from symmetry. The mean, median, and mode coincide for a symmetric frequency distribution (see Figure 2.7, A), but not for an asymmetric distribution. If the mode falls to the left of the median, the distribution is said to be *positively skew* (see Figure 2.7, B), if to the right of the median, to be *negatively skew* (see Figure 2.7, C). The usual measure of skewness incorporates the third moment of the distribution. The *fourth moment* of a frequency distribution is the mean of the fourth powers of the deviations of the observations about an arbitrary value; it measures the *kurtosis* of the distribution, i.e., its relative sharpness or flatness. On the basis of the fourth moment, distributions may be described as *platykurtic* or *leptokurtic* (see Figure 2.7, D and E). The various climatic elements usually have characteristic frequency distributions when examined for skewness and kurtosis.[19] (See Appendix 2.21.)

The values of higher moments than the fourth are not usually computed in climatology, because they become more and more dependent on extreme values. For observations of continuous quantities that have a symmetric frequency distribution, the second and third moments as computed from grouped observations will be too high, and must be reduced by means of Sheppard's correction.[20]

The magnitudes of the interquartile and interdecile ranges of a frequency distribution clearly are useful measures of the degree of dispersion of the original data, but they do not take into account individual observations near the extremes. The arithmetic mean gives no idea at all of the scatter of the observations, but merely locates the position of the center of gravity of the distribution. A new measure of dispersion is obviously needed, and is provided by the *standard deviation*, which is defined as the positive square root of the second moment of the distribution when its first moment is zero. The standard deviation (usually denoted by the symbol σ) should always be quoted in addition to the arithmetic mean if the simplest possible statistics are required for comparison: two stations may have identical arithmetic means, but the variability of their climates (as measured by their respective standard deviations) may differ considerably. There are other measures of dispersion or variability in the original set of data, for example, the *mean deviation* or *average variability*, the *coefficient of variation* or *coefficient of variability*, the *relative variability*, and the *intersequential variability*, but these are not employed as often as the standard deviation in climatology. The standard deviation of a normally distributed set of observations is a particularly useful statistic. (See Appendix 2.22.)

Standard deviations may be computed for mean values as well as for actual values,[21] but for both an important qualification must be borne in mind. The definition of standard deviation requires that the mean of the observations (i.e., the sample mean) be the same as the mean of the population, which is regarded as of infinite size. It is quite in order to compare two sets of data by means of their standard deviations if we are only interested in those particular sets of data. If, for example, daily temperatures during a ten-year period at stations A and B have standard deviations of 2.6°C and 1.4°C respectively, we may correctly infer that, during this period, daily temperatures at station A were nearly twice as variable as those at station B. But if we are using the ten years' data as a sample to estimate the long-term temperature distribution, we cannot compare the standard deviations for A and B directly, because we do not know the true mean of the whole population, and cannot be certain that the sample mean coincides with it. To remedy this deficiency, the *standard error* of the standard deviation should be calculated, and this correction applied to the standard deviation on the basis of the most likely difference between the sample mean and the population mean (see Appendix 2.23).

At this point it is appropriate to consider the purpose of a frequency analysis, or, for that matter, of any type of statistical analysis. The usual purpose of the analysis is to show that the observations confirm or refute some theory or hypothesis that has been formulated on nonstatistical grounds. In other words, we do not use the statistical evidence to formulate the hypothesis, but to test it. Any set of observations may contain "accidental errors," i.e., variations of the variable under consideration that are not due to the influences covered by the hypothesis under test. These errors may make it difficult to decide whether or not the observations do confirm or refute the hypothesis. The theory of mathematical statistics is useful here, for it enables limits to be set to the probable effect of accidental errors in introducing discrepancies between observation and hypothesis. These limits may be examined by means of *significance tests*.

The first step in significance testing is to set up a *null hypothesis*, i.e., the hypothesis that the accidental errors arise purely from sampling error, which is simply the error due to the fact that a sample cannot be completely representative of the whole population. The second step is to determine the probability that the null hypothesis is correct. This involves comparing the observed values of the variable with the values we should expect to occur if the theory or hypothesis under test (i.e., not the null hypothesis) is completely true. On the basis of the difference between observed and expected values, a *chi-square* (χ^2) *test* is carried out. The computed value of χ^2 is then compared with values given in chi-square tables. The tables are entered at the appropriate probability level and the appropriate number of degrees of freedom; usually the 5 per cent probability level ($p = 0.05$) is taken, but if special care is needed, the 1 per cent probability level ($p = 0.01$) is advisable. If the computed value of χ^2 is less than that given in the tables for the 5 per cent level, the data are consistent with the null hypothesis, but if it is greater than this value, the chances are more than 19 to 1 that the null hypothesis is incorrect, in which case, the observed distribution must differ significantly (in the statistical sense) from the distribution expected according to the theory or hypothesis under test. If the computed χ^2 value exceeds the 1 per cent value, the null hypothesis is almost certainly incorrect. If the computed χ^2 value lies between those given for the 1 and 5 per cent levels, the significance test is statistically inconclusive. (See Appendix 2.24.)

The contribution made by different influences to the variability shown by a set of data may be assessed by means of the *analysis of variance* procedure. Unlike standard deviations, variances can be added together. Hence the total variance of a series is given by the sum of the independent variances of the separate subvariates that combine to form the observed variate:

$$\sigma_T^2 = \sigma_1^2 + \sigma_2^2 + \sigma_3^2 + \cdots + \sigma_R^2,$$

where σ_T^2 represents the total (observed) variance, σ_1^2, σ_2^2, etc., represent the variances due to the different influences, and σ_R^2 represents the residual or "accidental" variance, i.e., that due to unknown influences. The fundamental idea behind the analysis of variance is that the total variability in the observations may be broken down into (1) the variation within each frequency group about the mean of all the observations within that group, and (2) the variation of each group or class mean about the grand (i.e., population) mean. These within-group and between-group variations are computed, and then tested for statistical significance by means of the *F*-test.[22] By this means, the percentage of the observed variation that may be attributed to the effect of different factors may be assessed, and the magnitude of the importance of the unexplained variance (i.e., any unknown factors) may be assessed.

PROBABILITY ANALYSIS

The concept of probability is at the heart of all climatological problems. In view of the complexity of the atmosphere, not to mention terrain effects, one can never be certain of any explanation or prediction in climatology. Data in climatology are almost always used to make some sort of decision or statement, and in any problems

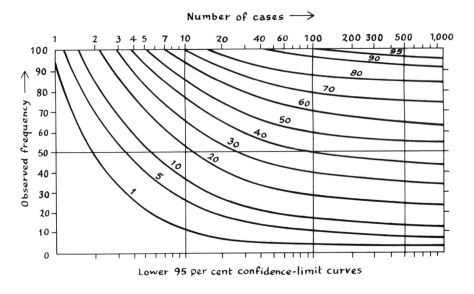

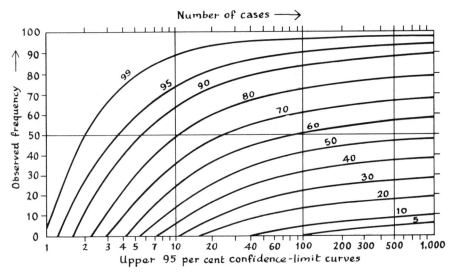

FIGURE 2.8.
Confidence limit curves of true relative frequency of an event that has an observed frequency as indicated (after I. I. Gringorten).

of decisionmaking under uncertainty, the mathematics of probability provides the logical guide.[23] It is impossible to go very far in dealing with climatic statistics without coming up against the concept of probability; in our discussion of standard deviations it was necessary to use the term.

Probability is a mathematical concept, which may be approached as follows. We may define the *relative frequency* of an event as the number of times the event actually happens during a specific period of time divided by the number of times it could happen during the same period. Thus if 10 days during a 40-day period have mean temperatures of 50°F or more, then the relative frequency of days $\geq 50°F$ is $\frac{10}{40}$ or 0.25. The *probability* of the event is defined as its relative frequency in the long run, i.e., in an infinite number of trials. The true probability of any event is usually unknown, and must be inferred from its relative frequency. According to Bernoulli (1713), as the number of trials increases, the probability approaches unity (i.e., certainty) that the relative frequency will differ by less than any desired amount from the true probability. In general, the relative frequency approaches the true probability as a limiting value as the number of observations or trials increases. It is therefore possible to specify *confidence limits* between which we can confidently expect the true probability to lie (see Figure 2.8).[24]

In practice, if a frequency curve has its frequencies reduced to percentages of the total sample size, it becomes a *probability curve*. The equation describing this curve (i.e., $f(x)$ in Figure 2.9, A) represents the *probability density function*, and the area under the curve represents cumulative probabilities when going from left to right. The integral of the probability density function,

$$F(x) = \int_{-\infty}^{x} f(t)dt,$$

is the *cumulative probability function*. The probability density function is used to prepare a graph of a theoretically derived curve; and the cumulative probability function is used to compare numerically the agreement between theoretical and observed curves. If it can be shown that the frequency curve describing a given set of observations closely approximates the curve representing certain types of probability density function, in particular the *normal curve*, then much useful data may be derived that would not be apparent from the original observations.

The frequency distribution for discontinuous elements or events—i.e., for elements, such as thunderstorms or tornadoes, which either do or do not occur, on a given day— is described in probability terms by the binomial theorem.[25] The expected frequency of a occurrences of such an event in N days out of a possible n days is given by:

$$N \frac{n!}{a!\,(n-a)!}\, p^a\, q^{n-a},$$

in which p is the probability the event will happen, and $q = 1 - p$ is the probability it won't. The limiting form of this expression for $p = q = 0.5$ plots as a smooth symmetric curve declining on either side of a central maximum (see Figure 2.7, A), if n is made indefinitely large. In other words, we may plot a frequency histogram of the above expression, in which the class intervals are represented by values of a ranging from 0 to n. This histogram obviously describes discontinuous data, but it can be made to represent a continuous quantity by assuming the latter to be made up of an infinite number of discrete (i.e., discontinuous) values spaced at equal intervals. Thus as we increase the size of n, so the "stepped" character of the histogram diminishes until, when n is infinitely large, the histogram becomes a smooth frequency curve. If this procedure is carried out for a binomial expression in which $p = q = 0.5$, then the resulting curve is smooth and symmetric; it may be shown that the equation for this curve is

$$y = \frac{Ne^{-x^2/2\sigma^2}}{\sigma\sqrt{2\pi}},$$

where y is the ordinate corresponding to abscissa x, σ represents the standard deviation of the distribution portrayed by the curve, and the latter is symmetric about the origin (see Figure 2.9, B). This bell-shaped curve is called the *normal frequency curve*, and was developed by Gauss, Laplace, and Maxwell. Areas under the curve represent probabilities: thus the area to the left of line *PQ* in Figure 2.9, B, is proportional to the probability that a value of x_a or less will occur. Tables of the *probability integral* conveniently give the portion of the area under the normal curve to the left of the ordinate distant x from the mean; these tables are therefore useful in the determination of cumulative probabilities (see Appendix 2.25).[26]

The normal frequency distribution described by the normal curve is completely determined by the mean and the standard deviation of the original data, and all of its odd moments are zero. Provided a distribution is normal, many types of standard error may be computed for it: for example, the standard error of the original observations, of the sample mean, of a variance, and of many other parameters. The standard error is a measure of the precision of the appropriate parameter; i.e., the standard error of any parameter computed from N values is an estimate of the standard deviation of a distribution composed of the values of that parameter which have been calculated on the basis of an infinite number of samples, comprising all possible values of N. An alternative measure to the standard error is the *probable error*. (See Appendix 2.26.)

The appropriate test of statistical significance for a theory or hypothesis intended to explain a set of observations whose frequency distribution is normal is not the χ^2 test, but the "*student's t*" test. Provided the data are normally distributed, the distribution of any parameter of the data will conform to the distribution represented by the "student's t", and tables giving values of the latter may be used to test the significance of the deviation of any observed value of this parameter from that postulated by the theory under test for the entire population (see Appendix 2.27).

To test whether a given frequency distribution is normal or not, the simplest procedure is to plot the data on *normal probability paper*.[27] This is a type of graph paper on which data for a normal distribution plot as a straight line. If the observations fall in a straight line on the graph, they may be regarded as being drawn from a population whose frequency distribution, and whose probability statistics, are the same as that of a population described by the normal probability curve. The great advantage is that if the distribution represented by some observations is a normal one, then the straight line may be used to interpolate or extrapolate data with confidence. An alternative to the graphic method for testing the normality of an observed frequency distribution is to apply *Cornu's criterion*.[28]

If a given frequency distribution plots as a zigzag curve instead of a straight line on probability paper, it can sometimes be separated into two or more components that are normally distributed. This may be done approximately, or more precisely by one of two more involved procedures.[29] Bimodal distributions, consisting of a combination of two normally distributed subvariables, are very common in climatology; for example, the frequency distribution of daily tropopause heights in middle latitudes has two modes, because the tropopause there may be either high and relatively cold or low and relatively warm, depending on whether tropical or polar influences are dominating. Figure 2.10 shows some typical frequency distributions for climatic elements.

If a nonnormal frequency distribution cannot be separated into normal components,

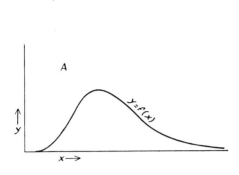

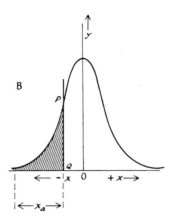

FIGURE 2.9.
Probability curves.

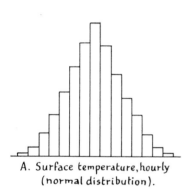

A. Surface temperature, hourly
(normal distribution).

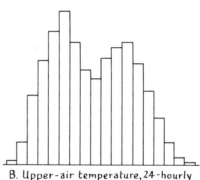

B. Upper-air temperature, 24-hourly
(Bimodal distribution).

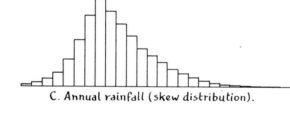

C. Annual rainfall (skew distribution).

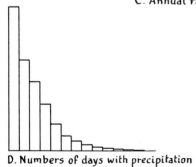

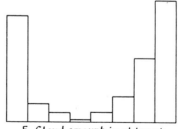

D. Numbers of days with precipitation
between specified amounts
(J-shaped distribution).

E. Cloud amount in oktas at
particular synoptic hour
(U-shaped distribution).

FIGURE 2.10.
Typical frequency distributions in climatology.

the usual procedure is to transform it in some way into a normal distribution. If the original distribution is positively skewed and leptokurtic, it may be made normal by a logarithmic transformation of the data; if it is negatively skewed, it can be normalized by a reversed log-normal transformation. Other transformations are also possible. (See Appendix 2.28.)

The actual process of fitting a normal curve to a set of observations can be completed from the original data, without the necessity for constructing a frequency histogram and then smoothing it. Often it is unnecessary to determine the normal curve. All that is necessary, after determining that the observed distribution is normal by graphing it on probability paper, is to use the probability integral tables after the standard deviation of the distribution has been computed (see Appendix 2.29).

The usual method of assessing probabilities is to base them on experience as revealed in historical data; in particular, we may obtain them by smoothing the historical frequencies. The question then arises: how large should the historical sample be? Experience is the most reliable guide here. Observations should be accumulated until a relatively stable frequency distribution (unimodal or multimodal) is obtained. If too many observations are used, the sample will be mixed, drawn from more than one "population," because of long-period climatic fluctuations and changes. The usual problem is that of scarcity of data. However, even if only a small number of observations is available, it is possible to construct a probability curve in an entirely logical way. The procedure is based on a cumulative frequency curve, which is fitted by eye to the original data (see Appendix 2.30).

Once the probability curve representing the population from which the observational sample is drawn has been constructed, probabilities may be read from it for any values of the variable. It may be shown mathematically that probabilities may be combined, and by means of probability algebra one can obtain much useful information from the figures. For example:

1. If the probability of an event is p (a number varying from 0.00 if the event is certain not to happen to 1.00 if it is certain to happen), then the probability that it will not happen is $q = 1 - p$. The expected number of times the event should happen in N trials is Np.

2. For mutually exclusive events (for example, the events rain or no rain, or dense fog, fog, mist, or none of them), the individual probabilities add up to unity; i.e., $p_1 + p_2 + p_3 + p_4 = 1$, where p_1, p_2, p_3, p_4, represent the probabilities of dense fog, fog, mist, and good visibility, respectively.

3. For events that are not mutually exclusive, i.e., that can all take place at the same time, the individual probabilities must be multiplied. If the probability of below-freezing temperatures at station X is p_x and that of below-freezing temperatures at station Y is p_y, then the probability of below-freezing temperatures at both X and Y is $p_x p_y$, and the probability of below-freezing temperatures at neither is $(1 - p_x)$ $(1 - p_y)$. The probability of below-freezing temperatures at X and above-freezing temperatures at Y is $p_x(1 - p_y)$, and so on. Three or more probabilities may be combined in exactly the same way.[30]

Probability algebra as described assumes that events are independent; e.g., below-freezing temperatures at station X are assumed to be completely independent of below-freezing temperatures at station Y. But climatic data are often interdependent, and the analysis then becomes one of *conditional probabilities*. The probability assigned to event A when it is known that another event B has occurred, or which would be assigned to A if it were known that B had occurred, is called the *conditional probability of event* A *given event* B, and is usually denoted by $P(A|B)$. The symbol for the ordinary or unconditional probability of event A is written $P(A)$. The probability that both events A and B will occur is termed their *joint probability* and is written $P(A,B)$. The mathematical definition of conditional probability is then given by

$$P(A|B) = \frac{P(A,B)}{P(B)},$$

where $P(B)$ is the unconditional probability of event B. If we know the conditional probabilities of event A given event B, we can obtain the unconditional probability of the joint occurrence of A and B by applying the multiplication rule,

$$P(A,B) = P(B)\,P(A|B).$$

This is the formula to be used for obtaining combined probabilities of events that are not statistically independent. In uncertain cases, it may be determined whether or not two events A and B are related, by applying the rule that if $P(A|B) = P(A)$, then events A and B are statistically independent. Very frequently in climatology, there is occasion to revise the probabilities calculated for some event. This involves distinguishing between *a priori* probabilities, i.e., the unconditional probabilities assigned to certain events before the additional data becomes available, and *a posteriori* probabilities, i.e., the conditional probabilities assigned to those events after the additional material has been incorporated into the main body of data. The new probabilities may then be assessed by means of *Bayes' theorem* (see Appendix 2.31).[31]

An interesting and useful form of probability analysis has to do with the maximum or minimum values of an element during a long period. Extreme values obviously give a rough indication of the range of a variable, and once an extreme has occurred it is not likely to occur again for some time. In extreme-value analysis, the basic statistic of the observed frequency is its *apparent return period*, which is defined as the reciprocal of the relative frequency. The return period is the average interval between recurrences of the event during a long period of time. Various probability statistics may therefore be expressed in terms of return periods by means of probability algebra (see Appendix 2.32). If a set of extreme values may be obtained from the observational data—for example, a set of "highest summer temperature" values may be obtained from a series of observations of the highest temperature of each summer day during a long period at a station—then by means of *Gumbel's theory of extreme values*, the following statistics may be inferred: (a) the return period for a specific extreme value; (b) the probable extreme value corresponding to any one return period; (c) the expected extreme value corresponding to any return period; and (d) the probability that the greatest observed extreme will depart from its expected value by an amount equal to or less than its actual departure. The mathematics behind the Gumbel theory is somewhat involved, but it is the basis of *extreme probability paper*, which enables rapid analyses to be made. If, when plotted on extreme probability paper, the observations lie in a straight line, then they obey the theory of extreme values, and the statistics (a) to (d) inclusive may be obtained directly from the graph. Actual observations usually plot as a zigzag line on the graph paper, in which case a confidence band centered on the computed " line of expected extremes " is usually determined.[32]

If the population from which the extremes were drawn has a double exponential type of distribution, a simplified method may be used for estimating extreme values. Some theoretical guide to the computation of extreme-value statistics is always desirable, because estimations based on relative frequencies alone can be most misleading. For example, the highest temperature that has occurred five times in 50 years has a relative frequency of 0.10, but the best estimate (at the 95 per cent confidence level) of its true probability as derived from theory is between 0.05 and 0.22. Thus its return period is not necessarily 10 years but may be between 4.5 and 20 years.[33]

As examples of frequency and probability analysis, studies of temperature, humidity, and precipitation analysis may be quoted. Particularly worthy of attention are analytical methods that enable more detailed information to be obtained from the observed data than it is at first apparent could be.[34]

THE RELATIONSHIPS BETWEEN SCALAR QUANTITIES

Frequency and probability distributions are often required for combined elements, i.e., for a combination of two or more sets of observations. The data may be conveniently displayed by a *contingency table* for a combination of two elements; the significance of the apparent relationship indicated by the figures may be assessed by a chi-square test, and the closeness of agreement between the two variables measured by the *correlation coefficient* (see Appendix 2.33). Correlation coefficients are meaningful for the samples from which they are computed, but it is very dangerous to make inferences about cause and effect in the total population from correlation alone. A general rule that works quite well is that a correlation coefficient must equal or exceed twice its standard error before it may be regarded as evidence of correlation between the two variables in the parent population. A more refined significance measure is

TABLE 2.3
Examples of Transformation of Precipitation Data for Pairs of Stations in Arizona

Station pairs	No. of years of data	Linear[a]	Type of Transformation		
			Square root	Cube root	Logarithmic
Tucson Natural Bridge	64	0.54	0.55	0.57	0.57
Tucson Phoenix	76	0.38	0.41	0.42	0.42
Tucson Flagstaff	50	0.44	0.47	0.46	0.38
Tucson Yuma	83	0.19	0.24	0.26	0.28
Natural Bridge Phoenix	64	0.61	0.65	0.65	0.69
Flagstaff Yuma	50	0.63	0.62	0.63	0.57
Natural Bridge Yuma	63	0.22	0.36	0.42	0.48

[a] Linear transformation values give the correlation coefficients calculated from the raw observations. All other values are correlation coefficients calculated from data obtained by transforming the raw observations as specified. Data from J. E. McDonald, *T*, 12 (1960), 176.

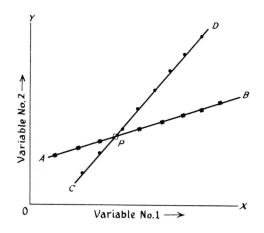

FIGURE 2.11.
Regression lines. *AB*, regression of *y* on *x*;
CD, regression of *x* on *y*; *P*, mean of
all the observations.

January

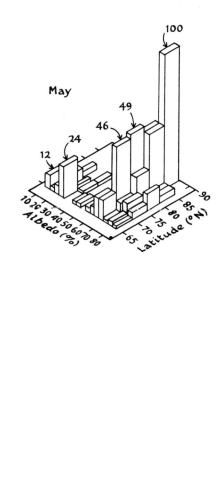

May

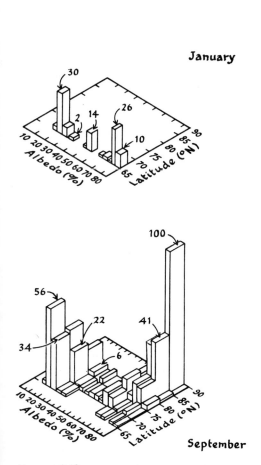

September

FIGURE 2.12.
Stereograms showing albedo variations over Arctic surfaces. Vertical scale is percentage of area in a given latitude belt. (After P. Larsson, *Geog. Rev.*, 53 (1963), 572).

provided by the "*t*-test." It is frequently the practice to normalize the data by means of logarithmic, square-root, or cube-root transformations before calculating their correlation coefficient, but unless a very precise measurement of the latter is justified, this is unnecessary, as Table 2.3 demonstrates. *Correlation fields* may be determined by mapping correlation coefficients, and useful information may be inferred from them: for example, they indicate that pressure systems are smaller vertically in the middle and upper troposphere, where their heights are more variable, than at sea level or in the stratosphere.[35]

It should be noted that, strictly speaking, a correlation coefficient should only be computed between two variables that are linearly related. Mathematically, the correlation coefficient *r* between two quantities *x* and *y* is defined by

$$r_{xy} = \frac{CV_{xy}}{\sigma_x \sigma_y},$$

where σ_x and σ_y are the standard deviations of the variables plotted along the *X* and *Y* axes, respectively, and CV_{xy} is the *covariance* between *x* and *y*. Covariances are used widely in many branches of climatology (see Appendix 2.34).

A correlation coefficient must lie between $+1$ and -1: $+1$ signifies perfect positive correlation; -1 signifies perfect negative correlation; 0 implies there is no correlation whatsoever between the variables. The meaning of these measures of correlation is illustrated by Figure 2.11. If all the data available for investigating the relationship between two variables are plotted on a rectangular coordinate graph, a *scatter diagram* results. If we assume that the relationship between the variables is linear, it is possible to draw two lines on the diagram that describe this relationship; these lines are *regression lines*, and the angle between them is an inverse measure of the correlation between the variables. For the plotted points in Figure 2.11 we can (a) divide the values of the variable plotted along *OX* into groups, and calculate the mean of all the *OY* observations falling into each group; similarly, we can (b) divide the values of the variable plotted along *OY* into groups, and then calculate the mean of the *OX* observations falling into each group. If we then fit a straight line to the mean values resulting from (a), we obtain one regression line. If we fit another straight line to the mean values resulting from (b), we obtain a second regression line. The smaller the angle between these lines, the closer is the relationship between the two variables. When the relationship is perfect, the two regression lines coincide; when there is no relationship, the two lines intersect at right angles (see Appendix 2.35).

The correlation coefficient in its simplest form, as described above, may only be used to determine the closeness of agreement between two variables that are represented by pairs of numerical values. If the variables are not represented in this form, a different type of correlation coefficient must be used. If observations of each variable can be ranked in order of magnitude, by assigning the rank one to the largest value, the rank two to the next largest, and so on, their degree of correlation may be assessed by computing the *rank correlation coefficient*. If one of the variables is expressed quantitatively, the other divided into two classes based on occurrence and non-occurrence, the appropriate measure of agreement is the *biserial correlation coefficient*. If both variables are divided into two classes for occurrence and nonoccurrence, the *tetrachoric correlation coefficient* should be used. If the variables are expressed qualitatively, and one or both of the series consists of more than two classes, the *coefficient of contingency* must be employed. For the biserial and the tetrachoric correlation coefficients, it is assumed that the variates are normally distributed. If the elements refer to derived quantities, rather than to directly observed ones, particular care is necessary in interpreting the correlation coefficient (see Appendix 2.36).

If the two variables are in the form of time-series, the correlation coefficient may yield extremely misleading results. For example, the correlation coefficient between two time-series may be zero, even though there is significant correlation between, say, the high-frequency components of each series, which is canceled by equally significant correlation of opposite sign between the low-frequency components. Because of this possibility, the correlation between two time-series is best examined by *cross-spectrum analysis* rather than by correlation coefficients. In effect, the cospectrum derived from a cross-spectrum analysis indicates to what extent fluctuations within a given range of periods contribute to the covariance between the two time-series. The quadrature spectrum then indicates how much would be contributed to the covariance if each harmonic of one of the time-series is shifted 90°. An example of cross-spectrum analysis is the correlation between zonal indices of 500-mb winds along the 25°, 40°, and 60°N parallels of latitude. Three-day moving averages of these indices, analyzed for cospectra and quadrature spectra, show that the zonal indices at 60°N and 40°N are negatively correlated, as are those at 40° and 25°N, although the former correlation is greater than the latter. Oscillations with periods of about 25 days are important at 40° and 60°N, but longer-period oscillations dominate at 25°N.[36]

If the two sets of observations to be correlated are not in a linear relationship to each other, the correlation coefficient will indicate only part of the total relationship between them, i.e., the part that depends on the extent to which there is a linear connection. The appropriate measure in this case is the *correlation ratio*. The correlation ratio between variables X and Y is defined as the ratio of variance of Y due to X to the total variance of Y, and its value varies from $+1$ for perfect correlation between the variables to zero if no correlation exists.[37]

To measure the relation between three or more scalar variables, *partial correlation* is used. If we have three variables numbered 1, 2, and 3, then the *partial correlation coefficient* between variables 1 and 2 when the effect of variable 3 has been removed is written as $r_{12.3}$, and is defined by

$$r_{12.3} = \frac{r_{12} - r_{13} r_{23}}{\sqrt{1 - r_{13}^2} \sqrt{1 - r_{23}^2}},$$

where r_{12}, for example, is the simple correlation coefficient (i.e., the *total correlation coefficient*) between variables 1 and 2. The regression lines between variables 1 and 2 may be obtained in terms of $r_{12.3}$ and the three total correlation coefficients, and then the closeness with which the regression equation fits the data may be found from the *multiple correlation coefficient*. Multiple correlation coefficients should be calculated only for data between which a physical relationship is known to exist, because the value of a multiple correlation coefficient is always positive, even when no physical correlation really exists between the variables (see Appendix 2.37).

Partial correlation coefficients may be computed and employed to produce regression equations between more than three independent variables, and the closeness of fit between the actual data and the equations may be tested by means of multiple correlation coefficients. However, the purpose of such a laborious procedure should

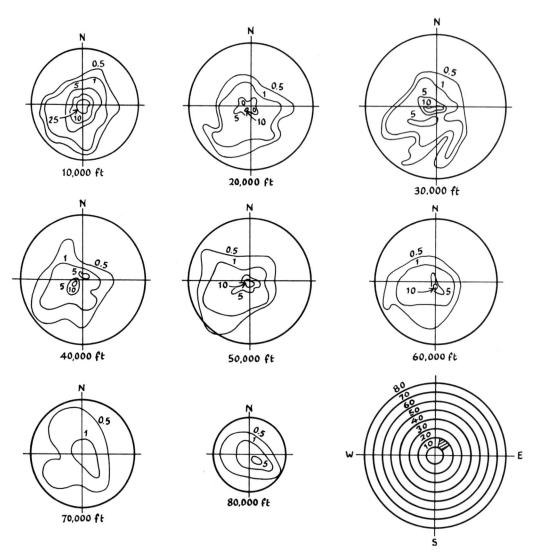

FIGURE 2.13.
Annual frequency surfaces for winds over the Argentine Islands (Grahamland, Antarctica), 1954-1958. On the radial scale, isopleths represent frequency densities (unit area shaded on diagram). After B. D. Giles, *Brit. Antarctic Survey Bull.*, no. 4 (1964), p. 39.

be carefully considered beforehand: if the purpose is prediction, then a graphic method, such as will be discussed later in this chapter, is preferable. A study of actual investigations into the relationships between scalar quantities will show that the problems are often very much more complex than a first perusal indicates. It is quite common for an investigation of the correlation between two variables to reveal that a third variable (or more) is important. For example, a regression analysis of temperature against altitude may show that the change in temperature with altitude is a function of temperature as well as of altitude, and that the function varies with the season.[38]

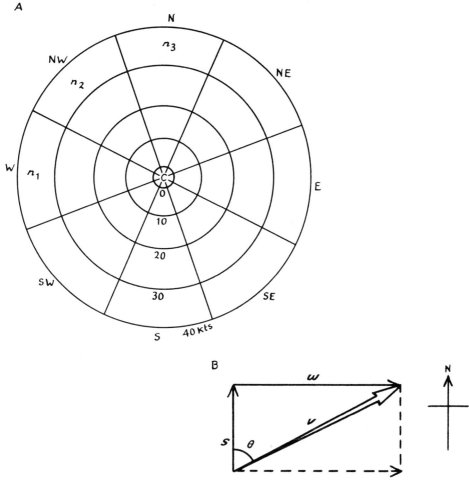

FIGURE 2.14.
Circular contingency table. In A, C represents the number of calms, and $n_1, n_2, n_3, \ldots$, represent the number of 30- to 40-knot winds reported from the west, northwest, north, . . . , respectively. In B, $\mathbf{V}$ represents the actual wind vector, θ being the actual wind direction. The resolved components of $\mathbf{V}$ are $\mathbf{S}$ and $\mathbf{W}$, given by $S = V \cos \theta$ and $W = V \sin \theta$, where S, V, and W are the modules of $\mathbf{S}$, $\mathbf{V}$ and $\mathbf{W}$, respectively.

Statistical Analysis of Vector Quantities

Vector quantities, such as air movement, must be expressed three-dimensionally if their statistical properties are to be determined satisfactorily. The frequency distribution of a vector quantity is usually represented graphically by means of either a *stereogram* (Figure 2.12) or a *frequency surface* (Figure 2.13). The stereogram is constructed, in effect, by erecting rectangular or square columns on each cell of a frequency table, the height of each column being proportional to the number of observations in each cell. If the total number of observations is increased, and at the same time the number of cells is increased, then in the limiting case the outline of the

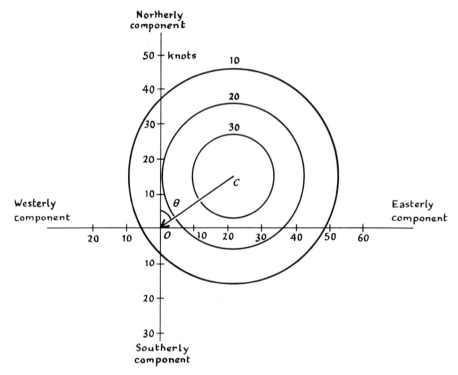

FIGURE 2.15.
Circles represent isopleths of constant frequency density for wind data. The vector **OC** represents the vector mean wind for a circular normal wind distribution, whose direction is given by angle θ and whose module is given by the length of OC according to the windspeed scale marked on the axes. Note: westerly and southerly components of the actual wind are given by **W** and **S** respectively, in Figure 2.14, B.

upper surface of the stereogram (numbered in Figure 2.12) will become a smooth curved surface, the frequency surface for the data (see Appendix 2.38).

Wind data are usually plotted as a circular contingency table (Figure 2.14), and values of *frequency density*, defined as the frequency of wind observations per unit area, are calculated from the values in each cell and plotted on the graph. These values are then regarded as either point values or, if greater accuracy is required, as area means, which are then converted to point values, and isopleths of constant frequency density are drawn. When the frequency-density isopleths have been determined, the "normality" of the wind data may be determined at a glance. If the data are drawn from a population that is normally distributed, then the isopleths will be concentric circles about a point that represents the vector mean of the observations (Fig 2.15), and the *circular normal probability function* may be used to obtain probability statistics, as the probability density function is used with scalar quantities.[39]

Wind data that plot as approximately circular isopleths are said to have *circular distributions*. Since homogeneous airflows in the upper atmosphere are approximately circular, data can be interpolated with a fair degree of confidence.[40] Data derived from different seasons or from different air masses will not necessarily be circular, and neither will winds that have been much influenced by topography, or surface

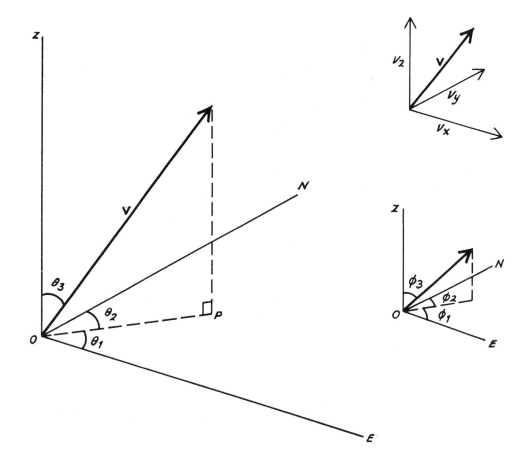

FIGURE 2.16.
Line *OP* is in the plane of *ON*, *OE*. Axes *ON* (in the north-south direction), *OE* (in the east-west direction), and *OZ* (in the vertical direction) are at right angles. The vector **V** represents the three-dimensional wind vector, which may be replaced by orthogonal components V_x, V_y, V_z in the *OE*, *ON*, *OZ*, directions. Then $V_x = V\cos\theta_1$, $V_y = V\cos\theta_2$, and $V_z = V\cos\theta_3$. The vector mean wind, $\mathbf{V_m}$, of a set of N wind vectors such as **V** is given by

$$\text{Module } V_m = \frac{1}{N}\sqrt{(\sum V_x)^2 + (\sum V_y)^2 + (\sum V_z)^2}.$$

Direction of $\mathbf{V_m}$ is given by angles φ_1, φ_2, and φ_3, where

$$\cos\varphi_1 = \frac{\sum V_x}{NV_m}, \cos\varphi_2 = \frac{\sum V_y}{NV_m}, \cos\varphi_3 = \frac{\sum V_z}{NV_m}.$$

winds in general. If mixing has occurred, for example, at the boundaries of inversions, between land and sea breezes, or near the tropopause, the air flow will be heterogeneous, and the wind data will not show a circular frequency distribution.

The standard deviation of the frequency distribution of a vector quantity is measured by the *standard vector deviation* of the distribution. As with scalar quantities, the standard vector deviation may be computed for any set of wind data, but it only has value in specifying the number of observations that fall within certain ranges if the data are normally distributed, i.e., if the frequency isopleths are circular and concentric. In order to obtain the standard vector deviation for a set of data, the wind observations are, in effect, represented graphically in three dimensions (Figure 2.16), and the *vector mean wind* is computed and described by its two components, its *module* and its *direction*. The three directions usually taken are north-south, east-west, and vertical. Since observed winds often refer to winds in the horizontal plane alone, here the vertical coordinate is zero, and the vector mean wind is given by a simple resolution of observed wind into north-south and east-west components; summing the components gives the *resultant wind.* For the computation of vector mean winds, winds should preferably be available in the form of vectors (i.e., the observations should specify both the direction and the speed of the wind), but the vector means may nevertheless be estimated from frequency tables of wind direction alone (see Appendix 2.39).

To understand the meaning of the standard vector deviation, some consideration of vector theory is necessary. Wind is a *bivariate* quantity; i.e., it may be represented by two quantities, in this case, orthogonal components from the north and east. If these components are independent, their normal distribution is described by the circular normal probability function already noted, but if the components are related, their normal distribution must be described in terms of the *elliptical bivariate normal function.* The difference between these two functions may be illustrated as follows.

The equation that describes a frequency surface includes as variables the number of observations of the vector, and the standard deviations of the orthogonal components of the vectors. Let the latter be σ_x in the OX (east-west) direction and σ_y in the OY (north-south) direction. The contours of the frequency surface form ellipses with centers on the common axis of the distribution (i.e., along the vector mean OP in Figure 2.17), and with their major and minor axes proportional to σ_x and σ_y (or vice versa) and at right angles. If the vector components are independent of each other, and if $\sigma_x = \sigma_y$, the frequency surface represents a normal circular wind distribution. The heights of any required frequency ordinate (i.e., the position of any given isopleth of frequency density) may then be found from the equation for the frequency surface, which simplifies in the normal case to

$$z = \frac{N}{\pi\sigma^2} \exp(-V^2/\sigma^2)\,, \qquad\qquad [2.1]$$

in which N is the number of vectors, V^2 is the sum of the squares of the deviations of the vector components from their vector mean, σ^2 is the sum of σ_x^2 and σ_y^2, and z is the height of the frequency surface at the point in question (see Appendix 2.40).

Equation 2.1 represents only one form of the normal distribution for vectors. If the vector components are not independent, i.e., if the correlation coefficient r_{xy} between the OX and OY components of the observations does not approximate zero, then a different equation is required, one in which σ_x and σ_y are not equal.[41] If a frequency surface follows this new equation, then the frequency distribution is said to be *elliptically normal,* and the frequency of vectors from any given point, sector, or area may be determined from formulae and graphs derived from the elliptical bivariate normal probability function. Many observed winds obey the elliptical distribution.

The quantity σ^2 in Equation 2.1, i.e., $\sigma_x^2 + \sigma_y^2$, defines the standard vector deviation (S.V.D.). Although by definition this quantity involves the deviations of the observed vectors from the vector mean, it is not usual to compute the S.V.D. from them. Instead, the S.V.D. is found from the modules of the individual vector observations.

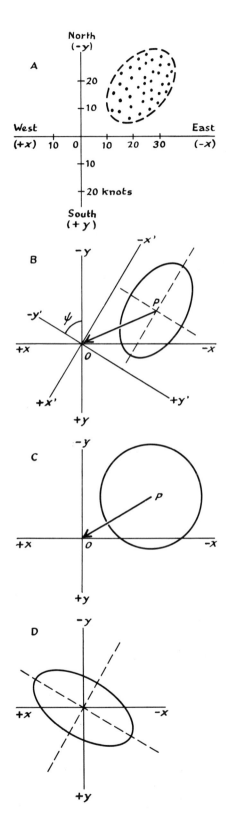

FIGURE 2.17.
Elliptical and circular wind distributions. In A, the plotted points represent individual wind observations, obtained by plotting vector components (at right angles) of each wind observation. If plotted points lie within an ellipse, as shown, the wind distribution is said to be an elliptical one. Note the signs of x and y on the axes: vector components from west and south are positive. B shows a generalized elliptical bivariate wind distribution where OP represents the vector mean wind. The ellipse is the projection of the bivariate frequency surface contour (which contains a specified percentage of the observations) onto the x, y, plane. Data for which σ_x and σ_y are not equal, and r_{xy} is not zero, obey this distribution. Rotation angle ψ (and hence the ellipse) may be obtained from OP, σ_x, σ_y. C and D present special cases. C shows a generalized circular wind distribution, obeyed by data for which $\sigma_x = \sigma_y$ and $r_{xy} = 0$. D results when vector mean and component means are zero.

It should be noted that the true vector mean can only be found by employing vector algebra.[42] For wind observations, the data usually are in the form of scalars, and the vector mean wind and the scalar mean wind are related by the *constancy* of the winds. The greater the constancy of the winds, the closer the scalar mean wind approaches the true vector mean wind. It is obviously useful to be able to determine standard vector deviations directly from scalar mean wind. Provided the wind distribution is the normal circular one, the relation is given by

$$\sigma = 1.13 \ V_s$$

for small values of constancy, where σ is the standard vector deviation and V_s is the scalar mean wind. (See Appendix 2.41.)

The standard vector deviation of, for example, the variation in height of a pressure surface, is in fact a quite complex parameter. Four different S.V.D.'s may be calculated for this quantity: the S.V.D. of the time variations of contour height at a point, the S.V.D. of the space variations in contour height at a given time, the S.V.D. of space variations for time-meaned contour heights, and the S.V.D. of time variations for space-meaned contour heights. Once these different standard vector deviations have been obtained, however, it is possible to obtain from them the mean amplitude, effective wavelength, and mean kinetic energy of the dominant pressure systems.[43]

For the correlation of vector quantities, the simple methods of correlation-coefficient analysis described for scalars are inappropriate, and must be extended dimensionally. If the fundamental definitions of linear correlation are applied to vectors, it may be shown that the multiple correlation coefficients of each component of one vector on all the components of the other vector are involved; a complex expression for the *total vector correlation* is regarded as the "true" measure of correlation. Alternatively, a linear treatment of vector correlation may be adopted, in which the correlation between two vectors is expressed in terms of two components, termed *stretch* and *turn*. The linear vector correlation coefficient R_{wz} between vectors W and Z is then defined by

$$R_{wz} = \sqrt{R_S^2 + R_T^2}$$

where R_s is the coefficient of stretch and R_T is the coefficient of turn; R_{wz} is usually called the *vector stretch correlation coefficient*, and is consistently larger than the total vector correlation coefficient (see Appendix 2.42). Provided the degree of wind connectedness in time or space is large ($R_{wz} > 0.4$ or thereabouts), the vector stretch correlation coefficient is the simplest and best measure of correlation, but for a low degree of connectedness ($R_{wz} \leq 0.3$), the more complicated total vector correlation coefficient must be used. Reference to actual examples will make clear the fact that the study of wind correlations in space and time needs very careful thought.[44]

Statistical Predictions in Climatology

The purpose of most statistical investigations in climatology is to provide an estimate of the probability of some event that either was not or cannot be observed. Provision of the probabilities of future events (future prediction), and provision of the probabilities of past events that were omitted either by accident or by design from the climatic record (past prediction), are both included in the concept *prediction*. The procedure of statistical prediction is essentially one of classification based on historical experience; i.e., one makes the assumption that if at some time it is observed that a specific event (the *predictand*) is associated with a specific combination of simultaneous or antecedent conditions (*the predictors*), then if at some other time (for which we have no observations of the event in question) this combination of conditions exists, we may expect that the event will be likely to occur. We must use the word "likely" because we never know whether or not we have all the significant conditions, and therefore the prediction must be on a probability basis. A distinction may be made between *underprediction*, which is a prediction based on insufficient predictors, and *overprediction*, which necessarily results when a selection from many possible predictors is made. Because of the danger of overprediction, the data to be analyzed should be divided into two sets of observations. One set, the *dependent sample*, should be used to estimate parameters and to determine relationships; the other, the *independent sample*, should be employed to test the predictions made by means of the relationships found from the dependent sample. If the relationships are both determined and tested by the same data, overprediction must result; i.e. the predictions will appear to be better than they really are.

The simplest form of statistical prediction is found for two variables that are related linearly. Suppose we have accurately measured variables X and Y, related as shown in Figure 2.18. We could predict unobserved combinations of values over the range between PQ and RS by subjectively drawing a straight line through the plotted points. However, this method is too imprecise to do justice to accurate observations, and it does not permit extrapolations to be made. The usual method to adopt is that of *least squares*; i.e. we fit a straight line to the plotted points in such a way that the sum of the squares of the distances of the plotted points from the line is a minimum (see Appendix 2.43). Since two sets of distances may be measured, two possible regression lines may be plotted by the method of least squares. That is, we may either (a) regard X as the predictand and Y as the predictor, in which case we determine the regression of Y on X, i.e., the best estimate of X for a given value of Y, or (b) regard Y as the predictand and X as the predictor, and compute the regression of X on Y, i.e. the best estimate of Y for a given value of X. The lines computed for these estimates pass through the point representing the means of X and Y.

If the two variables are not related linearly, the simplest procedure is to see if they can be transformed to linearity. For the *decay curve* and the *exponential curve* (Figure 2.19), linearity may be obtained by taking logs, and the method outlined in the previous paragraph then applies (see Appendix 2.44). For more complicated relationships, it is not possible to transform to linearity by a simple process. Although any curve, no matter how complicated, may be described by a polynomial expression if sufficient constants are introduced, often it will be apparent that more than two variables are involved. The standard methods employed in climatology for dealing with predictions from multivariable systems are those generally grouped under the heading of *objective weather forecasting*.

In general, a system of objective weather forecasting may be defined as one made without recourse to personal judgment by the forecaster; two or more forecasters, presented with the same data, would produce exactly the same prediction. The methods involved are all essentially climatological, because experience, as represented in the climatic record, provides the basis for them. Synoptic experience and a

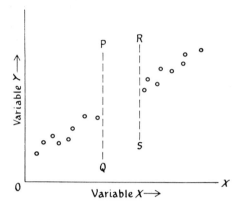

FIGURE 2.18.
Use of the method of least squares.

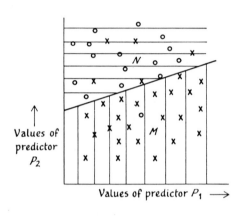

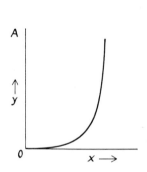

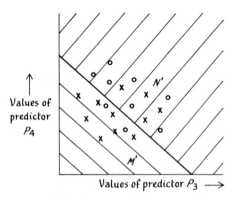

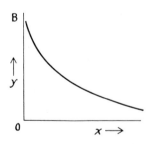

FIGURE 2.19.
A shows the exponential curve,
$y = ae^{bx}$. B shows the decay
curve, $y = ax^{-b}$.

FIGURE 2.20.
Graphic stratification (a hypothetical case).
Circles represent occurrences of event
(E); crosses represent nonoccurrences of
event ($\sim$E).

knowledge of atmospheric science is important in the selection of predictors. There are numerous methods of objective prediction; four of the most successful will be described.[45]

Simple stratification is appropriate when the predictand is simply whether or not some event happens, and the data are given in the form of categories. Suppose we denote the occurrence of the event by the symbol E, and its nonoccurrence by the symbol $\sim E$. The predictors we denote by A, B, C, and so on, these being the conditions that in the past have been associated with E. We first take condition A. If condition A does not prevail, we can predict $\sim E$, and we are finished; that is, if $\sim A$, then $\sim E$. If condition A prevails, we move on to condition B, and the process repeats. If we exhaust all the predictions without having predicted $\sim E$, then we may safely predict E. The method is thus one of elimination.

Graphic stratification is appropriate where the predictors are given in the form of continuous variables, and the predictand is the occurrence or nonoccurrence of an event E.[46] The predictors are taken in pairs, and a scatter diagram is plotted. Suppose we have predictors P_1, P_2, P_3, P_4, and so on. We first plot P_1 against P_2 and label the plotted points according to whether they represent E or $\sim E$; we then divide the diagram into two areas N and M, M containing a negligible number of occurrences of E, and N containing both occurrences and nonoccurrences (see Figure 2.20). We then plot the points that fall into area N from the next pair of predictors, P_3 and P_4, and divide area N into areas N' and M', M' containing a negligible number of E cases, and N' containing both E and $\sim E$ cases. The process of stratification is repeated until a very high percentage of E cases can be separated into one area of a scatter diagram. The prediction procedure is then as follows.

1. Enter the observed values of P_1 and P_2 in the diagram. If the plotted point falls within area M, forecast $\sim E$, and the prediction is then complete. If the plotted point falls within area N, proceed to stage 2.

2. Enter the observed values of P_3 and P_4 in the diagram. If the plotted point falls within area M', forecast $\sim E$ and stop; if it falls within area N', proceed to the next stage and so on. Not more than six predictors must be used, otherwise the accumulation of random influences decreases the efficiency of the scheme.

If both predictand and predictors are in the form of continuous variables, *graphic multiple correlation* should be employed. This is similar to graphic stratification, in that the predictors are plotted in pairs on scatter diagrams, but instead of marking areas of E and $\sim E$ on the latter, isopleths representing constant values of intermediate (fictional) variables are drawn by eye on the graph. For example, if we have predictors P_1 through P_8 to forecast predictand Z, intermediate variables V_1 through V_6 are defined as in Figure 2.21. Each stage is graphed separately, as in Figure 2.22. Stage 1 comprises four graphs, stage 2 includes two graphs, and stage 3 consists of the final graph. This technique has been used many times.[47] Once it is completed for a set of data it enables a prediction to be made in a very short time. When the isopleths are fitted by eye, some mutual adjustments are usually necessary; if great accuracy is desired, they may be fitted by least squares. Both methods assume that the relations between predictand and predictors are linear or at least curvilinear.

The *contingency table method of prediction* makes no assumption about the form of the relations between predictand and predictors.[48] The predictand data may be in any form; continuous variables, such as temperatures or rainfall amounts; attributes, such as the mutually exclusive classes of "clear days, partly cloudy days, overcast

128

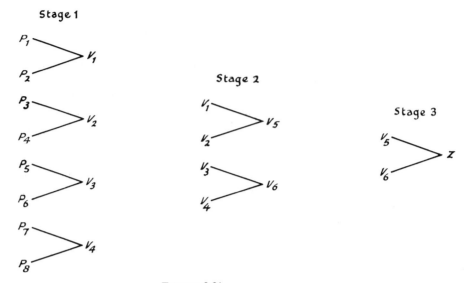

FIGURE 2.21.
Stages in graphing predictor values.

days"; or dichotomies, such as "day with thunderstorm, day without thunderstorm."
The predictors may be actual observations, derived elements, computed parameters,
tendencies, gradients, or weather types. The only prerequisite is that both predictors
and predictand must be organized into classes; if there are N observations, there
should not be more than n classes, where $2n^2 < N$. Usually no more than six predictors will be necessary. The procedure is in five stages, as follows.

1. Draw up a primary table of observed contingencies. In Table 2.4, which represents the relationship between the predictand and predictor number one, C_{11}
denotes the number of observations of predictor number one, class P_1, in which the
predictand fell into class Z_1, C_{53} denotes the number of observations of predictor
number one, class P_5, in which the predictand lay in class Z_3, and so on. The
number of observations required depends on the number of cells in the contingency
table; the number of observations should not exceed 50 times the number of cells.
For Table 2.4, more than 1,250 observations will not materially increase the reliability of the final prediction scheme. The cell with the fewest observations should
contain not less than 5 per cent of the total number of observations.

TABLE 2.4
Example of a primary contingency table.

Predictor classes	Predictand classes					Marginal Totals
	Z_1	Z_2	Z_3	Z_4	Z_5	
P_1	C_{11}	C_{12}	C_{13}	C_{14}	C_{15}	
P_2	C_{21}	C_{22}	C_{23}	C_{24}	C_{25}	
P_3	C_{31}	C_{32}	C_{33}	C_{34}	C_{35}	
P_4	C_{41}	C_{42}	C_{43}	C_{44}	C_{45}	
P_5	C_{51}	C_{52}	C_{53}	C_{54}	C_{55}	
Marginal totals						Grand total

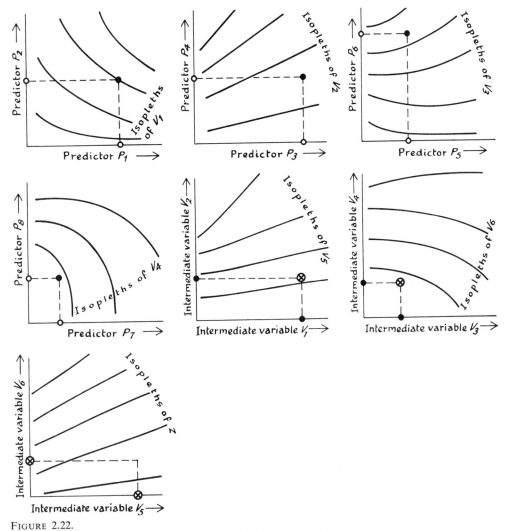

FIGURE 2.22.
Graphic multiple correlation (a hypothetical case). Values to be plotted on axes:

    ○ Observed value of predictor (stage 1);

    ● Derived value of intermediate variables V_1, V_2, V_3, V_4 (stage 1);

    ⊗ Derived value of intermediate variables V_5 and V_6 (stage 2).

Final predicted value is given by point P, read off from isopleths of Z (stage 3).

2. If each cell in the contingency table contains approximately the same number of observations, i.e., if each combination of predictand-predictor classes has the same probability of occurrence, compute the *contingency ratio* for each cell, which is the extent to which any observed class-combination frequency differs from the expected frequency (see Appendix 2.45).

3. If each cell in the contingency table does not contain approximately the same number of observations, i.e., if each class does not have an equal probability of occurrence, compute the *normalized contingency ratio* for each cell, and then normalize all the contingency tables to the same level of statistical significance (see Appendix 2.45).

TABLE 2.5
Example of a final contingency table.

Predictor		Predictand class										
Number	Category	Z_1	Pr	Z_2	Pr	Z_3	Pr	Z_4	Pr	Z_5	Pr
1	P_2	C_{21}		C_{22}		C_{23}		C_{24}		C_{25}	
2	P_4	C'_{41}	a	C'_{42}	e	C'_{43}	i	C'_{44}	m	C'_{45}	q
3	P_1	C''_{11}	b	C''_{12}	f	C''_{13}	j	C''_{14}	n	C''_{15}	r
4	P_1	C'''_{11}	c	C'''_{12}	g	C'''_{13}	k	C'''_{14}	o	C'''_{15}	s
5	P_2	C''''_{21}	d	C''''_{22}	h	C''''_{23}	l	C''''_{24}	p	C''''_{25}	t

Predictor number 1 is assumed to be represented by the primary contingency table, Table 2.4. Predictor number 2 is assumed to be represented by a similar table, with C_{11} replaced by C'_{11}. C''_{11} replaces C_{11} in Table 2.4 for predictor number 3, etc. Columns labeled *Pr* represent cumulative products, thus $a = C_{21} \times C'_{41}$, $b = C_{21} \times C'_{41} \times C''_{11}$, $h = C_{22} \times C_{42} \times C''_{12} \times C'''_{12} \times C''''_{22}$, etc.

4. Inspect each individual contingency table, and prepare a *final contingency table*, which is based on the antecedent conditions for the forecast situation. For example, suppose we wish to predict the monthly mean temperature at some station three months ahead. The predictand categories might be defined as follows: Z_1 denotes a monthly mean temperature of 30°F or less; Z_2 denotes a monthly mean temperature of 31–40°F; Z_3 one of 41–50°F; Z_4 one of 51–60°F; and Z_5 one of 61°F or more. Assume we have identified five predictors that are related to the predictand at a lag of three months; e.g., predictor number one for month m ($m = 1, 2, 3, \ldots,$ 12) is significantly correlated with predictand Z for month $m + 3$, etc. We wish to predict whether the predictand category will be Z_1, Z_2, Z_3, Z_4 or Z_5 in a given month on the basis of the predictor categories for three months ago. We therefore obtain the conditions, from the climatic record, that existed for three months previous to the forecast month. Suppose these antecedent conditions were that predictor number one fell into category P_2, predictor number two fell into category P_4, and predictors number three, four, and five fell into categories $P_1, P_1,$ and P_2 respectively. We then extract the rows corresponding to $P_2, P_4, P_1, P_1,$ and P_2 from contingency tables 1, 2, 3, 4, and 5 respectively, and assemble them in the final contingency table, Table 2.5.

5. Determine *cumulative contingency products* for each predictand class, as indicated for the columns labeled *Pr* in Table 2.5. The most probable predictand class is then the one with the highest cumulative product, and it should be predicted.

There are several points to be remembered when producing a contingency-table prediction scheme. Three predictors are the minimum number needed for a realistic forecast, and seven the maximum. Highly related predictors will not add much useful information to the tables, and should be reduced to one predictor by making a correlation analysis. If a large number of predictors are known to exist, compute the *information ratio* for each predictor, and select the six with the highest ratios.[49] When interpreting the cumulative contingency products, remember that a figure of 1.0 or thereabouts is the value to be expected by chance; a value greater than 1.0 indicates that the predictand class is more likely to occur than one would expect from chance associations between predictand and predictor classes. To determine what value of the cumulative contingency product one should regard as critical in making a definite prediction, test the contingency table system on an independent sample and compare the predictions with the actual events.

With all the prediction techniques described, it is necessary to test the prediction scheme on independent data. For instance, if we have ten years of observations, we may regard data for odd-numbered years as the dependent sample, and make predictions from a scheme based on them for the even-numbered years. We may then double-check by taking data for even-numbered years as the dependent sample, and then making predictions from them for the odd-numbered years.

The accuracy of the prediction scheme may be measured by calculating *skill scores* from a verification of predictions made for the independent sample. The skill score measures the advantage possessed by the predictions in comparison with estimates based on mean values, persistence, or pure chance (e.g., tossing a coin or a die). As a simple measure of skill, we may determine

$$\text{skill score} = \frac{F - D}{T - D},$$

where F is the number of correct predictions, D is the number of correct predictions based on mean values, persistence, or chance, and T is the total number of predictions made. The skill score is equal to unity if all predictions are correct, and to zero if the number of correct predictions is the same as that expected from mean values, persistence, or chance. There are other, more complex measures of skill; for probability forecasts, *sharpness* and *validity* are important considerations.[50]

Our discussion of methods for the interpretation of climatic data has so far dealt only with *univariate* quantities, i.e., with elements of which only one attribute has been measured. This is quite satisfactory when one is only concerned with a single attribute, e.g., temperature, but when we are concerned with prediction and have to describe that attribute in terms of predictors, we approach the study of *multivariate* quantities. Multivariate analysis is really the vector counterpart of univariate analysis, and deals simultaneously with "elements" for which more than one attribute has been measured. The standard observations made at a synoptic station in Britain at 9:00 A.M. may be regarded as observations of a single multivariate quantity, although each multivariate observation comprises a number of simultaneous univariate observations of pressure, temperature, dew point, cloud amount, visibility, etc.

The form of multivariate analysis that has found most application in climatology is *discriminant analysis*. This technique determines the linear function of a set of predictors that maximizes the ratio of the between-group sum of squares to the within-group sum of squares of the total variability in the observations. *Multiple discriminant analysis* extends the concept to a number of mutually uncoordinated linear functions. The essential difference between discriminant and regression analysis, and between multiple discriminant and multiple correlation analysis, is that the multivariate techniques involve vector concepts, usually expressed in the form of matrices. To every analytical expression in multivariate analysis, there is a geometric counterpart, so that every multivariate quantity may be represented by a point in *predictor space*. Predictor space is not ordinary three-dimensional space, but an *n*-dimensional space, in which each dimension is a predictor. The coordinates of the point are the values of the discriminant functions associated with that point.[51]

Application of discriminant techniques to climatic data is time-consuming. Special significance tests must be applied, the *Hotelling* T^2 *test* or *Mahalanobis* D^2 *test* corresponding to the "students' *t*" test of univariate statistics, and the *Wilks* Λ *criterion*

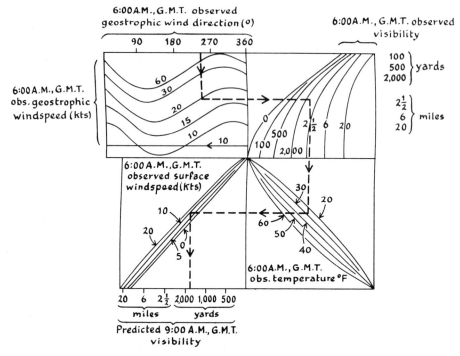

FIGURE 2.23.

Objective prediction of 9:00 A.M. visibility at London Airport. Plotted example shows procedure for observed surface wind of 10 knots, observed geostrophic wind of 240°, 20 knots, observed visibility of 3,500 yards, and observed temperature of 30°F (all at 6:00 A.M. G.M.T.), giving a predicted visibility at 9:00 A.M. G.M.T. of approximately 1,800 yards. (After M. H. Freeman, *QJRMS*, 87 (1961), 393.)

corresponding to the *F*-test for the analysis of variance. However, the discriminant techniques do provide convenient ways of selecting the best predictors; for example, for a prediction scheme for estimating the precipitation at Hartford, Conn., up to six hours in advance, the 175 predictors that were available were reduced to 16 by discriminant analysis.[52]

Many examples of statistical prediction schemes in climatology are available. They include simple "will occur/will not occur" schemes, graphic multiple correlation systems, linear regression schemes, and multiple linear regression systems, as well as the vector system just described.[53] Every method of statistical prediction has certain limitations, especially on the validity of the predictors. Usually, only those predictors that are known to have some physically demonstrable relationship with the predictand should be used. This presupposes a considerable knowledge of meteorological theory on the part of the investigator. If he lacks this knowledge, the relationships between the predictand and possible predictors should be investigated by correlation analysis, and only those predictors possessing a high positive correlation with the predictand should be used. For the statistical prediction of climatic data during short intervals of time or within small areas, synoptic experience forms a good basis for the selection of predictors. Figure 2.23 illustrates statistical prediction of visibility at London Airport some hours ahead: from 11 years' synoptic experience, 18 parameters were known to be significant in visibility forecasting; these were reduced to five, because there was much intercorrelation among the original 18 variables. A classic example involved the use of geographical parameters to estimate precipitation amounts in regions devoid of rain gauges.[54]

Cartographic Interpretation of Climatic Data

The production of climatic maps from statistical data is very difficult, if the distributions displayed are to have any physical meaning. The obvious difficulty is that the observed data refer to points in space, between which there are considerable gaps. If it may be assumed that the observed element is a variable with a linear rate of change between the points at which it is measured, then the number of data points may be increased by a simple triangulation procedure, and the problem of maxima and minima may be dealt with by a more elaborate procedure (see Appendix 2.46). If the original data are in the form of space averages instead of point-values, the space means may be converted to point-values, and an isopleth map obtained by a simple method. Sometimes point observations are required for computing space means, for example, area means of precipitation, in which case the *Thiessen polygon* method may be employed (see Appendix 2.47).

Almost always, if the results of a climatological investigation must be in the form of isopleth maps instead of statistical tables for the stations, one can produce some conventional scheme for transforming the data into maps. However, it must be realized that such schemes *are* mere conventions, and only represent the data in a logical manner. They most certainly do not tell us any more about the observed element than the tabular data do, although they may suggest lines for further investigation. The geographer's immediate reaction, "let's map the data," is to be deprecated for many climatological investigations. Mapping must be attempted only after very careful statistical analysis of the data. (See Appendix 2.48.)

A useful preliminary to climatic mapping is to make a *space-correlation analysis* of the data. For example, the correlation of observed pressures at one station with those at another results in the delimitation of *correlation fields* which show that pressure systems must vary more in height in the middle and upper troposphere than at sea level or in the stratosphere. Provided the original data at each station are normally distributed, or may be transformed to normality, one can produce probability maps with much more confidence than if one attempted to map the raw observations; such probability maps are especially useful for dealing with rainfall reliability.[55]

Despite the disadvantages of maps, climatic results are often wanted in map form. There are many ways to map data, but they all suffer from a fundamental difficulty. The construction of maps from climatic data involves *space-forecasting*, making use of techniques similar to those of objective weather prediction, which is concerned with *time-forecasting*. In all such forecasting, there is uncertainty about the initial state of the element in question. By means of statistical theory, extrapolation formulae may be developed that take into account the uncertainties about the initial state that are created by inadequate data; i.e., in space-prediction, the formulae take into account the lack of information between observing stations. A compromise must always be made: if the element to be mapped has been measured only at discrete points in space or time, then its true value is uncertain; if it is averaged over space or time to obtain a value which approximates to this true value, then detail is lost. It is possible to allow for these effects.[56]

It must be remembered that a given meteorological observation is part of *two* sets of measurements. It is part of a time-series of observations at one station, and it is also part of a space-series of observations along a geographical section or traverse that passes through the station in question. It is possible to convert a time-section into a space-section and vice versa.[57] Consequently, any spatial analysis of climatic data is incomplete if it does not take into account the parameters obtained from an analysis of the variations of the element as a function of time. Always, statistical space-predictions (or time-predictions, for that matter) should be made only from a knowledge of the physics and hydrodynamics of the climatic feature concerned.

Aids to Facilitate Climatological Statistical Analysis

Many of the methods described in this chapter are very laborious to apply, and any aids to lighten the drudgery involved are very welcome. For all but the very primitive analyses, logarithms or slide rules are essential. Equations may also be rapidly solved by means of a specially constructed alignment chart, or *nomogram* (see Appendix 2.49).

Various "dodges" may be used to perform very rapid statistical tests: some are simplifications of the standard methods, others are alternative ways of applying the standard methods that allow shortcuts in computation because of the properties of certain numbers. For example, the statistical significance of a mean value, or of the difference between two means, may be obtained very simply, and the association between two groups of figures may be found quickly. The existence of a significant trend in a set of figures may be obtained very rapidly, and even such a complex method of time-series analysis as the computation of power spectra can be carried out quite expeditiously by an approximation technique.[58]

Access to an electric desk calculating machine is essential if lengthy calculations are to be carried out without undue fatigue, for example, if an objective prediction scheme is to be produced that involves the examination of many predictors. Unless the same type of calculations are to be made many times over using different sets of figures, the time saved by using an electronic computer may not justify the time spent in writing the program, punching the tape, and so on. However, if data are to be processed on a routine basis, the preparation of a computer program should be well worth the effort. For example, processing rainfall data by computer allows a form of *quality control* that soon brings to prominence any doubtful observations. Contrary to popular expectation, electronic computers are of more value in the dynamic phases of climatology than as aids in statistical calculations.[59]

Since the raw data provided by meteorological instruments must nearly always be processed before the figures are suitable for climatological purposes, various types of *automatic data-processing systems* have been developed that compute the needed statistical parameters. For example, one machine provides continuous running means and standard deviations of the wind measurements made by an anemometer-bivane apparatus: the wind vector is measured in spherical coordinates (i.e., azimuth angle, elevation angle, and windspeed), but the data printout resolves into rectangular coordinates. Another device, an electronic analog system based on time-series theory, may be coupled to the output of a recording wind vane, and provides standard deviations automatically computed with an error of less than 5 per cent. A much more complex system can average 29 different meteorological variables over periods of 2, 10, or 60 minutes; this system can automatically average wind direction, which is a very difficult problem (see Appendix 2.50).[60]

Notes to Chapter 2

1. The classic textbooks dealing with statistical techniques in climatology, e.g., *HSM* and *MC*, are descriptive rather than analytical. For a clear statement of the difference between the descriptive and analytic approaches, see A. Court, *AGP*, 1 (1952), 75.

2. For an introduction to circular variables, see *ibid.* For a weatherman's guide to statistical analysis, see G. Reynolds, *W*, 22 (1967), 168, 188, and 439.

3. P. Lewis, *W*, 15 (1960), 121. On running means, see I. I. Gringorten, *T*, 20 (1968), 461, on probability distribution, and C. L. Godske, *T*, 18 (1966), 714, on dangers in the use of simple models for smoothed data. See J. S. Milner, *W*, 23 (1968), 435, for an example (rainfall in lee of south Pennines, England) of how misleading the idea of "normal periods" can be when using running means.

4. R. F. M. Hay, *MM*, 82 (1960), 121. On persistence, see R. Murray, *MM*, 96 (1967), 356, for monthly mean temperatures in central England, also R. P. Sarker and P. R. Mhaiskar, *QJRMS*, 93 (1967), 361, for temperatures in Bombay during the southwest monsoon. For a comparison of persistence and mean-value forecasts of upper winds, see H. S. Appleman, *JGR*, 67 (1962), 767.

5. For further details of this concept, see S. Goldman, *Frequency Analysis, Modulation, and Noise* (New York, 1948). Tables and graphs giving the Fourier integrals for a wide variety of "signals" (i.e., climatological time-series) are available in many references, e.g., G. A. Campbell and R. M. Foster, *Fourier Integrals for Practical Application* (New York, 1948). For examples of Fourier harmonic analysis, see: J. E. Carson, *JGR*, 68 (1963), 2217, on soil and air temperatures; O. S. McGee and S. L. Hasthenrath, *NOTOS*, 15 (1966), 79, on rainfall in South Africa; and S. L. Hastenrath, *AMGB*, 16, series B (1968), 81, for Central American rainfall. For a Fourier analysis

of kinetic-energy variables over North America, see E. C. Kung and S. Soong, *QJRMS*, 95 (1969), 501.

6. *HSM*, pp. 346–76, and *MC*, pp. 354–438.

7. A. Schuster, *Terr. Magnet.*, 3 (1898), 13; *Proc. Roy. Soc.*, 77, series A (1906), 136.

8. For details, see *HSM*, pp. 353–59.

9. See J. M. Craddock, *W*, 12 (1957), 252, and *J. Roy. Stat. Soc.*, 120 (1957), 387, on the use of band-pass filters in the analysis of the slower variations in temperature at Kew. For an example of the use of band-pass filters in the analysis of tropical stratospheric winds, see G. E. Edmond, *MM*, 94 (1965), 304.

10. For a comprehensive statement of this viewpoint, see J. L. Holloway, *AGP*, 4 (1958), 351.

11. D. L. Gilman, F. J. Fuglister, and J. M. Mitchell, *JAS*, 20 (1963), 182. For examples of spectral analysis, see G. I. Roden, *JAM*, 5 (1966), 3, on temperatures in the western United States, and G. W. Brier, R. Shapiro, and N. J. Macdonald, *JAM*, 3 (1964), 53, on rainfall in the United States. On the power spectrum of a red noise, see J. M. Mitchell, Jr., *JAS*, 21 (1964), 461. For the theory of spectral analysis of data distributed over a sphere, see W. M. Kaula, *Revs. in Geophysics*, 5 (1967), 83.

12. H. L. Griffiths, H. A. Panofsky, and I. Van der Hoven, *JM*, 13 (1956), 279. I. Van der Hoven, *JM*, 14 (1957), 160.

13. See H. A. Panofsky, *BAMS*, 36 (1955), 163, for details. On the use of the discrete Fourier transform in the estimation of power spectra and cross-spectra, see G. W. Groves and E. J. Hannan, *Revs. in Geophysics*, 6 (1968), 129.

14. For an example, see S. K. Pramanik and P. Jagannathan, *SPIAM*, pp. 86 and 106. For estimations of the relations between time series, using linear regressions, see B. V. Hamon and E. J. Hannon, *JGR*, 68 (1963), 6033. For studies of apparent trends, see J. Spar and P. Ronberg, *MWR*, 96 (1968), 169, on

annual precipitation in New York City; see also C. G. Abbott, *BAMS*, 45 (1964), 175, and S. L. Chorghade, *BAMS*, 46 (1965), 410, on precipitation in Nagpur, India, illustrating the importance of preliminary consideration of units of measurement.

15. For details of these methods, see *HSM*, p. 326. On the use of statistical control theory in the analysis of nonlinear time series, see R. H. Jones, *JAM*, 4 (1965), 701.

16. For an account of nonparametric statistics, see P. G. Hoel, *Introduction to Mathematical Statistics* (New York, 3d ed., 1962), pp. 329–49.

17. *HSM*, pp. 13–15.

18. See R. L. Schlaifer, *Probability and Statistics for Business Decisions* (New York, 1959), p. 99, for details and examples. This work provides an indispensable guide to the logical basis of probability analysis.

19. *HSM*, p. 129, give examples. On the information content of mean values, see N. C. Matalas and W. B. Langbein, *JGR*, 67 (1962), 3441. On the computation of the true means of climatological quantities, see J. D. Kalma, *W*, 23 (1968), 248, on daily mean air temperature and humidity, and A. Court and D. Waco, *MWR*, 93 (1965), 517, on daily mean relative humidity.

20. See *HSM*, p. 37, for details. For a discrete variate, Sheppard's correction should not be applied; see A. C. Aitken, *Statistical Mathematics* (Edinburgh and London, 1939), p. 44, for details.

21. For details, see *HSM*, p. 44. On nonnormal coefficients of variation (useful for studying arid-zone rainfall), see J. R. Hastings, *JAM*, 4 (1965), 475.

22. See *HSM*, p. 139, for details. See *TYS*, p. 157, for a discussion of the mathematical meaning of the *F*-test. On the derivation of significance tests when the data contain serial correlations, see J. Nordø, *T*, 18 (1966), 39. For significance tests applied to differences between the movements of typhoons and hurricanes, see W. T. Hodge, *JAM*, 6 (1967), 621.

23. H. L. Alder and E. B. Roessler, *Introduction to Probability and Statistics* (San Francisco, 4th ed., 1968), provide an excellent introduction to practical aspects of probability theory. On the use of decision-making techniques in meteorology, see H. R. Glahn, *MWR*, 92 (1964), 383.

24. H. L. Rietz, *Mathematical Statistics* (New York, 1927). See I. I. Gringorten, *JM*, 7 (1950), 388, for equations and graphs enabling confidence limits to be decided very rapidly.

25. See *TYS*, p. 50, for a discussion of the mathematical properties of the binomial distribution. On estimating probabilities of more than two mutually exclusive events by means of a nonlinear model, see R. H. Jones, *MWR*, 96 (1968), 383.

26. *HSM*, p. 85. See also *TYS*, p. 75, for a derivation of the normal distribution from the binomial distribution (i.e., as a consequence of the transforma-

tion of a discrete binomial variate into a continuous variate). Tables of values of the probability integral may be found, e.g., in Alder and Roessler, *op. cit.*, pp. 292–94.

27. Normal probability paper was introduced by A. Hazen, *Trans. Amer. Soc. Civil Engs.*, 77 (1914), 1539. See also N. Carruthers, *MM*, 83, (1954), 359.

28. See *HSM*, pp. 42 and 91, for details. On Cornu's criterion for testing normality for small samples, see J. J. O'Brien and J. F. Griffiths, *AMGB*, 16, series A (1967), 267.

29. The two procedures were introduced by K. Pearson, *Phil. Trans. Roy. Soc. London*, 185 (1894), 71, and by C. V. L. Charlier, *Lund Univ. Årsskrift*, vol. 1, no. 5 (1905). The procedures assume that either the means for the two components may be guessed, or the variances of the two components may be taken to be equal. A. Court, *op. cit.* (1952), gives details of both methods. For the approximate method see *HSM*, p. 130.

30. For further details of probability algebra, see Hoel, *op. cit.* (1962).

31. On estimating conditional probabilities for dichotomous random variables, see W. M. Brelsford and R. H. Jones, *MWR*, 95 (1967), 570. For the theory of joint/conditional probabilities as illustrated by point versus area precipitation, see E. S. Epstein, *MWR*, 94 (1966), 595. On the use of Bayes' theorem in estimating false-alarm rates in forecasts of rare weather events, see R. H. Olson, *MWR*, 93 (1965), 557. For measures of "goodness" of probability assessors, see R. L. Winkler and A. H. Murphy, *JAM*, (1968), 751.

32. See A. Court, *op. cit.* (1952), for a discussion of the basis of the mathematical theory, which first appeared in E. J. Gumbel, *Ann. Inst. Henri Poincaré*, 5 (1935), 115. See also E. J. Gumbel, *Statistics of Extremes* (New York, 1958). Extreme probability paper was introduced by R. W. Powell, *Civil Eng.*, 13 (1943), 105. See Court, *op. cit.* (1952), for plotting details and theory. On the use of extreme-value theory, see J. G. Lockwood, *MM*, 96 (1967), 11, on probable maximum 24-hour precipitation over Malaya, and *W*, 23 (1968), 284, on extreme rainfalls. For comparisons of the Gumbel theory with other extreme-value prediction techniques, see D. M. Herschfield, *JGR*, 67 (1962), 1535. For a plotting rule for extreme probability paper, see I. I. Gringorten, *JGR*, 68 (1963), 813. For an analytical theory involving "envelopes" for ordered observations, which may be applied to meteorological extremes, see I. I. Gringorten, *JGR*, 68 (1963), 815. For world-record climatological extreme values, see J. B. Rigg, *W*, 19 (1964), 241, and *W*, 20 (1965), 67; see also J. L. Paulus, *MWR*, 93 (1965), 331, for the world's highest point rainfalls.

33. Court, *op. cit.* (1952).

34. For details of descriptive use of statistics, see: E. N. Lawrence *MM*, 83 (1954), 195, on the frequency

of weather spells; H. C. Shellard and P. B. Sarson, *MM*, 91 (1962), 19, on the estimation of frequency distributions of hourly temperatures from monthly averages of daily maximum and minimum temperatures; G. I. Roden, *JAM*, 5 (1966), 3, on the analysis of temperatures from 1821 to 1964 in the western United States; A. B. Thompson, *MM*, 83 (1954), 293, and G. Reynolds, *W*, 11 (1956), 249, and 13 (1958), 151, on the statistical pattern of precipitation; J. C. Foley, *MM*, 81 (1952), 325, on the interpretation of rainfall variability in Australia; A. N. Dingle, *JM*, 12 (1955), 220, on secular change of American precipitation amounts; A. L. H. Gameson and R. D. Quaife, *MM*, 94 (1965), 173, on rainfall intensities; R. Murray and M. K. Miles, *MM*, 94 (1965), 1, on simple measures of the raininess of a month; and C. K. Stidd, *JM*, 11 (1954), 202, on the correlation between point-variable (e.g., precipitation) and space-variable (e.g., pressure) patterns.

For details of the analytic approach to climatic statistics, see: A. F. Jenkinson, *MM*, 85 (1956), 208, on the pattern of rainfall; E. N. Lawrence, *MM*, 86 (1957), 257 and 301, and J. McQuigg and W. Decker, *JAM*, 1 (1962), 178, on the estimation of runs of dry days; I. A. Lund, *JAM*, 5 (1966), 625, on the estimation of clear lines-of-sight (or sunshine) through the atmosphere; C. K. Stidd, *JAM*, 6 (1967), 255, and J. E. Kutzbach, *loc. çit.*, p. 791, on the use of eigenvectors; and I. I. Gringorten, *JAM*, 5 (1966), 606, on the use of stochastic models for the frequency and duration of weather events. On the use of Markov chains, see: A. M. Feyerherm and L. D. Bark, *JAM*, 4 (1965), 320, and 6 (1967), 770; B. Eriksson, *T*, 17 (1965), 484, on wet/dry days in Sweden; E. H. Wiser, *MWR*, 93 (1965), 511, on sequences of wet/dry days, using a modified Markov chain; W. P. Lowry and D. Guthrie, *MWR*, 96 (1968), 798, on Markov chains of order greater than one in the western United States; J. R. Green, *MWR*, 93 (1965), 155, on wet/dry day sequences; A. M. Feyerherm and L. D. Bark, *JAM*, 6 (1967), 770, on wet/dry day sequences in the north-central United States; E. A. Fitzpatrick and A. Krishnan, *AMGB*, 15, series B (1967), 242, on a first-order Markov chain for assessing rainfall discontinuity in central Australia; and J. E. Caskey, Jr., *MM*, 93 (1964), 136, on cold spells in London.

For examples of the descriptive use of precipitation statistics, see: R. Murray, *MM*, 96 (1967), 129, and 97 (1968), 181, on sequences in monthly rainfall in Britain; G. W. Brier, R. Shapiro, and N. J. Macdonald, *JAS*, 20 (1963), 529, and W. H. Portig, *JAS*, 21 (1963), 462, on rainfall calendaricities in the United States; J. M. Craddock, *W*, 20 (1965), 44, and G. Nicholson, *W*, 20 (1965), 322, on Thursday as the wettest day of the week at Teddington, but not at Kew, England; A. Bleasdale, *J. Inst. Water Engineers*, 17 (1963), 45, on the distribution of exceptionally heavy rains in the United Kingdom, 1863–1960; and

J. G. Lockwood, *NOTOS*, 13 (1964), 3, on estimation of probable maximum precipitation in South Africa. On statistical pitfalls in deducing cycles in rainfall data, see D. Hartley-Russell, *W*, 19 (1964), 121.

For examples of the analytic approach to precipitation statistics, see: H. C. S. Thom, *MWR*, 96 (1968), 883, on use of the Gamma function; S. Kotz and J. Neumann, *JGR*, 68 (1963), 3635, and 69 (1964), 800, on prediction of precipitation amounts for periods of increasing length, using the Gamma function; W. D. Sellers, *MWR*, 96 (1968), 585, on eigenvectors and American monthly precipitation patterns; and for an elementary analytical approach to wet and dry spells at Kew, see C. Chatfield, *W*, 21 (1966), 308. See also L. G. Veitch, *QJRMS*, 91 (1965), 184, on principal component analysis of atmospheric pressures in Australia.

35. B. Kinsman, *T* 9, 1957, 408. *MC*, p. 245. *HSM*, p. 220. J. E. McDonald, *T*, 12 (1960), 176. E. A. Bertoni and J. A. Lund, *JAM*, 2 (1963), 539.

36. H. A. Panofsky and P. Wolff, *T*, 9 (1957), 195, give details of cross-spectrum analysis, and perform such an analysis on the hemispheric westerly index of the general circulation.

37. See *HSM*, p. 241, for the mathematical definitions.

38. *HSM*, p. 251. J. T. Tanner, *JAM*, 2 (1963), 473. In many cases, such a study will reveal that the methods of pure mathematics rather than those of statistics are the appropriate ones for the investigations. See, e.g., J. G. Kemeny, H. Mirkill, J. L. Snell, and G. L. Thompson, *Finite Mathematical Structures* (Englewood Cliffs, N.J., 1959), for numerous examples.

For regression predictions of North Pacific typhoons, see H. Arakawa, *JAM*, 3 (1964), 524. For regression analyses on a time series, see S. Martinelle, *T*, 20 (1968), 179, on dependence of cosmic-ray intensity on air pressure, and G. W. Groves and E. J. Hannan, *Revs. in Geophysics*, 6 (1968), 129, on dependence of sea level on weather. On a regression technique for objective forecasts at 300 mb, see A. Woodroffe, *MM*, 95 (1966), 129. On "inflation" of regression forecasts, see H. R. Glahn and R. A. Allen, *JAM*, 5 (1966), 124, and E. S. Epstein, *JAM*, 6 (1967), 427.

On the use of multiple-correlation and multiple-regression techniques, see: B. I. Miller and P. P. Chase, *MWR*, 94 (1966), 399, on statistical prediction of hurricane motion; C. E. Jensen, J. S. Winston, and V. R. Taylor, *MWR*, 94 (1966), 641, on 500-mb height prediction from satellite infrared-radiation data; W. H. Klein, *JAM*, 5 (1966), 137, on objective prediction of surface temperatures in American cities.

On the possible influence of the moon on the Earth's weather and climate, see: P. J. Visagie, *JGR*, 71 (1966), 3345, on modulation of winter precipitationon the South African plateau by lunar phases;

G. O'Mahony, *QJRMS*, 91 (1965), 196, and G. W. Brier, *QJRMS*, 92 (1966), 169, on rainfall and moon phase at Sydney, Austr.; I. A. Lund, *JAS*, 22 (1965), 24, and 23 (1966), 633, on a statistically significant correlation between sunshine and moon phase in American data. On misleading results that may occur if running means are used, see E. L. Deacon, *JAS*, 23 (1966), 131.

39. See *HSM*, p. 161, and A. Court, *op cit.* (1952), for details of the mathematics.

40. The classic example of the application of this concept was the prediction of upper winds used in the compilation of the charts included in C. E. P. Brooks, C. S. Durst, and N. Carruthers, *Upper Winds over the World*, *MOGM*, vol. 10, no. 85 (1950). See C. E. P. Brooks, C. S. Durst, and N. Carruthers, *QJRMS*, 72 (1946), 55, for details of the method.

41. See H. L. Crutcher and L. Baer, *JAM*, 1 (1962), 522, for the equation.

42. See *HM*, pp. 219–25, for a summary of vector algebra.

43. For details, see W. L. Godson and M. A. MacFarlane, *T* 12 (1960), 259, and 13 (1961), 522.

44. E.g., see C. E. Buell, *JAM*, 1 (1962), 269. For examples of problems met with in the analysis of wind data, see: N. L. Canfield, O. E. Smith, and W. W. Vaughan, *JAM*, 5 (1966), 301, on circumventing limitations of upper wind records; B. F. Bulmer, *MM*, 96 (1967), 89, on 10-minute versus one-hour averages of wind; G. C. Holzworth, *MWR*, 93 (1965), 323 on the relation between average daily surface windspeed and hourly windspeed frequencies; and G. Lupton, *MM*, 97 (1968), 58, on estimation of monthly mean windspeeds from frequency distributions of windspeeds at fixed hours.

On the transformation of wind data from spherical to vector-mean Cartesian coordinates, see J. C. Kaimal and C. N. Touart, *JAM*, 6 (1967), 583. According to the theory of vector error distributions for upper winds, on which see D. N. Axford, *MM*, 97 (1968), 361, the frequency distribution of vector errors is not Gaussian, unlike the frequency distribution of scalar errors.

For examples of the elliptical frequency distributions characteristic of winds in the stratosphere, see R. A. Ebdon, *MM*, 97 (1968), 97. According to J. D. Tracy, *MWR*, 94 (1966), 407, vector errors in forecasting Atlantic hurricanes may be described by probability ellipses. For comparisons of methods for computing vector correlation coefficients, see R. L. Lamberth, *JAM*, 5 (1966), 736, and R. W. Lenhard, *JAM*, 6 (1967), 583.

45. For a review of methods, see I. I. Gringorten, *AGP*, 2 (1955), 57. For an example of least-squares prediction, see M. E. Graves and E. S. Epstein, *MWR*, 95 (1967), 375, on the estimation of 500-mb heights from 300-mb heights and temperatures. For an analysis of linear and curvilinear relationships for the

yearly distribution of rainfall intensity in Britain, see A. L. H. Gameson and R. D. Quaife, *MM*, 94 (1965), 173.

46. See S. Petterssen, *AMM*, 3, no. 15 (1957), 140, for this method.

47. E.g., D. L. Jorgensen, *MWR*, 77 (1949), 31, on the prediction of rainfall in central California, is a classic paper. See the bibliography in Gringorten, *op. cit.* (1955), for other examples.

48. For details of the theory involved in this method, see E. W. Wahl *et al.*, *AFSG*, no. 19 (1952), and I. A. Lund and E. W. Wahl, *AFSG*, no. 75 (1955). For examples of contingency-table predictions, see R. Murray, *MM*, 97 (1968), 141 and 303, for monthly and seasonal precipitation and temperature in England and Wales; see also R. F. M. Hay, *loc. cit.*, p. 278, on prediction of winter temperatures in Britain from autumn rainfall, and P. M. Stephenson, *MM*, 96 (1967), 335, on seasonal rainfall sequences in England and Wales. For a relatively simple method of predictor selection, see W. D. Mount and I. A. Lund, *JGR*, 68 (1963), 3619.

49. See Lund and Wahl, *op. cit.* (1955), for details.

50. G. W. Brier and R. A. Allen, *CM*, p. 841. F. Sanders, *JAM*, 2 (1963), 191. On probability forecasting, see: D. L. Jorgensen, *MWR*, 96 (1968), 887, on combining two probabilities by means of a scatter diagram; E. S. Epstein and A. H. Murphy, *JAM*, 4 (1965), 297, on the use of probability triangles for assessing probability scores, *JAM*, 6 (1967), 1002, on the theory of " hedging " in probability forecasting, and *loc. cit.*, p. 748, on verification of probability forecasts, on which also see F. Sanders, *loc. cit.*, p. 756; see also E. S. Epstein, *MWR*, 94 (1966), 487, on quality control theory for probability forecasts. On the measurement of forecast skill, see I. I. Gringorten, *JAM*, 4 (1965), 47, on prediction of continuous variables, and *JAM*, 6 (1967), 742, on the theory; and see J. G. Bryan and I. Enger, *loc. cit.*, p. 762, on maximization of skill score by use of probability forecasts.

51. For an elementary account of discriminant analysis, see P. G. Hoel, *op. cit.* (1962), p. 179. For the application of discriminant analysis to climatological data, see R. G. Miller, *AMM*, vol. 4, no. 4 (1962), who gives a full statement of the vectors and matrices. On multivariate analysis, see R. H. Jones, *JAM*, 3 (1964), 285, on linear prediction of multivariate time series. On discriminant analysis, see H. R. Glahn and R. A. Allen, *MWR*, 92 (1964), 509, on comparison with scatter-diagram analysis. For later developments from discriminant analysis, see M. J. C. Hu and H. E. Root, *JAM*, 3 (1964), 513, and H. R. Glahn, *loc. cit.*, p. 718, on the use of adaptive logic models in climatological short-period predictions of rain at San Francisco and ceiling height at Washington, D.C., respectively. See also H. R. Glahn, *JAS*, 25 (1968), 23 for canonical correlation and its relation

to discriminant analysis in the prediction of 500-mb heights over the United States.

52. For details, see Miller, *op. cit.* (1962).

53. See J. J. Papas, *JAM*, 1 (1962), 353, for a simple yes-no hail-forecasting scheme. I. A. Lund, *JAM*, 2 (1963), 517, described a system for predicting sea-level pressure changes for periods of less than 12 hours. F. L. Martin *et al.*, *loc. cit.*, p. 508, described a method for predicting the 24-hour movement of North American anticyclones, and K. W. Veigas and F. P. Ostby, Jr., *loc. cit.*, p. 24, explained the application of a moving-coordinate prediction model to North American east-coast cyclones. For other examples of statistical prediction schemes in climatology, see: J. W. Reed, *JAM*, 6 (1967), 360, on upper winds in the tropical Pacific; W. H. Klein, F. Lewis, and G. P. Casely, *loc. cit.*, p. 216, on maximum/minimum surface temperatures over North America by automated multiple-regression technique; W. H. Klein, C. W. Crockett, and J. F. Andrews, *JGR*, 70 (1965), 801, on daily precipitation and cloudiness from 700-mb circulation pattern in the United States; J. R. Borsting and F. F. Sheehan, *JAM*, 3 (1964), 132, on comparison of cyclone-prediction techniques for the northern hemisphere; J. A. Russo, Jr., I. Enger, and E. L. Sorenson, *loc. cit.*, p. 126, on surface winds in the New England–New York area by screening-multiple-regression technique. On the statistical prediction of time-series, see R. H. Jones, *loc. cit.*, p. 45, on linear prediction theory, and p. 285, on linear prediction theory for multivariate time-series.

54. "M. O. discussion," *MM*, 88 (1959), 207. See also M. H. Freeman, *QJRMS*, 87 (1961), 393. See W. C. Spreen, *TAGU*, 28 (1947), 285, for the precipitation study.

55. E. A. Bertoni and I. A. Lund, *JAM*, 2 (1963), 539. J. Glover, P. Robinson, and J. P. Henderson, *QJRMS*, 80 (1954), 602. For examples of probability maps, see R. R. Dickinson and J. Posey, *MWR*, 95 (1967), 347, on snow-cover probability in the northern hemisphere. For a detailed treatment of the problems inherent in the distributions displayed on a map of annual rainfall in central America, see S. L. Hasthenrath, *AMGB*, 15, series B (1967), 201. For anomaly-index maps, which measure the relative degree of local rainfall abnormality, for South Africa, see M. P. van Rooy, *NOTOS*, 15 (1966), 13. For climatological maps of interdiurnal pressure variability in the northern hemisphere, see K. Bayer, *AMGB*, supplement 1 (1966), p. 142.

56. See T. A. Gleeson, *JAM*, 1 (1962), 18, and 2 (1963), 202, on the extrapolation formulae.

57. T. Fujita, *AMM*, 5, no. 27 (1963), p. 138, outlines the procedure. The construction of *wind roses* provides a good example of the difficulties involved in apparently simple cartography in climatology. See: W. Slusser, *WW*, 18 (1965), 260, for an elementary wind-rose map for the United States; L. E. Truppi, *MWR*, 96 (1968), 325, for bias introduced into wind roses by anemometer starting speeds; and C. E. Wallington, *MM*, 97 (1968), 293, for the reduction of observing and procedure bias in wind-direction frequencies necessary before these should be used to construct wind roses. On the accuracy of estimations of area-mean rainfalls, see J. L. McGuinness, *JGR*, 68 (1963), 4763.

58. See W. E. Duckworth, *A Guide to Operational Research* (London, 1965), pp. 138–41, for details. See also M. J. Moroney, *Facts from Figures* (London, 1951), p. 66, on how to speed up calculations. See H. E. Cramer and F. A. Record, *JM*, 12 (1955), 146, for details of a simple quadratic-filter approximation to the power spectrum.

59. The desk calculator is, of course, quite distinct from an electronic computer or a punched-card sorting machine. See S. H. Hollingdale and G. C. Toothill, *Electronic Computers* (London, 1965), for details of the principles of operation of calculators and computers. *MC*, pp. 335–53, contains an account of the operation of punched-card sorting machines in climatology. On quality control, see A. Bleasdale and A. B. Farrar, *MM*, 94 (1965), 98. See also R. P. Waldo Lewis and B. Golding, *loc. cit.*, p. 109, for a discussion of errors in the traditional method of computing area means of monthly and annual rainfalls. On the computer analysis of aerological soundings, see N. E. Prosser and D. S. Foster, *JAM*, 5 (1966), 296.

60. F. V. Brock, *JAM*, 2 (1963), 755. F. V. Brock and D. J. Provine, *JAM*, 1 (1962), 81. H. Moses and F. C. Kulhanek, *JAM*, 1 (1962), 69. For a detailed discussion of the statistical analysis of each climatic element taken individually, see the standard textbooks, especially *MC*, pp. 17–60.

On computers in climatology, see: M. H. Freeman, *MM*, 97 (1968), 209, on a general-purpose computer program—METO—for statistical computations, simpler than either ALGOL or FORTRAN languages; P. M. Wolff and W. E. Hubert, *BAMS*, 45 (1964), 640, on computer economics; and F. W. Burnett, *WW*, 18 (1965), 196, and J. S. Sawyer, *Brit. J. Applied Physics*, 15 (1964), 379, for general accounts of the role of computers in meteorology.

On aids to statistical analysis, see: D. T. Acheson, *JAM*, 7 (1968), 548, on the theory of arithmetic averaging by means of electrical circuits incorporated in the measuring instruments; J. A. Turner, *JAM*, 7 (1968), 714, on estimation of standard derivation of wind direction from direct observation of a sensitive wind vane; and J. I. P. Jones, *Brit. J. Applied Physics*, 15 (1964), 467, for details of an electric circuit that, coupled to a new type of anemometer, provides continuous computation of frequency spectra of wind gusts, both along and at right angles to the mean wind direction.

CHAPTER 3

Climatological Models

Man uses many categories of phenomena or of behavior in his efforts to describe and understand what he believes to be reality. When he describes an object or event in terms of categories, he is describing a model of the object or event, and not the phenomenon as it actually is. It is easy to see that such a method of working must be common in fields such as physics, which deal with concepts not met in everyday life. It has not been until comparatively recently that the importance of models in such a science as geography has been realized.

Subjective geographical models are very common. These are the mental pictures of an unvisited place or region, which we build up in our minds from written accounts, sketches, paintings, photographs, and maps. We know that these subjective models are often far different from the reality. Scientific geographical models, on the other hand, must be definite, not woolly or ambiguous, and must be reproducible, so that they convey the same picture or idea to all users.

When a geographer produces a map by making observations of some element in the field at selected points, and then interpolating between the points on the assumption that the element has a uniform rate of change between adjacent points, he is making a scientific geographical model. The construction of a local relief map from surveyed spot heights will produce an accurate model, because the interpolations can be checked by eye in the field, and any gross errors discovered. A geological map is a good example of a less accurate geographical model, since it must be inferred from relatively few rock exposures.

The principle behind most geographical models is that, if we have observations of an element at point *A* and point *B*, then conditions along line *AB* will be most correctly specified by assuming the simplest possible rate of change with distance of the element along *AB*. This principle is not based on a law of nature, but on the famous maxim, known as Occam's Razor, of William of Occam (died about 1349), who said, "It is vain to do with more what can be done with fewer." Unfortunately, in climatology, as in all other disciplines, the simplest and most obvious assumption is rarely true. Occam's Razor does apply, but the fewest and simplest assumptions that are adequate to explain real phenomena are still many and complex.

Atmospheric science must, of necessity, describe the phenomena it studies in terms of categories. Five main categories may be recognized, the first being large-scale weather systems, including both primary (planetary) and secondary systems. The planetary systems include the major cyclones, anticyclones, and long-wave phenomena with a scale of 1,000 to 10,000 km; the secondary systems comprise the smaller cyclones and anticyclones with sizes between 100 and 1,000 km. All these phenomena are more or less horizontal, and are defined in geostrophic terms. The second category covers mesoscale weather systems, which are between 10 to 100 km, and often depend on topography. Third come small-scale weather systems on a scale of 1 to 10 km; these include showers, cumulus clouds, and thunderstorms, all more or less vertical phenomena and due to convection. Discontinuity systems, with a vertical scale of about 1 km, comprising fronts, inversions, and tropopauses, make up a fourth category. The final category embraces surface boundary systems—fogs, local winds, land and sea breezes, orographic and low turbulence clouds—with a vertical scale of 1 km.[1]

Excepting the fifth category, all these weather systems emerge from their environment, undergo a period of development, and ultimately merge into their environment again. Their development phase is self-generating, bound up with instability or divergence. They thus differ markedly in character from any other geographical phenomena.

One way of dealing with such phenomena is simply to observe them at fixed points on the Earth's surface, at fixed times, and then to summarize statistically the resulting figures. Maps or sections drawn in this way, showing average distributions over long periods of time, form *climatological mean models* of the phenomena. It is difficult to decide exactly what a climatological mean model shows, apart from a cartographic portrayal of certain statistics: perhaps it is best regarded as a historical document illustrating some aspects of atmospheric behavior during some given period of time, and according to some convention.

The mapping of rainfall distribution is an excellent example of the value of climatological models. Rainfall maps have in the past been constructed subjectively, on the assumption that precipitation distribution corresponds fairly closely to the pattern of orography. This assumption may be true on a very broad scale, but modern studies of precipitation dynamics indicate that it does not necessarily apply on middle or local scales.

If a rainfall map for a small area is constructed on the usual subjective basis, different distributions result, depending on the density of measuring points (see Figure 3.1). Even to increase the number of measuring points to almost fantastic proportions does not ensure that the true distribution of rainfall will be recorded, because of the serious inherent disadvantages of rain gauges discussed in Chapter 1.

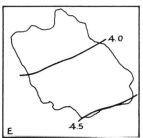

FIGURE 3.1.
Rainfall in central Illinois (after Huff and Street). A, 2.4 square miles per gauge. B, 4.75 square miles per gauge. C, 4.5 square miles per gauge. D, 19.5 square miles per gauge. E, 225 square miles per gauge, which is the standard United States Weather Bureau rain gauge density.

A reasonable compromise solution is to use *statistical models* of rainfall distribution. Such a model involves determining statistically the empirical relation between the recorded rainfall at specific locations and various topographic parameters at these locations. Rainfall distribution in the areas between the measuring points may then be calculated from the topography, on the assumption that the same statistical correlations exist as have been established for the rain-gauge sites. For example, Figure 3.2 gives the rainfall distributions for 1958 and 1959 over the Ystwyth Catchment in Wales, making use of two topographic parameters (altitude and exposure), the statistical model being the linear regression equation (see Appendix 3.1).

For maps of rainfall distribution during shorter periods of time (a month, a week, or less, for example), statistical models cannot be used, because the relevant topographic factors will not necessarily have a constant relation to precipitation. Also, atmospheric factors such as pressure, temperature, vertical motion, convergence, and divergence must then be considered. Accurate cartographic interpolation of rainfall amounts taking all these factors into account becomes very complex indeed.

Some examples of climatological mean models are given in Figure 3.3.[2] These models are based on observed behavior, show some degree of stability, and are purely descriptive. Since climatological mean models do not represent real phenomena of nature, inquiring into their causes is a waste of effort, and trying to use them to make

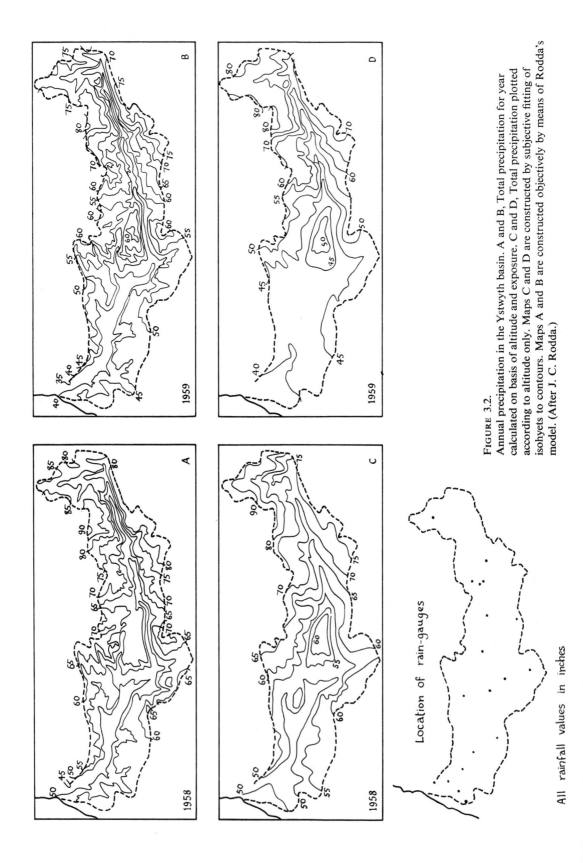

FIGURE 3.2.
Annual precipitation in the Ystwyth basin. A and B, Total precipitation for year calculated on basis of altitude and exposure. C and D, Total precipitation plotted according to altitude only. Maps C and D are constructed by subjective fitting of isohyets to contours. Maps A and B are constructed objectively by means of Rodda's model. (After J. C. Rodda.)

Location of rain-gauges

All rainfall values in inches

predictions is cheating, scientifically, although it is justified if the predictions are needed urgently, and correct answers are obtained.

Apart from climatological mean models, atmospheric science makes use of demonstration models, experimental models, and mathematical models. *Demonstration models* are useful as educational devices, in the broadest sense, but have no other application. *Experimental models* are very valuable in the discovery and testing of physical theories about the behavior of features of weather and climate. *Mathematical models* are essential for the accurate construction of climatic maps, as well as for scientific explanation and prediction in general.

Demonstration models are sometimes purely display techniques, but sometimes are true working models.[3] For example, a smoke-fog model made at Kew was demonstrated at a pollution exhibition held at the Royal Sanitary Institute in 1953. It consisted of a small wind tunnel, with a fan at one end drawing air through at a constant speed of 6 inches per second, and electric heaters at the other end to heat either the upper or lower air layers, so that temperature stratification could be set up. Smoke produced by burning sawdust in a small can was introduced into the tunnel by chimneys at the heater end. With only the upper heater switched on, foggy inversion conditions were simulated. With only the lower heater switched on, convection currents carried the smoke aloft, simulating a clear day with steep temperature lapse-rates.

Classic demonstration models of cyclones and tornadoes were made by Wilcke and Dines. Wilcke in 1780 at Stockholm rotated a thick steel wire a few inches long, mounted eccentrically, about the central axis of a large cylindrical water container. A vortex was generated along the central axis, with an outflow at the wire end and an inflow at the opposite end of the container. The motions were observed by placing burnt lime in the water.

W. H. Dines in 1896 made a model illustrating the formation of a tornado cloud.[4] A folding screen, consisting of six pieces of window glass, each 2 feet by 18 inches, arranged in two sets of three, and fastened together by twine, was surmounted by a horizontal wooden panel with a 7-inch diameter circular hole in its center. A hand-driven ventilating fan could be placed in the hole, or a portion of stove pipe with a gas jet burning in it instead of a fan. A shallow vessel of water was placed in the center of the screen and at its lower end, and was heated by a spirit lamp to provide water vapor for cloud formation.

When the fan was rotated, an upward current of air was produced in the center, and a cloud formed. By suspending a piece of cottonwool by a thread, and bringing it near the cloud, a distinct rotary motion in the cloud could be demonstrated, increasing in intensity as the center was approached. Once inside the cloud, the cottonwool was rapidly carried upward, indicating the presence of a strong updraft. If soap was dissolved in the water, the resulting bubbles were drawn in toward the cloud along a spiral path, carried up into the cloud in a heap, and then thrown violently outward by the centrifugal force within the whirl. The cloud column was distinctly hollow, indicating the great decrease in atmospheric pressure found in the center of tornadoes; when Dines made the experiment, it was not known from observation either that low pressure occurs at the center of a tornado or that its cloud column is hollow.

If the fan was rotated slowly, a cloud formed; if the rate of rotation was increased, producing a stronger updraft, the cloud disappeared but formed again soon afterward with greater intensity. If the speed of rotation of the fan was then reduced, the cloud disappeared, then reappeared again after a longer time than the first reappearance

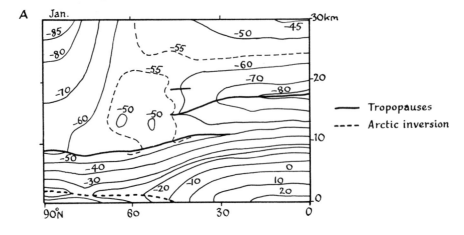

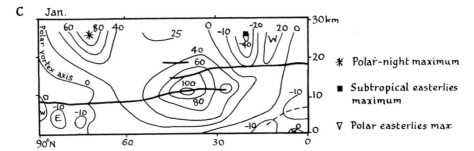

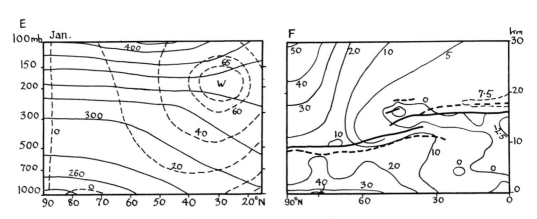

required. This corresponds with the variation in intensity of tornadoes as they move across the Earth's surface.

On occasion, "demonstration models" occur naturally. For example, model depressions have been observed on the Thames, near the stern of a moored ship, in very light westerly winds. An irregular layer of mist a few inches thick, covering the water, rendered the air movements clearly visible. Two streams of air, moving along the two sides of the ship, met at the stern and then continued side by side with little mixing, forming a "front." A wave two or three feet long was sometimes seen to appear on this front, and developed into a counterclockwise rotation, which finally broke away as an eddy. The mist towered two or three feet above these eddies, indicating that these "depressions" were the result of surface convergence.[5]

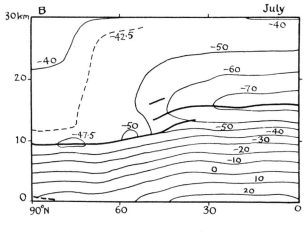

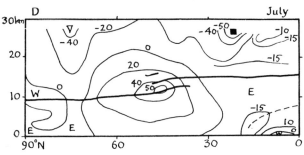

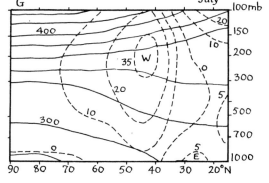

FIGURE 3.3.
Examples of climatological cross sections. A, B, Mean temperatures (°C) along longitude 80°W. C, D, Mean geostrophic zonal winds along 80°W. E, Potential temperatures (———) and zonal geostrophic winds (– –) averaged over all longitudes. F, Annual range of mean temperature (°C). ——— July tropopause – – – – – – January tropopause. G, Potential temperatures (———) and zonal geostrophic winds (– –) averaged over all longitudes. (After A. Kochanski and J. G. Moore.)

Models of this type sometimes provide a clue to causation. In the Thames model depressions, for example, the eddies had a life cycle of five or six seconds, and could be anticipated by several seconds. The first sign of a future wave proved to be movement of the mist toward the "front" from the south, the front appearing as a gently waving line of deeper, denser mist up to 20 yards long. Convergence often began three to four yards to the south of the front (not along it, as one would expect from the Polar Front theory), and no waves developed until the convergence reached the front.

Unlike demonstration models, experimental and mathematical models enable quantitative measurements and deductions to be made, and the greater part of this chapter will be devoted to them.

Experimental Models of Phenomena of Weather and Climate

Laboratory experiments offer many advantages to the climatologist. Most processes he studies are too large-scale for him to control in the field, but he can control measurable analogies to them in the laboratory. Experimental models provide vivid physical impressions of phenomena normally beyond the range of his experience, and they can indicate the most profitable times and places for carrying out field observations. They enable him to investigate quantitatively processes that would be too slow or too large for observation under ordinary conditions—climatic change, for example. Sometimes, if the mathematics involved in a problem appears to be intractable, it can be simplified by performing an experiment. However, the atmosphere is very complex, and it is very difficult, if not impossible, to design a model that will perfectly represent all the contributory factors.

The basic problem in designing an experimental model to represent any problem in meteorology or climatology is that of dimensional analysis. Both prototype and model must be dynamically and geometrically similar. According to the Π theorem of dimensional analysis, if n variables involved in a function require m dimensional characteristics for their expression, then the function can be rewritten in terms of $n - m$ dimensionless groups of these variables (see Appendix 3.2). Any geometric problem can be expressed in terms of ratios of lengths, and model and prototype (and the flow patterns involved in them) will be geometrically similar if corresponding length ratios have the same numerical values. In a kinematic problem—for example, a model of airflow in the free atmosphere—corresponding length and time ratios (i.e., velocities) must be similar in model and prototype. In a dynamic problem—for example, a model to illustrate apparent forces on a rotating Earth—corresponding length and time ratios and either mass or force ratios must be similar.[6]

The corresponding ratios will be dimensionless numbers, and in airflow problems, five of these numbers are of importance: the Reynolds number Re, the Richardson number Ri, the Froude number F, the Mach number M, and the Rossby number Ro.[7] Model and prototype will have similar flows if corresponding independent parameters in the general equation for similitude, of the form

$$E = f \left(\frac{b}{a}, \frac{c}{a}, \ldots F, Re, M \right),$$
[3.1]

have the same numerical values. In this equation, E represents the *Euler number* (see Appendix 3.3), which describes the flow characteristics of either prototype or model atmosphere, and $a, b, c, \ldots$ are a series of lengths representing the geometry of the terrain. In other words, although the analytic form of Equation 3.1 may not be known, provided that $\frac{b}{a}, \frac{c}{a}, \ldots, F, Re$, and M are numerically the same in both model and prototype, the flows will be comparable.

The *Reynolds number* is named after Osborne Reynolds, who in 1883 demonstrated from the flow of liquid in a narrow tube that laminar flow changes to turbulent flow when the ratio of the product of the velocity and density of the liquid and the diameter of the tube to the viscosity of the fluid exceeds a certain value. Re is defined by the equation

$$Re = \frac{Vd\rho}{\mu}, \qquad\qquad [3.2]$$

where V, ρ, and μ are the velocity, density, and kinematic viscosity of the fluid and d is the diameter of the tube. For water, the critical value of Re is approximately 2,000, depending on the shape of the entry pipe to the narrow tube. For the flow of liquid around a sphere or cylinder, the critical value of Re, separating laminar from turbulent flow, is in the range 2×10^5 to 4×10^5.

Apart from its significance in the design of models, Re is also important in placing limitations to the accuracy of balloon-ascent computations, and in deducing the maximum possible size of hailstones (see Appendix 3.4).

The Reynolds number applies when turbulence is generated in a fluid purely by its motion over solid bodies. If the density or velocity of the fluid varies, as they must if the atmospheric model is realistic, turbulence may also be generated by spontaneous action. This is denoted by the number Ri introduced by Richardson in 1920. It indicates the effect of vertical lapse-rate variations on turbulent convection.

The *Richardson number Ri* is defined in effect as the rate of loss of potential energy divided by the rate of gain of (turbulent) kinetic energy of the atmosphere (see Appendix 3.5). If Ri is less than unity, turbulence will tend to increase. This corresponds to an atmosphere thermally stratified so that its actual lapse-rate exceeds the adiabatic. In this unstable atmosphere, a vertically moving bubble of air will be accelerated by buoyancy forces, since at any level it will be lighter than its environment if moving upward and heavier than its surroundings if moving downward. Hence turbulence will be easily generated.

If Ri exceeds unity, turbulence will tend to decrease. This corresponds to an atmosphere in which the actual lapse-rate is less than the adiabatic. In such a stable atmosphere, an air bubble moving either upward or downward will be slowed down by buoyancy forces.

Whether or not a model will accurately simulate local weather effects depends on whether or not Ri in the model is approximately the same as Ri in the atmosphere. The difficulty is that atmospheric values of Ri cover a considerable range, the critical values varying from 0.04 to 1.0 in the lower atmosphere and from 10 to 1,000 in the free atmosphere.* Other values are also important; for example, clear-air turbulence often occurs with very low values of Ri (from 1 to 5) in the free atmosphere if there is large vertical wind shear.

The Reynolds number represents the effect of viscosity, the *Froude number* the effect of gravity, (i.e., the ratio of inertia forces to buoyancy forces), and the *Mach number* the effect of compressibility. The *Rossby number* takes into account the rotation of the Earth, and is more important than either F or M as a modeling criterion. (See Appendix 3.6.)

EXPERIMENTAL MODELS OF CLOUD FORMATION

In certain respects, cloud models are the easiest of all meteorological models to design. Classic work involved the recognition of Bénard cellular convection.[8]

Bénard cells, polygonal in form with ascending motion in their centers, downward

---

* The free atmosphere is not just the upper atmosphere, since part of the lower atmosphere can also be free, i.e., unaffected by conditions at the Earth's surface.

motion at their outer margins, and outward and inward motions at their tops and bottoms, respectively, develop in unstable layers of fluid in which there is no one general direction of motion. They may thus be expected to occur on occasion in the atmosphere, and there is radar evidence they do.[9]

By pouring a highly volatile liquid that contains tracers (e.g., benzene with thin gold flakes, as in cheap gold paint) into a small vessel to a depth of a quarter-inch or so, one can observe Bénard cells. Evaporation of the liquid leads to rapid cooling of its upper layer, resulting in marked instability. The cells forming apparently have clear liquid at their centers and outer boundaries, because the thin flakes of gold arrange themselves parallel to the motion, and therefore are visible at the top and bottom of each cell, but not at the center or periphery.

To form convection cells in air, a special chamber may be constructed, such as that developed by Chandra at Imperial College in 1938. This consisted of a shallow box containing parallel glass rods, on which wire was wound; on top of the box was placed a smooth metal plate, which could be heated by passing an electric current through the wire. A glass vessel was placed on top of the plate, and filled with water to such a depth that there was no likelihood of any great rise in temperature in the upper water layers. Layers of felt at the sides of the chamber prevented rapid heat leakage in or out. Cigarette smoke, injected by a two-way pump, rendered the air motion visible.

No motion was observed until the temperature difference between the top layer of the water and the base plate, as measured by platinum resistance thermometers, exceeded a certain value. If the water depth was 6 mm or less, no cells formed whatever the vertical temperature distribution. If the water depth was 10 mm, cells formed when the temperature difference was 11.4°C or more.

The effect of shearing motion in an unstable atmosphere was demonstrated when a long glass plate, forming the top of the chamber, was moved across it horizontally by means of a small electric motor. This resulted in a change in the speed of the water motion with depth, but no change in direction, analogous to a wind shear. No convection was found unless the temperature difference was above the limiting value with the top at rest, whatever the rate of shear. When this critical value was exceeded, ordinary convection cells resulted, with a low rate of shear; with a high rate of shear, the smoke took on the form of rolls. These rolls were longitudinal (along the direction of shear) for a relatively high rate of shear, and transverse (perpendicular to the shear direction) for relatively low rates of shear. Adjacent transverse cells rotated in opposite directions.

Atmospheric convection cells exhibit motions in the reverse sense, usually, to those observed in the laboratory; Bénard convection cells in the atmosphere show descending motion in their centers, of much smaller relative speeds. Therefore real clouds should form in regions analogous to regions of descending motion in the experimental models. Apart from this, the models are very realistic. Convectional clouds in the real atmosphere are aligned parallel to the isotherms when the latter crowd together with large wind shears; when the isotherms are farther apart, as occurs with small rates of shear, the clouds form approximately at right angles to the isotherms. Entrainment is difficult to simulate in the models.

Experiments with large air bubbles in water illustrate quantitatively many features of cumulus development. Normal air bubbles in liquids are small, 0.1 inch or so in diameter, and nearly spherical, their shape determined by surface tension. They are clearly unsuitable for use as analogies to cumulus convection. The shape of larger bubbles, however, is determined by buoyancy forces, not by surface tension, and the analogy is more realistic. Bubbles of 2 inches or so have a spherical cap and a ragged rear edge, very similar to cumulus bubbles.

At Imperial College, bubbles of this type were released into a tank of static water 4 feet long, 2 feet wide, and 3.5 feet deep.[10] The bubbles were created by partly immersing a spherical cup, pivoted about a horizontal axis through its center, and filled with a liquid denser than water up to the level of the water surface. The cup was then quickly overturned, creating cumulus convection in reverse, and the subsequent behavior of the resulting dome of heavy liquid was recorded on movie film.

For quantitative work, the excess density of the bubble over its surroundings must be known. A salt solution was used to make the heavy liquid, with dye or white precipitate added to make it easily visible. The bubbles were thus able to mix readily with their surroundings. The resulting motion was fully turbulent, viscous forces being unimportant, and all the resulting velocities were produced by buoyancy forces. Thus the ratio of inertia forces to buoyancy forces (i.e., the Froude number) was the same as in bubble convection in an otherwise calm atmosphere, and dynamic similitude was achieved. The only requisite was that the ratio of the extreme density values used should not be so large that the inertia forces became overly variable. This condition was satisfied if the maximum excess density of the bubbles over their surroundings was less than 15 per cent.

With water of uniform density, the bubbles were observed to have no wake, and to gradually turn inside out. The mixing responsible for the latter effect occurred mainly on the advancing (i.e., descending) cap of heavy liquid. The mixed material then flowed round to the back of the bubble, entering at its rear, the whole cycle then repeating itself. The result was that after every cycle the bubble became larger, more dilute, and slower moving. The bubble did not become eroded, as do real cumulus bubbles, but remained the same shape and grew by dilution.

With stratified water, corresponding to a stably stratified atmosphere, the bubbles became eroded, but remained approximately constant in size as they sank through their surroundings. This meant that the rate of erosion must have been equal and opposite to the rate of growth by dilution.

After a time, the motion inside each bubble was found to resemble that of a vortex ring. In the real atmosphere, the development of vortex rings is inhibited by the condensation taking place in cumulus towers, which continually create buoyancy. Another difference from the development of actual cumulus bubbles is that in the atmosphere the rate of erosion of a bubble is increased by evaporation. The evaporation of cloud droplets extracts latent heat from the air, producing negative buoyancy, so that a sinking shell of air heavier than its environment surrounds the cloud and increases its rate of erosion. Obviously, in the laboratory experiment it is impossible to simulate this sinking shell around the cloud, because of gravitational influence.

EXPERIMENTAL MODELS OF AIRFLOW OVER BARRIERS

The movement of air over orographic barriers, or over small-scale features such as windbreaks, has been extensively studied by means of wind tunnels and wave tanks.

Air currents for experimental work may be produced by a series of blowers directed as required in an open hall, but for meteorological purposes the resulting airflow is not uniform enough, and the turbulence is on too large a scale. (Despite these disadvantages, large models of mountains in hangars in the United States during the Second World War, the airflow being generated by aircraft propellors, proved to be very useful.) A wind tunnel with a uniform duct is therefore necessary, provided with a bell inlet, air being drawn through it by an exhaust fan or fans at the downstream end. The airflow velocity may be measured by means of pitot tubes (see Appendix 3.7) for speeds greater than 10 feet per sec or hot-wire anemometers for speeds of 5 to 10 feet per sec, and the air temperature by means of thermocouples. The air motion can be made visible by injecting chemical smoke (e.g., from titanium tetrachloride) or an oil fog into the tunnel, and using a dark background with top lighting slightly to the rear.

One of the earliest scale-model wind-tunnel experiments was made by Abe for a study of wind structure over Fujiyama.* A 1:50,000 scale model of Mt. Fuji was used for a series of experiments from 1929 to 1941, in which the pattern of airflow was made visible by miniature wind vanes, or by the use of smoke from an incense stick. Abe considered that he had achieved similarity between model and prototype Reynolds numbers by equating the molecular viscosity (a fluid property) in the model with the estimated eddy viscosity (a flow characteristic) in the prototype, on the apparently reasonable assumption that molecular motion on a small scale is analogous to eddy motion on a large scale. Using this assumption, a 22 miles per hour flow over Mt. Fuji and a 2 miles per hour flow in the wind tunnel both gave an *Re* value of 4,000. Most of Abe's model airspeeds resulted in almost wholly viscous flow, for two reasons: first, aerodynamic boundary-layer theory must also be taken into account; second, the ratio between kinetic and buoyant energy in both model and prototype must be considered if dynamic similitude is to be achieved.

Three types of airflow across a topographic barrier are theoretically possible, in each of which different dimensionless numbers must be used to obtain similitude.[11] In aerodynamic flow, there is no static stability; i.e., a statically neutral (adiabatic) layer exists next to the ground. Gravity does not enter into this type of flow, and the Reynolds number must be used as the critical modeling criterion. In barostromatic flow, vertical displacements are important, and the appropriate number is the *static stability number*, *S* (see Appendix 3.8). For airflow over a large barrier, the effect of the Earth's rotation must also be taken into account. This is geostrophic flow, and the modeling criterion is the Rossby number.

With aerodynamic flow of a uniform airstream over an orographic barrier, the amplitude of the resulting disturbance decreases with height; with barostromatic flow, it does not decrease with height. The static stability of an airstream extends the effect of an orographic barrier to greater heights than occurs with an adiabatic lapse-rate, and hence the stability number *S* must have identical values in both model and prototype; *S* is more often important than *Re* in wind-tunnel experiments. Its effect is to necessitate very low rates of airflow in the model (of about 70 cm per sec), which means that the control of the vertical profiles of both velocity and temperature becomes difficult. Also, of course, stability effects mean that the conditions that obtain when lee waves occur in the atmosphere are very difficult to model. Normal wind-tunnel flow is aerodynamic, representing adiabatic lapse-rate conditions. With barostromatic

---

* M. Abe, *Bull. Central Meteorol. Observatory* (*Tokyo*), vol. 7, no. 3 (1929).

flow, the atmospheric stratification is stable, and if an adiabatic layer (e.g., a layer of nimbostratus) is sandwiched between two stable layers, the airflow over the orographic barrier will be unsteady and contain overturning, not be steady and laminar as in the wind tunnel.

Boundary-layer separation introduces a great difficulty in wind-tunnel modeling. The pattern of separation varies considerably for different topographic configurations, and as the value of *Re* increases, at least four different types of airflow may occur.

At very low values of *Re*, the boundary layer is very deep, the viscous layer extending a great distance from the topographic surface. Under these conditions, the low-velocity airflow near the surface is quite stable. As the value of *Re* increases, stable eddies develop behind major irregularities of the topography, and the viscous boundary layer decreases in thickness. At moderate values of *Re*, the boundary layer becomes thin, but the flow remains laminar; eddies still form, but more irregularly, and soon merge into the general flow. At very high values of *Re*, the boundary layer becomes turbulent, and eddies form only in the region of the more angular surface obstructions.

The actual range of *Re* values appropriate for these different types of flow depends on the irregularity of the topographic boundary surface. In general, the more irregular the topography, the less important are viscous effects—at all but very small scales and very low air velocities—and the easier it is to obtain similitude. The more irregular the boundary surface becomes, the more the separation pattern is determined by the geometry of the surface. Therefore with the moderate to high Reynolds numbers characteristic of wind-tunnel airflows, if geometric similarity of topographic surfaces is achieved, similitude will result if the topography to be modeled is irregular. Perfect similarity is difficult to obtain for gently undulating topography, because extremely high air velocities are required to maintain *Re* at a sufficiently high value. However, the allowable increase in air velocity is limited, because as the Mach number approaches unity, marked compressibility effects appear. As the actual terrain becomes more streamlined, the angularity of the model terrain may be increased by distorting its vertical scale, but serious changes in the airflow pattern may be then introduced.

A 1:5,000 scale model of the Rock of Gibraltar made by Field and Warden in 1929 successfully solved some problems of the lee effects in the Bay when the Levanter is in operation that had caused many accidents to seaplanes.[12] Airflow patterns were studied by means of flags and streamers. The flags, consisting of short fine fibers of equivalent length 800 feet, were placed on wires at equivalent heights up to 7,000 ft, spaced at equivalent distances of 2,000 ft. The streamers, long strands of wool, were moved about until the major eddies were located. To study the effects of different wind directions, the model was skewed around.

The Gibraltar model indicated that at all levels up to at least 3,000 ft, an area of eddies and vortices can be expected to occur for 1.5 miles west of the Rock when an east wind is blowing, succeeded further west over the Bay by a wide region of very turbulent winds. This prediction was borne out remarkably well by actual observation: out of 360 double-theodolite pilot-balloon soundings, only 24 showed conditions different from the expected ones, despite a wide variation in the actual wind with both height and time.

Shelterbelt investigations have provided some of the most successful applications of wind-tunnel models.[13] Two effects have been observed in these experiments: the effect of airflow on (a) the pattern of snowdrift and soil (especially sand) erosion, and on (b) the ratios of horizontal air velocities to windward and leeward of model wind-breaks. Not only windspeeds, but also evaporation and transpiration, have been studied by this means.

The microclimatic conditions associated with shelterbelts and windbreaks fluctuate considerably in nature, and it is very difficult to site instruments in the field so as to minimize or isolate any shelterbelt feature such as height, breadth, or degree of penetrability. Therefore investigations with model shelterbelts in wind tunnels, where atmospheric conditions can be easily reproduced and controlled and different types of shelterbelt can rapidly be tested, provide an excellent basis on which to plan subsequent field investigations.

The resistance to the airflow experienced by a topographic obstacle such as a wind-break, shelterbelt, hedge, wood, house, or factory depends on both the rate of wind shear at the surface boundary, and the boundary-layer separation and consequent development of a turbulent wake. In nature, the boundary layer is fully developed and exceeds the height of the shelterbelt in thickness. In a wind tunnel, the boundary layer is mostly turbulent, but its depth is less than the height of the obstacle unless long test sections are employed.

Differences between model and real shelterbelts are therefore largely due to differences in the respective degrees of turbulence. The wind-tunnel model usually exhibits a much more extensive sheltered area in the lee of the shelterbelt than is found in nature, but this can be corrected by increasing the intensity of surface roughness in the tunnel. A rule found by experience is that model and prototype are qualitatively similar if distances measured from the model shelterbelt are multiplied by a factor K, the value of which depends on the ratio of Re to the height of the barrier; K often has a value of approximately 0.5.

Airflow velocities in these model experiments are based on pressure measurements. Pressure (p) and velocity (v) in fluid are related by the expression

$$p = \tfrac{1}{2}\rho v^2,$$

where ρ is its density. A body immersed in a moving fluid is acted on by two types of pressure: a *static pressure*, acting equally in all directions when the fluid is stationary, but persisting with some change in magnitude when the fluid is in uniform unacceler-ated motion; and an additional, *kinetic pressure*, set up by the impact of the molecules in the moving stream of fluid. The sum of the static and impact pressures is termed the *total head* of pressure (p_T), and the pressure due to the motion of the stream alone is the *velocity head* of pressure (p_o). The difference between p_T and p_o forms the *differential pressure*:

$$p_T - p_o = \tfrac{1}{2}\rho v^2.$$

The total head of pressure may be measured by pointing an open-ended tube upstream in the fluid, its other end being connected to a simple pressure gauge. The differential pressure is measured by a combined pitot-static tube, consisting of a double tube whose arms are connected to the two arms of a differential manometer. In very small wind tunnels, total-head tubes are used more often than pitot-static tubes, since the latter have to be very small. Usually, therefore, the pressures are sensed by means of small total-head tubes constructed from hypodermic tubing with an external diameter of about 1 mm. The static pressure then has to be found at some other point in the wind tunnel. The actual measurement of the pressure head is most conveniently done by means of Prandtl manometers. Both vertical and inclined manometers have been used: the latter type has a sensitivity of 0.002 inch of water at an inclination of 5° (see Appendix 3.9).

Details of some actual wind tunnels illustrate the possibilities of this type of work.

Finney used a test section 10 feet long, and with a 2-feet square cross section, in a wind tunnel to determine the effect of snow fences. Airflow velocities up to 45 miles per hour were produced by means of a 5-hp electric motor driving a 3-feet diameter propellor, and a small pitot tube was used to trace the resulting eddies. The tunnel was lined with coarse sandpaper, and flaked mica and fine sawdust were employed to indicate the snowdrift patterns forming in the lee of model snow fences built on a scale of 1 inch to 2 feet.

Nøkkentved in 1938–40 and Jensen in 1950–54 used open-circulation wind tunnels, driven by a 30-hp motor giving a velocity of up to 32 m per sec. Pitot-static tubes, one horizontal branch being 9 cm long, the other much smaller, recorded the pressures. Nøkkentved lined the tunnel floor with coarse paper and sawdust to show the pattern of snow drifting behind barriers 5 cm high and of varying degrees of penetrability. The variation was achieved by constructing barriers of one or two rows of spruce twigs, 1 cm apart, the individual twigs in a row being separated by 1.5 cm, or by using sheets with 4-mm diameter perforations. Measurements of pressures were then made to leeward of the barriers up to a distance of $27h$ (h being the height of the barriers), at distances of $0.16h$, $0.5h$, $1.1h$, and $1.5h$ above the tunnel floor. The results were then expressed as percentages of an undisturbed velocity of 10 m per sec at the same height.

Jensen employed model shelterbelts in the form of screens composed of vertical palings, horizontal cylindrical rods, and rectangular lattice. Measurements were made down to a distance of 1 cm ($0.2h$) above the tunnel floor, extending to a distance of $70h$ downwind from the barriers. The results were then expressed in terms of relative shelter effect, i.e., the reduction in velocity, due to the presence of the screen, as a percentage of the velocity of an unobstructed air current at the same point. Jensen also investigated the probable form of the boundary-layer over a wood by mounting corrugated paper (with corrugations 3 mm high and 8.8 mm long) on a horizontal flat plate 1.2 cm thick. The resulting model "woods" then represented widths of $20h$ to $205h$, and extended from 30 to 310 cm in the direction of the wind, occupying the entire width of the tunnel. As before, measurements were taken to a distance downwind of $70h$. In the same series of experiments, Jensen measured the effect of wind velocity on transpiration. He placed 20-cm-thick turves of clover and grass in boxes of zinc sheet, illuminated them artificially 12 to 16 hours per day, and periodically exposed them in the wind tunnel. They were weighed every so often to determine the water loss by evaporation, and the leaf area of each sample was measured after each experiment to assess the transpiration rate.

Blenk in 1952–53 used a test section 2.5 m long and 0.6 m wide, measuring pressures at heights of $0.5h$ and $0.17h$ above the tunnel floor by means of a 4-mm diameter pitot tube. Measurements were taken along a line extending a distance of $30h$ to both leeward and windward of the model, results being expressed as the ratio of the velocity determined at a given point with the model in position to the velocity at the same point in an unobstructed tunnel. The latter velocity was about 40 m per sec. The introduction of undulations in the form of a double sine curve along the tunnel floor enabled the effect of variations in the general topography to be studied. Blenk also measured the effect of airflow variations on evaporation by passing air from a jet mounted 60 cm above the ground over three cups of damp soil, sunk one behind the other in a slab level with the ground. The loss in weight of the soil after an exposure of 50 min to the airflow indicated the rate of evaporation. Model fences were then placed in front of the cups, and the resulting relative reduction in the rate of evaporation measured.

Woodruff and Zingg in 1952–53 introduced aluminum powder into their wind tunnel, and illuminated it so that the streamlines became visible. They employed a test section 12 feet long, commencing 40 feet downwind from the fan, and made measurements by means of four pitot tubes that could be lowered from the roof of the tunnel. The model barriers representing trees and shrubs were to a scale of 1 inch to 5 feet, and consisted of cedar twigs fitted into short lengths of 0.25 -inch diameter aluminum tubing mounted in a plywood base. The trees and shrubs were placed 36 to a row, individual trees being 1 inch apart, in rows spaced 2 inches apart, and their maximum height was 6 inches. A constant airflow velocity of 31 miles per hour was maintained at a height of $2h$ above the trees, and horizontal velocities were measured at 12 heights ($0.1h$ to $3.1h$ above the tunnel floor) and 23 locations, extending from $2h$ windward to $23h$ leeward of the models. Velocity profiles were then determined by plotting rectangular coordinate graphs of z/h against U/U_o, where z is the height above the datum level at which the velocities are measured, h is the height of the tallest tree in the models, and U, U_o, are the velocities measured in the tunnel with and without the models in position, respectively.

To determine the actual pattern of shelter at ground level requires a different approach to pressure measurements. Woodruff and Zingg utilized dune sand, 0.30 to 0.42 mm in diameter, placed on the tunnel floor in the lee of the barriers. Airflows with four different speeds were passed in turn over the models, and then the boundary of the sand area remaining at the end of each test indicated the approximate locality where the wind shear at ground level was reduced to a threshold value related both to the velocity gradient and to the grade of the sand or other erodible material (see Appendix 3.10).

Later experiments by Woodruff and Zingg involved a comparison of wind velocities aft of snow fences in both wind-tunnel and real conditions. The wind-tunnel work section, 16 feet long, began 36 feet downwind from the fan. The tunnel floor was covered with sieved gravel 0.17 to 0.25 inches in diameter, and four fences of galvanized tin, each 3 feet long, were placed at horizontal intervals of $15h$. Velocities were measured at 12 vertical heights in 18 separate locations in the lee of the barriers. The field plot, an ungrazed grass pasture in Kansas, had a uniform slope toward the southwest, with an unobstructed approach for several miles. The snow fences were spaced 60 feet ($15h$) apart, and velocities were measured by pitot-tube alcohol manometers at 0.5, 1, 2, 4, and 6 feet above the ground. The unobstructed wind velocity was measured by means of a normal three-cup anemometer. Measurements were taken in traverses, each complete traverse involving taking instantaneous velocities at 30-second intervals for 10-minute periods at 13 locations to the lee of a single barrier, and eight locations aft of a series of successive barriers.

The results indicated that very similar patterns were obtained for both model and prototype. This is interesting, because the Reynolds numbers were not similar, but because they were quite high, the drag coefficient was independent of Re, and so similarity of flow pattern was achieved. The sections compare wind-tunnel and atmospheric winds for single and for four successive snow fences by means of dimen-sionlessratios of height and velocity, respectively (see Appendix 3.11).

Caborn made 21 series of observations at Imperial College in 1955 using a 30-feet-long wind tunnel of octagonal cross section with a test section 66 inches long. A constant-speed airflow of 10.6 m per sec (34.7 feet per sec) was maintained by means of a two-bladed propellor driven by an electric motor controlled by a rheostat. Models made of uniformly spaced wire nails, projecting vertically 2 inches above their base, were employed, the longest model extending 30 inches along the test section. The nails were placed 1 cm apart in rows, 40 or so nails per row, the rows being 1 cm apart; an echelon arrangement, with the nails in one row opposite the spaces in the next, ensuring that a uniform decrease in penetrability occurred as the number of rows (i.e., the width of the shelterbelt) increased. The depths of the models were in multiples of their height: the $1h$ wide (width measuring the extent of the shelterbelt in the direction of the wind) model consisted of 4 rows of nails, the $2h$ wide model of 9 rows, and the $5h$ model of 24 rows. Shelterbelts of varying cross section were modeled by varying the height of the nails from, for example, 2 cm in the first row to 5 cm in the fourth row to represent a barrier whose upper surface sloped at 45°. A wind-pruned shelterbelt at Gosford in East Lothian was represented by a model with a scale of 1 cm to 8 feet, consisting of a solid plate 20 cm wide, increasing in height from less than 2 cm at the windward edge to 5 cm at its leeward edge. The sloping surface of the model, and its leeward edge, were covered with perforated galvanized paper. Paraffin-oil smoke introduced into the tunnel to make the airflow pattern visible indicated that the streamlines did not lift significantly when the air-stream passed over the wider models.

Pressure was measured at up to 13 horizontal distances downwind of the models, at heights of 1 cm ($0.2h$), 3 cm ($0.6h$), and 5 cm ($1.0h$) above the tunnel floor. Total-head tubes made from hypodermic tubing were used, a tap system enabling each tube to be connected in turn to one arm of an inclined-tube differential alcohol manometer. The static pressure was determined from a side-static hole in the wall of the tunnel, connected to the other arm of the manometer. Since this static pressure was not necessarily the same as the static pressure at the point where the sum of kinetic and static pressures was measured by the total-head tube, standard pitot-tubes were employed to determine the gradient of static pressure along the tunnel before the model experiments were commenced. This enabled a correction to be applied to the pressures recorded at the side-static hole during the experiments.

Measurements of differential pressure were made at a given point in the unobstructed tunnel, i.e., without any models in place and with the test-section base covered uniformly with plywood, and at the same point with the models in place. After thermometer and barometer were read to ensure the air density remained constant during the observations, the horizontal velocity ratios could be calculated (see Appendix 3.12).

Figure 3.4 presents some of the results of Caborn's experiments. They enable a number of generalizations to be made that have proved to be applicable to real shelterbelts too.

The two most critical factors deciding the effectiveness of a shelterbelt are its width/height ratio and its degree of penetrability. When the width is more than five times the height, the width of a shelterbelt is the limiting factor in determining its shelter effect, but the width/height ratio has to be much greater than for open

158

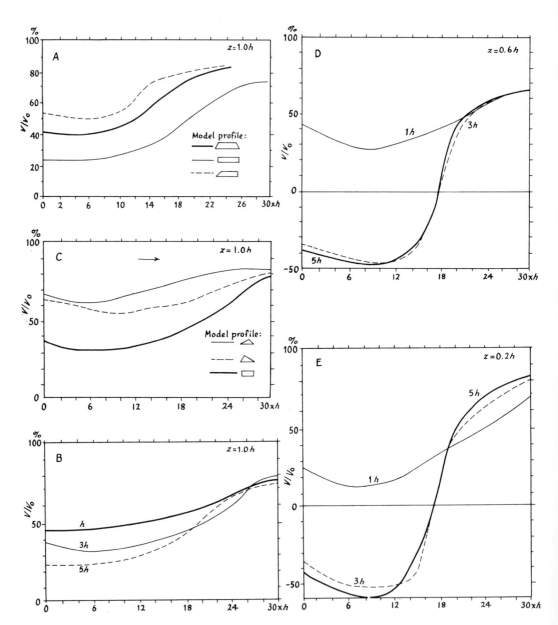

FIGURE 3.4.
Wind tunnel observations of Shelterbelt effects on windspeed (after J. M. Caborn). A, Relative velocities to leeward of models (5*h* wide) of varying cross sectional profiles. B, Relative leeward velocities for models *h*, 3*h*, and 5*h* wide. C, Relative leeward velocities for models 3*h* wide. D, Relative velocities to leeward for models *h*, 3*h*, and 5*h* wide, measured at a height of 0.6*h*. E, Relative velocities to leeward of models *h*, 3*h*, and 5*h* wide, measured at a height of 0.2*h*. In all graphs, absissae represent distances leeward of model shelterbelt in terms of shelterbelt height (*h*), ordinates represent velocity ratio V/V_0 as a percentage.

windbreaks. The width/height ratio is not, however, significant in deciding the extent and nature of the sheltered area unless the penetrability of the windbreak exceeds 20 per cent.

The zone of eddies to the lee of wide windbreaks is reduced by the airflow that moves over the top parallel to its upper surface. A rapid downward transfer of energy occurs when the airflow leaves the leeward edge of the windbreak, and the unobstructed wind velocity is resumed more rapidly than it is with narrow windbreaks; consequently, the sheltered area is reduced. An inclination of the windward edge of a windbreak has the same effect as an increase in its width: the extent of the sheltered area is reduced by an amount depending on the angle of inclination, and the effective penetrability of the belt is reduced because the greater part of the airstream is deflected over its top. The smaller the angle of inclination of the windbreak, the more pronounced is this upward deflection of the airflow. Windbreaks with vertical upwind and downwind edges are more effective in reducing the speed of the wind than those with sloping edges.

When two parallel windbreaks were placed $13h$ apart in the wind tunnel, the shelter effect at heights of $0.6h$ and $0.2h$ was slightly greater than behind a single windbreak; near the ground, it was quite pronounced, but no cumulative effect was detected at the level of the top of the windbreaks. The cumulative sheltering effect near the ground extended a distance $16h$ downwind of the rear barrier, but was not found beyond that.

An alternative way to study the effect of orographic barriers is, instead of moving air over a stationary model, to move a model in a stationary fluid. This was the procedure adopted by Long in an experimental investigation of airflow over the Owens Valley east of the Sierra Nevada in California.* He used a fluid channel 20 feet long, 2 feet high, and 6 inches wide, the fluid consisting of a mixture of salt and water such that its basic density distribution was linear with height. A model of the orography, 1 m long and 4 cm high, was moved along the bottom of the channel at a known constant speed. Aluminum particles in the fluid were registered as streaks on time-exposure photographs, made from a camera moving with the model on a parallel track, thus rendering the airflow pattern visible.

In fluid experiments of this type, it is impossible to obtain identical Reynolds numbers in both model and prototype, because of the effect of friction. The model flow is basically laminar, with Re in the region of 10^4; the atmospheric flow is basically turbulent, with Re typically around 10^{10}. The method of working must therefore be to compare mean motions in the model with mean-mean motions in the atmosphere at corresponding points. The appropriate equations are the Reynolds equations of turbulent motion for the atmosphere, and the Navier-Stokes equations for the model. These sets of equations are comparable only if either molecular and turbulent friction can be considered negligible in both model and atmosphere, or molecular friction is negligible in comparison with turbulent friction in the atmosphere, the latter being proportional to the Laplacian of the appropriate velocity component. Long found experimentally that friction was unimportant in the model, and could also be neglected in large-scale considerations in the atmosphere. Assuming the flow was two-dimensional, and that the Earth's rotation could be neglected, he finally deduced a condition for similarity of flow in model and prototype, making use of a modified Froude number (see Appendix 3.13).

---

* On wave tanks, see, e.g., R. R. Long, *ASM*, p. 1959.

Both single-fluid and multifluid model atmospheres were used. The former was 13 cm deep. At relatively high values of the Froude number, it exhibited a single wave crest over the Owens Valley, which moved downstream if the velocity of flow was increased. When the value of F was reduced, extra wave crests appeared: two wave crests developed over the valley with F in the range 0.09 to 0.17, and three when F was between 0.05 and 0.09, with confused flow beneath them. With moderate or small F values, the velocity of flow was reduced near the ground surface upstream of the wave, and increased just above the crest; with low values of F, several layers of alternatively high and low velocity flow were present.

The muitifluid model atmosphere, 26 cm deep, was made by placing several layers of immiscible fluid on top of the original salt-water mixture. The effect was to reduce the density in gradual stages through a considerable depth, rather than to reduce it suddenly at a free surface. For a given value of F (less than 0.060), the added layers were found to have negligible influence on the underlying model troposphere, all the fluid interfaces remaining undisturbed. At values of F around 0.105, the interfaces began to show signs of being slightly disturbed. At an F value of 0.219 the first interface became disturbed, and at 0.264 all the internal interfaces showed strong disturbances, with internal hydraulic jumps analogous to pressure jumps at some of them. Boundary-layer separation clearly showed up in some of the photographs.

EXPERIMENTAL MODELS OF THE GENERAL CIRCULATION OF THE ATMOSPHERE

The first experimental models of the general circulation were probably those of Vettin during 1857 and 1884, which involved the study of convectional systems in both rotating and nonrotating systems. Later models were developed by Exner in 1923 and Ahlborn in 1924, but the most comprehensive series of experiments were initiated by Fultz, at the University of Chicago, commencing in 1946. Other experiments have been conducted at Cambridge by Hide from 1953 onward, by Faller, and by Sabin at Massachusetts Institute of Technology. The earlier experiments were not quantitative, but later models, particularly those of Fultz and Hide, have involved much theoretical work.[14]

The original Chicago experiments were made by rotating an inverted hemispherical shell of liquid of relatively small thickness, contained between two glass bowls. A heat source maintained at the lower "pole" provided a convective interchange of liquid between pole and equator that indicated a broad similarity with atmospheric relative zonal motions. This type of model was later abandoned in favor of a horizontal circular disc (or a shallow cylinder) of liquid rotated about a vertical axis in a uniform field of local gravity. The horizontal disc system has the advantage over the hemisphere system that, for low rates of rotation, the disc approximates an equipotential surface for apparent gravity in a coordinate system that rotates with the pan which contains the liquid. The depth/diameter ratio of the liquid disc varies from 0.1 to 0.4, which is large in comparison with similar ratios in the actual atmosphere. The model follows the lead of Rossby, who in 1927 was the first to develop the ideas of von Helmholtz and apply realistic similarity concepts.[15] Rossby's model comprised a 2-m diameter pan, filled with water to a depth of 5 cm, which was rotated at 3 to 4 rpm. Instead of making use of cold or warm sources—Exner, for example, used a rotating disc heated at the rim by a ring of gas flames and cooled at the center by a cylinder of ice, and Vettin used a rotating disc with a piece of ice at the center to provide a cold source—Rossby produced density stratification by injecting solutions of colored salt from a tank at the center of the disc.

The Fultz experimental apparatus is complex, but its principle is simple. The rotating pan is similar in shape to an aluminum saucepan 30 cm in diameter. The liquid that is rotated is distilled water (which is a suitable liquid, since most of its physical properties are known with a fair degree of precision, except for its thermal conductivity, but there are still difficulties in using it, especially because its volume-expansion coefficient at 20°C is 81 per cent of its value at 25°C, which is at variance with the atmospheric values). A heat source is provided by means of a ring-shaped electrical resistance heating element around the rim of the rotating disc, and a cold source (or heat sink) is maintained by circulating a liquid (whose temperature may be controlled) to the surface that it is desired to cool. The liquid movements are made visible by sprinkling aluminum powder on the top surface of the rotating liquid; the aluminum particles float on the liquid surface and accurately follow the horizontal motions of the surface layer without disturbing them. To follow the subsurface motions, i.e., the "surface airflow," a malachite dye is dropped into the liquid, slowly falls to the bottom of the pan, and leaves a blue-green ribbon trail behind it that enables steamlines to be determined at any level within the model "atmosphere."

The experimental difficulties in the "Fultz dishpan" model are very great. For example, quite apart from dynamic similitude problems, the rotation rate must be accurately maintained as constant as possible, because the "atmospheric" motions to be followed are very sluggish (about 1 per cent of the absolute rate of rotation) in comparison with the rate of rotation of the container. The pan must be kept very steady during its rotation, because any slight vibrations or tilting of the axis of rotation may result in spurious wave motions in the fluid. Since the relative motions within the liquid are very small in comparison with its absolute rate of rotation, the velocity field relative to the rotating pan must be measured, and not the field of absolute velocity. This creates difficulties in observing and recording the flow, because the observer, or the recording camera, must be rotated too. A prism arrangement is necessary to overcome these difficulties (see Appendix 3.14). The aluminum tracers must be photographed by time-lapse cameras if synoptic charts of the motion are to be produced, and a high-speed flash unit must be used. Surface-tension effects influence the motions of the aluminum particles, giving rise to very high values of "wind shear" within the first millimeter below the surface of the liquid, and this must be counteracted by adding enough laboratory detergent to soften the surface film.

Vettin's rotating disc experiment resulted in the production of a simple meridional or "trade-wind cell" type of circulation throughout the whole of the model atmosphere, and Exner's experiment resulted in the production of very irregular motions, involving systems of vortices. The Fultz experiments show that three different types of flow regime may be produced. In the axially symmetrical or *Hadley regime*, a uniform rate of heating round the rim of the rotating disc and a constant cooling at its center result in the production of a single trade-wind cell (see Figure 3.5). This was the type of regime produced by Vettin in his experiments, and corresponds in effect to a relatively low rate of rotation of the model "Earth." As the rate of rotation increases, or more specifically as the thermal or kinematic Rossby number moves into the range from 10^{-3} to 0.2–0.6, the motions gradually begin to more closely resemble those of the actual atmosphere. That is, the motion of the "atmosphere" as a whole becomes irregular. Horizontal waves and vortices develop, the upper westerlies (i.e.,

162

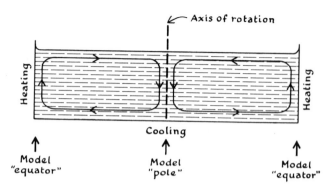

FIGURE 3.5.
Rotating dishpan model of
Hadley regime.

the flow at the top surface of the liquid) move outward from their Hadley-regime locations near the pole, and jetstreams develop. This is the *Rossby regime*, and represents a state of quasigeostrophic balance. Exner's experiments simulated conditions that can produce only a Rossby regime. When the rate of rotation increases still further, i.e., when the Rossby number decreases still further, the jets become more sharply defined and velocity shears increase, until finally the quasigeostrophic system is disrupted and the jets break down into small segments. Frontal-density discontinuities develop within the interior of the fluid, cyclonic-discontinuity shear lines appear, and families of "Polar Front" cyclones come into existence beneath upper-air jets as in the actual atmosphere.

The changeover from a Hadley to a Rossby regime and vice versa depends on the values of the rate of rotation of the model atmosphere (expressed geostrophically by the kinematic Rossby number) and the horizontal gradient of "air" density (represented by the thermal Rossby number), and on the nature of the vertical stratification within the liquid as measured by the internal Froude number or the Richardson number for synoptic-scale flow.

The third regime, the *annular Rossby regime*, involves the existence within the liquid of steady and regular baroclinic waves forming progressively along continuous westerly or easterly jetstreams. The annual Rossby regime was discovered by Hide, whose working fluid was an annulus contained between two concentric cylindrical surfaces that rotated together about a vertical axis. The inner cylinder acted as a cold source and the outer cylinder as a heat source. The annular Rossby regime occurred as a steady-state condition when the cylinders were rotated under conditions that produce the Hadley regime in the conventional rotating-pan model. For a given value of the ratio of the radius of the inner cylinder to the radius of the outer cylinder, a range of wave-numbers for the "upper-air" flow could be generated by varying the rate of rotation of the annulus or the value of the radial temperature gradient. When the Rossby number dropped below a certain range of values, the baroclinic waves broke down into irregular motions as before, and the transition from a Hadley to a Rossby regime also occurred, but much more sharply than in the Fultz experiments.

The importance of the Hide annular experiment as further developed by Fultz is that it succeeds in producing baroclinic "Rossby waves," i.e., baroclinic waves that behave in the steady and regular manner of the classic barotropic waves predicted theoretically and then discovered by Rossby.[16] The mode of development of the baroclinic waves is as follows. The annulus (a column of water 13 cm high) is rotated at a fixed rate of 20 rpm throughout the experiment, and the temperature of its outer wall is gradually raised while that of its inner wall is gradually lowered. Commencing with zero temperature difference between the inner and outer walls, a very slow symmetric flow pattern develops when the thermal Rossby number is about 0.001. This flow pattern has no actual counterpart in the real atmosphere. As the temperature differential is increased, a critical point is reached beyond which this symmetric flow becomes unstable and baroclinic waves develop. These waves are steady and regular, and at first a seven-wave pattern exists around the hemisphere. As the temperature differential is increased beyond the critical point, the seven-wave pattern transforms into a six-wave pattern, then into a five-, four-, three-, two-, and finally one-wave pattern. All these waves are quasigeostrophic, and they occur within the range of thermal Rossby numbers characteristic of the atmosphere (i.e., the range 0.01 to 0.1). If the temperature differential is increased beyond the value appropriate to a wave-number-one pattern, the Hadley regime results, i.e., an axially symmetric flow similar to that obtaining in the initial phase of the experiment, with a westerly current over the entire hemisphere in the "upper air" and an easterly flow everywhere at the surface, but the velocities of these currents are much greater than those of the corresponding flows in the initial phase.

If, in the Hide annulus experiment, the horizontal temperature gradient is gradually decreased after the Hadley regime has been achieved, then the transition points between the different regimes are found to occur at lower values of the thermal Rossby number than with an increasing temperature differential. Especially noteworthy is the fact that it proves impossible to recover the lower-velocity axially symmetric regime. If the rate of rotation of the annulus is varied, the initial instability at relatively high rotation rates manifests itself in the form of Bénard-like cells with alternating cyclones and anticyclones and no continuous jetstreams. If the rotation rate is increased still further, baroclinic waves embedded in an easterly (instead of the normal westerly) flow pattern result.

With a Rossby regime in which the Rossby number lies between 0.001 and 0.1, the annulus exhibits a hysteresis effect. It is then possible to generate any required wave-number state, for a given rate of rotation and a given horizontal temperature gradient, from a knowledge of the wave-number state in existence just before the annulus is brought to the required rotation rate and temperature gradient. Such a hysteresis effect is exhibited by the actual atmosphere: monthly mean weather patterns often resemble each other for a number of months in succession, and then change suddenly to a completely different set of patterns. When the annulus is near the transition zone between two wave-number states, or when it is in certain ranges of low Rossby number, *vacillation* may be observed, i.e., a cycle lasting for 10 to 100 revolutions (i.e., "days"), during which the flow pattern fluctuates from low-amplitude waves ("high index") to a system of closed upper cyclones in troughs ("low index"), and then to a pattern of southwest-northeast troughs, before returning to the first phase of the "index cycle" once more.[17]

The rotating disc or rotating cylinder experiments have been carried out in order to investigate specific phenomena, for example the Polar Front, convection, and jetstreams.[18] They are classic experiments, not only because of the way in which they have enabled precise measurements and quantitative estimates to be made of elements that are next to impossible to measure adequately in the actual atmosphere, but also because of their value in pointing the way to the most important controls of global-scale climatic fluctuation. For example, they emphasize the importance of the rate of rotation of the Earth and the equator-poles temperature gradient as primary controls of weather and climate. That is, they imply that the present-day pattern of our climatic zones is as it is because of a fortuitous combination of a certain rotation rate with a certain temperature gradient. If either of the latter change even very slightly—and geographical factors such as the distribution of sea ice, the distribution of air pollution, or the distribution of crops or of green vegetation, can easily bring about changes in heat balance that will affect the temperature gradient—then the pattern of the Earth's weather and climate could become vastly different. Another somewhat surprising implication of the experiments is that neither the existence of tropopause or stratosphere, nor momentum-balance considerations, appear to be necessary to account for the development of the basic atmospheric flow pattern in middle latitudes, although the details of the departures from this basic flow may be attributed to some extent to these factors.

Quite independently of the Fultz-Hide series of models, similar types of experimental model have been produced by other workers. For example, Long developed a model rather similar to Fultz' rotating cylinder in which two immiscible liquids were used to simulate vortices in a Polar Front model. This model indicated that a cyclone centered over the pole should be a stable feature, whereas an anticyclone should be unstable. Long also designed a rotating spherical shell model for studying the effects of large mountain barriers on zonal currents; it showed that a westerly airflow over a major north-south mountain barrier develops a strong anticyclonic circulation around the latter. The anticyclonic circulation then sets up planetary waves around the hemisphere, the wave-number frequency corresponding to an expression derived from the vorticity equation. The same apparatus was used by Frenzen, who took time-lapse movie films of the experiment that revealed unexpected circulation features. The frictional drag exerted on the air between two concentric spherical surfaces when the latter are rotated was used by Hubert to produce an experimental model of the formation of eddies by lateral shear. This experiment was particularly interesting in that air was the "working fluid," and the air motion was made visible by the injection of tobacco smoke.[19]

Mathematical Models of Phenomena of Weather and Climate

As examples of the uses of mathematical models in climatology, two general classes of model will be discussed: models derived from mathematical statistics, and numerical prediction models.

STATISTICAL MODELS

The *binomial model* is a good example of a simple mathematical model that is of value in the analysis of climatic statistics. It is applicable to the frequency distribution of noncontinuous elements, i.e., to events (such as a daily rainfall total of 0.1 inch or more) that either do or do not occur. If a given set of climatic figures on the frequency of some specific event conforms to the binomial model, then the following rules apply.

1. The probability that the event will occur x times out of a total of n possible occasions is given by

$$\frac{n!}{x!(n-x)!} p^x q^{n-x},$$

where p is the probability that the event will occur (as determined by the historical frequencies), and q is the probability that the event will not occur.

2. The probability of the event in a total of N occasions is

$$N \frac{n!}{x!(n-x)!} p^x q^{n-x}.$$

The value of the binomial model is that if the frequency distribution is symmetric, all the normal statistical parameters of the distribution may be obtained from a knowledge of N and the mean np only. A special form of the binomial model, the *Poisson distribution*, applies if the value of p (or q) is very small (say, 0.05 or less), but the possible number of occurrences is very large, so that the true value of p is unknown (see Appendix 3.15). The Poisson distribution shows, for example, that for a rare event that occurs on the average once in n days according to the observations, then the probability that the event will *not* occur during the next d days is $e^{-d/n}$. For a binomial model that produces a skew frequency distribution, with p (or q) less than 0.25 or so, but in which the possible number of occurrences is not large, the *Poisson exponential distribution* applies, and all the parameters of the distribution may be calculated simply from a knowledge of the mean (see Appendix 3.16).

The *Markov chain* is a model derived from the mathematical study of stochastic processes, and describes a sequence of events that takes place randomly.[20] For example, in the examination of a climatological time-series, the figures usually vary over a wide range, and this variation is taken as evidence of the existence of a genuine climatic fluctuation. However, if the figures can be generated by means of a Markov-chain process, then the variations may arise purely as properties of the model, and hence cannot be taken as evidence of the existence of a genuine climatological phenomena.

Markov chains of different orders may be defined. A first-order Markov chain is one in which the probability an event will occur on any one day depends only on conditions during the previous day. The probabilities obviously are conditional probabilities, and because they indicate the probability of moving from one state or value to another, they are referred to as *transition probabilities*.[21] If a set of figures, which classifies each day at a given station as either dry or wet, is being examined, two transition probabilities may be obtained:

(a) the transition probability for the occurrence of a wet day after a dry day, represented by $P(W|D)$; and (b) the transition probability of a wet day after a wet day, $P(W|W)$.

The probability of a wet day $[P(W)]$ is then given by $P(W|D)| (1-d)$ where

$$d = [P(W|W) - P(W|D)],$$

and the probability of a dry day $[P(D)]$ is given by

$$P(D) = 1 - P(W).$$

If the data follow a Markov chain, then six types of probabilities and transition probabilities may be obtained:

(a) $P(D_t)$, the probability that the tth day of the sequence will be dry;

(b) $P(W_t)$, the probability that the tth day of the sequence will be wet;

(c) $P(D_t \mid D_{t-1})$, the transition probability that the tth day will be dry given that the $(t-1)^{th}$ day is dry:

(d) $P(W_t \mid D_{t-1})$, the transition probability that the tth day will be wet given that the $(t-1)^{th}$ day is dry;

(e) $P(D_t \mid W_{t-1})$, the transition probability that the tth day will be dry given that the $(t-1)^{th}$ day is wet;

(f) $P(W_t \mid W_{t-1})$, the transition probability that the tth day will be wet given that the $(t-1)^{th}$ day is wet.

If the set of observed data may be represented by a first-order Markov chain, the data are not necessarily completely random, but the model provides an excellent first approximation for calculating the probability of occurrence of any particular sequence of wet and dry days. The model has been successfully applied to precipitation data, for example, for Tel Aviv, Israel, the Canadian prairie provinces, and certain stations in the United States.[22]

NUMERICAL PREDICTION MODELS

Mathematical models of the atmosphere describe the relationships that should exist between the various atmospheric parameters according to the known laws of dynamics and physics. The "atmosphere" used for these purposes is a theoretical atmosphere, usually defined by reference to a Standard Atmosphere and incorporating some features of the latter. Standard Atmospheres picture the actual atmosphere as far as it is known at a given time, based on all available observations in which the parameters have been computed according to the most up-to-date physical constants and definitions. Consequently, Standard Atmospheres must be revised every few years; one that has achieved worldwide application is the U.S. Standard Atmosphere of 1962, with supplementary atmospheres for the tropics and subtropics. At the level at which atmospheric density is least variable (i.e., the *isopycnic level* at 7 to 8 km), actual densities do not depart from the U.S. Standard value by more than 1 or 2 per cent in any area or season. It is convenient to express observed densities as percentages of the U.S. Standard value (see Figure 3.6). The largest deviations from the latter are found at the 70-km level north of latitude 60°, where the mean density varies from 65 per cent of Standard in January to 130 per cent in July. Figure 3.6 presents mean monthly cross sections for a new Standard Atmosphere that is more detailed than the U.S. Standard Atmosphere of 1962.[23]

Numerical prediction models were initially designed for short-range weather-forecasting purposes, but they have distinct uses in climatology. They may be used to predict climatological mean values for periods in which the amplitude of fluctuation of the general circulation is relatively low; they provide a basis for objective predictions of climatic conditions in areas devoid of observing stations; and they may be used to make probability forecasts.[24]

Essentially, numerical prediction models represent solutions of the equations of motion for the atmosphere by means of finite differences. The equations of motion

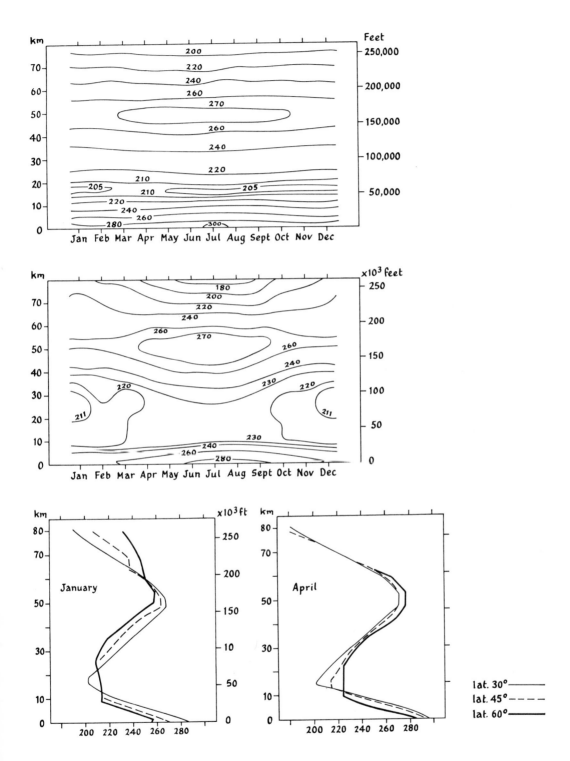

FIGURE 3.6.
Mean monthly atmospheric temperatures (°K): A, at latitude 30°N; B, at latitude 60°N. (After
A. J. Kantor and A. E. Cole.)

are nonlinear differential equations for which no analytic solution in terms of known functions (such as sin, exp, or $\log_e$) exist, and which must therefore be solved by replacing derivatives by finite differences; for example, the derivative $\left(\dfrac{\partial h}{\partial x}\right)_5$, which represents the instantaneous rate of change of 500-mb height with distance in the east-west direction, is replaced by $\left(\dfrac{\Delta h}{\Delta x}\right)_5$, which is the difference in height between two points on the 500-mb surface divided by the distance between them in the east-west direction. The normal x, y, p, t coordinate system used in dynamic meteorology is often used for numerical prediction, but there are difficulties with it in mountainous regions, where the lower limit of the atmosphere is no longer a coordinate surface. It has been suggested that the vertical coordinate p (i.e., atmospheric pressure) should be replaced by the independent variable σ, defined by $\sigma = \dfrac{p}{\pi}$, where π is the pressure at ground level, σ varying from zero at the top of the atmosphere to unity at the surface of the Earth.[25]

Although the principles behind numerical predictions of the instantaneous or mean state of the general circulation have been understood for many years, it has only been since the introduction of the electronic computer that the principles could be applied routinely.[26] In particular, computers with an adequate memory were necessary, because atmospheric computations generate many intermediate quantities before the final result is achieved. Classic experiments in the numerical prediction of 24-hour forecasts made at Princeton required 1,960,000 multiplications for the preparation of each forecast. These necessitated issuing 42,300,000 orders to the computer, in order to obtain a forecast in just over one hour. Each order (e.g., "multiply" or "divide") required 65 microseconds (i.e., 65 millionths of a second), and each multiplication or division took 500 microseconds to complete.

The starting point for a numerical prediction is usually the 500-mb and 1,000-mb charts for some instant in time. The geostrophic vorticity field at future instants in time is then calculated by the computer. Two general classes of models are employed in this work: barotropic and baroclinic.

The *barotropic model* is based on Charney's form of the vorticity equation for a barotropic atmosphere, which is solved by means of finite differences (see Appendix 3.17). The replacement of the strictly accurate differentials by approximately accurate finite differences results in an error, which is termed the *truncation error*.[27] In the application of the vorticity theorem, the equation describes the motion over an area bounded by a simple closed curve in terms of (a) the specification of the height of the appropriate pressure contours (usually 1,000 and 500 mb) everywhere along the boundary, and (b) the vorticity on that part of the boundary at which the air is entering the region bounded by the closed curve. Since these assumed conditions along the boundary are unrealistic for the actual atmosphere, a *boundary error* results. To reduce the boundary error, the area for which the forecast is required is located at the center of a much larger region over which the computations are carried out; i.e., the boundaries are chosen so that they are remote from the forecast area.[28]

In the classic application of the barotropic model made at Princeton by Charney, Fjörtoft, and von Neumann, forecasts of the height of the 500-mb surface in middle latitudes were made 24 hours ahead.[29] The initial (i.e., observed) 500-mb chart was sampled at the intersections of grid lines in a system of 15 × 18 lines spaced at distances of 736 km (i.e., 8° of longitude in latitude 45°) apart, and the computer made successive predictions for steps of three hours each. The total number of grid points at which the 500-mb contour height was measured was determined by the internal memory capacity of the ENIAC computer. Forecasts of the 3:00 A.M. G.M.T. 500-mb chart for North America were produced for January 5, 30, and 31, and February 13, 1949. Comparison of predicted and actual charts showed that the forecasts were in serious error in areas where potential energy was being converted into kinetic energy, i.e., because the model did not take into account the fact that vorticity need not be conserved but may be created or destroyed. The results suggested that baroclinic effects are not manifested as a steady and widespread conversion of potential into kinetic energy, but rather as sporadic and violent local overturnings accompanying the sudden development of baroclinic instability. For example, a result of the latter was the development of a major trough over eastern North America on January 30 (but not on the following day), which was completely omitted in the forecast made from the model. Tests of the model were also made by applying it to European conditions.[30]

By means of automation, 31 successive forecasts of the 500-mb surface were made as a test of the most suitable forecast interval. The over-all accuracy of the forecasts decayed steadily as the forecast period increased up to 72 hours, and the errors for the entire forecast series were related to the geographical positions and orientations of the "centres of action" in the general circulation and their seasonal variations in intensity. For forecast periods of 72 hours or more, the barotropic model can only predict phenomena with the same order of magnitude as the centers of action.[31]

Many subsequent tests have revealed deficiencies and sources of error in the barotropic model. For example, it often predicts excessive anticyclogenesis. Errors arise from lack of smoothing, from lack of data in the initial chart, from the use of the geostrophic assumption in the estimation of vorticity, and from differences in the initial analysis.[32] Some of these errors are quite interesting. For example, as the paucity of data in the initial chart increases, the smaller synoptic features in the forecast charts disappear, and the larger features become smoothed and change their speeds of movement; the fast-moving systems decelerate and the slow-moving systems accelerate, so that the mean flow remains constant in velocity. Use of the geostrophic assumption results in errors in the prediction of changes of intensity in the disturbances, tending to displace depressions too far north over the United States, and does not permit the development of fronts. Certain large-scale errors tend to persist in some geographical areas: for example, the model usually predicts excessively high 500-mb heights off the southeastern coasts of continents in the northern hemisphere as far south as latitude 13°N, and excessively low values off the northwestern coasts. These errors largely result from incorrect forecasts of the phase of the long waves with wave-numbers one to four inclusive. The delineation of mean-error fields, defined as the mean error of the preceding 10 to 15 forecasts, enables a correction to be applied to the next 24- or 48-hour forecast.[33]

An improved barotropic model incorporating the effect of the stratosphere has been developed, and also a two-parameter barotropic model that includes the total thickness field and provides better forecasts of strong development than the classic barotropic model.[34]

The barotropic model has been used to provide hemispheric forecasts, and routine forecasts based on it have been made in the United States since 1955. The form of the barotropic model used for the routine forecasts contains terms that allow for divergence arising from flow over mountains and for the spurious retrogression of ultra-long waves (i.e., wave-numbers one, two, and three) that was found in the original model.[35]

The barotropic model appears to work quite well in the Caribbean as far south as 15°N, except in the vicinity of the center of the subtropical ridge. It may be used for the prediction of five-day mean patterns, provided large-scale vorticity sources and sinks, surface friction, and nonadiabatic heating are taken into account, but not for forecasts a month in advance.[36]

One major defect of the barotropic model is that it predicts excessive synoptic development, i.e., anticyclones that are too intense and cyclones that are too deep. The barotropic model is essentially a Eulerian concept, and this particular defect may be remedied to some extent by making use of a Lagrangian model. An alternative method is to incorporate baroclinicity into the barotropic model, i.e., to produce a *baroclinic model*, for example, by incorporating the atmospheric thickness field into a barotropic model.[37]

The barotropic model assumes that atmospheric flow is two-dimensional and non-divergent. The so-called $2\frac{1}{2}$-*dimensional model* is the simplest type of model that takes into account the transformation of potential into kinetic energy without making use of the three-dimensional equations of motion.[38] It is thus applicable only to a two-dimensional atmosphere, but it allows for the three-dimensional structure of the real atmosphere by making use of two horizontal fields of distribution, i.e., one pressure-contour field and one temperature field. The $2\frac{1}{2}$-dimensional model assumes that temperature changes are caused by horizontal advection only, and that the thermal wind has a similar variation in magnitude with height at all places, but the direction of the thermal wind is assumed to be independent of height. An alternative model on these lines was developed by Eady, and a baroclinic model that takes the three-dimensional structure of the atmosphere more completely into account was devised by Arnason.[39]

A baroclinic model used for the preparation of routine short-period forecasts in Britain is the *Sawyer-Bushby model*. This model describes a baroclinic atmosphere in which the thermal wind is constant in direction in any vertical column, but not necessarily parallel to the wind direction at any level, and assumes that the speed of the thermal wind is proportional to the pressure difference through the layer concerned. Geostrophic motion and a parabolic variation of vertical motion with regard to pressure is assumed, and negligible air movement is assumed across the two pressure surfaces that bound the model atmosphere (see Appendix 3.18). The model gives realistic 24-hour predictions of the general state of the atmospheric circulation over western Europe and the eastern Atlantic (see Figure 3.7), and, although omission of the effects of nonadiabatic heating from the model causes appreciable errors over sea areas, the boundary error is not serious in the region of the British Isles. The Sawyer-Bushby model is more complex than the barotropic model, and applying it by means of a computer takes up to four hours to produce a 24-hour forecast. Minor terms in the vertically average vorticity equation used in the model may result in gross forecast errors under certain conditions.[40]

The Sawyer-Bushby model is essentially a two-parameter model; i.e., the atmosphere is regarded as comprising two layers, in each of which the motion is baroclinic. It may be extended to become a three-parameter model. Three-parameter models more

FIGURE 3.7.
Early predictions with the Sawyer Bushby two parameter baroclinic model (after F. A. Bushby and M. K. Hinds). A–D, 3:00 A.M. G.M.T., January 28, 1952. E–H, 3:00 P.M. G.M.T., January 28, 1952. A, E, Actual 500 mb charts. B, F, Predicted 500 mb charts. C, G, Actual total thickness charts. D, H, Predicted total thickness charts. I–M, 500 mb charts for March 15, 1949. I, Actual 500 mb chart, March 14, 1949 (initial chart for prediction experiment). J, Predicted chart, 3:00 A.M. K, Actual chart, 3:00 A.M. L, Predicted chart, 3:00 P.M. M, Actual chart, 3:00 P.M.

accurately describe the actual atmosphere and thus provide more detailed forecasts of the circulation. For example, the model devised by Wiin-Nielsen realistically predicts the development of slowly moving baroclinic waves that become unstable for large values of the vertical wind shear. Multiparameter models are also available.[41]

A two-layer baroclinic model developed by Lorenz incorporates energy transformations, including the tendency for static stability to increase as kinetic energy is released, and conserving the sum total of kinetic and available potential energy. This model also includes the Coriolis parameter as a variable, which reduces spurious anticyclogenesis, and may be regarded as the simplest, fully consistent baroclinic model of the general circulation that is now available. Examination of the Lorenz model shows that the distribution of stability in a developing baroclinic wave must be such that the air becomes more stable than average when moving downstream from a ridge to a trough, and more unstable than average when moving from a trough to the next ridge downstream.[42]

Other baroclinic models have been developed by Aubert and by Berkofsky, and the Norwegian Meteorological Service makes use of a two-parameter baroclinic model (in addition to a barotropic model) for the preparation of operational forecasts. Baroclinic models provide good predictions of the 1000-mb and 500-millibar circulations over the Arctic. The Arctic atmosphere seems especially amenable to numerical prediction, because it contains fewer complicating factors than middle-latitude or tropical atmospheres; for example, orographic effects are relatively minor, and nonadiabatic effects are relatively uniform except around the fringes of the Arctic.[43]

Comparisons of forecast charts produced by both barotropic and baroclinic models indicate that both types of model account for 500-mb height changes equally well. Typical root-mean-square errors in predictions of 24-hour changes of contour height average 230 feet per day, the correlation coefficient between actual and predicted height changes being on the order of $+0.75$ in areas subject to orographic influences, but these change to less than 200 feet per day and more than $+0.90$ in areas such as the southeastern United States, where no major orographic effects are present. For forecasts of air motion at the 1,000-mb level, however, the baroclinic model gives the best results. Both barotropic and baroclinic models predict excessive movement of momentum into the jetstream; the baroclinic model forecasts vertical motions that become excessively cellular after 12 hours and in serious error after 24 hours. Neither barotropic nor baroclinic models provide detailed predictions for planetary-scale motions, because they predict quasistationary waves at this scale. For his numerical experiment in predicting the state of the planetary-scale circulation, Phillips combined models of amplifying baroclinic waves with those of damped barotropic waves. The models will not provide accurate forecasts of local-scale circulations, but for intermediate-scale circulations on a continental scale they provide predictions whose accuracy is indeed remarkable in view of the mathematical complexities and atmospheric simplifications involved.[44]

Much effort has been devoted to the improvement of these models. One line of approach has been to examine the initial parameters by means of which the models are developed. For example, the standard barotropic and baroclinic models employ parameters that are derived directly from the height of certain pressure levels, whereas an alternative procedure may be adopted in which the atmospheric variables are expanded in terms of empirical orthogonal functions. Another line of approach is to analyze the known sources of error in the standard models. For example, the primitive

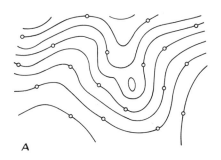

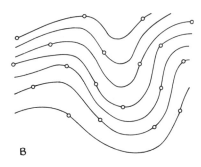

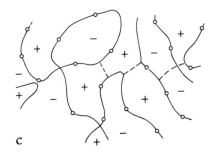

FIGURE 3.8.
Numerical forecasting initial analysis-error field (after P. D. Thompson).
A, Hypothetical "true" contour pattern. B, Reconstruction of the contour pattern by
interpolation between given "true" height values at circled points ("stations").
C, Initial analysis error, obtained by subtracting A from B, giving "grain patterns."

equations of motion (i.e., the Eulerian equations in hydrodynamics, modified by assuming hydrostatic balance) are very sensitive to the initial ($t = 0$) fields of wind and pressure if they are used as a basis for predictions at a time $t = t'$. Because of this sensitivity, the resulting forecast chart will consist very largely of spurious systems or "noise" in the form of high-frequency gravity waves, so that the significant synoptic-scale low-frequency waves will be hidden. The sensitivity may be minimized, particularly for baroclinic systems, by mutually adjusting the distributions of wind and pressure at time $t = 0$, i.e., by making the geostrophic assumption.[45]

Uncertainty about the state of the atmosphere at the commencement of the forecast procedure, i.e., a lack of sufficient observing stations, is an important source of error. Because of this error, the error in the forecast winds may be twice the observed-wind error after a forecast period of 48 hours, and the errors after six to seven days may rise to the proportions of pure guesswork.[46] The distribution of analysis error has a characteristic scale or "grain size" (see Figure 3.8), which depends on the distances between observing stations. The magnitude of the analysis error increases as the average distance between stations increases; doubling the existing over-all density of upper-air stations in the northern hemisphere would eliminate the increase in inherent prediction errors for forecast periods up to several days. Fortunately for the climatologist, zonally averaged wind fields appear to be inherently more predictable than the actual (instantaneous) wind observations.

The boundary error inherent in numerical prediction models is usually reduced by increasing the size of the forecast region, so that the area for which the forecasts are required occupies only a small central portion of the forecast region. However, in making forecasts for an entire hemisphere, this procedure results in new sources of error not connected with boundary conditions. Arbitrary boundary assumptions cause the generation of distortions and spurious irregularities in the short-wave components of the circulation, and these may be reduced by smoothing and filtering operations.[47]

Certain atmospheric parameters and influences omitted from the simple barotropic and baroclinic models are usually incorporated into the operational models. Paramount among these are nonadiabatic effects and orographic influences. The nonadiabatic effects are many: the release of latent heat of condensation, for example. The effect of nonadiabatic heating in general may be included in the numerical models simply by the addition of a new term to the basic equation of the model, provided the heat source only affects the motion field very slowly. However, if the nonadiabatic heating is due to release of latent heat of condensation of the water vapor in the atmosphere, the effect on the fields of temperature, vertical motion, and moisture is immediate, and this simple procedure cannot be applied. Instead, the numerical model must be very appreciably modified to enable it to include precisely the effects of latent heat release. In fact, the vorticity equation is not sufficient in itself as a model, and the final model comprises additional equations representing the first law of thermodynamics, continuity and vertical motion, plus a vertical-motion equation.[48]

The numerical computation of pressure tendencies at a very early stage in the development of the barotropic model indicated that the Rocky Mountains had a profound influence on airflow that must be allowed for in some manner. Charney and Eliassen were able to prove theoretically that both large-scale mountains and surface friction must be important influences in the generation of stationary planetary-scale waves. Smagorinsky was able to compute dynamically the combined influence of surface friction and large-scale heat sources and sinks on these waves, and Wiin-Nielsen demonstrated that the broad features of the distribution of mean atmospheric temperature on January normal maps may be computed from the vorticity equation, provided the effects of surface friction and orography are included in the model.[49]

The incorporation of orographic and surface-friction effects into barotropic or baroclinic models may be done quite satisfactorily, provided an orographic model of the appropriate part of the Earth's surface is available; such a model is usually produced from generalized contours. Orographic effects are capable of producing four types of atmospheric motion: (a) *small-scale turbulent motions*, especially in the lower troposphere, which arise from variations in the drag on the atmosphere exerted by the Earth's surface, and which are capable of influencing large-scale motions predicted by a dynamic model, even though the latter does not provide a means for transmitting these motions upward from the friction layer (see Appendix 3.19); (b) *gravity oscillations*, which are induced by the macroorography, may extend upward as far as the stratosphere, and may be predicted by orographic-wave theory; (c) *ageostrophic motions*, for example, quasihorizontal wind oscillations with wavelengths of a few hundred km, due to the horizontal deflection of an airstream by a mountain of about 100 km in width, and quasihorizontal wind oscillations with periods of about 18 hours as observed by transosonde flights; and (d) *quasigeostrophic motions* of both small and large amplitude. Type (a) motions are capable of generating an upward vertical velocity of 1.3 cm per sec in an airstream of velocity 20 m per sec flowing parallel to a mountainous coast, which compares with the 5 cm per sec that is the maximum computable vertical velocity in the two-parameter baroclinic model. Type (b) motions are due to the *form-drag* of the Earth's orography, which is on the order of 7.6 dynes per cm^2 for hills 300 m high spaced at 10-km intervals (see Appendix 3.20). Type (c) motions are very complicated and are usually not included in the dynamic models. Type (d) motions of small amplitude can be appreciable; because of their existence, a vorticity charge on the order of 3.5×10^{-10} per sec^2 may be generated by a ridge 300 m high and 1,000 km wide if an airstream of 20 m per sec crosses it. Motions of types (a) and (b) act as resistances to larger-scale motions and dissipate energy. Large-amplitude type (d) motions, i.e., over mountains that are too high for all the airstream to flow over them, are very difficult to describe dynamically. For example, it may be shown dynamically that northerly airstreams should cross the Alps without significant deflection, but that airstreams from other directions should be diverted.[50]

Orographic and surface-friction effects have been fairly successfully incorporated into the American operational barotropic forecasts by including a mountain term and a surface-friction term in the barotropic model equation (see Appendix 3.20). The mountain term is introduced on the assumptions that the lower boundary of the atmosphere is everywhere at 1,000 mb, and that a vertical motion is induced at this lower boundary depending on the magnitude of the flow of a fictitious 1,000-mb wind up or down the slope of the generalized ground surface. The mountain term also allows for the fact that windspeeds at ground level are higher than those given by the fictitious 1,000-mb wind, and assumes that vertical velocities decrease upward from the standard pressure at ground level. The surface-friction term includes the *skin-drag coefficient* of the Earth's surface, which is the sum of two components: (1) the form-drag of the major relief, which depends on the properties of the airstream as well as on the macroorography, and (2) the surface friction at the ground, which is relatively constant, whatever the type of airstream, for a given surface windspeed. (See Appendix 3.21.) A model is also available that separates the vertical motion due to surface friction from that due to orography.[51]

Orographic and frictional effects have been incorporated into the Sawyer-Bushby model by assigning to each grid point, at which pressure-contour height is measured, a value for orography that is equal to the mean value of the height of the ground over an area one grid-length (approximately 300 km) square surrounding the point in question. It is assumed that all the air at sea level is forced to rise over an obstacle at the grid point whose height is given by this mean value of the orography, and the vertical velocity induced by the latter is incorporated into the model (see Appendix 3.22).

The methods of incorporation of the effect of orography into the barotropic and baroclinic models just described are two-dimensional, in that the effect is included via the lower boundary condition of the models, but since the effect of terrain on the atmosphere varies with height, a three-dimensional form of the vertical-motion equation must be solved. If the model atmosphere is one in which static stability is a function of pressure alone, the decrease with height of the orographically induced vertical motions at the lower boundary may be obtained quite easily.[52]

The barotropic and baroclinic models essentially are dynamic models of the general atmospheric circulation over regions the size of a continent. There are numerous other available mathematical models that are of use to the climatologist, but only very brief mention of these is possible here. They may be classified under four headings.

1. *Models of Temperature and Precipitation Distributions on Hemispheric or Regional Scales.* Here we have models that predict precipitation amounts for the northern hemisphere on a routine basis 12, 24, or 36 hours ahead, and models that provide objective forecasts of five-day mean surface-temperature patterns. Both prediction systems are based on initial forecasts provided by barotropic and baroclinic models. A numerical prediction model with an entirely different basis, i.e., the equations of conservation of thermal energy in the troposphere and at the surface of the Earth, provides forecasts of large-scale mean temperatures for an entire hemisphere. Of considerable interest to the climatologist is the fact that this model satisfactorily predicts the normal pattern of temperatures for a given month or season. The implication is that the parameters that have been incorporated in the model must be the factors that control both the average and the "actual" mean distribution of temperature in the real atmosphere. These factors prove to be very different from those usually listed in elementary textbooks of climatology. Hydrodynamic models for long-range weather forecasting have been developed in the U.S.S.R.[53]

2. *Models of Elements of the Atmospheric Circulation.* These vary from models of mean asymmetries in the perturbations of the circulation that enable the total effect of the inhomogeneities of the Earth's surface to be calculated, to models of individual features such as cyclones and low-level jetstreams. Models have been developed that describe the main features of the mean subtropical flow over Australia in summer and the mean flow shown in a cross section along the northern portion of the Pacific trade winds.[54]

Jetstreams have been a favored topic for dynamic models, which show, for example, that the breakdown of the polar-night jet is probably a baroclinic instability feature, and that the main Polar Front jet of the troposphere in December is dynamically unstable, both barotropically and baroclinically, with one mode of maximum instability at a wavelength of 3,000 to 4,000 km and a secondary maximum at a wavelength of 10,000 km. Another mathematical model of a jetstream shows that a jet in latitude 45° should have a half-width of 1,000 km and a maximum wind of 60 m per sec. When there is no divergence, the jet is most unstable at a wavelength of 5,500 km; for divergent flow the most unstable wavelength is still 5,500 km, but this wave takes more than seven times as long to amplify.[55]

Tropical cyclones are also popular subjects for mathematical models. One model of the low-level inflow layer of a mature hurricane shows that the maximum wind and central pressure experienced in the hurricane depend on the inflow angle: moderate hurricanes require an inflow angle of 20° and intensive hurricanes an inflow angle of 25°. Other models indicate that the vertical velocity field within a hurricane is primarily determined by the coefficient of friction in the radial direction from the hurricane center, and that the subsidence within the eye of the hurricane is thermally controlled by lateral diffusion of eddies from the warm core. The latter model also demonstrates that the low-level indraft is related to vertical exchange processes in the tangential wind system near the sea surface. Another model provides a criterion for the development of typhoons that depends on the vertical perturbation of the pressure and vertical velocity distributions. Hurricanes are difficult to model, in that two motions at different scales must be described, because all tropical cyclones are in essence very intense vortices embedded in a large-scale flow. They may be modeled by assuming, for example, that a hurricane is an axially symmetrical vortex that moves with a constant, invariant shape and intensity, in which case the velocity of the vortex proves to depend on the vorticity gradient in the large-scale flow, or by assuming that a hurricane may be represented by a steady-state model with inflow and outflow at different levels, maintained by energy supplied locally from the sea surface.[56]

3. *Models of Atmospheric Processes.*   Models of convection are available, both for isolated cases and for turbulent convection. One model predicts that a moist thermal should have the shape of a tall, slender column instead of the spherical shape usually assumed in the bubble theories of convection. Another shows that entrainment of surrounding air sometimes does not explain the growth of a thermal bubble, and that there must be turbulent exchange of air into and out of the bubble.[57]

Heat exchange between the atmosphere and the Earth's surface on the global scale may be described by a model that assumes heating is proportional to the difference between the temperature of the air at the bottom of the atmosphere and the temperature of the underlying surface. The model shows that the rate of heating should depend on the motion of the atmosphere, and predicts that for stationary disturbances of the zonal flow in middle latitudes, heat sources in the lower part of the atmosphere should be located 30° of longitude to the west of temperature maxima at the Earth's surface. Another mathematical model describes the meridional heat budget of the layers of the atmosphere below 25 mb between latitudes 20°N and 70°N. The exchange of matter between atmosphere and sea may also be studied by means of models.[58]

The effect of a heated land surface on the overlying atmosphere is a basic problem in climatology, and may be examined by some useful models. Airflow over a small flat island is a convenient starting point, because orographic factors are eliminated, and, if the island is about 2 to 10 miles in length, it is large enough to affect the overlying atmosphere, but not large enough to necessitate taking the curvature and rotation of the Earth into account. If a larger island is taken, say, one 20 to 100 miles in length, then the Earth's rotation must be considered, but the curvature of its surface may be neglected. By combining studies from both smaller and larger islands, the effect of the Earth's rotation on the airflow may be eliminated.[59]

Three distinct scales of motion, each different physically from the other two, combine to determine the airflow over a small, heated island. These are: the basic airflow, which persists for at least one day, includes disturbances with a scale of some hundreds of kilometers, and is essentially described by the geostrophic wind; the small-scale eddies or thermal turbulence set up at the ground, which persist for not more than 30 minutes and have a scale of about 50 to 150 meters; and organized convective motions, approximately the same order of size as the island, which develop and decay as the diurnal rate of heating increases and decreases. One mathematical model had as its aim the prediction of the main features of these organized convective motions, which are important in the development of cloud streets. The model was tested by observations made in the Caribbean, where stationary cloud streets had been reported as streaming out in the leeward of small islands for many tens of miles, despite a basic airflow of 10–20 miles per hour. The differential equations making up the model (see Appendix 3.23) were solved by linearizing them as follows. The interaction

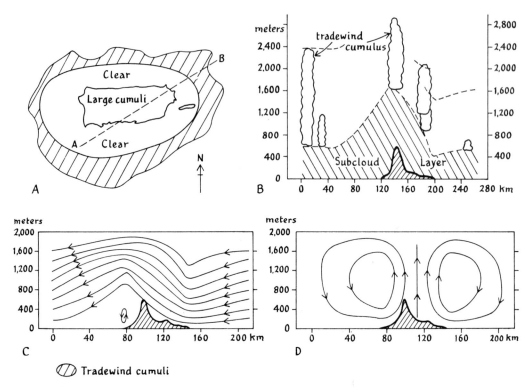

Tradewind cumuli

FIGURE 3.9.
Some effects of the island of Puerto Rico on the trade-wind airstream. (After J. S. Malkus.)

between the basic airflow and the mean convective motions was made linear by selecting for the field observations an island small enough so that the turbulent heat energy it injected into the atmosphere was negligible in comparison with the energy of the basic flow. The observed motions could then be regarded as the sum of the geostrophic wind plus the motion due to the organized convective systems. The final result of the application of the model was a proof that the observed motions could be described by two equations: (a) a heat-diffusion equation, which is important only near the island, and (b) an equation describing the flow of air over a fictitious *equivalent mountain*, whose amplitude and shape depend on the temperature profile along the surface of the island, the coefficient of eddy heat exchange directly above the latter, and the speed and stability characteristics of the basic airflow. The equivalent mountain grows and decays during a single day, and since its amplitude is inversely proportional to the speed of the basic current, the mountain corresponding to a given island may vary considerably in height from day to day.

Application of the equivalent-mountain model to Puerto Rico showed that a symmetric orographic-convection cell (see Figure 3.9) is present over the island on sea-breeze days, with cumulus forming in ascending air over the island and no clouds within the ring of compensatory subsidence surrounding the island. The equivalent mountain proves to be at least 1.3 times as high as the actual mountains (600 m) on the island for an island-sea temperature differential of 1.5°C. In other words, the model demonstrates that if surface air temperatures on Puerto Rico average 1.5°C higher than the average surface temperature of the surrounding sea, then the resulting air motions over the island will correspond to those over mountains of about 2,500 feet in height. The model also indicates that the sea-breeze must reverse at a height that depends on the stability and velocity of the basic current (see Appendix 3.24), and demonstrates that, in general, the balance between physical processes in the trade-wind atmosphere is in a critically sensitive state. There need be only a very small degree of subsidence within the subcloud layer to prevent the formation of cumulus, although the cloud layer itself may be completely unaffected by subsidence.

A model of airflow over a heated land mass 200 km in width and of infinite length predicts the appearance of wave-like velocity perturbations that move outward from the center of the land mass, indicating that the air over the latter continually rises and falls back again, producing air waves rather similar to the expanding waves produced in a water surface when a stone is dropped into it. The perturbations appear some time after the commencement of the heating, whether or not the overlying atmosphere is stationary. They rise to a greater height if there is no wind; if a wind is blowing, the vertical velocity decreases with increasing height less rapidly, and a miniature cold front develops and moves across the island during the heating period. The model also shows that subsidence (with associated inflow of air) occurs even during heating.[60]

4. *Models of Weather Phenomena*. Numerical prediction models of fog have been developed, based on the equations of turbulent diffusion, and also models that predict the total cloud cover that will develop over the northern hemisphere (on a synoptic scale) during 24-, 36-, and 48-hour periods in specific atmospheric layers. Other models demonstrate that mother-of-pearl clouds may be formed by three-dimensional orographic waves set up by individual mountain peaks rather than by an entire mountain range, and that the shape of a cumulonimbus cloud depends on its internal dynamics. Models of cumulus clouds naturally are closely connected with convection models. For example, the appearance of lines of cumuli may be predicted by means of a model of buoyant convection from an instantaneous thermal-line source, and the characteristics of a convective cloud in the updraft stage may be predicted by means of a steady-state model that not only includes heat and momentum transfer by lateral diffusion as well as by entrainment, but also takes into account the drag of the condensed water on the ascending air. Another model shows that disturbances with dimensions of the magnitude of a cumulus cloud must develop in a conditionally unstable atmosphere if the stratification of the latter is such that the air is unstable for ascending motion but stable for descending motion throughout an extensive area, and small random perturbations exist in the airflow, for example, due to the air flowing over relatively uniform but slightly irregular topography. The existence of stable stratification in a region of descending motion increases the critical lapse-rate, narrows the regions of ascending air, and thus results in the latter becoming further apart and separated by regions of descending air of ever-increasing width. This result is very significant for precipitation distribution: the regions of ascending air prove to be about the same size as the rain areas in tropical storms, although the latter are usually regarded as arising primarily from nonconvective processes.[61]

The diurnal variation of wind near the Earth's surface is at first sight an obvious subject for a mathematical prediction model. One such model, based on the equations of motion, employs two boundary conditions: the wind is assumed to vanish at the lower boundary (i.e., the ground surface), and to become geostrophic at the upper boundary (i.e., the top of the friction layer). The model then demonstrates that the planetary-scale boundary-layer motions must be spiral in character, and proves that diurnal pressure variations (i.e., thermal pressure effects) have no significant influence on diurnal wind variation within the atmospheric boundary layer.[62]

A model of airflow over an infinitely long mountain ridge enables vertical wind oscillations at different levels to be computed for an atmosphere in which temperature and stability vary with height more realistically than in the Scorer orographic-wave theory. Mathematical models are also available for atmospheric solitary waves, pressure jumps, and, in fact, for numerous other phenomena that give rise to distinct weather events.[63]

Although a perusal of the literature on mathematical models may be somewhat disconcerting to the nonmathematical climatologist, these models are of vital importance to climatology. Not only do they provide concise explanations of climatological phenomena in an intellectually stimulating manner that no simple qualitative verbal description can hope to equal, but they also provide a means for generating climatic data in regions where observations are lacking or for phenomena that may be observed only indirectly. The precision of the mathematical model is of considerable value to the climatologist. Since many of the observed fluctuations of a climatic element are often of the same order of magnitude as the observational error, it is more accurate to estimate the values of the required element indirectly, by means of an appropriate model into which is fed observations of an atmospheric parameter, such as pressure, that can be measured precisely.

Notes to Chapter 3

1. R. C. Sutcliffe, *W*, 14 (1959), 163.

2. The cross-sections are based on A. Kochanski, *JM*, 12 (1955), 95, and J. G. Moore, *MM*, 85 (1956), 167. For weather models of Polar Front weather devised by Bergeron, see *WAF*, II, 198–205.

3. For a demonstration model, see H. H. Lamb's description, *MM*, 89 (1960), 319, of a model of the general atmospheric circulation; also see K. H. Stewart, *MM*, 83 (1954), 114, for details of a smoke-fog model.

4. For a reprint of Dines' paper, see *W*, 9 (1954), 249.

5. B. C. V. Oddie, *MM*, 84 (1955), 354.

6. For a general account of modeling criteria applied to atmospheric phenomena, see H. Rouse, *CM*, p. 1249. See also D. Fultz, *CM*, p. 1235. For a review of laboratory simulations of atmospheric phenomena, see G. M. Hidy, *BAMS*, 48 (1967), 143.

7. For details of these numbers in a meteorological context, see D. Fultz, *JM*, 8 (1951), 262; see also *MM*, 81 (1952), 369. On the Froude number employed as a similarity criterion for laboratory models of wind stress at the air-water interface, see Jin Wu, *JAS*, 26 (1969), 408.

8. For a review of experimental modeling of clouds, see D. Brunt, *CM*, p. 1255. For laboratory models of cloud development, see: H. F. Eden and B. Vonnegut, *JAM*, 4 (1965), 745, on cumulus in a gaseous medium; R. List, *BAMS*, 47 (1966), 393, on similarity problems; and A. J. Faller, *JAS*, 22 (1965), 176, on cloud rows formed by large eddies in the atmospheric boundary layer. For a large-scale experiment in the development

of clouds, see J. Simpson *et al.*, *Revs. in Geophysics*, 3 (1965), 387, for experiments on tropical cumuli in the Caribbean, using aircraft observations and cloud-seeding techniques. For a natural experiment in cloud development, involving mesoscale eddies in the wake of islands resembling von Kármán vortex streets, from TIROS photographs, see K. P. Chopra and L. F. Hubert, *JAS*, 22 (1965), 652. For a review of experimental laboratory models of condensation and evaporation, see J. S. Turner, *W*, 20 (1965), 124. For experimental laboratory models of optical phenomena, see J. O. Mattsson, *W*, 21 (1966), 14.

9. H. Foster, *JM*, 9 (1952), 437, and R. F. Jones, *MM*, 81 (1952), 152.

10. F. H. Ludlam, *W*, 9 (1954), 169; R. S. Scorer and C. Ronne, *W*, 11 (1956), 151.

11. R. S. Scorer, *QJRMS*, 79 (1953), 70.

12. J. H. Field and R. Warden, *MOGM*, vol. 7, no. 59 (1933).

13. The main source for the following discussion of wind tunnels is J. M. Caborn, *Shelterbelts and Microclimate, Forestry Commission Bulletin*, no. 29 (1957). See also " M. O. Discussion," *MM*, 82 (1953), 18, and N. P. Woodruff and A. W. Zingg, *TAGU*, 36 (1955), 203. For details of wind-tunnel research in the British Meteorological Office, see G. E. W. Hartley, *MM*, 95 (1966), 144. See also: E. J. Plate and A. A. Quraishi, *JAM*, 4 (1965), 400, on a low-speed wind tunnel for modeling velocity distributions inside and above tall crops; A. S. Thom, *QJRMS*, 94 (1968), 44, on measurement of mass and heat exchange between an artificial leaf and the airflow in a wind tunnel; A. C. Chamberlain, *loc. cit.*, p. 318, on wind-tunnel measurements of water-vapor transport to and from surfaces of differing roughness; R. C. Malhota and J. E. Cermak, *JGR*, 68 (1963), 2181, for a theory of wind-tunnel modeling of atmospheric diffusion; and A. H. Schooley, *loc. cit.*, p. 5497, on measurement of vertical profiles of wind above waves generated in a short-fetch wind tunnel. On the generation of wind shear in a wind tunnel, see A. Lloyd, *QJRMS*, 93 (1967), 79.

14. F. Vettin, *Ann. Phys.* (Leipzig), 100 (1857), 99, and 102 (1859), 246; *Meteorol. Z.*, 1 (1884), 227 and 271. F. M. Exner, *Akad. Wiss. Wien*, 132 (1923), 1. F. Ahlborn, *Beitr. Phys. Freien Atmos.*, 11 (1924), 117. D. Fultz *et al.*, *AMM*, vol. 4, no. 21 (1959), gives full details on his experiments. R. Hide, *QJRMS*, 79 (1953), 161. A. J. Faller, *JM*, 13 (1956), 1. On Sabin, see Fultz, *op. cit.* (1959), p. 72.

15. C.-G. Rossby, *MWR*, 54 (1926), 237; *Beitr. Phys. Freien Atmos.* 14 (1928), 240; *BAMS*, 28 (1947), 53.

16. See D. Fultz, *DC*, p. 71, and D. Fultz and R. Kaylor, *ASM*, p. 359.

17. For the dynamics of vacillation, see E. N. Lorenz, *JAS*, 19 (1962), 39, and *JAS*, 20 (1963), 448.

18. D. Fultz, *JM*, 9 (1952), 379. J. Corn and D. Fultz, *AFGP*, no. 34 (1955). See also D. Fultz, *DC*, p. 71. H. Riehl and D. Fultz, *QJRMS*, 83 (1957), 215, and 84 (1958), 389.

19. R. L. Long, *JM*, 8 (1951), 207, and 9 (1952), 187. P. Frenzen, *BAMS*, 36 (1955), 204. L. F. Hubert, *TAGU*, 33 (1952), 817.

For *rotating annulus* models, see the following. On vacillation: *MWR*, 95 (1967), 75, on two kinds of vacillation; and W. W. Fowlis and R. L. Pfeffer, *JAS*, 26 (1969), 100, on amplitude vacillation. On Rossby waves, see A. Ibbetson and N. Phillips, *T*, 19 (1967), 81. On asymmetric waves, due to the simultaneous presence of a 4-wave and a 5-wave circumpolar vortex, see R. L. Pfeffer and W. W. Fowlis, *JAS*, 25 (1968), 361. For an Antarctic circulation model, see T. T. Gibson and D. A. Douglas, *W*, 24 (1969), 309. On thermal convection, see M. Bowden and H. F. Eden, *JAS*, 22 (1965), 185; see also W. W. Fowlis and R. Hide, *loc. cit.*, p. 541, on the effect of viscosity on the symmetry of flow. On the theory of rotating annulus circulations, see: P. E. Merilees, *JAS*, 25 (1968), 1003, on the transition from axisymmetric to nonaxisymmetric flow; H. A. Snyder and E. M. Youtz, *JAS*, 26 (1969), 96, on the effect on flow when differential heating of the side walls of the container is varied sinusoidally; R. Hide, *JAS*, 24 (1967), 6, on the effect of horizontal temperature gradient on vertical stability; G. P. Williams, *loc. cit.*, p. 144 and 162, and 25 (1968), 1034, on thermal convection; also R. Hide, *BAMS*, 47 (1966), 873, for a review of the dynamics of rotating fluids. For details of a *rotating paraboloid-shaped annulus* model, which produces inertial oscillations in a circumpolar vortex, exciting Rossby waves which are then propagated and result in instability of the vortex, see D. Fultz and T. S. Murty, *JAS*, 25 (1968), 779.

For *hurricane* models, see J. S. Turner, *QJRMS*, 94 (1968), 589, for a rotating water tank with ring convection, air bubbles used as tracers, and circulation driven entirely by convection in hurricane-eye wall; and R. K. Hadlock and S. L. Hess, *JAS*, 25 (1968), 161, for one with simulation of latent heat release.

20. A stochastic process may be regarded as any sequence of experiments that can be subjected to a probability analysis. See J. G. Kemeny *et al.*, *Finite Mathematical Structures* (Englewood Cliffs, N.J., 1959), pp. 146–50, for Markov chains and other stochastic processes.

21. For mathematical details, see *ibid*.

22. See A. M. Feyerhern and L. D. Bark, *JAM*, 4 (1965), 320, for full details on such calculations. J. Neumann and K. R. Gabriel, *QJRMS*, 88 (1962), 90. J. W. Hopkins and P. Robillard, *JAM*, 3 (1964), 600. L. L. Weiss, *MWR*, 92 (1964), 169.

For *statistical models*, see the following. For a simple model of wet and dry spells, see C. Chatfield, *W*, 21 (1966), 308; see also *W*, 22 (1967), 78. For a stochastic model, incorporating Markov chains and Bayes' theorem, of the frequency and duration of weather events at Minneapolis and extremely high temperatures at Dharan, Saudi Arabia, see I. I. Gringorten, *JAM*, 5 (1966), 606. See also the references in the second and fourth paragraphs of note 34 of Chapter 2.

23. For details of the U.S. Standard Atmosphere 1962, see A. E. Cole, A. Court, and A. J. Cantor, pp. 2.1–2.22 in S. L. Valley, ed., *Handbook of Geophysics and Space Environments* (AFCR Labs., Office of Aerospace Research, USAF, 1965). A. E. Cole and A. J. Kantor, *JAM*, 2 (1963), 90, and *AFSG*, no. 157 (1964). A. E. Cole and A. Court, *AFSG*, no. 151 (1962). For details of the computations involved in the new Standard Atmosphere, see A. J. Kantor and A. E. Cole, *JAM*, 4 (1965), 228, who give mean monthly cross-sections, profiles and data of atmospheric properties, surface to 80 km, for latitudes 30, 45 and 60°N. On standard atmospheres, see A. E. Cole, *MM*, 95 (1966), 236, for a comparison of temperature-height curves observed in the stratosphere over Britain with corresponding curves for standard atmospheres.

24. R. S. Scorer, *T*, 6 (1954), 23. R. M. Endlich and J. R. Clark, *JAM*, 2 (1963), 66. T. A. Gleeson, *JAM*, 1 (1962), 497.

25. N. A. Phillips, *JM*, 14 (1957), 184. The normal system was introduced by, especially, R. C. Sutcliffe, *QJRMS*, 73 (1947), 370.

26. See L. F. Richardson, *Weather Prediction by Numerical Process* (Cambridge, Eng., 1922); for a retrospective review of the book, see G. W. Platzman, *BAMS*, 48 (1967), 514; Platzman, *T*, 4 (1952), 168, outlined a logical design for a digital computer suitable for meteorological investigations, and produced the basic program and flow diagram for computation of the geostrophic vorticity field corresponding to a pressure-contour analysis.

27. A. Wiin-Nielsen, *T*, 14 (1962), 261, discusses this error.

28. For examples, see B. Bolin, *T*, 7 (1955), 27, who shows that, over an area of 9,000 × 12,000 km, the boundary influences are unimportant in the center of the area for 24-hour forecasts, but result in serious errors in 72-hour forecasts.

29. For details, J. G. Charney, R. Fjörtoft, and J. von Neumann, *T*, 2 (1950), 237.

30. *T*, 4 (1952), 21, and 6, (1954), 139.

31. B. R. Döös, *T*, 8 (1956), 76. J. Namias, *T*, 8 (1956), 206.

32. W. L. Gates, *JM*, 14 (1957), 332. R. Fjörtoft, *T*, 7 (1955), 462. E. O. Jess, *T*, 12 (1960), 21. J. Charney, *T*, 7 (1955), 22. E. Charasch, *T*, 10 (1958), 95. A. Bring and E. Charasch, *T*, 10 (1958), 88. W. H. Best, *T*, 8 (1956), 351. T. Berggren, *T*, 10 (1958), 289.

33. D. E. Martin, *T*, 10 (1958), 451. S. Williams, *T*, 10 (1958), 216.

34. B. Bolin, *T*, 8 (1956), 61. S. J. Smebye, *T*, 5 (1953), 219.

35. J. F. Blackburn and W. L. Gates, *JM*, 13 (1956), 59. For a review of the success of the forecasts, see E. B. Fawcett, *JAM*, 1 (1962), 318. Also see P. Bergthorsson *et al.*, *T*, 7 (1955), 272.

36. C. L. Jordan, *JM*, 13 (1956), 223. For details, see P. F. Clapp, *T*, 5 (1953), 80, and *JM*, 13 (1956), 341. For examples of barotropic forecasts for the tropical Pacific area, see J. Vederman, G. H. Hitara, and E. J. Manning, *MWR*, 94 (1966), 337. For experiments in barotropic forecasting of hurricane tracks in the Caribbean and North Atlantic, see F. Sanders and R. W. Burpee, *JAM*, 7 (1968), 313, and R. W. James, *JAM*, 3 (1964), 277. On error estimates for numerical prediction of hurricane trajectories, see A. Kasahara and G. W. Platzman, *loc. cit.*, p. 110.

For additional references on barotropic models, see: N. J. Macdonald and R. Shapiro, *loc. cit.*, p. 336, on the relation between persistence and forecast accuracy; F. G. Shuman and J. D. Stackpole, *MWR*, 96 (1968), 157, on finite-difference equations incorporating a map-scale factor; and D. L. Williamson, *T*, 20 (1968), 642, on a spherical geodesic grid, involving equal-area equilateral triangles, instead of the normal rectilinear grid, for integrating the barotropic vorticity equation).

37. A. Wiin-Nielsen, *T*, 11 (1959), 180. See also T. N. Krishnamurti, *JAM*, 1 (1962), 508. H. Sigtryggson and A. Wiin-Nielsen, *T*, 9 (1957), 296.

38. For a general account with mathematics, see A. Eliassen, *T*, 4 (1952), 145. See also B. Bolin, *T*, 5 (1953), 207.

39. E. T. Eady, *T*, 4 (1952), 157. G. Arnason, *T*, 4 (1952), 256, and *T*, 5 (1953), 386.

40. F. H. Bushby and M. K. Hinds, *QJRMS*, 81 (1955), 396, and 80 (1954), 165. D. E. Jones and P. Graystone, *QJRMS*, 88 (1962), 250.

41. F. H. Bushby and C. J. Whitelan, *QJRMS*, 87 (1961), 374. A. Wiin-Nielsen, *T*, 13 (1961), 320. For a review of multiparameter models, see B. Bolin, *T*, 5 (1953), 207.

42. E. N. Lorenz, *T*, 12 (1960), 364. W. L. Gates, *T*, 13 (1961), 460. A. Wiin-Nielsen, *T*, 15 (1963), 1.

43. E. J. Aubert, *JM*, 13 (1956), 207. L. Berkofsky, *loc. cit.*, p. 102, and *JM*, 14 (1957), 93. H. Økland, *T*, 15 (1963), 280. M. A. Estoque, *T*, 12 (1960), 41. For additional information on *baroclinic* models, see H. W. Ellsaesser, *JAM*, 7 (1968), 153, on faults of early baroclinic models. On the operational British Bushby-Timpson ten-level model for predicting frontal rainfall, which includes the effect of latent heat but ignores frictional and topographic effects, see F. H. Bushby and M. S. Timpson, *QJRMS*, 93 (1967), 1, and 94 (1968), 12; also see G. R. R. Benwell and F. P. Bretherton, *loc. cit.*, p. 123.

44. P. D. Thompson and W. L. Gates, *JM*, 13 (1956), 127. W. L. Gates and C. A. Riegel, *T*, 15

(1963), 406. A. P. Burger, *T*, 10 (1958), 195. N. A. Phillips, *T*, 6 (1954), 273.

45. I. Holmström, *T*, 15 (1963), 127. N. A. Phillips, *T*, 12 (1960), 121.

46. P. D. Thompson, *T*, 9 (1957), 275.

47. A. Wiin-Nielsen, *T*, 11 (1959), 45. C. E. Wallington, *QJRMS*, 88 (1962), 470.

48. R. G. Fleagle, *T*, 12 (1960), 127. L. Berkofsky, *CD*, p. 85.

49. B. Bolin and J. Charney, *T*, 3 (1951), 248. J. G. Charney and A. Eliassen, *T*, 1 (1949), 38. J. Smagorinsky, *QJRMS*, 79 (1953), 342. A. Wiin-Nielsen, *T*, 13 (1961), 127.

50. J. S. Sawyer, *QJRMS*, 85 (1959), 31. For details on ageostrophic motions, see P. Queney, *BAMS*, 29 (1948), 22, and M. Neiburger and J. K. Angell, *JM*, 13 (1956), 166. H. Reuter and H. Pichler, *T*, 16 (1964), 40, discuss the Alps.

51. G. J. Haltiner, L. C. Clarke, and G. E. Lawniczak, Jr., *JAM*, 2 (1963), 242.

52. For details, see L. Berkovsky, *JAM*, 3 (1964), 410. For additional information on numerical prediction models in general, see the following.

For simple, *graphically integrated models*, for forecasting the 1,000-mb chart, see: M. A. Estoque, *JM*, 14 (1956), 292, for one that incorporates orographic effects; R. J. Reed, *JM*, 15 (1960), 1, for one suitable for student use, and *MWR*, 88 (1960), 209; H. S. Muench, *JAM*, 3 (1964), 547, for one that incorporates ageostrophic effects: and D. A. Lowry and E. F. Danielsen, *MWR*, 96 (1968), 86, for one that incorporates variable stability.

For models making use of the *primitive equations of motion*, see: F. G. Shuman and L. W. Vanderman, *MWR*, 94 (1966), 329; L. W. Vanderman and W. G. Collins, *MWR*, 95 (1967), 950, on a barotropic model for the tropics; Y. Kurihara, *MWR*, 93 (1965), 399, on a spherical grid system for global use for numerical integration of the primitive equations; M. Grimmer and D. B. Shaw, *QJRMS*, 93 (1967), 337, on long-term global integrations using spherical polar coordinates; and A. Grammeltvedt, *MWR*, 97 (1969), 384, on finite-difference schemes for the primitive equations in a barotropic atmosphere. See F. G. Shuman and J. B. Hovermale, *JAM*, 7 (1968), 525, for a six-layer baroclinic model that became operational at the National Meteorological Center in the United States in mid-1966.

For the use of *spherical harmonic (spectral) series*, instead of finite-difference grid methods, for integrating the vorticity equation for numerical prediction purposes, see: G. W. Platzman, *JM*, 17 (1960), 635, for a spectral form of the vorticity equation, and *JAS*, 19 (1962) 313, on the analytical dynamics of the spectral vorticity equation; H. W. Ellsaesser, *JAM*, 5 (1966), 263, on methods for expanding northern hemisphere meteorological data in surface spherical harmonic Laplace series, and *loc. cit.*, p. 246, on a comparison of spectral and grid methods for northern hemisphere data; and A. Robert, *JAM*, 7 (1968), 83, for an application to global 500-mb charts, and *loc. cit.*, p. 730, for a method that incorporates moisture/precipitation aspects of the general circulation.

For the operational use of the *balance-wind equation*, to provide accurate nondivergent wind fields for determining the initial stream functions at 500 mb for the barotropic model, and at other levels for baroclinic models, see R. Asselin, *T*, 19 (1967), 24. On the reduction of truncation error in balance-equation models, see A. Sundström, *MWR*, 97 (1969), 150. On *grids*, see B. Culbertson and P. T. Willis, *JAM*, 3 (1964), 541, for the effect of grid spacing on the accuracy of predictions made by numerical models, and G. E. Hill, *JAM*, 7 (1968), 29, on "grid telescoping," which enables numerical predictions to be made on a very fine (local) scale. On *initial conditions*, see R. H. Jones, *JAS*, 22 (1965), 685, for statistical techniques that, in effect, employ earlier numerical predictions to obtain best estimates of initial conditions for a forecast. On the incorporation of *radiation effects* into numerical models, see D. B. Danard, *MWR*, 97 (1969), 77, on longwave radiation in tropospheric models, and J. C. Gille, *BAMS*, 48 (1967), 714. On the incorporation of *wind profiles* into numerical models, see H. W. Ellsaesser, *MWR*, 96 (1968), 277.

On *multilevel models*, see K. Miyakoda, *et al.*, *MWR*, 97 (1969), 1, on a nine-level model for two-week predictions of precipitation distribution over the northern hemisphere that incorporates orographic and sea-surface temperature effects. As an alternative to multilevel models, see L. Berkofsky, *JGR*, 68 (1963), 4187, for *vertically integrated numerical models*, which assume that the variations with height of meteorological quantities are known, and then introduce these assumptions into equations for the vertically averaged atmospheric flow. For a two-level quasigeostrophic model with variable Rossby parameter (β), see A. Huss, *AMGB*, 16, series A (1967), 105. On the generation of *monthly snow-cover frequencies* within numerical prediction models based on North American data, see P. F. Clapp, *JAM*, 6 (1967), 1018.

53. J. Vederman, *MWR*, 89 (1961), 243. W. H. Klein *et al.*, *T*, 12 (1960), 378. J. Adem, *MWR*, 92 (1964), 91. E. N. Blinova and I. A. Kibel, *T*, 9 (1957), 453. For a simple mathematical model, based on the energy balance of the Earth-atmosphere system, which predicts average annual sea-level temperature in 10° latitude belts, see W. D. Sellers, *JAM*, 8 (1969), 392. For prediction models for the distribution of monthly and seasonal temperature and precipitation over the northern hemisphere, see J. Adem, *MWR*, 93 (1965), 495.

54. B. Saltzman, *JAS*, 20 (1963), 226. J. Adem and P. Lezama, *T*, 12 (1960), 255. H. Wexler, *T*, 13 (1961), 368. J. F. de Lisle and J. F. Harper, *T*, 13 (1961), 56. J. S. Malkus, *T*, 8 (1956), 335.

55. J. G. Charney and M. E. Stern, *JAS*, 19 (1962), 159. G. J. Haltiner, *T*, 15 (1963), 230. F. B. Lipps, *JAS*, 20 (1963), 120.

56. J. S. Malkus and H. Riehl, *T*, 12 (1960), 1. T. N. Krishnamurti, *T*, 13 (1961), 171, and 14 (1962), 195. M. A. Estoque, *T*, 14 (1962), 394. S. Syōno, *T*, 5 (1953), 179. A. Kasahara and G. W. Platzman, *T*, 15 (1963), 321. H. Riehl, *JAS*, 20 (1963), 276.

For whole-atmosphere models, see: E. N. Lorenz, *T*, 17 (1965), 321, for a 28-variable model to test the predictability of the atmosphere; M. Sankar-Rao, *MWR*, 93 (1967), 213, on the stationary harmonics of atmospheric motion; and B. Saltzman, *GPA*, 57, no. 1 (1964), 153, and *GPA*, 69, no. 1 (1968), 237, for a model of an axially symmetric, time-averaged atmosphere that predicts time-averaged climatic variables.

For planetary-scale phenomena, see: R. J. Deland and Yeong-jer Lin, *MWR*, 95 (1967), 21, on traveling planetary waves at 500 mb; P. A. Gilman, *T*, 16 (1964), 160, on mean meridional circulation in presence of steady-state, symmetric circumpolar vortex; B. Saltzman, *MWR*, 93 (1967), 195, on forced, stationary perturbations of the average winter atmosphere, caused by internal heating and cooling of the atmosphere and flow over topography; and T. W. Kao, *JGR*, 70 (1965), 815, on blocking in a stratified atmosphere. For models of the stratospheric circulation, see: Li Peng, *GPA*, 61, no. 2 (1965), 191, on the lower stratosphere; J. M. Wallace and J. R. Holton, *JAS*, 25 (1968), 280; and R. J. Reed, *QJRMS*, 90 (1964), 441, on the tropical 26-month oscillation. See also I. Tolstoy and T. J. Herron, *JAS*, 26 (1969), 270, for a mathematical model predicting jetstream winds from ground-level pressure fluctuations.

For tropical phenomena, see: W. J. Koss, *MWR*, 95 (1967), 283, on tropospheric waves in equatorial latitudes; T. N. Krishnamurti, *QJRMS*, 95 (1969), 594, for a comprehensive model predicting the I.T.C.Z. and associated disturbances and tropical storms, and incorporating cumulus convection; T. N. Krishnamurti and D. Baumhefner, *JAM*, 5 (1966), 396, on an easterly wave associated with an upper cold-core cylcone; Man-Kin Mak, *JAS*, 26 (1969), 41, on large-scale tropical eddies; C. S. Barrientos, *JAM*, 3 (1964), 685, on transverse circulations in a steady-rate hurricane; and K. Ooyama, *JAS*, 26 (1969), 3, on the life-cycle of tropical cyclones. On numerical prediction of tropical cyclones, see S. L. Rosenthal, *MWR*, 92 (1964), 1. For a model of the surface boundary layer of a hurricane, see R. K. Smith, *T*, 20 (1968), 473.

For subtropical phenomena, see S. L. Rosenthal, *MWR*, 95 (1967), 341, for a mathematical model of synoptic-scale disturbances over the subtropical oceans. For mathematical models of cyclogenesis, see: H. Økland, *MWR*, 93 (1965), 663, on the northern hemisphere; also, for one incorporating latent-heat release, M. B. Danard, *JAM*, 5 (1966), 85 and 388,

on North America. For a model of a cut-off low in the northwestern United States, see D. W. Stuart, *JAM*, 3 (1964), 669. For mathematical models of fronts, see H. Økland, *T*, 21 (1969), 359, on frontogenesis in a deepening cyclone, and J. Holmboe, *T*, 18 (1966), 830, on the growth of a young frontal wave.

57. Y. Ogura, *JAS*, 19 (1962), 492, and 20 (1963), 407. J. M. Richards, *loc. cit.*, p. 241. J. R. Herring, *loc. cit.*, p. 325. J. Warner, *loc. cit.*, p. 546. For mathematical models of convection, see: H. J. Lugt and E. W. Schwiderski, *JAS*, 23 (1966), 54 and 808, on local heating of a horizontal surface; M. A. Estoque, *JAS*, 25 (1968), 1046, on penetrative convection, enabling environmental lapse- and hydrolapse-rates to be estimated; E. C. Nickerson, *JAS*, 22 (1965), 412, on buoyant convection; A. I. Barcilon, *JAS*, 25 (1968), 796, on buoyant jets in stagnant surroundings, somewhat analogous to thunderstorm updrafts; H. L. Kuo, *JAS*, 23 (1966), 25, on convective vortices; and R. C. J. Somerville, *JAS*, 24 (1967), 665, on an atmosphere unevenly heated from below, producing an asymmetric, direct solenoidal circulation, resembling a sea breeze or Hadley cell and related to Bénard cells.

58. B. R. Döös, *T*, 14 (1962), 133. P. A. Davis, *JAS*, 20 (1963), 5. P. Welander, *T*, 11 (1959), 348.

59. For a general discussion, see J. S. Malkus, *SP*, 43 (1955), 461.

60. R. C. Smith, *QJRMS*, 81 (1955), 382, and 83 (1957, 248. For a model of diurnal temperature variation in a non-grey atmosphere, driven by a harmonically oscillating ground temperature, see P. J. Gierasch, *JAS*, 26 (1969), 65. For a model of meridional temperature gradients contrary to radiational forcing, see Li Peng, *GPA*, 62, no. 3 (1965), 173.

For a model of airflow over a heated coastal mountain (the Santa Ana Mountains in southern California), see M. A. Fosberg, *JAM*, 8 (1969), 436. For a model of vertical motion due to airflow over irregular terrain, see L. Berkofsky, *JAM*, 3 (1964), 410. For a model of orographic precipitation incorporating mountain-wave theory, which indicates that precipitation due to the Western Ghats of India extends 40 km to leeward of the crest of the range, see R. P. Sarker, *MWR*, 95 (1967), 673.

For a model of a viscous boundary layer in a stratified, rotating atmosphere and its effect on transient motions, see J. R. Holton, *JAS*, 22 (1965), 402. For a model of momentum flux due to mountain waves in a stratified, rotating atmosphere, see W. Blumen, *loc. cit.*, p. 529.

61. E. L. Fisher and P. Caplan, *JAS*, 20 (1963), 425. J. E. McDonald, *loc. cit.*, p. 476. C. E. Jensen, *JAM*, 2 (1963), 337. E. Hesstvedt, *T*, 14 (1962), 297. P. Squires and J. S. Turner, *T*, 14 (1962), 422. D. K. Lilly, *T*, 14 (1962), 148. G. J. Haltiner, *T*, 11 (1959), 4. G. J. Haltiner and E. M. Chase, *T*, 12 (1960), 393. H. L. Kuo, *T*, 13 (1961), 441.

For a model of a cloud, see H. L. Kuo, *JAS*, 22

(1965), 40, and for its use in estimating the area covered by convective clouds in the Caribbean from observed moisture convergence, see T. N. Krishnamurti, *JAM*, 7 (1968), 184. For a model of the initiation of cumulus over mountainous terrain, see H. D. Orville, *JAS*, 22 (1965), 684, and, for grid-interval effects, *JAS*, 25 (1968), 1164. For a model of precipitation from winter storms in the eastern United States, incorporating latent-heat release, see F. Sanders and D. A. Olson, *JAM*, 6 (1967), 229; see also F. K. Schwarz, *JAM*, 7 (1968), 297.

62. G. J. Haltiner, *T*, 11 (1959), 452, and 13 (1961), 438.

63. B. R. Döös, *T*, 13 (1961), 305. R. R. Long, *T*, 8 (1956), 460. A. J. Abdullah, *JGR*, 71 (1966), 1953.

For a model of diurnal wind variation in the planetary boundary layer in the northern hemisphere, see K. Krishna, *MWR*, 96 (1968), 269. For a model of diurnal oscillations of tropospheric winds above a low-level jet, see G. L. Darklow and O. E. Thompson, *JAS*, 25 (1968), 39.

PART II

Some Applications
of the Basic Techniques

CHAPTER 4

Radiation Climatology

Solar radiation is the ultimate control of weather and climate, because it is the only source of the Earth's energy. It is most complicated to study, since a fundamental law of nature is that a body, on receiving radiant energy, itself begins to radiate energy. The Earth's surface receives energy radiated from clouds and from the atmospheric particles themselves, and itself radiates energy back to the atmosphere and to space. There is also a continual exchange of radiation between all elements of the landscape, inorganic and organic. It is difficult to separate these different radiations, in particular to distinguish "meteorological" radiation, the net balance between all solar and atmospheric radiations, from "geographical" radiation, the net balance between the streams of energy radiated and absorbed by the various features and objects of the Earth's surface.

There is no single instrument that will satisfactorily measure all significant variations of either meteorological or geographical radiation, but there are many radiation instruments. We must consider radiation instrumentation in detail, because the published climatological data on radiation cannot be compared unless the limitations of the instruments from which the data were derived are clearly understood.

Figure 4.1 presents some climatological radiation distributions on the macroscale. Microscale distributions of radiation are functions of many local radiative sources, rather than of solar, atmospheric, and general terrestrial sources; it is not feasible to construct maps of them. Figure 4.1 definitely indicates that the geographical distribution

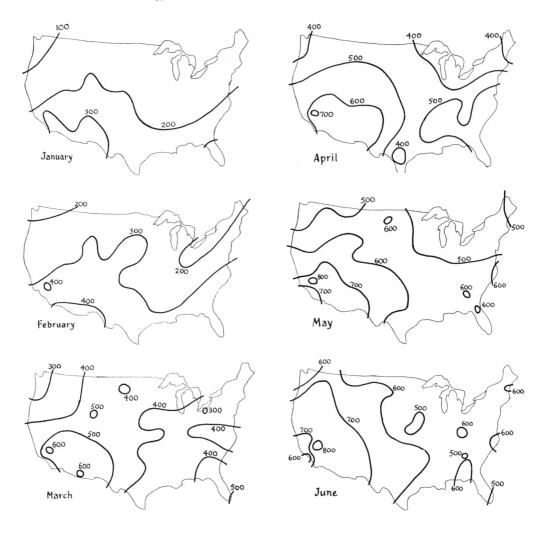

of radiation varies considerably in time and space, just as the other climatic elements do. In other words, although the relative positions of Earth and sun follow their well-established annual pattern, the distribution of solar radiation on the Earth's surface is by no means as regular. The effect of atmospheric absorption and reradiation of solar energy, and terrestrial absorption and emission of both direct and diffuse solar radiation, is so variable both geographically and temporally that the actual pattern of radiation distribution at the Earth's surface is almost as turbulent as the patterns for the other elements of weather and climate.

Most radiation maps and tables in standard textbooks are based on theoretical calculations, usually those of Simpson or of Baur and Phillips, and not on actual observation. Radiation theory has still to be employed in many problems, but observations are now beginning to accumulate.

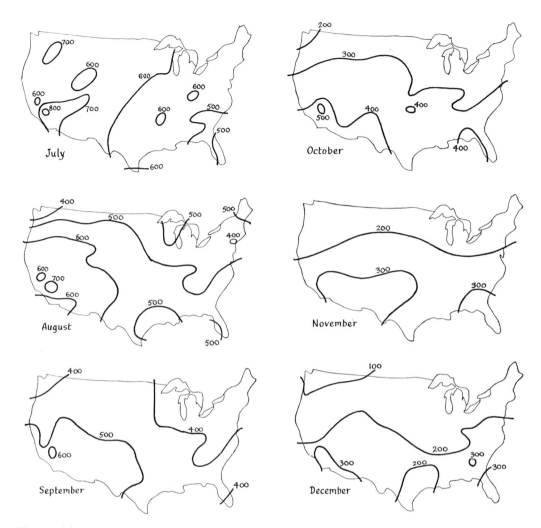

FIGURE 4.1.
Monthly mean radiation patterns for the United States in 1960. Average daily solar radiation (direct plus diffuse) received on a horizontal surface, in langleys (cal per cm²). (After the U.S. Weather Bureau.)

The Physical Nature of Radiation

Radiation is the only important form of heat transfer in interplanetary space, and is thus the process that transports electromagnetic energy from the sun to the Earth. One can calculate from astronomical theory exactly how much solar radiation is received at any point on the outer edge of the Earth's atmosphere, how much would be received at any point on the Earth's surface if there were no atmosphere, and how much seasonal variation in these amounts there is (see Appendix 4.1).

Known radiation covers a wide range of wavelengths, from above 10^{10} cm for electricity in power lines to well below 10^{-14} cm for cosmic rays. The wavelengths of electromagnetic energy are usually measured in terms of the micron, μ (10^{-4} cm), or the Ångström unit, Å (10^{-8} cm). Visible light occupies the range from 0.4 to 0.7 μ;

and the appreciable solar and terrestrial radiation has wavelengths mainly between 0.15 and 120 μ. The most important part of the solar radiation, from a meteorological viewpoint, is from 0.15 to 30 μ.

Every body of matter radiates heat energy, the amount of heat radiated being proportional to the fourth power of the absolute temperature of the body (see Appendix 4.2). This relationship, the *Stefan-Boltzmann Law*, applies to a black body, i.e., a body that radiates the maximum possible amount for a given temperature. Such a body is said to have a continuous spectrum. Some objects act only partly as black bodies, not emitting radiation at certain wavelengths, and are said to have discontinuous spectra, with the gaps in their spectra being described as absorption lines. Snow and ice surfaces and clouds absorb long-wave radiation fairly uniformly, but reflect short waves. The gases and vapors in the atmosphere have discontinuous spectra. The solar spectrum shows numerous fine lines; those produced by absorption of solar radiation by gases and vapors in the sun's atmosphere are termed *Fraunhofer lines*.

The sun acts effectively as a black body for the range of radiation from 0.1 to 30 μ. Long observations of solar radiation reaching the Earth indicate that the incoming radiation at the outer limits of the atmosphere varies only slightly, if at all, from 2.00 cal per cm^2 per min, the *solar constant*. From this, the sun should have a temperature of 5,750°K, which is its effective temperature; the temperature it should have on the basis of its color is 6,108°K, as determined from Wien's law (see Appendix 4.3).

According to Planck's law, a black body at the temperature of the sun should radiate approximately 99 per cent of its energy between the wavelengths 0.15 and 4 μ. One-half of this radiation will be between 0.38 and 0.77 μ, forming visible radiation; the remainder will be invisible, forming short-wave (ultraviolet) or long-wave (infrared) radiation. The geographical distribution of visible solar radiation is well-known, but that of ultraviolet or infrared radiation is less well-mapped.

Terrestrial temperatures lie within the general range 200 to 330°K, giving radiation from 4 to 120 μ. Solar and terrestrial radiations thus do not overlap; the former is generally regarded as short-wave radiation and the latter as long-wave radiation. Earth radiation is, of course, invisible, and lies in the infrared region of the electromagnetic spectrum.

The amount of solar radiation that would be received at a point on the Earth's surface if there were no atmosphere depends on four main factors. First, latitude determines the angle of incidence of the solar beam, and the duration of the solar radiation (i.e., the length of day). Second, the ellipticity of the Earth's orbit around the sun means that the Earth is closer to the sun in January (91,300,000 miles) than in July (94,500,000 miles), and although this must have some effect on radiation received at the Earth's surface, the exact effect is uncertain.

Third, the revolution of the Earth around the sun, because the Earth's axis is inclined at an angle of 66½ degrees to the plane of the ecliptic, results in the seasons. On or about March 21 and September 23, the sun is vertically overhead at the equator, at which time (the vernal and autumnal equinoxes), days and nights are equal in length all over the Earth. At the time of the solstices, the sun's vertical rays reach their extreme in poleward migration, and days and nights are the most unequal in length. At the summer solstice (on or about June 21), the sun is vertically overhead at the Tropic of Cancer; at the winter solstice (on or about December 22), it is vertically above the Tropic of Capricorn.

Fourth, the amount of radiation emitted from the sun varies with time. Some of the variations are irregular; for example, the solar constant seems to fluctuate by 1 or 2 per cent. Other variations are periodic, for example, those associated with sunspots and solar storms.

Let us consider what the radiation distribution on the Earth would be if there were no atmosphere. In the Earth's winter hemisphere there would be a region of nil radiation, and maximum radiation (more than 1,000 cal per cm^2 per day on a horizontal plane) would fall in the polar regions of the summer hemisphere. The December maximum in the southern hemisphere would be greater than the June maximum in the northern hemisphere, the Earth being closer to the sun in December. At the time of an equinox, the latitudinal distribution of radiation would be symmetric, with nil receipt at the poles, and maximum receipt (about 900 cal per cm^2) at the equator. At the solstices the distribution would be very asymmetric. On June 21, the radiation receipt would steadily increase as one moved north, from nil between the South Pole and latitude 68°S, to a maximum of more than 1,100 cal per cm^2 at the North Pole. There would be a distinct drop in radiation receipt for some distance north of latitude 60°N, where the increasing obliquity of the solar beam offsets the increasing length of day as the pole is approached.

The presence of the atmosphere weakens the intensity of solar radiation reaching the Earth's surface. Three processes are important here. First, the solar radiation is scattered by small particles in the atmosphere (molecules of dust and of air itself) whose diameter is less than the wavelength of the radiation (see Appendix 4.4); the clear daylight sky appears to be blue (when one looks across the solar beam) because all shorter wavelengths in the solar beam have been scattered by the air molecules and so reach the eye. Second, all wavelengths are diffusely reflected by particles in the atmosphere (dust specks and cloud droplets) whose diameter is greater than the wavelength of the radiation. Since the reflection is nonselective, the sunlight reflected by a cloud contains all the original wavelengths in equal amounts, so the cloud appears to be white if the sun is behind the observer. Third, some of the longer wavelengths are absorbed by water vapor. If the atmosphere contains a relatively high percentage of water vapor, much of the heat radiated by the Earth during the night will be absorbed by the atmosphere and not lost into space; nocturnal cloud cover prevents the development of very low temperatures at the Earth's surface. Ozone and oxygen also absorb solar radiation, and ozone absorption is of great importance in the upper atmosphere.[1]

Radiation Measurements

The fundamental instruments for radiation measurement were not invented until the early twentieth century. They include the Ångström compensation pyrheliometer, standard in Europe and Africa; the Abbot silver-disc pyrheliometer, standard in North America, Australia, and most of Asia; and the Michelson bimetallic actinometer, which is easy to read, but must be checked at regular intervals and is not suitable as a standard reference instrument.

The goal at first was to measure direct solar radiation at normal incidence. Instantaneous measurements presented many problems, but monthly means and other climatological data could be provided fairly accurately. The IMO in 1905 recommended the Ångström pyrheliometer as standard for all work of this sort; this entailed the adoption of the *Ångström scale of radiation*, introduced near the end of the last century. In Europe, the standard instrument was established by the Swedish meteorological service at Uppsala, and other Ångström pyrheliometers are still calibrated with it today.

In 1913, the Smithsonian Institute introduced a radiation scale based on the Abbot pyrheliometer, the standard instrument being established in Washington. The Smithsonian Institute produced an improved scale in 1932, and announced that their 1913 scale was nearly 2.5 per cent too high. For very clear atmospheric conditions, the ratio of the 1913 Smithsonian scale to the Ångström scale is 1.035 to 1, although occasional differences of up to 6 per cent have been observed. The Ångström and 1932 Smithsonian scales differ by only 1 per cent in clear conditions. The WMO at Davos in 1956 recommended the universal adoption by January 1, 1957, of a new *International Pyrheliometric Scale*. All Ångström scale readings have to be increased by 1.5 per cent to bring them into conformity with the international scale, and all 1913 Smithsonian readings must be reduced by 2.0 per cent. Radiation comparisons based on climatological data must therefore be approached very warily.

In the 1930s, it was discovered that, although diffuse sky radiation is less intense than direct solar radiation, its climatological sum, over a year, for example, is about as great. The reason, of course, is that diffuse radiation goes on throughout the day unchecked, whereas direct radiation only reaches the ground when the sun is not obstructed by clouds.

The commonest radiation measurement today is of *global radiation*, which is the sum of the diffuse radiation and the vertical component of direct radiation. Diffuse radiation is usually measured by means of a horizontal thermopile, under a glass globe opaque to long-wave radiation, and shielded from the sun by a disc. The disc must be so arranged that it moves with the sun. If the disc is omitted, the thermopile measures the global radiation.

Total radiation means the sum of all long- and short-wave radiation. Of the 318 stations and meteorological institutes measuring radiation in 1954, 90 per cent recorded the total radiation from sun and sky on a horizontal surface. During the IGY, radiation networks were expanded. In the Soviet Union, for example, 200 stations were set up to record solar radiation, but only one-quarter were self-recording.

There are, besides visible global radiation, also invisible radiation fluxes, in particular of long-wave infrared radiation from both the Earth and the atmosphere during both day and night. Such radiation can be measured easily only during the night, when there is no visible radiation to complicate matters. Atmospheric long-wave radiation is especially difficult to measure.[2]

Radiation-Measuring Instruments

In general, *actinometers* measure direct solar radiation, *pyranometers* measure scattered and global radiation, and *balance-meters* measure net radiation balance. Pyranometers are usually divided into three groups: pyranometers, for measuring short-wave radiation; effective pyranometers, for measuring combined long- and short-wave downward fluxes; and infrared effective pyranometers, for measuring long-wave downward flux. Radiation balance-meters measure the difference between all upward and downward radiation fluxes, and infrared balance-meters the difference between upward (terrestrial) and downward (atmospheric) long-wave radiation. The radiation balance is counted as positive if it represents a gain in heat for the ground. *Pyrgeometers* determine the long-wave terrestrial radiation (see Appendix 4.5).

Radiometer is a general term for radiation-measuring instruments. General problems in radiometer design include: the choice of a suitable absorbant blackening material (usually paint) with which the radiation receiver surface is coated; the choice of an appropriate "window" material, to protect the receiver surface from rain, yet expose it to the range of wavelengths of radiation that are to be measured; and, finally, artificial ventilation. Most radiometers can be used in calm weather only, because heat transfer from the receiver surface to the air varies with windspeed. This difficulty can be overcome by artificial ventilation, making the velocity of air movement over the receiver surface constant; by heating the receiver surface, usually electrically; or by using filter plates or caps to shelter the receiver from the wind.

Radiometers present special problems for observers. There are the questions of standardization and installation of the instruments; there is the problem of choosing the reference surface above which the radiation balance is to be measured.

Long-wave radiometers have perhaps created the most difficulties. For many years, it was possible to measure long-wave radiation only in the absence of solar radiation. The covers or "windows" of glass or quartz, which, in radiometers measuring direct solar or diffuse sky radiation, protect the receiver surface from wind effects, cannot be used, because they are not transparent to long-wave terrestrial or atmospheric radiation. Instead, compounds such as polyethylene must be used. By 1956, there were twenty different types of instrument for measuring long-wave radiation, and no one type could really be regarded as standard.

There are so many different types of radiation-measuring instrument that it is almost impossible to present a complete yet readable account. It is nevertheless essential to know something about these types and their relative merits and disadvantages, not only because without this knowledge it is impossible to interpret published radiation data and maps intelligently, but also because such knowledge brings home the realization that the basic driving force behind climate, i.e., the variation of solar energy at the Earth's surface and in its atmosphere, is still incompletely charted.

For meteorological purposes, local measurements are not as necessary with radiation as with other elements, because once data for the given solar elevation are known for clear-sky conditions (and such data can be regarded as constant for a comparatively large area), one need only know how much cloud cover there is in the locale to derive a fair estimate of its average radiation conditions. For climatological purposes, however, the local radiation needs to be known as accurately and minutely as possible, particularly in applied, as distinct from dynamic or synoptic, climatology investigations.

The following account divides radiometers into three groups: first, classical and later instruments that either have been adopted as standard instruments for calibration purposes, or are used for continuous recording at observatories; second, instruments that have been used mainly for research purposes or for specific investigations, but not for continuous long-period records; third, instruments used for measuring atmospheric infrared radiation.*

---

* For an account of the measurement of nocturnal radiation at stations in India, using the Ångström pyrgeometer, see Y. Viswanadham and R. Ramanadham, *GPA*, 78, no. 3 (1967), 214.

STANDARD RADIATION-MEASURING INSTRUMENTS

The first standard instrument, of which large numbers are in use today, was probably the *spherical pyranometer* invented by Bellani in 1836. This converts radiation into heat, which causes the distillation of a volatile liquid into a graduated condenser; the volume of liquid condensed gives the total, or integrated value, of radiation received. Several refinements have been added to the original instrument (see Figure 4.2).[3]

Bellani suspended the instrument with its receiver and condenser in full sunlight; the condenser was subject to considerable temperature variation, and distillation was erratic. Gunn protected the condenser from sunlight with a light metal shield, reducing temperature fluctuation, and Pereira kept the ambient temperature of the condenser constant by suspending the instrument in a vertical steel cylinder sunk into the ground, so that the top of the bulb was at ground level. The original instrument had a glass receiver, which Gunn in 1945 replaced by a copper one, in the form of a spherical bulb coated with lamp black and sealed into an evacuated glass tube leading to the condenser. The instrument is very simple and comparatively inexpensive. Its efficiency is 70 to 75 per cent if alcohol is used as the distillate (see Appendix 4.6).

A water-filled Bellani-Gunn radiometer is very suitable for reading once a day, and gives the (relative) daily integrated total of short-wave solar radiation at ground level. It also gives integrated evaporation totals that compare quite well with values obtained from evaporation tanks. The instrument is very suitable for use in the tropics, particularly at relatively high altitudes, where seasonal radiation changes are at a minimum

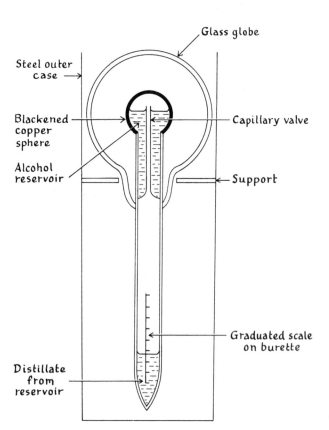

FIGURE 4.2.
Bellani radiometer.

and effective ambient temperatures are controlled very largely by soil temperatures and are independent of the season. In temperate latitudes, weekly totals from the alcohol-filled instruments are reliable, but daily totals may vary greatly, since there is a temperature-dependent threshold, below which there is negligible distillation (see Appendix 4.7).

Most standard actinometers are designed to measure total (i.e., direct solar plus diffuse sky) radiation in all wavelengths over the range 0.3 μ to 4 μ, the radiation intensity being expressed as the amount of radiant energy falling on a unit area of receiver surface in a unit time. Two types of plane surface are usually employed: one is horizontal; the other at "normal incidence," i.e., perpendicular to the line joining the place of observation to the instantaneous position of the sun; the latter type restricts radiation to that coming from directions lying within about 5° of the direction of the sun's center. Of the standard instruments of the latter type, the Ångström pyrheli-ometer, the Abbot silver-disc pyrheliometer, and the Moll-Gorczynski thermopile, the first two are suitable for spot readings only, and hence are mainly used for calibration purposes with their "angle of view" centered on the sun. The thermopile instrument can be made to record continuously by linking it to a recording galvanometer. By fitting it to a heliostat driven by a clockwork or synchronous electric motor, so that the thermopile element is rotated about an axis perpendicular to the plane of the celestial equator, the sensitive surface of the instrument can be kept perpendicular to the line joining the thermopile to the sun. The axis of rotation must, of course, be continually adjusted as the angle of declination of the sun changes.

The *Ångström pyrheliometer* is the standard "absolute" instrument for measuring direct solar radiation at normal incidence, and is the only actinometer in which all relevant factors can be determined (see Appendix 4.8). It consists essentially of two identical strips of thin manganin or platinum, coated black on their upper surfaces, and with similar mountings; one strip is exposed to solar radiation, the other shielded from it. Insulated copper-constantan thermocouple junctions are attached to the underside of each strip, connected in mutual opposition, in series with a sensitive galvanometer. The strips are mounted in a case fitted with diaphragms to limit the area of sky "seen" by the exposed strip.

In operation, an electric current from a battery is passed through the shielded strip until the temperatures of the two strips are equal, and is measured by an accurate milliammeter. The rate at which heat is developed in the shielded strip because of its electrical resistance equals the rate at which radiant energy is absorbed by the exposed strip. The rate at which heat is generated in the shielded strip can be calculated theoretically from the dimensions and resistance of the strip (see Appendix 4.9). By taking into account the absorption of the blackening material, one can accurately calculate the received radiation intensity. Absorption by the blackened strip may be imperfect, and the exposure of the strips may differ very slightly, so that in practice, although the instrument is theoretically absolute, it is usual to compare individual pyrheliometers with the standard one at Uppsala every five years or so.

The *Abbot silver-disc pyrheliometer* is a simpler instrument, but provides only relative measurements. It measures the direct solar radiation that falls on a massive, blackened silver disc supported by fine wires inside a copper case, the latter being placed inside a wooden box to reduce temperature changes. Diaphragms limit the angle of acceptance of the radiation to a cone of semiangle 5°, and the radiation can be com-pletely cut off from the disc by means of a shutter. A mercury thermometer with its bulb attached to a cavity in the disc allows the latter's temperature to be read. The instrument is first set up with its shutter closed, and thermometer readings are taken after, say, 20 and 120 seconds. The shutter is then immediately opened, and the readings repeated. The intensity of radiation received can then be found from the differences in temperature during the time interval chosen (see Appendix 4.10). The instrument maintains its calibration for many years, and is a secondary standard in America.

The *Michelson actinometer* is also a useful secondary standard instrument, and is very portable, needing no external source of power. Its main feature is a thin, blackened bimetallic strip enclosed in a narrow chamber in a metal case, with a fine quartz fiber attached to an extension of one end of the strip, the other end of which is fixed. Radiation is allowed to fall perpendicularly on the strip, which changes its curvature as its temperature rises. The resulting movement of the fiber is observed by means of a microscope with a graduated scale in its eyepiece. The bimetallic strip has a lag of 20 to 30 seconds, and so will not respond to instantaneous radiation changes. It is also influenced by temperature, so that zero-setting screws are necessary. It is not, of course, an absolute instrument, and must be calibrated against a standard Ångström pyrheliometer.

The *Eppley pyrheliometer* has been used extensively in the United States. It measures hemispherical radiation of wavelengths less than 3μ on a flat surface. The receiving surfaces consist of a central white disc surrounded by a concentric black surface, mounted horizontally in the center of a clear, spherical glass bulb. The surfaces are made by coating copper rings with magnesium oxide and with soot in shellac, respectively, and 10- or 50-junction thermopiles are attached to them. A tripod base with spirit level and leveling screws is provided. The difference in temperature between the black and white surfaces is recorded by a potentiometer or galvanometer.

Errors arise in its readings because the rings have variable temperature coefficients, the thermopile output varies with the direction of gravity and with air convection inside the glass bulb, and radiation may come from the rear of the instrument or be reflected by the glass cover. The absorptivity of the black ring depends on the angle of incidence of the radiation beam: with low sun and a clear sky, the instrument underestimates the radiation intensity, probably because of reflection from its receiving surface. It records a higher intensity of radiation when temperatures are low than when they are high, the error being 1 or 2 per cent for each 10°C change in ambient temperature. The Eppley pyrheliometer therefore provides relatively coarse measurements. It is not really suitable for use in high latitudes, and measurements taken with it on bare soil or other relatively absorbent surfaces have a greater proportionate error than those taken on ice or similar reflecting surfaces. It measures both short-wave and diffuse solar radiation.[4]

Thermopile instruments generally involve exposing a thin blackened surface, supported inside a relatively massive polished case to maintain temperatures uniformly over the surface, to the solar radiation. The blackened surface rises in temperature until its rate of heat loss by all processes equals its rate of heat gain by radiation. The difference in temperature between the receiving surface and a reference point inside the case is then measured by thermojunctions in series. If the temperature rise is to be a function of radiation intensity only, the blackened surface must have a small lag coefficient, be independent of changes in ambient temperature, and be shielded from wind currents, and convection currents must not arise inside the case.

For measuring direct solar radiation at normal incidence, the Moll-Gorczynski large-surface thermopile has been used for many years at Kew, in the *Gorczynski pyrheliograph*.[5] The thermopile consists of 80 thermojunctions on an equatorial mounting driven by a clock. A heliostat keeps the thermopile surface normal to the direct rays from the sun, and three metal diaphragms ensure that only radiation from the sun and from a narrow annulus of sky falls on the thermopile. The latter is protected by a glass cover, and its output is carried to a recording millivoltmeter, which is calibrated against an Ångström pyrheliometer.

The *Moll-Gorczynski solarimeter* measures both total and diffuse radiation falling on a horizontal surface. The sensitive surface consists of thin strips of constantan and manganin, arranged alternatively and blackened. The active thermal junctions run along the center line of this surface; the inactive junctions are in thermal contact with supporting posts that are electrically insulated from the base plate of the instrument. The receiving surface is covered by two concentric hemispheric glass domes, which shield it from rain and wind, and reduce convection currents inside the case. The remainder of the instrument is mounted in the center of a circular metal plate, the top surface of which is flush with the sensitive surface. The guard plate prevents direct solar radiation from reaching any part of the instrument other than the blackened strips. When radiation falls on the latter, their temperature rises, setting up an EMF that is measured by means of either a galvonometer (on land) or a potentiometer (at sea). For diffuse-radiation measurements, a shade is placed over the glass domes to cut off direct rays. A shutter can also be placed over the glass domes, so that a meter reading for zero radiation can be obtained at any time during the day (rather than having to be taken at night) and heat lost from the sensitive surface by conduction through the air to the glass domes be allowed for. The instrument is calibrated by comparing it with a standard instrument under normal conditions of use. Laboratory calibration will give wrong results if the radiation transparency of the glass hemispheres is not the same as that of the standard for all wavelengths.

The *Linke-Feussner actinometer* is somewhat similar to the Moll-Gorczynski solarimeter but has no glass domes. The thermopile is mounted within a shell consisting of a series of concentric, milled copper rings, highly polished so that they absorb only a very small amount of heat when exposed to direct solar radiation. The receiving surface is directional, only accepting radiation over a cone of semiangle 5°, so that it can be used to measure radiation from selected parts of the sky as well as for direct solar radiation at normal incidence. Filters allow different wavelengths of radiation to be investigated, and if a very sensitive galvanometer is used, the instrument is also suitable for measuring long-wave earth radiation.

The *Robitzsch actinograph*, designed for continuous recording of sun and sky radiation received on a horizontal surface, was introduced in 1932.[6] Its receiving element consists of three horizontally mounted bimetallic strips, arranged so that the two outer strips—painted white so as to absorb as little radiation as possible—act opposite to the center (blackened) absorbing strip. The end of the latter moves a pen, via a lever system, across a recording chart attached to a clockwork-driven drum that usually rotates once every 24 hours. A glass hemisphere protects the elements from rain and dust.

The *M.O. bimetallic actinograph*, the standard United Kingdom instrument, was developed from the Robitzsch design.[7] In this instrument, all three strips are attached to a rod at one end; at the other end, the whitened strips are fixed to the frame of the instrument, but the blackened strip is attached to a lever mechanism and recording pen. The movement of the pen arm is proportional to the temperature difference between the strips, and hence proportional to the total incoming radiation received. A heavy case is provided to eliminate effects of ambient temperature changes. The instrument is simpler than the thermopile instruments, but has a fairly large lag (5 to 10 minutes, compared with the 10 seconds of the solarimeter thermopile), and this, plus the relatively great friction between the pen and the chart, renders it unsuited for recording instantaneous values. It is very useful for recording daily totals of incoming sun and sky radiation.

Although sky radiation (i.e., the portion of incoming solar radiation that is received not directly from the sun but from scattering of the direct beam by the atmosphere and its constituents) can be rather intense, records of sky radiation are very much less numerous than those of direct radiation. Devices for providing standardized measurements of sky radiation are now available. These involve some means by which the direct solar beam can be prevented from falling on the receiver of the instrument, which is usually a pyrheliometer. The shading device may be rotary, driven through an equatorial mounting that casts a shadow on the pyrheliometer at any required time of day or year; or it may be fixed, adaptable to particular astronomical or geographical conditions. Examples are the Robinson shading ring, the Drummond adjustable ring, and the Blackwell shading frame.[8]

In the Robinson device, the radiation instrument is located at the center of a "sphere," 40 inches in diameter, on the surface of which moves an opaque strip cut from the surface of a sphere equal in size to that on which the strip travels: the mean radius of the strip varies so that the effective occulting area remains constant whatever its position on the sphere. Different times of year require strips with different radii of curvature. The basic "sphere" is represented by two metal rings, one mounted in the horizontal plane, the other mounted vertically in the north-south plane. The two points at which these rings are fixed together represent the north and south points on the horizon. The occulting strip is attached to two points on the horizontal ring, and touches the vertical ring to form a segment of the surface of the basic sphere. The appropriate radius of curvature of this strip is calculated from astronomical data (see Appendix 4.11).

All three of these devices must be corrected because part of the sky is obstructed by the shading strip or frame. Assuming the sky is isotropic (which it sometimes is not), standardizing corrections may easily be applied (see Appendix 4.12).

The instruments so far described are the major standard ones. When interpreting data obtained from them, the relative standard of accuracy of each type of instrument should be kept in mind. Percentage accuracies are as follows:

Ångström pyrheliometer,	± 1.5–2.5 per cent;
Silver-disc pyreheliometer,	± 0.5–1.1 per cent;
Michelson actinometer,	± 1.0–1.5 per cent;
Moll-Gorczynski actinograph,	± 5 per cent for daily values;
Eppley pyrheliometer,	± 3 per cent for longer periods;
Robitzsch actinograph,	± 10 per cent for daily values, ± 5 per cent for monthly means.

The figure for the Ångström instrument is especially important, since it is the only "absolute" standard in the list.[9]

NONSTANDARD AND EXPERIMENTAL INSTRUMENTS

These will be described in the following order: (1) instruments measuring incoming short-wave direct solar radiation and diffuse sky radiation; (2) instruments measuring long-wave radiation, both incoming (atmospheric) and outgoing (earth); (3) instruments measuring net radiation, i.e., the balance between incoming (short-wave) and outgoing (long-wave) fluxes of radiative energy.

A classic experimental instrument was the Dines radiometer, first set up at Benson in the United Kingdom in 1920, and later operated at Kew from 1930 to 1940. It consisted essentially of a receiver, a blackened thermopile of 120 pairs of copper-constantan junctions covering a circular area of 10 cm diameter, placed 10 cm from the inner end of a blackened metal cylinder 65 cm long (see Figure 4.3). The receiver was opposite an aperture, 10 cm in diameter, which could be covered by a shutter, and a spherical mirror could be adjusted to reflect radiation through this aperture. A small tank of water beneath the mirror formed a standard radiator. The metal cylinder was immersed in a constant-temperature water tank, and the radiometer was mounted near a north-facing wall so that the mirror could be directed at the northern sky. The radiation absorbed by the thermopile was recorded by a moving-magnet galvanometer, and the instrument was calibrated against a black-body radiator over the normal range of atmospheric temperatures.[10]

A simple experimental way to measure insolation is to use a standard Campbell-Stokes sunshine recorder as an integrating actinometer. If the weight of cardboard burnt in the sunshine card is weighed both before and after exposure to sunlight, and a series of these weighings is calibrated against an Eppley or other standard pyrheliometer, then a regression equation may be obtained that will convert daily sunshine duration into daily insolation totals. The resulting accuracy is about ± 11 per cent (see Appendix 4.13).

A very useful and cheap integrator for daily total solar radiation on a horizontal surface has been designed by Whillier. This instrument employs an ordinary commercial D.C. ampere-hour meter (similar to the household electric meter, which measures alternating current) connected directly to silicon cells. It depends on the principle that the short-circuit current of a silicon cell is linearly proportional to the intensity of the solar radiation impinging on it. Three round silicon cells are normally used, 12.5 cm^2 of cell area being needed to provide the required starting current of 0.05 ampere at the minimum solar intensity it was desired to record (0.2 cal per cm^2 per min). The resulting data—read from the dials of the meter—are within ± 4 per cent of the figures obtained from conventional pyranometers, and are probably quite as accurate as those from Eppley and other pyrheliometers.[11]

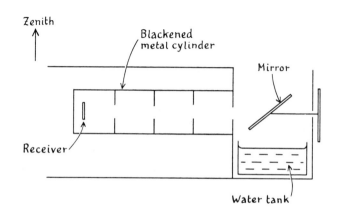

FIGURE 4.3.
Dines radiometer, side view
(after G. D. Robinson).

A more complicated instrument, which determines the direct component of total solar radiation, is the panradiometer. This consists of three spheres, each 6.5 cm in diameter, one painted black, one painted white, and one highly polished. These are exposed to sunlight, and the quantity of heat that must be supplied to each of two spheres to bring them to the temperature of the third is measured by means of thermocouples. The procedure is repeated with the spheres shaded from direct sunlight, giving the sum of radiation reflected from the ground plus radiation scattered by the atmosphere. Subtracting the latter sum from the first value gives the direct solar radiation.[12]

More involved actinometers have been developed in Australia by Albrecht. The triple actinometer consists of three thermopiles (each of 30 thermocouples of copper-constantan), connected in parallel and mounted side by side at the bottom of a brass tube two inches in diameter. The active joints of the thermopiles are mounted in a detecting plate of thin cellophane; the passive joints are free in the air underneath. The brass tube contains four diaphragms, each with three openings, one above each thermopile, with different colored filters in the different diaphragms. The instrument is first directed at the sun, and zero on the galvanometer found for each of three positions, with the end of the tube covered. The cover is then removed, the thermopiles exposed to the sun for two minutes, and the galvanometer read. The cover is then replaced and the galvanometer read. The cover is then replaced and the galvanometer read for a third time. The solar radiation intensities are then found by determining the differences between the "open" readings and the mean of the two "closed" readings. The response time of the instrument is about 30 seconds.[13]

A somewhat differently constructed triple actinometer was used in the Antarctic in 1954. In this model, the metal tube contained three smaller tubes of varying diameter, one covered by clear glass, one by a yellow and one by a red filter, with a bolometer fitted to the bottom of each internal tube. The bolometers were connected to a galvanometer by a selector switch; the operational sequence was to take three readings (tube covered; tube open; tube covered) for each of the three internal tubes. The instrument required calibration against a standard pyrehliometer or actinometer.[14]

An optical actinometer, designed by Albrecht and used at Melbourne, consisted of a diaphragm tube, whose effective area covered 5° around the moon, fitted with various filters. A segmented disc, mounted between the bottom of the tube and a photoelectric cell, was rotated by an electric motor. Thus direct solar radiation alternately fell upon the cell and was prevented from reaching it. The circuit was arranged to give a full-scale deflection of the galvanometer at a light intensity equal to one-tenth of that of a half-moon in a cloudless sky. The direct solar beam could be measured fairly precisely by this method.[15]

Global radiation, the sum of the incoming direct solar beam plus the sky radiation received on a horizontal plane, depends on the height of the sun, the water-vapor content of the overlying atmosphere, altitude above sea level, and the albedo of the ground surface at the measuring site. Provided the albedo and the amount of radiation absorbed by the water vapor is known, the global radiation for a clear sky may be calculated indirectly. It may also be estimated from average cloudiness figures (see Appendix 4.14), and directly measured by atmometers or, more expensively, by remote-reading telepyranometers. The atmometer method is cheap, and quite sufficient where only monthly totals of global radiation are required. The telepyrameter method provides almost instantaneous values.

The atmometer apparatus includes two Livingstone spheres, 5 cm in diameter and made of porous porcelain, which are fixed by spring clips to cross arms mounted on a vertical brass rod. Two arrangements for replenishing the water lost by evaporation from the spheres under radiation are possible: type (a) is replenished from a supply vessel connected to a glass top; type (b) is replenished directly through the top of a burette. The apparatus must be assembled very carefully, since air must be excluded from it, and valves must be introduced to prevent the absorption of precipitation. The instrument needs refilling with water at least once a day, especially during high winds.[16]

Albrecht's telepyranometer allows radiosonde soundings of sun-plus-sky radiation to be made at 200 to 1,000 miles away from the base station, depending on ionospheric propagation conditions. The instrument employs condensers whose dielectrics are affected by outside temperatures, and stable high-frequency oscillators whose frequency depends on the effect of temperature on its tuning capacitance (see Appendix 4.15). Two radio-frequency oscillators (on different frequencies) produce, after mixing, an intermediate frequency, which is amplified and then transmitted as a radio frequency signal by an aerial system. The lower-frequency oscillator, which has a tuning capacitance of very high temperature coefficient, has its condenser painted black and exposed to solar radiation; its frequency varies with the temperature of the condenser. The higher-frequency oscillator has several condensers (painted white and exposed to the instrument temperature, but not to solar radiation), whose temperature coefficient is such that the change in frequency (in kilocycles per second) exactly equals that of the first oscillator if the same temperature difference is applied. Thus the intermediate frequency remains constant whatever the temperature of the instrument, but changes whenever the balance between the oscillators is disturbed by incoming solar radiation.

Neither the atmometer nor the telepyranometer give continuous, integrated records. One can construct an electrical circuit that, coupled to a pyrheliometer, will automatically produce such a record.[17]

Daily totals of global radiation may be recorded optically by a photographic apparatus designed by Albrecht. A box camera using ordinary film is mounted so that it points vertically upward, its lens being replaced by a slit covered with a sheet of milky glass. The slit must be not less than 0.1 mm wide, to avoid the formation of diffraction images. Polarizing filters are fitted underneath the slit to reduce the light sensitivity, and either the camera itself is rotated or a revolving mirror is used to swing the incoming light beam. The response of the instrument depends on the motion of the film; usually it is ± 1.7 minutes (of time) for a 0.1 mm slit.[18]

Incoming long-wave radiation from the atmosphere (sky back radiation) may be measured by the infrared radiometer devised by Stern and Schwartzmann. This instrument may be operated at any time of day or night. It employs an infrared window of highly polished thallium bromoiodide in the form of a hemisphere 8 mm thick, which also screens the receiving element from wind. The element is a nickel disc sensor, $\frac{1}{8}$ inch in diameter, 0.001 inch thick, located just below the optical center of the hemisphere; a thermocouple is attached to its lower surface, and its upper surface is gold-blackened. The hemisphere projects above the top of a cylindrical case whose temperature can be controlled, and the thermocouple is connected to a rapid-response potentiometer. The instrument must be calibrated against a uniformly radiating hemisphere. For use during the daytime, a plexiglass cover, opaque to all radiation of wavelength greater than 3.5 μ, is placed over the hemisphere before the potentiometer is read. The cover is then removed and the potentiometer read again. The difference between the two potentiometer readings is proportional to the back radiation from the sky.[19]

Outgoing long-wave radiation from the Earth's surface may be measured by Albrecht's pyrgeometer. The sensor consists of three pairs of plates with different surfaces whose temperature is measured. The surfaces are black/white, white/white (heated), and metallic/white; the latter pair can also be heated, to remove dew. The temperatures of the surfaces are read from a sensitive galvanometer, via a selector switch. In use, the instrument is first set up at the appropriate site so that it has an unrestricted view to the horizon on all sides. The plates are directed at right-angles to the sun, and are shaded by a screen. A series of readings are then taken. The screen is then readjusted so that infrared radiation, but not direct sunlight, reaches the plates, and the sensor exposed for ten minutes, after which the readings are repeated. The instrument can also be used at night.[20]

Instruments for measuring radiation balance, i.e., the net difference between in-coming short-wave and out-going long-wave radiation, operate on different principles from the experimental devices so far discussed. They may be divided into two groups: (a) instruments usually used to determine radiation balance at or near the Earth's surface; (b) instruments usually used to measure radiation balance at some height in the free atmosphere.

For certain locations, there is a definite relationship between daily totals of solar radiation and daily totals of net radiation, which may be expressed by an equation (see Appendix 4.16). Thus net radiation may be estimated from readings of a pyrheli-ometer. However, before the equations can be computed, the radiation balance must first be measured.

Net radiative energy at the Earth's surface expresses the difference between the upward and downward fluxes of energy in the range $0.17\,\mu$ to $0.80\,\mu$. Solar radiation reaches its maximum at about $0.5\,\mu$, terrestrial radiation at around $10\,\mu$. Important instruments for measuring the net difference between these two radiation streams have been produced in the United Kingdom, in Australia, and in the United States. The net radiometer measures the net flux of radiation through a horizontal plane. A blackened thermal transducer is placed in the radiational field and the difference in temperature between its upper and lower surfaces is measured, usually by means of a potentiometer. The blackening material should obey the cosine law; i.e., its absorptivity should be independent of the angle of incidence of the radiation, and of the wavelength of the incident radiation. In practice, the radiometer must be equally sensitive to all wavelengths from 0.2 to 60 μ. The net radiation R_n is defined by

$$R_n = R_s\,(1 - r) + R_k\,(1 - r) - R_e,$$

where R_s is the incoming solar radiation, R_k the incoming sky radiation, R_e the out-going earth radiation, and r the reflectivity of the surface. Also, $r = 1 - E$, where E is the emissivity of the surface. The effect of thermal convection on the transducer surfaces must be minimized for the balance equation to be applicable.

A British instrument, the MacDowall fluxmeter, measures the net flow of radiation through a horizontal surface. The radiometer element is a square plate, three inches on each edge, mounted horizontally, and blackened on both sides. An electric blower directs a jet of air equally across both top and bottom surfaces, thus ensuring that energy exchanges between the plate surfaces and the air are maintained at a relatively rapid rate that does not change appreciably when the windspeed varies. The plate is a bakelite frame, in which are two aluminium sheets, 0.01 inch thick, separated by thin polyethylene sheets. A thermopile, of 60 turns of constantan wire on bakelite strips, is incorporated in the bakelite frame. The sensor has a field of vision covering 97 per cent of the hemisphere, and is equally responsive to all wavelengths from 0.3 to 40 μ. It does not function properly in rain or fog, because the wetted surfaces act as a "wet bulb" instead of a radiometer. The instrument must be calibrated twice, first for long-wave, second for short-wave radiation, and may be made continuous-recording by being coupled to a potentiometer system. Its response remains constant at 2 per cent in winds of up to 20 miles per hour.[21]

The Australian CSIRO radiometer devised by Funk consists of a sensor element, a 250-junction thermopile, shielded from wind and precipitation by sealed hemispheres of very thin polyethylene. The latter are kept inflated and free from internal condensation by a circulating dry gas. A heating ring prevents the formation of dew and rime. Diagonal strips of selectively reflecting paint equalize the long- and short-wave sensitivities of the sensor, which has a response time of about a minute and a wind dependence of less than 1 per cent in windspeeds of up to 15 m per sec. By employing two CSIRO radiometers, one mounted some distance above the other, one can measure how much cooling in the lowest few meters of the atmosphere is caused by radiation. The instruments are first set side by side at the lower height, connected in opposition, and the more sensitive instrument of the two is shunted to give a nil reading on a galvanometer, thus equalizing the sensitivities. One instrument is then moved to the higher level, and its output EMF measured by a potentiometer; this gives the difference in radioactive flux between the two heights. By this means, radiative cooling rates as high as 12°C per hour were found near Melbourne. The CSIRO radiometer is available commercially, as is the Gier and Dunkle radiometer and the Thornthwaite radiometer, which are American instruments operating on similar principles.[22]

An experimental American instrument, the Suomi net radiometer, measures the net flux of radiation through a surface parallel to the Earth's surface. In general principle, it is similar to the MacDowall fluxmeter, although it is nonrecording. A blackened plate is mounted parallel to the ground surface and the resulting temperature difference through it, after exposure to radiation, is measured. If the thermal conductivity of the material out of which the plate is constructed is known, the net radiation required to produce the observed temperature difference can be calculated (see Appendix 4.17).

The plate of the Suomi net radiometer includes a thermopile made of a rounded-edge glass microscope slide, 7.5 cm long and 2.5 cm wide, with 120 turns of wire 0.2 mm in diameter wound around it. The plate is attached to a vane, which is placed in the nozzle of an air blower, and an electric heater is fitted to the plate. The instrument is calibrated either by comparing it with the calculated flux between two plane, ideal radiators at different temperatures, or by taking readings with the plate in shadow and then in full sunlight, in which case the change in net radiation is set equal to the vertical component of direct solar radiation as measured by an actinometer. The sensitivity of the instrument is 1.3 mV per cal per cm² per min, as against 30 for the CSIRO radiometer.

The Suomi net radiometer can also be used as a total radiometer, giving estimates of downward flux of long-wave radiation (see Appendix 4.18). For climatological purposes, if the instrument is to be used in an area of great contrasts in relief and vegetation, it is often very useful to determine radiation conditions above several different surface types simultaneously, rather than to observe solely at a single site. A totaling system using a relaxation oscillator circuit makes this possible.[23]

The Fritschen net radiometer is a small and comparatively inexpensive instrument that provides continuous records. Its element is only 1 inch in diameter, unlike the foot-square plate of the Suomi instrument, and it may be used in all weathers, whereas the polyethylene shells of the latter can be penetrated by moisture. However, its shielding arrangement renders it insensitive to sudden changes in radiation intensity, so that it cannot be used to record instantaneous values of the radiation balance. Its transducer element, both surfaces of which are painted black, is mounted between four sheets of 0.002-inch-thick mica to form a capped cylinder shielding the sensor from wind and precipitation. It is calibrated in a special box against an Eppley pyrheliometer; both sides of the transducer require calibration. In use, it is connected by a selector switch to a sensitive recording potentiometer. The switch permits several instruments to be connected to a single recorder, so that conditions over a variety of surfaces can be recorded at a single point. If at least two radiometers are used, with different spectral absorptions, both incoming and outgoing radiation fluxes may be measured.[24]

Later instruments by Fritschen made use of four polystyrene radiation windows instead of the mica sheets of the original instrument (see Appendix 4.19). Mica is not completely satisfactory as a radiation window, because it strongly absorbs radiation between 8.8μ and 10.3μ and transmits poorly between 10.3μ and 15μ. The outer two windows are hemispherical, so rain will run off easily, and although dew or condensation on the hemispheres is not prevented—as it is, for example, in the CSIRO and Thornthwaite radiometers—its effect appears to be negligible compared with sampling and instrument errors. Indeed, the presence of a heater to eliminate condensation may produce larger errors than the condensation. A further defect of the original instrument, its limited output, required a stable D.C. voltage for amplification, and limited its field use. A thermistor sensor provides a larger output, which does not require amplification.[25]

Measuring radiation balance at some point in the free atmosphere is more difficult than measuring it at the Earth's surface. The normal method for many years was indirect: to measure the water-vapor content of the air at appropriate heights, and then to apply the radiative transfer equations, on the assumption that water vapor is the main radiating constituent of the atmosphere. Adaptation of the Suomi net radiometer for airborne use by Kuhn later led to the production of the radiometersonde for direct measurements; this is essentially a net radiometer suspended beneath a balloon.[26]

The sensor system of the Suomi-Kuhn radiometersonde consists of alternating layers of styrofoam insulating material, 12.7μ polyethylene films acting as both convection shields and radiation windows, and 6.4μ mylar sheets, the inner surfaces of the latter being aluminized to reflect long-wave radiation, and their outer surfaces blackened. Thermistors measure the temperatures of the uppermost and lowermost aluminized mylar surfaces, while the spaces between all surfaces are occupied by air at external pressure. The whole unit is not more than 5 cm high and 30 cm long, and is attached to the side of the ordinary radiosonde. The instrument measures upward, downward, net, and total components of the infrared radiation in the atmosphere. It can also measure the temperature change of the air caused by radiation in any layer,

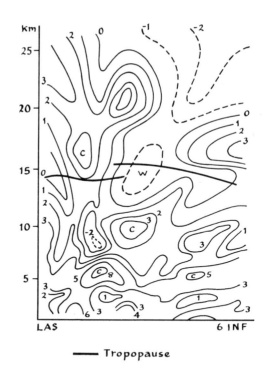

FIGURE 4.4.
Net radiation divergence: cross section for 6.00 A.M., G.M.T., July 29, 1959, from Las Vegas (LAS) to International Falls (INF). Units are vertical divergence in hundredths of a cal per cm² per minute per 100 mb. *W* is warming; *C* is cooling. (After P. M. Kuhn and V. E. Suomi.)

— Tropopause

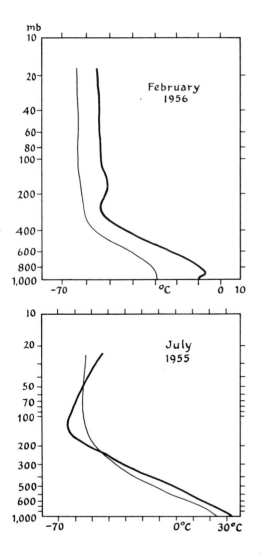

FIGURE 4.5.
Black ball radiometer soundings over Minneapolis. Heavy line is average air temperature; lighter line is average equivalent radiation temperature. (After J. L. Gergen.)

and the rate of emission of radiation by water vapor at different heights. Figure 4.4 illustrates some soundings made over the United States with the radiometersonde.[27]

A version of the radiometersonde developed by Clarke in the Caribbean has resulted in a relatively inexpensive instrument that, mounted in an aircraft flying at 120 knots, can record average radiation flux for a distance of 4 miles with a probable error of only ± 0.007 langleys per minute. This enables the radiational cooling at a point in the atmosphere to be computed with an accuracy of $\pm 1°C$ per day. Aircraft observations of atmospheric net radiation above the United Kingdom used Moll-Gorczynski solarimeters for long-wave radiation (0.3 μ to 3 μ), their outputs connected across millivoltmeters and photographically recorded, and photometers for short-wave radiation in the visible band (0.43 μ to 0.7 μ), with potentiometer recorders.[28]

An alternative instrument to the radiometersonde for obtaining radiation-balance soundings in the free atmosphere is the black-ball radiometer devised by Gergen, which gives results that are accurate to 1 per cent up to 50 mb. The black-ball thermistor replaces the humidity element of a normal radiosonde circuit; the black ball is essentially a convection-shielded spherical absorber, blackened, and suspended beneath the balloon. The thermistor measures a temperature (the equivalent radiation temperature) that is proportional to the total (upward and downward) incident radiation. The observed temperature must be corrected, because convection of heat prevents the black absorber from attaining true radiative equilibrium. Both the black-ball temperature and the air temperature are measured; the method then assumes that the atmosphere will tend to warm (by infrared radiation) when the black ball is warmer than the air, and tend to cool when the black ball is cooler than the air. All soundings must be made at night. Figure 4.5 illustrates some soundings over Minneapolis.[29]

Independent observations appear to indicate that the black ball does not measure the true heating or cooling rate of the atmosphere. It is nevertheless very useful for estimating the radiation temperature of solid quasiblack bodies in the high atmosphere, and so for verifying the validity of the assumed absorption coefficients on which all earlier indirect attempts at estimating atmospheric net radiation were based.[30]

An Australian black-ball instrument developed by Funk is simpler than the American instrument, but is intended primarily for use near the ground. The black ball consists of a spherical, 40-watt, 230-volt electric light bulb, copper-plated and blackened, with 13 copper-constantan thermojunctions to measure the temperature difference between the air and the black-ball surface. The heating current is supplied from the lamp filament.[31]

The Suomi-Kuhn net radiometer and the Gergen black ball have been compared by mounting them under the same sonde balloon, together with a disc radiometer and a blackened silver sphere. The disc radiometer consisted of two spheres, separated by an aluminium sheet blackened on both sides, in which a thermistor was incorporated. Double polyethylene shields reduced convection. The Suomi-Kuhn instrument and the disc radiometer were found to give good agreement, but the black ball differed from the disc radiometer (probably because of convection and conduction inside the black ball). The black ball indicated that the atmosphere was being warmed by radiation when the Suomi-Kuhn radiometer showed it was being cooled. Thus, although soundings from the two systems are relatively valuable when studied separately (in particular when the data have been reduced to mean values), they cannot be expected to yield accurate, absolute measurements of instantaneous conditions. It is certainly not possible to combine readings from the two instruments.[32]

Standardization is now satisfactory for instruments intended to reveal the broad features of the distribution of incoming solar radiation at the Earth's surface, and efficient networks are in operation. However, the mapping of local radiation patterns is much poorer. If the climatologist is concerned with radiation balance or long-wave radiation, he must decide precisely what data he needs, select his instrument accordingly, and go make a series of short-period observations on location. These observations must be compared statistically with the nearest permanent stations to determine long-period estimates. An alternative approach is to take some element that is relatively easy to measure, e.g., water-vapor content or sunshine duration, and then compute the radiation data indirectly.

Indirect Estimation of Radiation from Theory

Radiation theory is extremely useful for calculating how much long-wave radiation from the atmosphere will be received at the Earth's surface. In fact, the effects of long-wave radiation in the free atmosphere were, until very recently, known only through these theoretical calculations, not from actual measurements.[33]

Atmospheric long-wave radiation is heat radiation, its energy being derived from the kinetic energy of the air molecules. This energy is radiated to the Earth's surface by those constituents of the atmosphere that have absorption bands in the range from 4μ to 100μ, i.e., mainly water vapor, carbon dioxide, and ozone. The basic equations of radiative transfer of energy (see Appendix 4.20) describe the intensity of long-wave radiation at the Earth's surface (or at any height in the troposphere) in terms of the mass of absorbing material in the overlying atmosphere. These equations are based ultimately on Planck's law. They involve absorption coefficients and transmission functions for water vapor, carbon dioxide, and ozone taken from the results of laboratory experiments and assumed to be valid in the atmosphere as a whole. The derivation of these equations is complex, but the final product is an expression that may conveniently be made into a chart, which can be used to solve the equations. The chart is quite fundamental, even though various assumptions about radiation transmission within narrow intervals of the energy spectrum must be made because the laboratory experiments whose results are taken require that radiation transmission be measured over homogeneous air paths through which pressure, temperature, and humidity mixing-ratio are constant, conditions which do not necessarily hold in the real atmosphere. The Curtis-Godson approximations allow laboratory-derived transmissions over homogeneous paths to be converted into figures appropriate to the real atmosphere. Atmospheric transmission of radiative energy depends on the pressure and temperature of the atmosphere, the wavelength of the radiation, and the mass of absorbing material. The Godson procedure is to use laboratory transmission data for pressure and absorbing-mass variations, and to employ theoretical considerations to allow for the temperature dependence.[34]

Three charts for solving the radiation equations have gained wide recognition; these are the Möller chart, the Elsasser chart, and the Yamamoto chart.[35] The Möller chart, introduced in 1943, uses optical depth as abscissa, and a variable quantity P (a function of both optical depth and absolute temperature T) as ordinate. Isotherms and isopleths of constant T are permanently shown on the chart. The Elsasser chart,

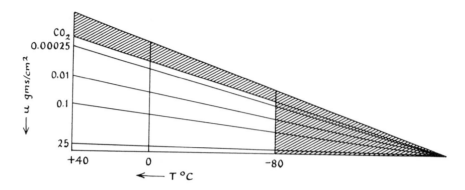

FIGURE 4.6.
The Elsasser chart. Shaded portions are omitted from the operational chart, and the information they contain is given in tabular form.

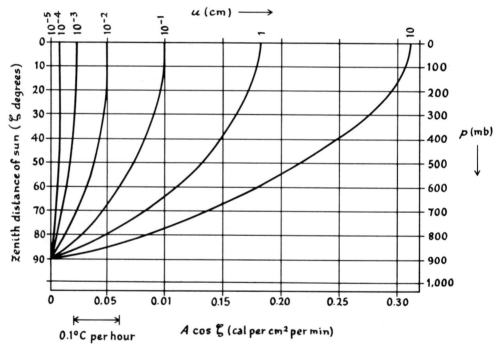

FIGURE 4.7.
Yamamoto radiation chart. A is total absorption of solar radiation. (After G. Yamamoto and G. Onishi.)

introduced in 1942, uses T as the independent variable (abscissa) and Q (also a function of optical depth and T, but not the same as P) as ordinate. Lines of constant Q are by definition lines of constant optical depth, and the ordinate axis is labeled in terms of optical depth values. The original graph is transformed so that the x-axis equals aT^2 and the y-axis equals

$$\frac{Q}{2aT}$$

where *a* represents area. Thus areas are preserved in the transformation and the operational chart (see Figure 4.6). The chart is triangular, but in practice, its right-hand portion is usually omitted, and the values that would be taken from it are given in a table. The Yamamoto chart, introduced in 1952, employs black-body flux as the independent variable. The chart is rectangular (see Figure 4.7), abscissa proportional to σT^4 and ordinate proportional to

$$\frac{Q}{T^3}.$$

The horizontal axis is normally labeled in terms of both precipitable water and $A \cos \zeta$, where ζ equals zenith distance, and the vertical axis in terms of both ζ and pressure. (See Appendix 4.21.)

To use the Yamamoto chart, the relation between pressure and precipitable water at various heights in the atmosphere must first be found from aerological soundings.[36] The incident angle of the beam of radiation whose absorption is to be calculated is chosen, as is a point on the graph that corresponds to the given values of zenith distance and optical depth. This point is moved vertically to the ordinate that corresponds to the pressure for the height in question; the abscissa then gives the absorption within a vertical air column whose base is at that height and which contains the given amount of precipitable water. Similar points are determined for the complete aerological sounding, and the curve that joins these points describes how absorption by water vapor and carbon dioxide varies with pressure within the whole column. The rate of temperature increase within the air column caused by this absorption of incoming solar energy is proportional to the difference in absorption between two points on the curve separated by the appropriate pressure interval. These rates of heating are marked on the graph (usually in units of 0.1°C per hour) for pressure intervals of 100 mb.

The Elsasser and Möller charts are similar in principle—on each one plots a curve from an ordinary aerological sounding—but their details differ. With the Elsasser chart, for example, the initial step is to calculate values of optical depth for given heights in the atmosphere from the values for specific humidity and pressure given in the sounding.[37]

The interpretation of a radiation chart can best be illustrated for the Elsasser chart (see Figure 4.8). The line *COB* represents the aerological soundings as plotted on an Elsasser chart. Above level P_o, no appreciable water is present in the atmosphere, so here the curve follows a line of constant optical depth. The Earth's surface, when radiating as a black body, radiates in the same manner as an isothermal surface of infinite optical thickness. If the Earth's surface has a temperature T_E, then the energy it radiates is given by area *ABC* (usually in cal per cm³ per 3 hours), and the total radiation from the atmosphere received at the ground at the point in question is given by area *x*. The net loss of radiation from the Earth's surface is thus area *ABC* minus area *x*, i.e., area *y*.

For conditions at some level *P* within the atmosphere, where the air temperature is T_A, optical depth is plotted against temperature (going upward from level *P*) to represent absorption within the overlying atmosphere: area *x'* thus gives the downward radiation flux to *P*. Optical depth is then plotted against temperature going downward from *P* to represent absorption within the underlying atmosphere,

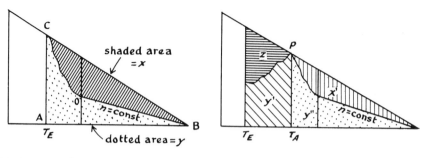

FIGURE 4.8. Interpretation of the Elsasser chart.

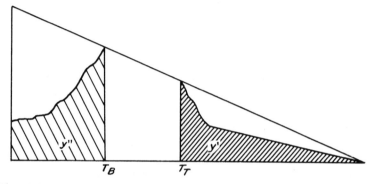

FIGURE 4.9. The effect of clouds.

terminating at the Earth's surface (temperature T_E). Area z is the portion of the downward flux of radiation due to the atmospheric layers below level P, and the net upward flux to level P is area $y' + y''$.

Figure 4.9 illustrates the effect of clouds. If T_T represents the temperature at the top of the cloud, then area y' is the net radiative loss from the cloud top. If T_B represents temperature at the base of the cloud, then area y'' is the net gain in radiation at the cloud base due to radiation from the ground plus radiation from water vapor in the atmosphere below the cloud.

A comparison of radiation data calculated with the aid of the Elsasser chart, and those derived from black-ball radiometer measurements,[38] shows that atmospheric temperatures produced by radiational cooling as calculated from the chart are always warmer than the "true" (i.e., black-ball) temperatures, particularly in winter, when the water-vapor path length is less at all altitudes. Thus Earth and atmosphere radiate less infrared energy to space than the chart suggests.

From work at Kew, Robinson found that the Elsasser chart gave results 6 to 14 per cent too high for moderate amounts of radiation, but results correct to ± 3 per cent for high amounts. The reasons for the inaccuracies were: (a) the proportion of carbon dioxide in the atmosphere varies, although the chart assumes it is concentrated in a limited band centered near 15 μ; (b) the chart ignores the effect of emission of radiation by particulate matter; (c) Elsasser's original theory made insufficient allowance for the variation of emissivity of water vapor with temperature, and assumed emissivities that are too great for moderate amounts of moisture, but are approximately correct for larger amounts.[39]

The Möller chart is interpreted somewhat differently from the Elsasser chart. The radiation from an isothermal atmospheric layer is given by the area bounded laterally by the lines, $y = 0$, $y = m$ (where m is the radiating mass of the layer), $x = 0$, and $x = T$. The radiation from a nonisothermal atmospheric layer may also be found, by a graphic integration.[40]

Neither the Elsasser, the Möller, nor the Yamamoto chart is of use for the stratosphere, where the absorption of radiation by water vapor is negligible, but a Russian radiation chart, details of which are available in English, may be useful for it, because it takes into account absorption by ozone as well as by water vapor and carbon dioxide. Charts and numerical tabulations made recently by Elsasser and Culbertson allow accurate radiation calculations to be made up to 25 or 30 km, and approximate calculations from 30 to 40 km. An alternative to using charts or tables is to design a computer evaluation; a program available for radiation computations employs actual or mean aerological soundings as input data.[41]

Some other radiation charts deserve mention. Robinson's Kew chart is an empirical chart that shows emissivity (in per cent) and water-vapor path length (in cm) against temperature (in °C). The downward radiation from the atmosphere received at the Earth's surface is usually computed by this chart to be less than the observed radiation, but the discrepancy decreases as the amount of radiation increases. The error of the Kew chart is 0.05 ± 0.55 mw per cm^2, as against 1.8 ± 0.8 mw per cm^2 for the Elsasser chart. When surface temperatures are very high (i.e., 40°C or thereabouts), the Elsasser chart gives values 30 to 40 per cent below those of the Kew chart.

Brooks's radiation chart is more rapid both to use and to interpret than Elsasser's. First introduced in the form of tables giving the rate of change of temperature caused by radiation, the Brooks method was based on empirical measurements of the emissivity of water vapor. The data were later made into a chart that allowed incoming atmospheric radiation to be estimated from detailed temperature and moisture soundings. Brooks's results indicated that on occasions one-half the incoming radiation at the ground came from the lowest 100 feet of the atmosphere. His chart is intended for use in studying conditions near the Earth's surface, although it is not suitable for true microclimatological work (see Appendix 4.22). In the working chart the abscissae represent 4th powers of the absolute temperature of the air at the surface, and the ordinates are atmospheric emissivities on a linear scale. An area of one square unit on the chart represents one unit of heat-flow caused by radiation.[42]

Radiation at the Earth's surface may also be estimated theoretically, without recourse to charts. Thus Funk makes use of a theoretical equation that is especially suitable for microclimatological work. The equation gives the divergence of radiation (i.e., the decrease in radiative energy, which leads to cooling of the atmosphere) within a layer of the atmosphere near the ground in terms of the mean air temperature in the layer, the optical thickness of the atmosphere above the layer, and the ground temperature.[43]

Other investigators have preferred statistical methods. Thus Swinbank used Dines' classic experimental observations at Benson to prove that, over a wide range of temperatures and humidities, and in a wide variety of locations,

$$R = -17.09 + 1.195\sigma T^4,$$

where T is the absolute temperature of the air at the Earth's surface, R is the incoming long-wave radiation from a clear sky in mw per cm^2 per °K, and σ is the Stefan-

Boltzmann constant. To a first approximation $R = 5.31 \times 10^{-14} T^6$. This result follows the same lines as the classic early inductions concerning nocturnal cooling at the Earth's surface by Brunt and Ångström, who showed that

$$\frac{R}{\sigma T^4} = a + b\sqrt{e} \text{ and } \frac{R}{\sigma T^4} = \alpha - 3.10^{\gamma e},$$

where e is vapor pressure in mb and a, b, α, β, γ are empirically determined constants for specific sites. Similar statistical estimates for an ocean location have been made by Martin and Palmer using multiple regression coefficients.[44]

Mapping the Distribution of Radiation

There are three main methods for mapping the distribution of solar radiation and radiation balance at the Earth's surface: (a) directly, using actual measurements of radiation, and interpolating between the observing stations by subjective judgments; (b) indirectly, by making use of the known relations between radiation and other climatic elements; (c) by employing measurements made from artificial Earth satellites.

Until very recently, the world distribution of stations recording direct observations of solar radiation intensity has remained very patchy; there were only a few more than 200 in all the continents in 1958 for example, and most of these were in North America and the Soviet Union west of the Urals. As a result, for many years all published maps of world radiation distribution were based on theory and subjective estimations, rather than on direct measurements, including the classic world maps of Simpson, Baur and Phillips, and Landsberg. Detailed work by Budyko and others in the Soviet Union resulted in the production of new maps based on closer networks of stations, and later world maps have been developed by Black and Landsberg, among others.[45]

A few examples will illustrate the problems involved in radiation mapping. In a few areas, notably the United States and the western Soviet Union, the station networks are close enough, and cover enough areas for small-scale maps showing the broad features of radiation distribution to be drawn by direct interpolation.[46] Elsewhere, or if more regional detail is desired, the relationship between solar radiation and such elements as sunshine duration, cloud amount, and cloud type (which are recorded at many more stations than record solar radiation) must be determined for that area, and the distribution of solar radiation inferred from it. The steps involved in such calculation are as follows.

First, the amount of radiation that ought to be received at a specific location is estimated from the latitude for a particular time and date. This amount is modified to take into account the local transmission characteristics of the atmosphere, which are determined from radiation observations at the nearest available stations. These transmission characteristics vary with time of year, solar zenith distance, altitude above sea level, dust content of the air, and amount of precipitable water in the atmosphere. Second, the average cloudiness is used to calculate the diffuse sky radiation and the depletion in insolation caused by absorption or reflection by the cloud cover. Third, the local albedo of the terrain is used to determine how much radiation is reflected from the ground surface. (See Appendix 4.23.) This procedure is lengthy, and can obviously be simplified. Attempts at simplification usually commence by concentrating on two climatic elements only: sunshine and cloudiness.

Empirical equations, derived by Ångström in 1922, Kimball in 1928, Savinov in 1933, and Ukraintsev in 1939, involve the correlation between the average amount of solar radiation and both the average duration of sunshine and the average cloudiness (see Appendix 4.24). According to Ångström, the total incoming radiation, Q, received on a horizontal square centimeter and the number n of hours of sunshine as measured by a sunshine recorder are related by

$$\frac{Q}{Q_o} = a + b\left(\frac{n}{N_o}\right),$$

where Q_o is the maximum total amount of solar radiation impinging on a horizontal square centimeter on a perfectly clear day, N_o is the maximum number of hours of sunshine recorded under cloudless conditions, and a and b are empirical constants for a specific site and month. For Athens, for example, a varies from a minimum of 0.25 in August to a maximum of 0.44 in December, with an average value of 0.34; b varies from a minimum of 0.58 in September and December to a maximum of 0.68 in April, with an average of 0.63. Over much of the United States, average values of a and b for the year as a whole are 0.35 and 0.61, respectively.[47]

More recently, Black showed that the relation between mean monthly incoming solar radiation, Q, and sunshine as recorded on a Campbell-Stokes recorder could be expressed by

$$\frac{Q}{Q_A} = a' + b'\left(\frac{n}{N}\right),$$

where Q_A is the maximum possible radiation received in the absence of an atmosphere, N is the maximum possible duration of bright sunshine, and n is the monthly amount of sunshine.[48] The constants a and b have values around 0.23 and 0.48, respectively, for the northern hemisphere, and show no systematic variation with latitude. By substituting length of day for N, one can construct monthly mean radiation maps quite easily from sunshine records even if few radiation records are available. Maps for western Europe and the eastern North Atlantic have been prepared in this manner.[49]

A slightly different expression,

$$\frac{Q}{Q_o} = \alpha + (1 - \alpha)\left(\frac{n}{N_o}\right),$$

was used by Drummond and Vowinckel to construct seasonal radiation maps for South Africa. Here, α is a positive constant less than one.[50]

For the radiation-cloudiness correlation, Black developed a world map, based mainly on North American and European data, that proved

$$\frac{Q}{Q_A} = 0.803 - 0.340C - 0.458C^2,$$

where Q_A is the radiation received at the outer limit of the atmosphere, Q is the total (sun plus sky) radiation on a horizontal surface, and C is the cloudiness. The maps were based on values of Q for each 5 degrees of latitude and longitude, and agreed well with observed amounts of radiation outside North America and Europe.[51]

Efimov (1939) and Kuzmin (1950) introduced equations that allowed for the radiational influence of clouds at different heights; Berliand (1952) took into account the effect of outgoing long-wave radiation from the Earth's surface; and Budyko added the effect of the difference in temperature between the underlying surface and the air on effective outgoing radiation (see Appendix 4.25). These and similar equations work well for monthly or longer-period mean values, but they are useless for daily or instantaneous radiation calculations; even for monthly totals, it is often necessary to allow for differences in atmospheric solar absorption.

When accurate estimates, which must include receipt of both direct and diffuse solar radiation at the ground, are needed, there is evidence that cloud type must definitely be taken into account. In Arctic regions, for example, the transmissivity of a cloud type varies greatly with the seasons, being particularly high in winter.[52] Cloud transmissivity varies also from place to place, so much so that the geographical variation in the transmissivity of any one cloud type is greater than the differences in transmissivity between various types of low and medium cloud at a specific location. In general, the transmissivity of a particular cloud type increases northward and with the elevation of the sun in the sky; the increase is small in middle latitudes, but greater in high latitudes.

In middle latitudes, high clouds cut off about 20 per cent of the incoming solar radiation, middle clouds between 50 and 60 per cent, and low clouds or fog from 65 to more than 80 per cent. For both overcast skies and cloudless skies, insolation may be expressed as a function of air mass, by means of empirical coefficients that depend on geographical location and cloud type. Provided that the insolation received from a cloudless sky is known for a specific site, the incoming radiation to be expected with differing cloud amounts at that site may then be calculated.[53]

For stratiform clouds, regression equations adequately describe the relation between mean annual or monthly amounts of sunshine and amounts of cloudiness as percentages. Cloudiness figures are available for many more localities than are records of sunshine, which may conveniently be estimated from the former. For United States data, the regression line for months with relatively small amounts of sunshine in cloudy areas is displaced on the correlation graph well below that for months with relatively small amounts of sunshine in less cloudy areas.[54]

Although basing climatic maps of radiation elements on estimation from sunshine and/or cloudiness averages is often convenient, it is really a very approximate method, because one is combining quantities that have been measured (or guessed at) with greatly varying degrees of accuracy. A direct method is much to be preferred, and one is now available on a global scale since the advent of artificial Earth satellite transmissions.

Early satellites attempted to measure the infrared radiation given off by the Earth and atmosphere, and transmit the data back to the ground. Explorer VII sent back observations of terrestrial radiation at night that revealed definite associations between radiation centers and centers of high and low pressure at the surface. Its radiation sensor consisted of a bolometer in the form of a thin hemispherical shell thermally insulated from a large mirror. The satellite was spinning about an axis parallel to the mirror surface, so that the bolometer acted as a spherical mirror-backed hemisphere spinning in space in the shadow of the Earth. Since the heat-balance equation for such an object can be deduced from theory, the upward radiation through it (and hence the nocturnal terrestrial radiation) can be computed (see Appendix 4.26).

TIROS I took many thousands of photographs showing the distribution of clouds across the face of the Earth, but since its cameras responded only to reflected solar radiation, it could operate only in daylight. TIROS II carried radiometers in addition to cameras, and could gather data on radiation balance conditions both day and night. By measuring the effective temperature of the radiation sensor, the radiation streams through it could be calculated (see Appendix 4.27). The sensor had a 5° angle of view, subtending an area of 60 km in diameter on the Earth's surface when pointing directly downward. There are difficulties in interpretation—for example, the limb-darkening effect, especially marked with high humidity in the stratosphere, causes radiometer observations of the horizon to give lower results than observations of an area directly below the satellite—but nevertheless numerous radiation maps have been produced from the signals transmitted from TIROS II and later satellites; TIROS III, for example, provided data for maps covering five spectral regions. These maps have also proved useful for deducing the cloud types and heights above the darkened side of the Earth. Some of the maps show that large-scale variations in long-wave Earth radiation are related to temporal variations in kinetic energy and available potential energy: as the strength of the northern hemisphere westerlies increases, the outgoing radiation decreases, particularly in low latitudes. Seasonal changes in terrestrial infrared radiation are clearly superimposed on the pattern of maximum radiation emission in regions of little or no cloud, and of minimum emission in cloudy areas.[55]

TIROS II provided the first direct means of mapping the Earth's infrared radiation, and proved particularly useful for finding variations in large areas and over long periods. During the period November 26, 1960, to January 6, 1961, the latitudinal profile of terrestrial outgoing radiation was at a minimum in land areas between 5°N and 15°S, and in oceanic areas between 5°N and 10°N. The minimum was displaced southward over the continents because of the cloudiness associated with the inter-tropical convergence zone. Maximum outgoing radiation on land was at 30°S in the summer hemisphere, and at 10°N in the winter hemisphere, reflecting the southward displacement of the subtropical anticyclones. Southern hemisphere land areas showed greater variations in infrared emission than corresponding areas in the northern hemisphere, and land areas in general were more variable as radiation sources than oceanic areas. These and similar facts substantiate deductions made on theoretical grounds a long time ago, but the TIROS II data are the first actual observations.[56]

TIROS II radiation observations have proved useful in mapping frontal systems and the structure of frontal zones, and for estimating the surface temperature of the Earth, and its diurnal variation, in large areas. From observations of actual radiation and how it differs from the radiation expected without a cloud cover, it has proved possible to calculate vertical motion in the atmosphere at 700 mb.[57]

TIROS III observations have revealed a maximum total outgoing radiation of 340 w per m^2 from the North African desert, and a minimum of 190 w per m^2 from cloudy areas of the tropical Atlantic and the western United States. Albedos of 30 per cent were measured for the desert under clear skies, 15 per cent for heavily vegetated areas with clear skies, and 55 per cent for overcast areas. TIROS III data also made synoptic maps of albedo over continent-sized areas possible for the first time.[58]

Satellite radiation information is very complex to interpret, and the initial processing must be left to the resources of national and international meteorological organizations. However, tabulations of TIROS data now form an important source for studies in radiation climatology; and the well-established fact that the radiative

FIGURE 4.10.
Radiation balance (after J. London).

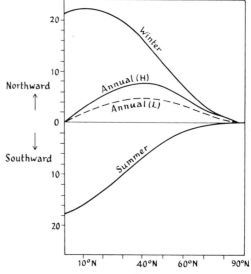

Total heat flux as required by
radiation balance

Units are 10^{16} cal per minute.
H : according to Houghton (1954)
L : according to London (1960)

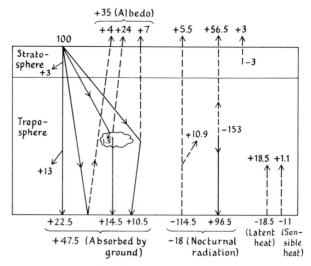

Annual radiation balance

⟵——— incoming radiation
- - - ▷ outgoing radiation
+ indicates an energy gain, − indicates an energy loss.
100 units are equivalent to 0.5 langleys per minute.

properties of atmospheric gases may be inferred from remote radiometer observations opens up new vistas in the study of atmospheric radiation balance.[59]

The Climatology of Radiation Balance

The heat energy available at the Earth's surface, and hence the temperature of the atmospheric environment of all life on Earth, is determined by the balance between incoming and outgoing streams of radiation. For many years, it has been customary in climatology texts to summarize the partition between the various elements of the radiation balance concept in terms of what happens (on the average, over the whole Earth) to 100 units of solar radiation impinging on the outer limits of the atmosphere.

From the known distribution (both horizontal and vertical) of temperature, pressure, cloudiness, and gases that emit and absorb radiation, London calculated, by means of the Elsasser chart, the infrared flux for latitudinal strips 10 degrees wide from equator to pole, for each season, and then summed the figures to give an over-all picture in terms of a mean annual heat balance that can be assumed to be correct for periods of about 10 to 20 years.[60] In general (see Figure 4.10, A), an annual surplus of energy is received in latitudes between the equator and 40°N, and an annual deficit of radiative energy from latitude 40°N to the north pole.

Figure 4.10, B, shows the over-all balance. Assuming that the mean rate of receipt of solar radiation at the outer limit of the atmosphere equals 100 units (it is actually $S/4$, where S is the solar constant, or 0.50 ly per min), then 35 units are returned to space as short-wave radiation; 114.5 units of long-wave energy are radiated upward from the ground surface; 153 units are emitted by water vapor and carbon dioxide in the troposphere; and 3 units are radiated to space by ozone and carbon dioxide in the stratosphere in the 10 μ and 15 μ bands, respectively. To complete the energy balance, nonradiative effects must be taken into account: 11 units are carried upward from the Earth's surface to the atmosphere by turbulence, forming the residual necessary to balance the over-all budget, and 18.5 units are injected into the troposphere as latent heat of condensation during the formation of precipitation. In general, the amounts of energy received and reradiated by the various sources are: stratosphere, 3 units; troposphere, 135 units; Earth's surface, 144 units. The "greenhouse effect" of the atmosphere, absorbing and radiating back to Earth the energy that would otherwise be lost to space, results in the surface temperature of the Earth being 33°C higher than it would be if there were no atmosphere, and this increases the infrared emission from the surface sufficiently to bring about an over-all balance.

The first detailed geographical consideration of radiation balance was by Simpson, and was based on analytical connections between the known distribution of water vapor and the computed distribution of long-wave radiation for the world. Simpson assumed that temperatures at the tops of clouds were the same all over the Earth, and found that the resulting outgoing infrared flux varied uniformly for both place and time, with a balance between outgoing and incoming radiation during every month of the year. Houghton, employing the Elsasser chart and all available pyrheliometric data for North America, found that the actual amounts of both incoming solar and outgoing terrestrial radiation were much higher than Simpson's calculations suggested.[61]

Later workers have preferred to express their results in the form of atmospheric

220

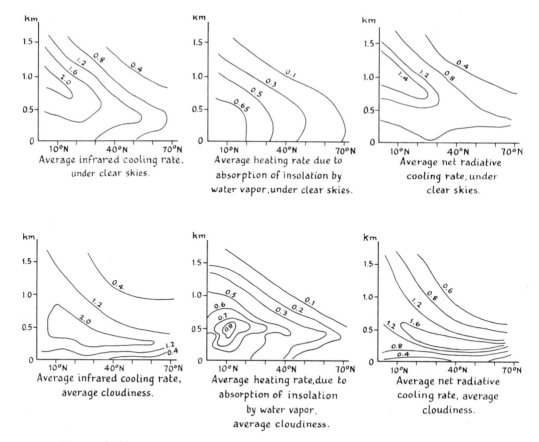

FIGURE 4.11A.
Radiation balance in the troposphere in March, in °C per day.

cross sections, which represent the distribution of radiation balance as determined from theoretical calculations plus the known distribution of pressure, temperature, water vapor, cloudiness, ozone, carbon dioxide, and oxygen. London (Figure 4.11) found that the entire troposphere is a source of radiative energy, with cooling concentrated in the middle troposphere in all latitudes because the middle cloudiness layer is located at 3 to 4 km, where net heat loss due to radiation is 1.5 to 2.0°C per day; this heat loss is counterbalanced mainly by the release of latent heat of condensation. Roach's cross sections give the 24-hour heating rates resulting from radiation absorption by water vapor and carbon dioxide up to 10 mb in a cloudless atmosphere for the northern hemisphere. Yamamoto also took absorption by molecular oxygen into account. Manabe and Möller calculated the seasonal distribution of temperature variations caused by water vapor, carbon dioxide, and ozone absorption, and from this obtained the annual mean rate at which temperature is changed by radiation.[62]

Davis computed the radiation budget for the atmosphere alone below 25 mb in latitudes 10°N to 70°N, after eliminating the effects of land and water distributions. His sections show that radiation causes the least heating in the region of the tropopause, and the most heating in the lower layers of the atmosphere (except in the Arctic winter region), but, in general, air columns extending from the Earth's surface to

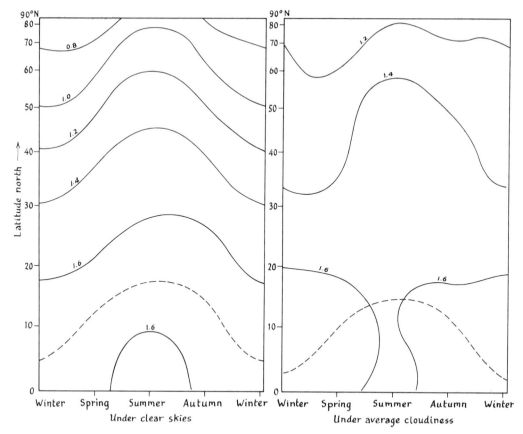

FIGURE 4.11B.
Average infrared cooling rate of the entire troposphere, in °C per day.

25 mb exhibit net cooling by radiation almost everywhere. Maximum heating by latent heat is in middle latitudes. Thermal deficits are balanced by convergence of large-scale eddies transporting heat north of 50°N, and by the mean meridional circulation in the subtropics, especially in winter.[63]

It is obvious that the sea must be a great storage reservoir for energy; how do variations in its heat storage influence the radiation balance of the atmosphere? Assuming that local energy storage in both the sea and the solid land undergoes an annual cycle such that the net annual accumulation of deficit is nil, Baur and Phillips calculated that if during any one year, 1 per cent of the total effective solar radiation (i.e., 0.003 cal per cm^2 per min) is stored instead of being radiated back to space, then the mean temperature of the atmosphere would increase 6.3°C. The observed increase would be only about half as much, because the atmosphere's capacity to absorb water vapor from the sea would also increase. According to Rossby, however, if only the surface layer of the oceans (i.e., the uppermost 50 to 100 meters) is taken into account, the increase in mean temperature of this layer after a 1 per cent increase in storage of incoming radiation would only be a few tenths of a degree, since the heat stored in the surface layers of the southern hemisphere oceans is decreased at the same time as heat storage in the northern hemisphere oceans is increased. Although this temperature increase would be insignificant on a global scale, it could nevertheless be a decided influence on local radiation balance. Furthermore, if the effect of the deeper layers of the oceans is also included, the immediate influence on the atmosphere of a slight increase in heat storage becomes even less: as much as 1 per cent of the total incoming radiation could be stored within a layer 1,000 m deep in the depths of the sea without producing a temperature increase of more than 0.015°C per year. Thus radiation anomalies may be stored and temperature effects isolated within the oceans, their influence not being felt in the atmosphere until decades or even centuries later. The Earth as a whole cannot therefore be in precise radiation balance with outer space, even over periods of several decades.[64]

Despite this delayed-action effect of the oceans on global radiation balance, it is possible to investigate fairly accurately the radiative heat exchange between sea surface and overlying atmosphere within limited areas. In the North Atlantic, for example, the geographic patterns of heat exchange alter markedly near 700 mb, below which oceanic or continental influences on radiation balance are strong and the meridional gradient very pronounced, above which atmospheric influences and zonal gradients dominate the radiation balance.[65]

Although theoretical computations for infrared radiation indicate that the tropopause could be maintained primarily by radiative processes within the atmosphere, radiometersonde observations suggest that it is in fact maintained by processes other than radiation.[66]

The radiation balance of the stratosphere and higher atmospheric levels is much bound up with the effects of ozone and other gases that are comparatively rare near the Earth's surface. A very prominent warm layer in the atmosphere around 55 km, first discovered by Lindeman and Dobson in 1922, forms a high altitude heat source that is capable of damping down winds in the troposphere. This warm layer is caused by the absorption of short-wave solar radiation at the upper boundary of the ozone layer, which results in temperatures of 50° to 80°C, much greater than those at the surface. Within the ozone layer, situated between the tropopause and 60 km, the ozone concentration at a particular point reaches a maximum in early spring and a minimum in late autumn. The total amount of ozone (although very small, corresponding to a pressure of 0.3 cm at normal temperatures and pressures) is at a minimum at the equator and at a maximum near 60°N, but, particularly in middle and high latitudes, it varies considerably from day to day; the daily variation is in fact as large as the annual variation. Absorption of infrared radiation from the troposphere by ozone heats the lower stratosphere: the flux of infrared radiation reaches maximum or minimum after the solstices, but the amount of ozone reaches maximum or minimum somewhat earlier, and hence the temperature extremes are in between them. The seasonal temperature variation in the upper part of the ozone layer is about 50°C, because the density of the atmosphere is two to three times greater in summer than in winter at these altitudes in middle latitudes, largely as a result of the great summertime increase in the total amount of ozone.[67]

Figures 4.12 and 4.13 illustrate some cross sections of radiation balance in the stratosphere. In examining the basis on which these and similar sections are constructed, one must remember that, unlike tropospheric sections (Figures 4.10 and 4.11), the stratospheric ones cannot be compiled with the aid of Elsasser or other radiation charts: the concentrations of absorbing gases are too low in the stratosphere.

Ohring's model sections (Figure 4.12) show that differences in radiation balance

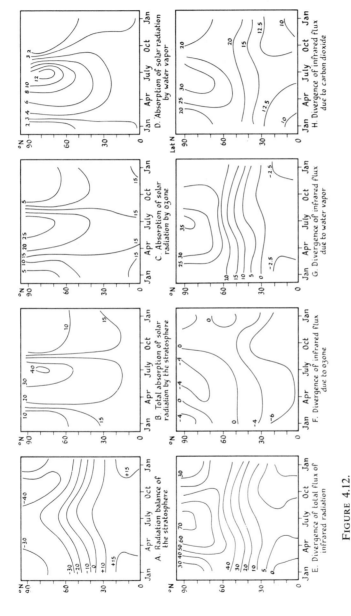

FIGURE 4.12.

Radiation balance of the stratosphere. Units are 1,000 langleys per minute. All sections cover the stratosphere from the tropopause to 55 km. + indicates that absorption of insolation exceeds emission of infrared radiation. − indicates that absorption of insolation is less than emissions of infrared ratiation. (After G. Ohring.)

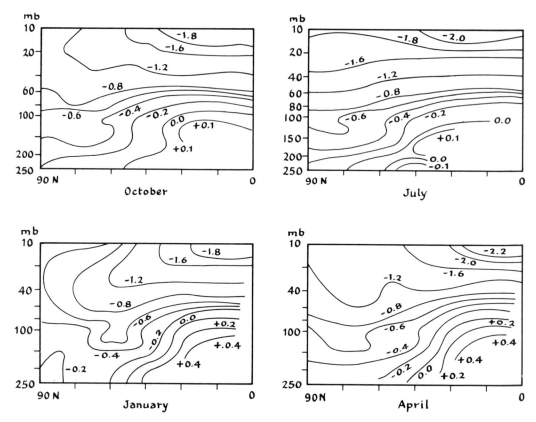

FIGURE 4.13.
Radiative temperature change due to carbon dioxide, in °K per day (after D. L. Brooks).

should make the stratosphere warmer in high latitudes than in low latitudes, and actual observation confirms this. Within the stratosphere, low latitudes are a source of radiative energy, and high latitudes a sink. Carbon dioxide is more important than water vapor for cooling the stratosphere in this model, and ozone produces a convergence of infrared energy. In low latitudes (0° to 40°N) the tropopause is heated by this long-wave radiation, and at the top of the stratosphere (50 km) a region of net infrared heating is found. During autumn and winter, there must be a poleward movement of energy, at least in the upper stratosphere, to account for the computed radiation balance, and an equatorward flow in the late spring and summer. An advection of radiative energy into and out of the northern hemisphere stratosphere must also be present at its lower (tropopause) and southernmost (equator) boundaries. Radiative equilibrium is not attained at any point in the stratosphere, or in the entire stratosphere, in any season.[68]

Murgatroyd and Goody computed sections for latitudes from 30° in the winter hemisphere to 60° in the summer hemisphere, and showed that the upper atmosphere (between 30 and 90 km) closely approaches radiative equilibrium at all heights above the stratosphere; the heating rates due to radiation seldom fall outside ±2°K per day. However, the upper atmosphere above the polar regions does depart from equilibrium, with excess heating in summer and excess cooling in winter. Sources and sinks of radiative energy in the 30–90 km layer appear to cancel each other out; there must be comparatively little vertical heat transfer by nonradiative processes in this layer.[69]

Houghton's sections give the calculated rate of heating of the atmosphere between 5 and 50 km because of absorption of solar radiation. Brooks computed sections (Figure 4.13) to show the heating rates caused by radiation from carbon dioxide alone for the layer from 10 to 250 mb.[70]

It is possible to calculate theoretically the 12-hour rate of local temperature change in the stratosphere caused by radiation alone (see Appendix 4.28). Values obtained thus for the southeastern United States indicate that at 200 and 100 mb, horizontal advection is more important than radiation in deciding the local rate of temperature change; at 50 mb, advection and radiation are of equal importance; at 200 mb, vertical exchange processes (convective and eddy transfer of energy) are as important as radiation. The cross sections in Figure 4.12 and 4.13 give heating rates caused by radiation alone; they would have to be modified considerably to indicate total heating rates.

For the troposphere, London's cross sections of radiation balance show that, in the climatological sense, the entire troposphere always loses heat by radiation. There is an average radiation balance, but not necessarily one at any one time or place. In March, the heat lost from the entire troposphere by radiation amounts to 200 cal per cm^2 per day; 70 per cent of this heat loss is balanced by the transfer, by eddy conduction from the Earth's surface to the atmosphere, of sensible heat; the remaining 30 per cent by the release of latent heat of condensation.[71] Within limited areas, the radiation balance of the troposphere can be expressed empirically as a function of the total daily receipt of solar radiation on a horizontal surface, the upper-air temperature, and the cloud amount (see Appendix 4.29).

The geographical pattern of tropospheric radiation balance can successfully be studied synoptically. There is a definite correlation, during periods of a few days, between outgoing terrestrial radiation and the distribution of cloudiness. The pattern of radiation balance caused by water vapor at 700 mb over the eastern North Pacific changes from day to day as the airflow pattern changes, but there seems to be no simple relationship between airflow and radiation balance, which also conforms with neither the winds nor the 700-mb contour pattern. The pattern of radiation balance between the surface and 400 mb over western Europe, for short periods, can depart considerably from the mean pattern.[72]

Consistent relationships are found between infrared radiation flux to space and synoptic weather conditions. Centers of maximum (minimum) infrared emission usually coincide with high (low) pressure centers, and a definite, although nonlinear, relationship exists between long-wave flux and the height of the 700-mb pressure surface. Figure 4.14 gives some examples of infrared synoptics. Variations in cloud cover and cloud type prove to be the main determinants of long-wave radiation fluctuations, as would be expected from theory, but direct observation of the infrared variations has brought some interesting features to light: the variation in infrared emission from a relatively small geographical area on any one night is at least as large as the variation in mean long-wave loss to space between the equator and the poles during the year; and infrared emission varies by a factor of two over very small distances, a fact that must be significant for the formation of local weather systems. On a broader scale, the rate of emission of long-wave radiation to space from the atmosphere and the Earth's surface combined is mainly a function of the frequency of

May 26, 1959

FIGURE 4.14.
Infrared radiation synoptics,
showing constant power loss to
space in milliwatts per cm²
(after J. L. Gergen and
W. F. Huch).

May 27, 1959

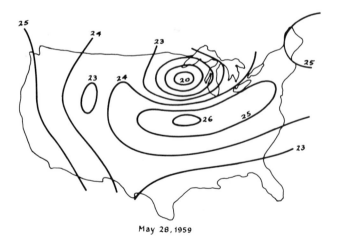

May 28, 1959

high clouds, and is about as variable as the kinetic and potential energy of the atmosphere.[73]

Radiation balance at the Earth's surface is quite complicated to study, because of the variegation of the Earth's topography and the resulting complex pattern of local variations in reflectivity and radiative properties. Some distance above the surface, however, the pattern becomes simpler: aircraft observations over southeastern England and the English Channel show that there the Earth's surface at night radiates as a black body at a temperature between that of screen air and that of grass or the sea surface, despite topographic complexities.[74]

Because the climatological picture of the troposphere is essentially one of radiative cooling, it is obviously of interest to know where radiative heating becomes more important, as it must be at the Earth's surface (on the average, and despite nocturnal cooling) if the average temperature of the surface is to remain approximately constant. For southern England, observations suggest that radiative heating, which is very strong close to the ground, decreases rapidly with height at first, but only changes to radiative cooling at 850 mb. In cloudless summer anticyclones, radiative cooling and convective heating almost balance each other in the layer between the Earth's surface and 850 mb, but the development of clouds completely upsets this balance.[75]

Local radiation emission from short turf is close to black-body radiation at the temperature of a grass-minimum thermometer. The difference between the mean surface temperature of the turf and the mean temperature of the overlying air has itself a seasonal variation that complicates surface radiation balance. On clear days at Rothamsted, surface heating is proportional to net radiation, and the diurnal variation in the net long-wave balance is closely related to changes in surface temperature. The heating coefficient (i.e., the increase in infrared radiation loss per unit increase in net total radiation) is 0.41 for dry bare soil under these conditions; for tall crops it is 0.08, and for grass, potatoes, and sugar beets it is between 0.15 to 0.22, depending on windspeed. There may thus be large local differences in radiation balance because of differences in use of land or in type of crop.[76]

Daily totals of net radiation as a percentage of incoming solar radiation are as follows: bare soil, 37; short grass, 41; tall crops, 46; water surfaces, 53. Not only do different surfaces reflect incoming radiation differently, but different crops have appreciably different reflection coefficients, which must be known to within ± 0.02 if the net radiation is to be calculated to an accuracy of ± 5 per cent (see Appendix 4.30). Maximum reflection coefficients determined at Rothamsted for grass, lucerne, potatoes, sugar beets, and spring wheat during the months May to September range from 0.25 to 0.27. In Ontario small maize plants have coefficients of 0.12 to 0.15 in moist soils and of 0.21 in dry soils; for fully developed maize plants, growing 10 inches apart in 36-inch rows, the reflection coefficient is between 0.17 and 0.19.[77]

In general, reflection coefficients are lower for the early stages of crop and vegetation development, because the leaves are smaller, shade each other more, and reflect less. The highest reflectivity for all agricultural crops seems to be about 0.26. Also, leaves of different species have approximately the same reflection coefficient, which is that of the crop as a whole when the leaves are developed enough to shade the entire area beneath the crop. Deserts absorb less radiation than agricultural crops, and forests absorb more. For fresh green vegetation, and rural areas generally, the reflection coefficient is about 0.20.[78]

The reflection coefficient of a surface varies with the elevation of the sun as well as with the nature of the surface. Thus at Rothamsted, the reflectivity for grass was found to vary from 0.23 with a solar elevation of 60° to 0.28 with an elevation of 20°; that for bare soil varied from 0.16 to 0.17 for the same sun angles.[79]

When reflectivities of about the same value are in force over extensive areas, the resulting effect on radiation balance can be immense. For example, the snow-covered Antarctic plateau loses enough heat by radiation to cool the uppermost 10 meters of snow 100°C per year, if the energy balance of the snow surface were controlled by radiation alone.[80]

Local variations in atmospheric aerosols also influence radiation balance at the Earth's surface. In cloudless summer weather in southeastern England, for example, the presence of aerosols in the atmosphere decreases the total incoming radiation by 10 per cent, on the average, and increases the ratio of diffuse to total radiation by 8 per cent of the total radiation amount.[81] Absorption by man-made pollutants can be very great: it is negligible in rural areas in summer, is about 5 or 10 per cent in winter in some country districts, but reaches 30 per cent or more in industrial areas, depending on the smoke concentration in the atmosphere.

Diffuse sky radiation can vary greatly even within small areas, yet it is often neglected in the study of radiation balance. For surfaces within a small area and with approximately equal reflection coefficients, the ratio of mean monthly incoming radiation to mean monthly total (direct plus diffuse) radiation depends on the local amount of cloudiness (see Appendix 4.31). The variations within continental areas can be very large even when averaged out, as they are, for example, in South Africa: although in summer the diffuse sky radiation is uniform over the whole country at 200 to 250 ly per day, in winter it ranges from 50 or 100 ly per day in the southern part of the plateau to 200 or more in the more equatorial regions.[82]

In high latitudes, the study of surface radiation balance is complicated by special reflectivity and other effects. In polar regions, for example, radiation instruments operating at the Earth's surface suffer from two important errors: one is due to the effect of the angle of incidence on incoming radiation, the other is a temperature effect.[83] The former causes the instruments to record too little radiation when the sky is clear and the sun is low, because radiation is reflected from the receiving surfaces of the sensors. The temperature effect causes the instruments to record more radiation when the air temperature is low than when it is high; the recording is 1 to 2 per cent higher than the true value for each 10°C drop in the ambient temperature. In the sub-Arctic, these errors are not as large, and there it is possible to estimate mean values of the net radiation flux during the daytime from observations of total (direct plus diffuse) radiation alone (see Appendix 4.32).

It has proved possible to bring some order into the quantitative analysis of surface radiation balance by ignoring the complexities of the distribution of the topographic elements, and concentrating instead on their radiative properties. Since it is easy to measure incoming radiation, emphasis has been laid on estimating the outgoing long-wave component of the terrestrial radiation balance by means of fundamental radiation theory. Noteworthy contributions to this theory have been made by Brunt, Wexler, Groen, and Fleagle.

Brunt integrated the classical heat conduction equation by assuming that the net outgoing radiation remains constant, and proved that the temperature decrease at the surface during nocturnal radiation is proportional to the square root of the period of time during which the radiation is in progress; the resulting temperature therefore has no lower limit (see Appendix 4.33). Later additions to the theory were made by Frost and by Knighting, who introduced turbulence parameters and took into account the fact that the air itself supplies some of the radiated heat. Wexler assumed that the net outgoing radiation decreases as the Earth's surface cools, reaching zero when a certain critical difference between air and ground temperature is reached; if so, for any given air temperature, there must be a value below which the ground temperature cannot fall. Groen assumed that both net outgoing radiation and the heat flow through the ground approach zero as the minimum surface temperature is approached; he then showed that the actual temperature approaches this minimum value asymptotically as the net radiation decreases. He finally obtained the outgoing terrestrial radiation as a function of surface windspeed and temperature, the water-vapor content of the lower atmosphere, and the thermal properties of the local soil type. Fleagle integrated the heat conduction equation to demonstrate that horizontal obstructions cause radiative cooling to be slower than it would be under unobstructed conditions. The lowest temperature that the local ground surface may achieve depends on the thickness of the surface layer of the ground that undergoes the temperature change, the thermal conductivity of this layer, and the presence of obstructions above the horizontal   · plane.[84]

In practice, a combination of theory and empiricism gives quite good results in the estimation of the rate of cooling of the Earth's surface at night. The starting point is usually Brunt's equation,

$$\frac{R}{\sigma T^4} = a + b\sqrt{e},$$

but other expressions have been devised by Loennquist and Robitzsch (see Appendix 4.34). A more recent one by Monteith shows that the annual mean net long-wave radiation (L) for short grass throughout the British Isles is approximately given by $178c - 200$ cal per cm^2 per day, where c is the fractional cloudiness. The annual net (total) radiation of 29 kcal per cm^2 varies very little over the British Isles. In North America, variants of the Brunt equation derived by Jacobs, and empirical expressions derived by Young, Donnel, and Kangieser are usually preferred.[85]

The rate of nocturnal cooling at the surface may also be calculated from the radiative flux divergence within the lowermost layer of the atmosphere, either by direct observation using two CSIRO radiometers, or indirectly by computation using Funk's equation. By subtracting the calculated flux divergence from the observed temperature change, then integrating the resulting differences, profiles of vertical heat flux may be obtained.[86] The results show that radiative flux divergence may produce large rates of temperature change during the day as well as during the night, but those during the day do not much influence the rate of eddy heat flux in the lowest few meters of the atmosphere.

Another approach to nocturnal cooling, particularly where night minimum temperatures must be predicted from climatological information, was first introduced by Saunders.[87] This approach is based on the study of the local thermograph records for clear sky conditions. The night-cooling curve on a thermograph record is not smooth, but consists of two gradual declines separated by a steep drop, termed the evening temperature discontinuity. The latter is primarily a boundary-layer feature, and probably represents the time of first deposition of dew. It is very pronounced near the ground, less pronounced at screen height, delayed about one hour at a depth of 2 inches in light soils, and difficult to distinguish both above and below these levels. The time that the evening temperature discontinuity occurs is unaffected by windspeed, air temperature, and humidity: it is thus a very convenient quantity to forecast from climatic records. An empirical expression that gives T_r, the screen temperature at the time of the discontinuity, with clear skies, to an accuracy of $\pm 1°F$, is

$$T_r = \tfrac{1}{2}(T_{max} + T_d) - t,$$

where T_{max} is the maximum temperature during the afternoon preceding the night in question, T_d is the dew point at the time of T_{max}, and t is an empirical constant for the location that varies with the presence or absence of an inversion (see Appendix 4.35). For low-lying inland airfield sites in the United Kingdom, t equals 2.0°C if there is an inversion below 850 mb, and 1.0°C if there is not.[88] When T_r has been determined, graphing may conveniently be employed to determine the night-cooling curve from the values of T_r, T_{max}, T_d, the time of the evening temperature discontinuity, the time of minimum temperature (taken to be the time of local sunrise for clear skies), and the geostrophic windspeed.[89]

Fluctuations in Solar Radiation and Their Climatic Significance

The solar constant is the standard index for measuring the intensity of solar radiation received by the Earth and its atmosphere. Determination of the solar constant for a specific day is a lengthy procedure. The Smithsonian method involves (a) measuring with a pyrheliometer several curves of solar spectral irradiance, on absolute scales, for different solar altitudes or air masses (the latter approximating the secant of the sun's zenith angle) during the morning; (b) fitting the measured irradiance values for a specific wavelength to Bouguer's law (see Appendix 4.36), and then determining the irradiation corresponding to zero air mass by extrapolation; (c) carrying out similar extrapolations for various measured wavelengths, thus allowing a curve of zero-air-mass solar spectral irradiance to be constructed. When corrected for ultra-violet and infrared radiation not included in the measurements, the area under this curve, referred to mean solar distance, gives the value of the solar constant for that day.[90]

The Smithsonian Institute's value for the solar constant, based on 30 years' observations, is 2.00 cal per cm^2 per min, with a probable error of ± 2 per cent, and is currently accepted as the standard value. Another modern estimate, by Allen, gives the value as 1.99 ± 0.02 cal per cm^2 per min. Apparent variations in the numerical value of the solar constant are caused by slight differences in the method of calculation or in the periods of observation; there is as yet no experimental proof that the constant does vary in magnitude. Nevertheless, by means of a relatively simple mathematical model one can calculate what effects a slight variation in the solar constant would have. With a dry atmosphere, a 1 per cent change in the constant would cause a change of 0.75°C in equatorial temperatures; with a moist atmosphere, a similar change would cause an increase in temperature of 0.3 to 0.7°C in tropical latitudes, and a slightly greater increase in high latitudes. The general result would be a decrease in the zonal atmospheric circulation.[91]

Conditions during an eclipse of the sun provide a good example of how weather and climate on Earth can be influenced by changes in how much solar radiation is received. A solar eclipse causes a relatively rapid shutting off and restoring of the supply of solar energy to the Earth's surface, and so causes a drop in temperatures, a rise in humidities, and changes in winds and clouds.

Observations made during the path of the solar eclipse of August 31, 1932, across North America furnish some interesting information.[92] For each decrease of 0.1 cal per cm^2 in the solar radiation received at the ground, temperatures decreased an average of 0.7°F, which is less than the normal rate of decrease in afternoon temperature during August (1.1°F) for the same radiation drop. The actual rate of decrease in radiation was four times the normal afternoon decrease. In the zone of total eclipse, the fall in temperature averaged 6°F for stations with clear skies; in the zone of 40 per cent obscuration temperatures did not drop more than 2 to 3°F; and no temperature drop was recognizable outside the 40 per cent zone. Changes in temperature also depended on cloudiness, water, and altitude. No temperature falls were detected on the surrounding oceans or above 500 meters in the free atmosphere. Ground temperatures, at places that had been in full sunlight, dropped at least 30°F, and even shaded or clouded stations recorded falls of 3 to 7°F.

There were marked effects on elements other than temperature. Pressures dipped 0.3 mb in the zone of total eclipse, and 20 minutes later than the time of total eclipse, at which time there had been a slight hump in the barogram of not more than 0.1 mb. The dip in pressure was the result of the release of free air from surface drag as convection ceased, with a consequent attenuation of the atmosphere within the shadow zone; the pressure hump was due to the inflow of air aloft during the eclipse-induced cooling, which caused the pressure surfaces to bend downward. Convection and pressure were minimal a few minutes after the minimum temperature and the maximum of accumulated air from the outflow, so that maximum pressure began to precede the time of total eclipse.

Wind velocities and gustiness decreased during the eclipse. The normal tendency of air to turn toward the left over North America diminished, and in some places an "eclipse wind" developed. Changes in absolute humidity were small and irregular, but vapor pressures dropped during the first half of the eclipse, and rose immediately after totality; the drop was caused by the fact that the rate at which the atmosphere became charged by evaporation decreased more rapidly than the rate at which vapor was concentrated by the decrease in convection; the rise was due to evaporation increasing while convection was still decreasing. Cumulus and even cumulonimbus decreased during the eclipse (an increase in the amount of high clouds has been noted with other eclipses), and an altocumulus band, moving with the wind, represented a disturbed layer lying within the shadow zone parallel to the solar crescent.

Observations of another eclipse made in a sheltered locality on the southwest coast of Sweden showed a fall in temperature from 61°F to 55° at the time of mid-eclipse, the last 2°F during a 3-minute period. Following totality, the temperature rose rapidly to 60°F at 3:00 P.M. despite rapidly increasing lower cloud.[93]

The general question of how the sun influences terrestrial weather and climate has attracted attention for many years.[94] The main difficulty with trying to answer it is the fact that, although many workers have been able to demonstrate statistical correlations between solar activity (usually measured in terms of numbers of sunspots) and climate at the surface of the Earth, many of these correlations are relatively short-lived—for example, the supposedly well-established relationship between sunspots and the level of Lake Victoria broke down after 1955. A further difficulty is that, although actual, specific observations unmistakably show that the amount of solar radiation received at the Earth's surface does vary considerably, these variations of the solar constant are within the limits of experimental error. The measured variations in the solar constant, less than 0.2 per cent, seem to be due to climatic phenomena, i.e., to variations in atmospheric transparency and so on, not vice versa.

The answer to the problem clearly must be sought in the high atmosphere, in the ozonosphere in particular. The ozone layer is very likely the link between the sun's fluctuations and the reactions of surface weather and climate. But here we get another difficulty: the observed variation in ozone concentration in the atmosphere does not conform to either the sunspot cycle or the occurrence of solar flares. Nevertheless, local connections do exist. For example, ozone concentrations in the upper stratosphere at Arosa, Switzerland, correlate positively with solar radiation flux.[95]

In general, during times of increased solar activity in the ultraviolet, solar radiation at wavelengths near 0.2 microns increases greatly, but there is little change near 0.3 μ. The consequent increase in radiation absorption by ozone is at a maximum near the equator in the winter hemisphere, and in turn intensifies the stratospheric westerlies. The energy is injected into the bipolar wave, i.e., wave-number two of the stratospheric circumpolar vortex, and then causes accentuation of wave-number two of the tropospheric middle-latitude westerlies, thereby increasing both the eddy available potential energy and the poleward flow of water vapor. During the winter, therefore, the greatest temperature increase is in high latitudes, while in both middle and high latitudes precipitation is increased. There is more of an increase in ozone absorption during the summer than during the winter, but the summer north-south gradient is smaller and may even be reversed; as a result, the temperature increase is small and general at all heights in the atmosphere.

Both the Earth's magnetic fields and the electron density of the upper atmosphere reflect the sunspot cycle. Both vary with the variations in ultraviolet solar radiation, which, being electromagnetic energy, travels at the speed of light, and causes ionization in the high atmosphere, and in solar corpuscular radiation, which consists of slower-moving solid particles that cause the formation of storms in the ionosphere. Neither are direct causes of surface climate, although there are indirect relations. Thus days of maximum emission of particles by the sun (normally in winter, when sunspots are relatively inactive) are generally followed by marked rises in atmospheric pressure in high latitudes and a slight fall in lower-middle latitudes—the maximum differential pressure change between latitudes 40° and 70°N is about 2.2 mb—which result in antiycclogenesis in polar regions and increased storminess in middle latitudes about five days after the solar disturbance. These high-level blocking anticyclones of the winter season develop only over North America, not over Asia, and are most frequent in the vicinity of the Yukon Valley.[96]

The stratospheric circulation responds to changes in input of solar energy as much as 40 days late, as the northern hemisphere reverses from its normal westerly circulation in winter to its summer easterlies. As a consequence, there is a global westerly circulation twice yearly for about 40 days; the westerlies begin to develop in the winter hemisphere at the same time that the zonal westerlies begin to die down in the summer hemisphere before changing to an easterly circulation.[97]

Although the mechanism of the connection between solar activity and climate is not definitely known, and the solar constant shows no proven variation, empirical studies indicate definite connections between solar events and terrestrial weather and climate. Yearly means of the monthly amounts of solar radiation received at normal incidence at the Earth's surface clearly vary, and these variations must have a physical cause. Actual monthly amounts also show considerable variation; in the United States, the effects of dust and haze in the high-pressure areas, and of forest fires, is very evident in the radiation curve.[98]

The work of C. G. Abbot of the Smithsonian Institution on the variations of the solar constant has been a major effort carried on for many decades. Abbott's graphs seem to show periodic repetitions of dent-like depressions in the solar-constant curve with periodicities forming harmonics of 273 months. Although the variations are within the limits of experimental error, the periodicities have proved to have some limited value in climatological forecasting.[99]

The variations in the following three factors cause the rate at which the sun heats any given area of the Earth's surface to fluctuate: (a) the altitude of the sun during different months of the year; (b) the amount of solar radiation reaching the atmosphere; (c) the transparency of the atmosphere. If it can be assumed that maxima and minima in the numbers of sunspots correspond to maxima and minima in the total output of solar energy, then the well-known sunspot cycle should produce changes in factor b comparable in magnitude to those of factor a, but with a periodicity of just over 11 years.

The evidence from volcanic eruptions is important for factor c. A general increase in winter temperatures in the northern hemisphere from 1900 to 1940 is definitely established: mean January temperatures at Spitzbergen, in particular, have increased 24°F over this period. According to Wexler, an important factor in this increase may be a general clearing from the atmosphere, with consequent increases in the amounts of solar radiation reaching the ground, particularly in high latitides, of the volcanic dust from the great eruptions of 1883, 1900, and 1902. The Krakatoa eruption on August 27, 1883, ejected 13 cubic miles of ash, lava, and mud into the atmosphere, one-third of which (mainly pumaceous plates) took years to settle out of the atmosphere. The ejected material reduced the solar radiation reaching the Earth's surface by nearly 10 per cent for some time. The dust took three months to reach western Europe in amounts large enough to produce optical effects; in late November, it reduced the solar radiation recorded at Montpellier by 25 per cent, and the radiation amounts remained below average for three years.[100]

In general, volcanic dust in the atmosphere can decrease direct solar radiation at the ground by up to 20 per cent, but it also increases the diffuse radiation because the dust acts as a scattering medium. The net decrease in total (direct plus diffuse) radiation is therefore just below 10 per cent. The rate of terrestrial infrared radiation into space is not affected by more than 1 or 2 per cent.[101]

The increase in the amount of industrial aerosol in the atmosphere during this century represents an increase in the rate of production, not in net accumulation, of pollutants. The latter either fall out of the atmosphere, or are washed out, within one or two weeks after their injection into the air. The pollutants consist of relatively large particles, and cause a net increase in the downward flux of infrared radiation, which balances the slight decrease in the amount of insolation reaching the ground because of absorption by the aerosol.[102]

When naturally vegetated land is converted to agriculture, the results include increased evaporation and therefore a higher vapor content in the atmosphere. The latter must have some effect on radiation balance, but it is probably not enough to be a major cause of fluctuations in the amount of solar radiation received at the ground.

Detailed tables are available, covering sunspot activity from the year 649 B.C. up to the present. The data suggest that the fundamental rhythm in the production of sunspot cycles has a mean period of 11.1 years, and is itself a component of a major cycle of seven sunspot cycles (i.e., 77.5 to 78 years) in length. A cycle is usually longer during aurorally weak periods, and its maxima tend to be earlier in the fundamental period during aurorally active periods.[103]

The basic 11.1-year cycle is extremely irregular. It varied in length from 7 (the mean for 1830 to 1837) to 17 (the mean for 1787 to 1804) years during a 200-year period, and its periodicity can be longer (or shorter) than normal for several cycles in succession. The four maxima between 1761 and 1787 averaged 8.67 years apart; the four maxima between 1787 and 1830 averaged 14.33 years apart. Although there is no consistent correlation between numbers of sunspots and temperatures or amounts of precipitation, generally the zonal climatic belts shift latitudinally as the sunspot activity changes, but not always, for example, not between 1820 and 1830.

Some quite large differences in pressure, temperature, and precipitation in the northern hemisphere have been found by comparing figures for times of sunspot maxima and minima.[104] Standardized climatic data, examined in terms of numbers of sunspots, have revealed the existence of two cycles, superimposed on the basic 11.1-year cycle, that appear to be very significant for surface climate: the 80- to 90-year cycle, and the 20- to 24-year double-sunspot cycle.

The double-sunspot cycle, whose existence was first proved by Hanzlik in 1931,[105] coincides with the following developments in terrestrial climate. During the period from sunspot minimum to the major sunspot maximum, polar continental anticyclones dominate in high latitudes in winter, with warm dry summers in the interior of the continents. During the period from the sunspot minimum to the minor sunspot maximum, the prevailing storm tracks in middle latitudes are located south of their average positions in the northern hemisphere, bringing cool, wet weather to the continents in lower-middle latitudes. At the time of the minor sunspot maximum, the zonal circulation pattern tends to shift poleward, bringing warm, dry conditions back to continental interiors in lower and middle latitudes.

The 80- to 90-year cycle, first recognized by Schove and by Willett in 1961, is characterized by a sharp break from peak sunspot activity during the final quarter of the cycle to minimum activity during the first quarter of the following cycle.[106] The the break is associated, in the atmospheric circulation, with a marked change from a low-index blocking pattern (during the final quarter) to a low-latitude zonal pattern with cool, wet conditions in low and middle latitudes. The second and third quarters of the cycle, coinciding with increasing sunspot activity, are associated with a poleward shifting of the zonal circulation pattern, with resultant warming in most latitudes, and with an intensification of continental-maritime contrasts equatorward of latitude 50°N during the third quarter.

The 80- to 90-year cycle is most significant in lower-middle and subtropical latitudes, particularly in summer and autumn; it may result from an actual fluctuation of the effective (but not necessarily of the intrinsic) solar constant. The double-sunspot cycle is most significant in middle and high latitudes, especially in winter; it probably results from change in the transmissive properties of the atmosphere.

Sunspot numbers also apparently reveal a 25- or 26-month periodicity that seems to excite the oscillation of the stratosphere. The 24- to 26-month cycle in equatorial stratospheric winds may be due to a variation in solar ultraviolet radiation with this period.[107]

Some correlations between sunspot activity and surface weather conditions are worth noting even though they have been established only for limited areas or periods. For example, in the vicinity of South America, the general circulation of the atmosphere tends to be in the low-index stage during sunspot maxima, and in the high-index stage during sunspot minima.[108] At the time of a sunspot maximum, pressures decrease in low latitudes and increase in high latitudes, the latter because of a weakened subtropical high-pressure belt and an extended southern circumpolar vortex and trade-wind cell.

During years with not too many sunspots, sea-level pressures in North America show a pronounced decrease in persistence two weeks after large increases in terrestrial magnetic activity. This decrease could be brought about by a funneling of charged solar particles by the Earth's magnetic field: the radiation would enter the atmosphere in the auroral zones, and would result in a sudden change in the wave-number of the circulation between 100 mb and 80 mb; it should take about two weeks for such a change to be dynamically propagated to sea level.[109]

A decrease in sunspot activity is associated with a northward shift of the atmospheric circulation over the Bay of Bengal, and an increase with a southward movement. The northward shift advances the seasonal progression of climatic features over Madras; the southward one retards them. Sunspot maxima are periods of weak easterlies and infrequent cyclogenesis over the Bay of Bengal; sunspot minima are times of active cyclogenesis and numerous westward-moving storms. During the northeast monsoon, days of excessively heavy precipitation in southeast Madras are more frequent during sunspot minima, with intervening periods of subnormal precipitation. Drought conditions are often approached during sunspot maxima.[110]

For the Mediterranean region as a whole, the 35 coastal and island stations can be grouped according to whether one solar cycle corresponds to one, two, or three oscillations of precipitation; the grouping holds equally for winter, summer or annual rainfall totals. Sunspot influence is particularly marked in winter: more than 80 per cent of the stations have their principal maximum of winter precipitation between the fourth and sixth year of the sunspot cycle, at which time summer rainfall exhibits its principal minimum for 24 of the stations.[111]

Data for 1889-1938 suggest that there may be a monthly cycle in both precipitation and sunspots. The highest maximum of both, on July 12, is more than 50 per cent higher than the lowest minimum, six months earlier.[112]

For the tropics as a whole, there has been a secular change in the relation between temperatures and sunspot activity. The correlation between sunspot numbers and surface temperatures was negative before 1920, but has been zero or even positive since then. This change seems to be associated with a minor pressure oscillation (or a secular change in the Southern Oscillation) that has resulted in an increase in the upper westerlies above northern Australia.[113]

High surface temperatures definitely appear to be related to sunspot maxima. For at least 30 consecutive months, from April 1947 onward, mean air temperatures in many parts of the world were the highest ever recorded, and annual sunspot means attained their highest values at the beginning of the two warmest spells during this time. Summer 1947 was a time of heat waves in western Europe; during this period, there were seven sunspot maxima, which occurred regularly every 27 days, with temperature minima on the 21st day. There is generally a negative correlation between solar activity and heat waves, with a phase difference of either $+21$ or -6 days being most usual.[114]

Correlations between atmospheric events and solar phenomena other than sunspots have been demonstrated. A rise in atmosphere temperatures of up to 1°C at 100 mb follows ionospheric storms caused by solar corpuscular radiation. This temperature increase is not evident at 50 mb, and has little or no effect on the height of the 500-mb surface: hence it quite probably has no effect on surface weather and climate in general. However, the 500-mb troughs over the Gulf of Alaska and Aleutian Islands were definitely affected by solar corpuscular emissions during the winters of 1946–48. Three days after the arrival of a solar corpuscular cloud, as manifested by a very bright auroral display, these troughs became amplified. During the winter of 1956–59, the 300-mb troughs over the same area also underwent amplification, greater cyclonic isobaric curvatures developing two to four days after the corpuscular emissions; the troughs reached maximum development some days later.[115]

Eruption of solar flares is associated with cyclogenesis in the high troposphere (at about 30,000 feet) over the Marshall Islands, the cyclone appearing between six and ten days after the flare. Less than 36 hours after the solar explosion, there is a sudden rise in temperature in the layer between the tropopause and 300 mb and above the equatorial convergence zone; this rise is maintained at the rate of 7°C per day at 100 mb. The mechanism of this association probably involves absorption, by water vapor, of the additional ultraviolet radiation that must penetrate the ozonosphere near the equator during the eruption of a solar flare.[116]

Weather conditions in the central midwestern United States correlate to some extent with the appearance of the solar corona. Excessive precipitation in spring, summer, and autumn is associated with a high ratio of green to red in the coronal emission, and subnormal precipitation in these months with a high red-to-green ratio. Rainfall is usually heavier, and more frequent, near the times of the central meridian passage of the centers of heightened green emission.[117]

Notes to Chapter 4

1. The total daily radiant energy available at the top of the atmosphere may be described in terms of a rapidly convergent harmonic series in the day of the year, whose coefficients are a function of geographical latitude and certain well-known astronomical constants; see E. C. McCullough, *AMGB*, 16, series B (1968), 129. For speculation on the effects of the eccentricity of the Earth's orbit on its climate, see R. H. Olson, *WW*, 21 (1968), 190. On Wien's displacement law, see W. K. Widger, Jr., *BAMS*, 49 (1968), 724 and 1142.

2. A standard reference on basic radiation instruments is *MOHSI*, pp. 350–51. See also A. Ångström, *CM*, pp. 50–56, and W. Mörikofer, *SPIAM*, pp. 521–25. For a review of the development of radiation scales, see: L. G. Aldrich and C. G. Abbot, *SMC*,

vol. 110, no. 5 (1948); C. G. Abbot, *SMC*, vol. 110, no. 11 (1948); A. J. Drummond, *BWMO*, 5 (1956), 75; W. Mörikofer, *BWMO*, 6 (1957), 25. According to T. G. Kyle, *T*, 19 (1967), 240 (see also A. Ångström, *T*, 20, 1968, 198), an error must exist in the Ångström radiation scale because of thermal conductivity effects within the blackened strips in the pyrheliometer. On the introduction of new types of radiation instrument into Europe's observing network, see R. Dogniaux, *BWMO*, 17 (1968), 203. For the relation between net and solar radiation at selected stations, see J. A. Davies, *QJRMS*, 93 (1967), 109. For details of radiation measurements at particular stations, see: A. Ångström and A. J. Drummond, *T*, 18 (1966), 801, on high-altitude mountain stations; H. Wörner, *GPA*, 57, no. 1 (1964), 193, on short-wave insolation along longitude 10°E; S. E. Jensen and H. C. Aslyng, *AMGB*, 15, series B (1967), 127, on net and net long-wave radiation at Copenhagen; M. K. Elnesr and N. A. Hegazy, *GPA*, 59, no. 3 (1964), 267, on total solar radiation at normal incidence, at Giza, U.A.R.; M. K. Elnesr and A. M. Khalil, *GPA*, 60, no. 1 (1965), 217, on total solar radiation on vertical and inclined surfaces during cloudless days in Egypt; and K. Bullrich, R. Eiden, and W. Nowak, *GPA*, 64, no. 2 (1966), 220, on sky radiation in Greenland. For the relations between net radiation, solar radiation, and albedo over natural surfaces at St. Paul, Minnesota, see S. B. Idso, D. G. Baker, and B. L. Blad, *QJRMS*, 95 (1969), 244. For the relation between net and solar radiation over irrigated field crops, see L. J. Fritschen, *Agric. Meteorol.*, 4 (1967), 55.

3. D. L. Gunn and D. Yeo, *QJRMS*, 77 (1951), 293. H. C. Pereira, *QJRMS*, 85 (1959), 253.

4. D. Fuquay and K. Buettner, *TAGU*, 38 (1957), 38. S. Fritz, *PM*, p. 159.

5. J. M. Stagg, *MOGM*, no. 86 (1950). L. Jacobs, *MM*, 90 (1961), 284.

6. F. H. W. Albrecht, *GPA*, 29 (1954), 115.

7. L. Jacobs, *op. cit.*, and *MOHSI*, p. 330.

8. For details, see: N. Robinson, *BAMS*, 36 (1955), 32; N. Robinson and L. Stoch, *JAM*, 3 (1964), 179; A. J. Drummond, *AMBG*, 7 (1956), 413.

9. A. Ångström, *CM*, p. 50. On the effect of instrumental aperture on actinometric measurements, see A. Ångström and B. Rodhe, *T*, 18 (1966), 25. On the use of colored glass filters, which enable the Eppley pyranometer to isolate specific spectral bands of solar radiation, see A. J. Drummond and J. J. Roche, *JAM*, 4 (1965), 741. According to R. A. Hamilton and R. H. Collingbourne, *QJRMS*, 93 (1967), 186, either the Moll-Gorczynski solarimeter is defective to some extent, or there is an unknown absorbing gas present in the atmosphere, which is most concentrated at noon. For a motorized occulting disc that enables records of diffuse radiation to be obtained, see C. J. Sumner, *JAM*, 7 (1968), 145. See also: M. C. Anderson, *JAM*, 6 (1967), 941, on solarimeter compari-

sons; B. G. Collins and E. W. Walton, *MM*, 96 (1967), 225, on the Moll-Gorczynski pyranometer; J. A. Davies, *JAM*, 4 (1965), 547, on use of the Gunn–Bellani radiometer in West Africa; and E. C. Flowers and N. F. Helfert, *MWR*, 94 (1966), 259, on Eppley pryanometers and pyrheliometers. On the importance of the geometry of the receiving surface of a radiometer on the measurement of radiation balance, see W. E. Reifsnyder, *Agric. Meteorol.*, 4 (1967), 255. On the use of the Gunn–Bellani radiometer, see J. S. G. McCulloch and F. J. Wangati, *Agric. Meteorol.*, 4 (1967), 63.

10. G. D. Robinson, *QJRMS*, 73 (1947), 127.

11. A. Whillier, *Solar Energy*, vol. 8, no. 3 (1964).

12. A. M. Stoll and J. D. Hardy, *TAGU*, 36 (1955), 213.

13. F. H. W. Albrecht, *GPA*, 29 (1954), 115.

14. H. J. Albrecht and R. Dingle, *GPA*, 38, no. 3 (1957), 222.

15. H. J. Albrecht, *GPA*, 29 (1954), 155.

16. T. C. O'Connor, *GPA*, 30, no. 1 (1955), 130.

17. See D. P. Brown and R. A. Harvey, *BAMS*, 42 (1961), 325, for details.

18. F. H. W. Albrecht, *op. cit.*

19. S. C. Stern and F. Schwartzmann, *JM*, 11 (1954), 121.

20. F. H. W. Albrecht, *op. cit.*

21. J. MacDowall, *MM*, 84 (1955), 65.

22. J. P. Funk, *BWMO*, 11 (1962), 83, and *QJRMS*, 8 (1960), 382. J. T. Gier and R. V. Dunkle, *Proc. Am. Inst. Elec. Engrs.*, 70 (1951), 339. See also K. P. Mackay, Jr., *JAM*, 4 (1965), 112, for a description of a modification of the commercially available Gier–Dunkle instrument (which is the Beckman and Whitley total hemispheric radiometer) that makes an external reference junction unnecessary. On the Thornthwaite radiometer, see D. M. Gates, *AMM*, 6, no. 28 (1965), p. 23.

23. B. C. Goodall, *JGR*, 67 (1962), 1383.

24. L. J. Fritschen and W. R. van Wijk, *BAMS*, 40 (1959), 291. For details see C. B. Tanner, J. A. Businger, and P. M. Kuhn, *JGR*, 65 (1960), 3657.

25. L. J. Fritschen, *JAM*, 2 (1963), 308. For details see G. S. Campbell, G. L. Ashcroft, and S. A. Taylor, *JAM*, 3 (1964), 640.

26. P. M. Kuhn, *MWR*, 89 (1961), 285.

27. V. E. Suomi and P. M. Kuhn, *T*, 10 (1958), 160. See also: V. E. Suomi, D. O. Staley, and P. M. Kuhn, *QJRMS*, 84 (1958), 134; P. M. Kuhn, *JAM*, 2 (1963), 368; P. M. Kuhn, V. E. Suomi, and G. L. Darkow, *MWR*, 87 (1959), 129; P. M. Kuhn and V. E. Suomi, *JGR*, 65 (1960), 3669.

28. D. B. Clarke, *JGR*, 68 (1963), 235. W. T. Roach, *QJRMS*, 87 (1961), 346.

29. J. L. Gergen, *RSI*, 27 (1956), 453, and *JM*, 14 (1957), 495.

30. F. Möller, *JM*, 16 (1959), 87.

31. J. P. Funk, *JM*, 18 (1961), 701.

32. R. L. Aagard, *JM*, 17 (1960), 311. J. A. Businger and P. M. Kuhn, *JM*, 17 (1960), 400. For the relation between insolation and sunshine at West African stations, see J. A. Davies, *QJRMS*, 91 (1965), 359. For the variation of mean daily sunshine duration with wind direction at Durham, see A. J. W. Catchpole, *MM*, 93 (1964), 149. On pyranometers for use in field crops, see M. Bringman and N. Rodskjer, *AMGB*, 16, series B (1968), 418. On integrating pyranometers, see J. P. Kerr, G. W. Thurtell, and C. B. Tanner, *JAM*, 6 (1967), 688, for climatological stations, also C. A. Federer and C. B. Tanner, *JGR*, 70 (1965), 2301, for daily solar radiation amounts. On radiometers, see: A. C. Combs *et al.*, *JAM*, 4 (1965), 253, on infrared radiometers; O. Vittori, F. Prodi, and V. Vicentini, *GPA*, 77, no. 2 (1967), 186, on Italian instruments for long-wave radiation; J. P. Funk, E. L. Deacon, and B. G. Collins, *GPA*, 64, no. 2 (1966), 213, on a radiosonde radiometer; J. M. Norman, B. L. Blad, and D. G. Baker, *JAM*, 5 (1966), 730, on rainfall correction for nonshielded net radiometers; and B. G. Collins and T. G. Kyle, *GPA*, 63, no. 1 (1966), 231, on polythene-shielded radiometers. On the calibration of balance meters, see H. S. Paulsen, *AMGB*, 15, series B (1967), 156; also C. L. Palland and L. Wartena, *AMGB*, 16, series B (1968), 95. See also: T. G. Kyle, *GPA*, 66, no. 1 (1967), 126, on a crystal pyrradiometer measuring both long and short-wave radiation; D. G. James, D. W. S. Limbert, and J. C. McDougall, *MM*, 95 (1966), 161, on a British adaptation of the Suomi–Kuhn radiometersonde; A. Mani, C. R. Spreedhaven, and V. Srinivasan, *JGR*, 70 (1965), 4559, on infrared soundings over India; B. J. Lieske and L. A. Stoschein, *AMGB*, 15, series B (1967), 67, on radiative flux divergence in Alaska from radiometer measurements; and E. J. Williamson and J. T. Houghton, *QJRMS*, 91 (1965), 330, on radiometric measurements of emission from stratospheric water vapor. For a solarimeter designed in New Zealand to measure photosynthetically active radiation, see K. J. McCree, *Agric. Meteorol.*, 3 (1966), 353.

33. For a historical review, see F. Möller, *CM*, p. 34.

34. W. L. Godson, *AMBG*, 12 (1962–63), 1 and 196, and *JM*, 12 (1955), 272 and 533.

35. For details, see: F. Möller, *CM*, p. 34; W. M. Elsasser, *HMS*, no. 6 (1942); W. M. Elsasser and M. F. Culbertson, *AMM*, vol. 4, no. 23 (1960); G. Yamamoto and G. Onishi, *JM*, 9 (1952), 415.

36. *Ibid.*

37. For details on use of the charts, see *HM*, pp. 364–71, and W. D. Sellers, *Physical Climatology* (Chicago, 1965), pp. 47–64.

38. J. L. Gergen, *JM*, 15 (1958), 350.

39. G. D. Robinson, *QJRMS*, 73 (1947), 127, and 76 (1950), 37.

40. F. Möller, *op. cit.*

41. K. Y. Kondratiev and H. J. Niilisk, *GPA*, 49, no. 2 (1961), 197, and 46, no. 2 (1960), 216 and 231. W. M. Elsasser and M. F. Culbertson, *AMM*, vol. 4, no. 23 (1960). R. A. McClatchey, *JAM*, 3 (1964), 573.

42. D. L. Brooks, *JM*, 7 (1950), 313. F. A. Brooks, *JM*, 9 (1952), 41.

43. J. P. Funk, *JM*, 18 (1961), 388.

44. W. C. Swinbank, *QJRMS*, 89 (1963), 339. W. H. Dines and L. H. G. Dines, *Mem. Roy. Meteorol. Soc.*, 2, no. 11 (1927), 1. D. Brunt, *QJRMS*, 58 (1932), 389. A. Ångström, *SMC*, vol. 65, no. 3 (1918), and *Medde. Stat. Meteorol. Hygrog. Aust.*, vol. 6, no. 8 (1936). F. L. Martin and W. C. Palmer, *JAM*, 3 (1964), 780. For further information, see: F. L. Martin and J. B. Tupaz, *MWR*, 96 (1968), 416, for a numerical procedure for computing outgoing terrestrial flux from the Elsasser–Culbertson theory; B. H. Armstrong, *JAS*, 25 (1968), 312, on Curtis–Godson approximations; W. G. Zdunkowski and F. A. Lombardo, *AMBG*, 15, series B (1967), 141, on the Möller radiation chart; G. M. Jurica, *MWR*, 94 (1966), 573, on a computer version of F. A. Brooks's 1950 tables; W. G. Zdunkowski, R. E. Barth, and F. A. Lombardo, *GPA*, 63, no. 1 (1966), 211, on a correction to the Elsasser–Culbertson radiation tables; C. R. Rodgers and C. D. Walshaw, *QJRMS*, 92 (1966), 67, on new computation schemes for infrared cooling, on which see also P. A. Davis and W. Viezee, *JGR*, 69 (1964), 3785; M. A. Atwater, *JAM*, 5 (1966), 824, on comparison of infrared cooling estimations for the boundary layer; W. G. Zdunkowski and D. V. McDonald, *GPA*, 65, no. 3 (1966), 185, on infrared flux divergence in an idealized valley; O. Lönnqvist, *T*, 15 (1963), 382, on a chart for estimation of surface temperatures from outgoing terrestrial radiation; W. G. Zdunkowski and F. G. Johnson, *JAM*, 4 (1965), 371, on estimation of infrared flux divergence from tables derived from radiometersonde soundings; and G. D. Robinson, *QJRMS*, 92 (1966), 263, on direct versus indirect measurements of radiation absorption over southern England.

45. M. I. Budyko, *BWMO*, 7 (1958), 166. G. C. Simpson, *Mem. Roy. Meteorol. Soc.*, vol. 3, no. 21 (1928). F. Baur and H. Phillips, *Gerlands Beitr. Geophys.*, 47 (1936), 218. H. E. Landsberg, *HM*, p. 937. J. N. Black, *AMBG*, 7 (1956), 165. H. E. Landsberg *et al.*, *World Maps of Climatology* (Berlin, 1963). *HBES*, pp. 97–138, gives a convenient account of these maps.

46. T. H. MacDonald, *MWR*, 91 (1963), 658.

47. G. J. Macris, *MWR*, 87 (1959), 29.

48. J. N. Black, C. W. Bonython, and J. A. Prescott, *QJRMS*, 80 (1954), 231. Q_A may be determined theoretically, or taken from Angot's table given in *PDM*, p. 112.

49. G. J. Day, *MM*, 90 (1961), 269.

50. A. J. Drummond and E. Vowinckel, *JM*, 14 (1957), 343.

51. J. N. Black, *AMGB*, 7 (1956), 165.

52. E. Vowinckel and S. Orvig, *JAM*, 1 (1962), 552.

53. B. Haurwitz, *JM*, 2 (1945), 154, 3 (1946), 123, and 5 (1948), 110.

54. R. L. Fox, *MWR*, 89 (1961), 543.

55. E. G. Astling and L. H. Horn, *JAS*, 21 (1964), 30. W. R. Bandeen, V. Kunde, W. Nordberg, and H. P. Thompson, *T*, 16 (1964), 481. P. F. Clapp, *MWR*, 90 (1962), 287. J. S. Winston and P. K. Rao, *MWR*, 90 (1962), 307. P. K. Rao, *W*, 19 (1964), 88.

56. Astling and Horn, *op. cit.*

57. On frontal systems and zones, see R. S. Hawkins, *JAM*, 3 (1964), 564. The surface temperature of the Earth is inferred from the radiative flux emitted from it. The calculation of the radiative flux from remote radiometer data is very involved; see D. G. Wark, G. Yamamoto, and J. H. Lienesch, *JAS*, 19 (1962), 369, for details. S. Fritz, *JAM*, 2 (1963), 645, explains the derivation of surface temperatures from the radiative flux. On vertical motion, see W. E. Shenk, *JAM*, 2 (1963), 770.

58. W. Nordberg, W. R. Bandeen, B. J. Conrath, V. Kunde, and J. Persano, *JAS*, 19 (1962), 20. S. Fritz, P. K. Rao, and M. Weinstein, *JAS*, 21 (1964), 141.

59. See W. R. Bandeen, *JAS*, 21 (1964), 573, and J. I. F. King, *JAS*, 20 (1963), 245. For radiation maps, see: W. H. Terjung, *AMGB*, 16, series B (1968), 279, for world maps of isanomalies of global and net radiation; A. Mani, O. Chacko, V. Krishnamurthy, and V. Desikan, *AMGB*, 15, series B (1967), 82, for the Indian Ocean; J. P. Kerr *et al.*, *MWR*, 96 (1968), 232 and 237, for global radiation in Wisconsin and its mesoscale variability; and C. C. Wallén, *T*, 18 (1966), 786, for global radiation in Sweden. On local differences in attenuation of solar radiation over Britain, see J. L. Monteith, *QJRMS*, 92 (1966), 254; for solar radiation on the south coast of England, see A. R. Rees, *QJRMS*, 94 (1968), 397; for estimated daily values of solar radiation at stations in England and Wales, see S. M. Taylor and L. P. Smith, *MM*, 90 (1961), 289. On sunshine duration and cloudiness in Egypt, see M. K. Elnesr and A. F. El-Sabban, *GPA*, 59, no. 3 (1964), 256. For the empirical relation between hourly short-wave total solar radiation at the sea surface and cloud categories (OWS "Juliett"), see F. E. Lumb, *QJRMS*, 90 (1964), 43. For radiation distributions derived from satellite observations, see J. S. Winston, *JAM*, 6 (1967), 453, on the Pacific sector, and *MWR*, 95 (1967), 235, for global charts for middle and tropical latitudes; also T. Fujita and W. Bandeen, *JAM*, 4 (1965), 492, on the resolution capabilities of the NIMBUS satellite infrared radiometer, and W. R. Bandeen *et al.*, *T*, 16, (1964), 481, for TIROS-derived radiation maps of a tropical hurricane. See also E. Raschke and W. R. Bandeen, *JAM*, 6 (1967), 468, for monthly quasi-global maps of tropospheric water-vapor content as inferred from TIROS IV radiation measurements. P. M. Kuhn and J. D. McFadden, *MWR*, 95 (1967), 565, show that atmospheric water-vapor profiles may be inferred from satellite radiation data. On the estimation of surface insolation over the tropical oceans, see W. H. Quinn, W. V. Burt, and W. M. Pawley, *JAM*, 8 (1969), 205. For the correlation between sunshine duration and total cloud amount (Mildenhall, England), see F. B. Webster, *MM*, 98 (1969), 87.

60. J. London, *A Study of the Atmospheric Heat Balance* (New York, 1957).

61. G. C. Simpson, *Mem. Roy. Meteorol. Soc.*, vol. 3, no. 21 (1928), on which see also C.-G. Rossby, *ASM*, p. 10. H. G. Houghton, *JM*, 11 (1954), 1.

62. J. London, *JM*, 9 (1952), 145. W. T. Roach, *QJRMS*, 87 (1961), 364. G. Yamamoto, *JAS*, 19 (1962), 182. S. Manabe and F. Möller, *MWR*, 89 (1961), 503.

63. P. A. Davis, *JAS*, 20 (1963), 5.

64. C.-G. Rossby, *ASM*, pp. 10–14.

65. R. V. Godbole, *JAM*, 2 (1963), 674.

66. L. D. Kaplan, *SPIAM*, p. 583. D. O. Staley, *QJRMS*, 91 (1965), 282.

67. See F. W. P. Götz, *CM*, p. 275, and R. A. Craig, *CM*, p. 292, for detailed accounts.

68. G. Ohring, *JM*, 15 (1958), 440.

69. R. J. Murgatroyd and R. M. Goody, *QJRMS*, 84 (1958), 225.

70. J. T. Houghton, *QJRMS*, 89 (1963), 319. D. L. Brooks, *JM*, 15 (1958), 210.

71. J. London, *JM*, 9 (1952), 145.

72. F. Möller, *SPIAM*, p. 561. A. H. Thompson and M. Neiburger, *JM*, 10 (1953), 167.

73. For details, see: P. T. Gannon, *JAM*, 2 (1963), 175; J. L. Gergen and W. F. Huch, *N*, 186 (1960), 426; J. L. Gergen, *JAS*, 22 (1965), 152; P. M. Kuhn and V. E. Suomi, *JGR*, 65 (1960), 3669.

74. J. T. Houghton, *QJRMS*, 84 (1958), 448.

75. G. D. Robinson, *CPRMS*, p. 26.

76. J. L. Monteith and G. Szeicz, *QJRMS*, 87 (1961), 159, and 88 (1962), 496.

77. J. L. Monteith, *QJRMS*, 85 (1959), 386. W. G. Graham and K. M. King, *QJRMS*, 87 (1961), 425.

78. J. L. Monteith, *QJRMS*, 85 (1959), 386, and 88 (1962), 508.

79. Monteith and Szeicz, *op. cit.* (1961).

80. K. J. Hanson, *JGR*, 65 (1960), 935.

81. Monteith, *op. cit.* (1962).

82. A. J. Drummond and E. Vowinckel, *JM*, 14 (1957), 343.

83. See S. Fritz, *PAS*, p. 159, for a discussion of these effects.

84. D. Knighting, *QJRMS*, 76 (1950), 173. H. Wexler, *MWR*, 64 (1936), 122. P. Groen, *JM*, 4 (1947), 63. R. G. Fleagle, *JM*, 7 (1950), 114.

85. O. Loennquist, *Arkiv. Geofys.*, 2 (1954), 151. J. L. Monteith, *QJRMS*, 87 (1961), 171. W. C. Jacobs,

BAMS, 21 (1940), 297. P. C. Kangieser, *MWR*, 87 (1959), 19.

86. J. P. Funk, *QJRMS*, 86 (1960), 382, and *JM*, 18 (1961), 388. For the procedure, see W. P. Elliott, *QJRMS*, 90 (1964), 260.

87. W. E. Saunders, *QJRMS*, 75 (1949), 154, *QJRMS*, 78 (1952), 603, and *MM*, 89 (1960), 65.

88. For specific examples, see, *inter alia*: W. D. Summersby, *MM*, 82 (1953), 210, for Northolt; W. J. Bruce, *MM*, 85 (1956), 90, for London Airport (Heathrow); T. H. Parry, *MM*, 82 (1953), 368, for Shawbury; E. D. Roberts, *MM*, 84 (1955), 48, for Mildenhall; W. E. Saunders, *MM*, 84 (1955), 76, for St. Eval; K. Pollard, *MM*, 87 (1958), 12, and W. E. Saunders, *MM*, 87 (1958), 87, for Wittering; W. E. Richardson, *MM*, 84 (1955), 301, for high-level stations in Cumberland, and *MM*, 85 (1956), 218, for Achnagoichan (Scottish Highlands, 1,000 ft); W. J. Bruce, *MM*, 84 (1955), 105, for Wahn, Germany; and I. Penzar, *Zagreb Geophys. Inst. Papers*, 3d series, no. 5, for Zagreb.

89. See J. A. Barthram, *MM*, 93 (1964), 246, for details. On reflection and absorption of visible solar radiation in the United States, see K. J. Hanson, T. H. Vonder Haar, and V. E. Suomi, *MWR*, 95 (1967), 354. On surface radiation balance, see: R. L. Charnell, *JGR*, 72 (1967), 489, on long-wave radiation of the sea-surface near Hawaii; E. Vowinckel and S. Orvig, *AMGB*, 15, series B (1967), 1, on that of the Polar Ocean and its climatic change implications; K. J. Buettner and C. D. Kern, *JGR*, 70 (1965), 1329, on infrared radiation from terrestrial surfaces; G. Stanhill, G. J. Hofstede, and J. D. Kalma, *QJRMS*, 92 (1966), 128, on that of natural and agricultural vegetation in Israel; E. A. Fitzpatrick and W. R. Stern, *JAM*, 4 (1965), 649, on that of irrigated cotton fields in western Australia; W. G. Zdunkowski and L. B. Stowe, Jr., *T*, 20 (1968), 293, on the influence of CO_2 on radiative flux divergence in the lower few meters of the atmosphere. See also: W. S. Hering, C. N. Touart, and T. R. Borden, Jr., *JAS*, 24 (1967), 402, on the climatology of ozone heating in the stratosphere; K. Y. Kondratiev *et al.*, *GPA*, 58, no. 2 (1964), 187, on vertical profiles of radiation balance over the Soviet Union to 30 km; S. L. Hastenrath, *GPA*, 66, no. 1 (1967), 140, on the tropospheric radiation budget in the Caribbean–Gulf of Mexico area under clear skies; F. Möller, *T*, 17 (1965), 350, on backscatter of global radiation by the sky over snow-covered ground; V. V. Salomonson and W. E. Marlatt, *JAM*, 7 (1968), 475, on directional properties of reflected solar radiation; P. M. Kuhn and V. E. Suomi, *JAM*, 4 (1965), 246, on radiative effects of invisible clouds; E. B. Tinney and P. Menmuir, *MM*, 97 (1968), 165, on night cooling in England; J. Gordon and S. E. Virgo, *MM*, 97 (1968), 161, on night-minimum temperature prediction based on radiation

considerations; and L. P. Steele, P. A. T. Stroud, and S. E. Virgo, *MM*, 98 (1969), 107, on the accuracy of Saunder's method of night-minimum temperature prediction. For a review of the receipt and disposal of radiation at the Earth's surface, see D. M. Gates, *AMM*, 6, no. 28 (1965), 1. On the climatonomy of short-wave radiation balance at the Earth's surface, see H. Lettau and K. Lettau, *T*, 21 (1969), 208. For observations of thermal radiation (1.55-cm wavelength) emitted by the sea surfaces, see W. Nordberg, J. Conaway, and P. Thaddeus, *QJRMS*, 95 (1969), 408. For a climatology of infrared radiative cooling in the middle atmosphere (30–110 km), see W. R. Kuhn and J. London, *JAS*, 26 (1969), 189. For the infrared radiation transmissivity of high cirrus clouds, see P. M. Kuhn and H. K. Weickmann, *JAM*, 8 (1969), 147. For a climatology of radiative heating due to water vapor in the troposphere and lower stratosphere of the northern hemisphere, see G. M. Jurica, *MWR*, 94 (1966), 573.

90. See S. Fritz, *CM*, p. 13, for a general account; F. S. Johnson, *JM* 11, (1954), 431, described in detail the determination of the solar constant at the Smithsonian Institution.

91. C. W. Allen, *QJRMS*, 84 (1958), 307. C. V. Cunniff, *MWR*, 83 (1955), 101. See also S. Fritz, *CM*, pp. 13, 50–53. For the model, see J. S. Sawyer, *USCC*, p. 233.

92. C. F. Brooks *et al.*, *HMS*, no. 5 (1941).

93. S. M. Bower, *W*, 10 (1955), 357.

94. For general reviews, see: R. A. Craig and H. C. Willett, *CM*, p. 379; D. H. McIntosh, *MM*, 82 (1953), 11; C. G. Abbot, *SMC*, vol. 146, no. 3 (1963); W. L. Godson, *USCC*, p. 323; G. B. Tucker, *W*, 19 (1964), 302; and H. C. Willett, *JAS*, 22 (1965), 120. See also: W. L. Webb, *JAS*, 21 (1964), 582, for stratospheric solar response; H. Wexler, *BAMS*, 32 (1951), 10 and 48, for the effects of volcanic dust on insolation; and H. C. Willett, *BAMS*, 33 (1952), 255, for atmospheric reactions to the emission of solid solar particles.

95. W. L. Godson (1963), *op. cit.*

96. H. C. Willett (1952), *op. cit.*

97. W. L. Webb (1964), *op. cit.*

98. C. V. Cunniff, (1955), *op. cit.*

99. C. G. Abbott (1963), *op. cit.*

100. H. Wexler (1951), *op. cit.*, and quoted in W. T. Roach, *W*, 9 (1954), 379.

101. Wexler (1951), *op. cit.*

102. W. L. Godson (1963), *op. cit.*

103. D. J. Schove, *JGR*, 60 (1955), 127.

104. H. C. Willett, *JM*, 8 (1951), 1. S. W. Visser, *JM*, 10 (1953), 232. H. Wexler, in H. Shapley, ed., *Climatic Change* (Cambridge, Mass., 1953), p. 73.

105. S. Hanzlik, *Gerlands Beitr. Geophys.*, 29 (1931), 138.

106. D. J. Schove, *ANYAS*, 95 (1961), 107. H. C. Willett, *loc. cit.*, p. 89.

107. P. Westcott, *JAS*, 21 (1964), 572. R. Shapiro and F. Ward, *JAS*, 19 (1962), 506.

108. K. S. Ramamurti, *JM*, 10 (1953), 474.

109. R. Shapiro, *JM*, 13 (1956), 335.

110. P. K. Sen Gupta, *W*, 15 (1960), 52, and 12 (1957), 322.

111. L. N. Carapiperis, *SPIAM*, p. 139.

112. H. P. Gillette, *BAMS*, 35 (1954), 374.

113. A. J. Troup, *GPA*, 51, no. 1 (1962), 184.

114. P. Tetrode, *W*, 7 (1952), 14. S. W. Visser, *QJRMS*, 75 (1949), 411.

115. F. W. Ward, Jr., *JM*, 17 (1960), 130. F. W. Ward, Jr., and R. Shapiro, *JM*, 18 (1961), 247 and 256. N. J. Macdonald and W. O. Roberts, *JM*, 18 (1961), 116.

116. C. E. Palmer, *JM*, 10 (1953), 1, and 13 (1956), 315.

117. E. D. Farthing, *BAMS*, 36 (1955), 472. For the change in visible solar radiation (amount of sunshine) in British Columbia, 1901–1960, see J. M. Powell, *QJRMS*, 91 (1965), 95. On ozone variations during a solar eclipse, see D. Stranz, *T*, 13 (1961), 276, and R. D. Bojkov, *T*, 20 (1968), 417; also B. G. Hunt, *T*, 17 (1965), 516, for the theory. For the effect of the eruption of Mt. Agung, Bali, on solar radiation at the South Pole, see H. J. Viebrock and E. C. Flowers, *T*, 20 (1968), 400. For speculations on correlations between sunspots and terrestrial climates, see E. N. Lawrence, *W*, 20 (1965), 334 and 355, 21 (1966), 367, on smog, and 22 (1967), 69, on British summers; see also D. V. Deshpande, *W*, 23 (1968), 78, on Indian rainfall. See: T. Németh, *GPA*, 63, no. 1 (1966), 205, for a theory of the sunspot cycle in terms of the conjunction of the planets; C. J. Bollinger, *T*, 20 (1968), 412, for a theory of climatic cycles and trends in terms of solar tides caused by the planets; M. Bossolasco *et al.*, *GPA*, 62, no. 3 (1965), 207, for the relation between sunspots and the solar constant; K. Toman, *JGR*, 71 (1966), 3285, for a Fourier transform of the sunspot cycle; and D. Shaw, *JGR*, 70 (1965), 4997, for the relation between sunspots and terrestrial temperatures in terms of spectra. For a possible physical connection between sunspots and terrestrial climates, via the equatorial electrojet of the ionosphere over India, see K. N. Rao, D. R. K. Rao, and K. S. Raja Rao, *T*, 19 (1967), 337. According to H. L. Stolov and J. Spar, *JAS*, 25 (1968), 126, atmospheric pressure provides no evidence that the troposphere responds to solar chromospheric flares. For a model of heat transfer at the Earth's surface, applied to climatic changes resulting from solar radiation fluctuations, see K. Takahashi, *TJC*, 4 (1967), 94. For the relation between solar flares and terrestrial precipitation, see K. Takahashi, *TJC*, 4 (1967), 102.

CHAPTER 5

Temperature

Temperature is a continuous element; that is, it always has some value, and is always an important ingredient of weather and climate. Physically, temperature is a measure of the heat of a body. Heat is a form of energy, and temperature is that condition of a body which determines its ability to communicate heat to other bodies, or to receive heat from them. In a system of two bodies, that which loses heat to the other is said to be at the higher temperature.

In the measurement of temperature, the thermometer is first brought to the same temperature as the object whose temperature is to be measured, and then the temperature of the thermometer is found. Thermometers are calibrated by reference to fixed points, relating to certain easily achieved changes of state that take place at temperatures which can be determined from first principles. There are four fixed points of meteorological significance: the temperatures of (1) equilibrium between solid and gaseous carbon dioxide; (2) equilibrium between liquid water and its vapor (boiling point), (3) melting ice (melting point, 0.00°C); (4) transition of sodium sulphate (32.38°C). Interpolation between the fixed points is determined by the use of a platinum resistance thermometer (see Appendix 5.1).

Calibration in this manner gives the Celsius (centigrade) scale. The pressure of a perfect gas on this scale is zero at −273.16°C, absolute zero. The degree Celsius is defined as the variation in temperature that produces one hundredth of the increase in pressure undergone by a perfect gas at constant volume, when its temperature passes from that of melting ice (0°C) to that of boiling water (100°C) at 760 mm pressure; °C plus 273.16 gives the absolute (Kelvin) scale of temperature, on which the melting point of water is 273.16°K and boiling point is 373.16°K, at a pressure of 1013.25 mb.

The Kelvin scale avoids negative temperatures, and so has long been used for upper-air temperatures. The Fahrenheit scale is very convenient for meteorological temperatures near the Earth's surface, because it gives greater precision than the Celsius scale without use of decimals. However, the Celsius scale has now been universally adopted for all meteorological work.

Accuracy in quoting meteorological temperatures is limited by the fact that air temperatures vary considerably within short distances and times. The second-by-second variations are on the order of several degrees Fahrenheit. Hence it is usually inadvisable to quote beyond the nearest degree F, except if the temperatures are to be used for calculating humidities, in which case accuracy to 0.1°F is necessary, or if the fine structure of the atmosphere is to be measured. Soil temperatures vary much less than air temperatures, and an accuracy of 0.1°F near the ground surface, or of 0.02°F below a depth of 4 feet, should be the object. With water temperatures, the desirable degree of precision is 1°F.

Types of Thermometer

The basis of any thermometer is a substance with a physical property that changes as a function of temperature. Four properties are of particular importance: thermal expansion, thermoelectric effects, the dependence of electrical resistance on temperature, and the dependence of the speed of sound propagation in air on air temperature.

Liquid-in-glass thermometers use the differential expansion of a liquid in its glass container. Mercury-in-glass thermometers are standard for surface air, sea, and soil temperature measurements, for temperatures above the freezing point of mercury (−37.9°F). For temperatures below −38°F, ethyl-alcohol-in-glass thermometers are usual. All these thermometers are usually enclosed in a glass sheath to protect the scale markings and form a robust instrument.

Maximum thermometers have a constriction in the stem, an inch or so above the bulb. As temperature rises, the mercury forces itself past the constriction. When temperature falls, the mercury cannot all flow back, but breaks at the constriction. They are usually read and reset at 9:00 A.M. and P.M. G.M.T., the reading in each case giving the maximum temperature reached during the preceding 12 hours.

Minimum thermometers are alcohol thermometers in which a small metal or glass index has been introduced inside the liquid. The index is free to slide along inside the liquid thread. As temperature falls, the index is dragged toward the bulb by surface tension at the meniscus. When temperature rises, the alcohol moves past the index, which remains at the lowest temperature reached. These thermometers are also read and reset at the same times as the maximum thermometers.

Grass minimum thermometers are normal minimum thermometers mounted horizontally on pegs, with the bulbs just touching the tips of a short grass cover. They are read at 9:00 A.M. and reset at 9:00 P.M. G.M.T. Their measurements indicate Earth radiation better than those of screen-mounted minimum thermometers.

Errors in readings from liquid-in-glass thermometers arise mainly from parallax errors on the part of the observer, or from unavoidable secular changes in the glass. After several years, the glass in the bulb contracts slightly, raising the zero, which hence must be redetermined.

For sea-surface temperatures, a sample of water is obtained by lowering a bucket overboard, and its temperature is measured by a porcelain-mounted mercury thermometer. There are several sources of error, the major one being that the initial temperature of the bucket usually differs from that of the sea, and the water in the bucket may have changed its temperature by the time it reaches the deck, owing to heat exchange and evaporation. An alternative type of bucket, which is towed along in the sea for 30 seconds, the water circulating through it, gives better results. A special thermometer is mounted in a duralumin tube so that readings can be taken without removing the thermometer bulb from the water in the bucket. Accuracy for this type of measurement is 0.1°F.[1]

The thermometers so far discussed are all indicating, not recording instruments. The most common recording thermometers depend on deformation of various metals.

The *bimetallic thermograph* consists essentially of a metal strip formed by welding together pieces of invar and either brass or steel. The strip bends as its temperature changes, because of the different rates of expansion of the two metals. The strip is usually in the form of a helical coil of several turns: one end is attached to a pen arm, which makes an ink trace on a paper chart wrapped round a clockwork-driven drum. The other end is clamped to an arm, which may be adjusted to make the instrument read correctly. The metal element exhibits elastic hysteresis, and must be "aged" before use by subjecting it to a number of complete temperature cycles; otherwise the zero position may tend to drift. Seven-day charts are normally used.

The *Bourdon thermograph* uses a curved metal tube of elliptical cross section, which is filled with a liquid. When the temperature rises, the liquid expands and the tube straightens somewhat, which motion is transmitted to a recording-pen arm by a lever system. The Bourdon tube is easily damaged, and calibration is soon upset by moderately high temperatures. Secular changes occur both in the metal and in the liquid, so that this instrument is used less than the bimetallic thermograph.

The *mercury-in-steel thermograph* is a special form of the Bourdon thermograph, in which the liquid expansion takes place in a separate container, the change in pressure being transmitted to the Bourdon tube by means of a narrow-bore capillary tube. The entire system is made of steel, and filled with mercury. Up to 150 feet or so of capillary tube may be used, so that the instrument can be used for remote recording of sea or soil temperatures. If a long capillary tube is used, its temperature may vary, causing errors in recording. This is compensated for by enclosing some invar in the capillary, usually as a wire nine-tenths the diameter of the capillary. The instrument is very accurate and robust, but is obviously limited to temperatures above the freezing point of mercury.

The thermometers described above can be regarded as standard instruments. A number of other types are used for research purposes: these include electrical-resistance thermometers, thermocouples, and sonic thermometers.

In *resistance thermometers*, a suitably mounted wire is exposed to the air, a current passed through it, and its resistance measured. Resistance is proportional to temperature, and, since the wire is insensitive to solar radiation, very accurate measurement is possible. The wire must be kept from contact with anything solid, and a very small measuring current must be used, or the instrument will function as a hot-wire anemometer. For very accurate work, a potentiometer is used to measure resistance; for normal purposes, a Wheatstone bridge is used. The thermometer is calibrated by determining its resistance at the fixed points of 0°C, 100°C, and −78.51°C (the temperature of sublimation of carbon dioxide), and then intervening resistances may be found by interpolation (see Appendix 5.2). Temperatures can then be read to any degree of accuracy, depending on the sensitivity of the galvanometer in the potentiometer or Wheatstone-bridge circuits.

For very accurate measurement, platinum wire 0.05 to 0.10 mm in diameter is used, wound in a coil on a mica or other insulating frame, and protected by a weatherproof sheath. For less accurate work, nickel or copper wire is used. Commercially made resistance elements are also available, various types of thermistor, for example, often with a bead- or rod-like base of ceramic material. Thermistors can be made very small, and since they consist of mixtures of various metallic oxides whose resistance changes rapidly with temperature, the error in measurement induced by the leads to the thermometer element is much reduced (see Appendix 5.3).

The *thermocouple* depends on the principle that if two or more different metals are joined to form a continuous circuit, an electromotive force will be set up if the metals are at different temperatures. Thus a current will be set up in the circuit, the intensity of the current being proportional to the difference in temperature, if only two metals are used to form the circuit (see Appendix 5.4). One metal junction is kept at a constant temperature as a reference junction, using either melting ice (0°C) or, in tropical climates, melting diphenyl ether (26.6°C).[2] The other junction forms the measuring junction: it can obviously be made very small, using wire a few thousandths of a millimeter in diameter; a series of such junctions enables vertical temperature gradients to be measured very accurately. Upper and lower plates are usually used to shield the measuring junctions from direct solar heating, and to prevent the instrument from radiating heat to the sky at night, while permitting free ventilation.

Either copper-constantan or iron-constantan thermocouples are used, the latter giving a greater emf per degree difference in temperature. The current in the circuit is measured by either a potentiometer or a millivoltmeter. In the former, calibration of the instrument is independent of both the resistance of the leads and the sensitivity of the galvanometer. In a millivoltmeter, deflection is not necessarily a linear function of the current, and the entire circuit has to be calibrated at several temperatures throughout its range.

The fact that the speed of sound in air is a function of temperature alone can be used to measure temperature whatever the radiation conditions (see Appendix 5.5). A transmitter generates a pulse of high-frequency sound that is picked up by a receiver; the voltage generated by the latter, on receipt of the signal, is amplified and fed back into the transmitter, which generates another pulse. The time between pulses equals the time taken for the sound to travel from transmitter to receiver plus the time taken within the electrical circuit from receiver back to transmitter. The latter time is practically constant, while the former depends on the temperature of the air between generator and receiver. The distance between transmitter and receiver is adjusted so that successive pulses produce a constant-frequency signal. The frequency is then a measure of air temperature.

The propagation of compressional sound waves from explosions is a function of temperature and windspeed, and if the latter is known, the former can be inferred. Early experiments involved dropping 500-lb bombs at varying distances from a fixed

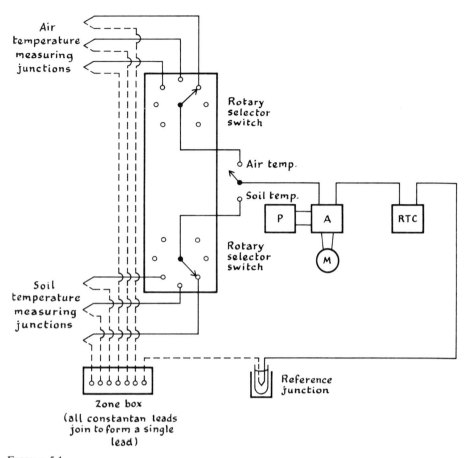

FIGURE 5.1.
Thermocouple circuit for simultaneous soil- and air-temperature measurement (after D. J. Portman).
Solid lines are copper wire; broken lines are constantan wire. *P*, power supply; *A*, amplifier,
M, meter; *RTC*, reference temperature compensator.

recording site over the sea, and setting off a 200-lb bomb at ground level at a single site, observing the explosion from several recording stations nearby. Temperatures were estimated up to a height of 60 km by these methods.[3]

Air temperatures within a few centimeters of a surface can be measured by optical means. Fleagle used a field-artillery range finder, in a vertical position and pointed at targets at different heights, to measure the vertical deflection of a horizontal light beam. This revealed a temperature anomaly some 10 cm above a cold surface, which could not be discovered with the normal temperature probes.[4] Acoustical and optical methods have the advantage that the thermometric substance whose properties they measure is the air itself, and therefore they measure temperatures directly, unlike ordinary thermometers.

Two types of temperature measurement have been much neglected: ground-surface and soil temperature. The standard mercury-in-glass soil or earth thermometer measures the temperature of the air around the bulb, not the true soil temperature. Ground-surface temperatures are difficult to measure accurately, because it is almost impossible to locate the sensor at the true ground-air interface, and also because they fluctuate greatly.

Rider made use of copper-constantan thermocouples at depths of 1, 5, 10, 20, and 40 cm, mounted in such a way that the soil and vegetation was not disturbed to any extent. A 1.5-inch-diameter tufnol tube was driven into the ground, with three thermocouples mounted at each depth. The thermocouple leads were brought up the center of the tube, then carried to a recording galvanometer via junction boxes and Post Office relays. The completed instrument enabled a record of temperature at each depth to be made once every ten minutes or so to an accuracy of 0.2°F.[5]

Attempts to measure ground-surface temperatures by laying alcohol or mercury thermometers on the ground are inaccurate, because the bulbs are affected by wind, radiative heating during the day, and radiative cooling at night. If the bulb is partly buried, it will merely record the mean temperature of the upper soil layers. If a fine wire thermocouple is laid on the surface, it will be affected by drifting soil, and subject to cooling and warming by the wind if it is not in direct contact with the soil.

If a metal plate, with a thermometer element built into it, is laid firmly on top of the soil, flush with the ground surface, it will be subject to the same radiation, wind, and mosture effects as that surface. Harrell and Richardson used a copper plate 12 square inches in area and 0.05 inch thick, with holes drilled through it to keep the soil from being shielded against precipitation and evaporation. Thermistors were sealed into the underside of the plate, enabling continuous records to be made with an accuracy of about 0.5°F.[6]

By using thermocouples at several heights above the ground and at several depths in the soil at the same time, in conjunction with a rotary switch as a selector, one can obtain average temperature profiles during periods of four or five minutes to an accuracy of $+0.05°C$ (see Figure 5.1). The observer switches rapidly from one thermocouple to another, taking a meter reading each time; within five minutes, ten or so readings can be obtained for each height and each depth.[7]

Errors in Temperature Measurement

Before temperature can be measured accurately, the thermometer must attain thermal equilibrium with its environment, which it does by convection, conduction, and radiation. Air is very transparent to radiation and has a poor conductivity, so the radiation exposure of a thermometer must be minimized. This is done by using concentric polished metal shields for thermocouples, and Stevenson screens for ordinary thermometers and thermographs.

The Stevenson screen is a wooden box painted white, with louvered sides, a ventilated double floor and roof, and its north-facing side hinged to form a door. Radiation cannot proceed in a straight line from outside to inside the screen; it is blocked by solid matter before it reaches the thermometers. The white paint ensures that the maximum amount of solar radiation is reflected. Stevenson screens are normally mounted with their floors 4 feet above the ground in English-speaking countries, and 1.7 to 2.0 m above the ground in continental European countries. They should be erected on level ground with a short grass cover, and the sites should be freely exposed, not sheltered by trees or buildings.

A screen does not ensure that true air temperature is recorded, but it does ensure uniformity, which is necessary for comparative purposes. Screen temperatures are up to 2°F too high on sunny afternoons, and up to 1°F too low on clear, calm nights, because of conduction from the outer walls of the screen. Portable, inexpensive thermometer shields give maximum temperatures 2 to 3°F higher, and minimum temperatures up to 1°F lower, than screen temperatures.[8]

Three mercury thermometers are normally mounted in a Stevenson screen: a maximum thermometer nearly horizontally, and two others vertically as dry- and wet-bulb thermometers, the latter with a muslin wick wetted with distilled water wrapped round its bulb. In addition, a minimum thermometer is mounted nearly horizontally in the screen, which also holds a thermograph. There is also space for a hygrograph, usually.

In the upper air, uniformity of exposure is more difficult to obtain. Boundary-layer effects from a radiosonde balloon, its equipment, and its supporting framework can cause errors at least as great as the instrumental errors, giving a warming effect during the day and a cooling effect at night. The temperature of a floating balloon at night is approximately that of a black body, but during the day, a balloon at 5 to 10 mb warms 10°C, bringing it almost to air temperature.[9] Fine wires are insensitive to solar radiation and sunrise effects, but radiation shields usually are mounted on radiosondes to guard against radiation from other parts of the equipment reaching the thermometer element. At high levels, unfortunately, the air reaching the element is heated during daytime by the shield itself.

All types of thermometer suffer from an error due to the time it takes for the instrument to reach the temperature of its environment. This is "lag error." When a thermometer is placed in an environment with a temperature different from its own, it does not reach the new temperature instantly, but approaches it exponentially. From Newton's law of cooling, the rate of heat loss or gain from a body is proportional to the difference in temperature between the body and its surroundings. The constant of proportionality (λ) is the lag coefficient (see Appendix 5.6).

Different types of thermometer have different lag coefficients. Values of λ vary from 50 or 60 seconds for mercury-in-glass to between 100 and 200 seconds for mercury-in-steel thermometers, and from 20 seconds for bimetallic thermographs to 5 seconds for electrical resistance thermometers and between 1 and 3 seconds for thermocouples. Thus, of all the thermometers described, thermocouples are the most sensitive, and mercury thermometers the least sensitive. If the temperature of the air is changing at a constant rate, the thermometer reading will lag behind the true temperature by a constant amount. The thermometer will indicate what the true temperature was λ seconds previously.[10]

The Physical Picture of Temperature

The temperature at a given place at a given time depends on the balance between incoming and outgoing heat. Hence a consideration of heat balance provides a good physical basis for understanding the geographical pattern of temperature.[11]

The heat-balance equations are special cases of the law of conservation of energy. They apply to a vertical column extending through the whole of the geographical medium, both Earth and atmosphere, and summarize all the streams of heat energy flowing between the Earth's surface and the ambient space. The basic equation states that, to a first approximation.

$$R = LE + P + A.$$ [5.1]

Here, R is the radiational flux of heat, P is the turbulent flux of heat between the Earth's surface and the overlying atmosphere, A is the nonturbulent heat flux between the surface and the lower layer of the atmosphere, L is the latent heat of evaporation, and E is the rate of evaporation or condensation; R is regarded as positive if it designates an inflow of heat to the Earth's surface; and all other quantities are positive if they designate expenditure of heat.

The equation ignores several items that are insignificant in comparison with the five elements in it. These include the expenditure of heat for photosynthesis, the gain of heat from oxidation of biological substances, the heat used to melt snow or ice, the heat released when water freezes, the heat generated by dissipation of the mechanical energy of winds in waves, tides, and currents, and the positive or negative heat flux transferred by the fall of precipitation with a different temperature from that of the underlying surface.

On exceptional occasions, for example when heat is generated by forest fires, leading to a rapid discharge of heat accumulated by photosynthesis, these additional factors must be taken into account. Normally they can be ignored, except for short-period considerations, when heat loss or gain due to phase changes of water may be important. Advection of heat has no effect on heat balance at the Earth's surface (see Appendix 5.7).

TABLE 5.1
Heat balance of the Earth's surface

Latitude	Oceans					Continents				Earth				
	Q[a]	R	LE	T	C	Q	R	LE	T	Q	R	LE	T	C
50–65°N	88	34	−34	−18	18	93	23	−19	−4	91	28	−25	−10	7
40–50°N	109	54	−51	−15	12	119	38	−22	−14	114	46	−36	−15	5
30–40°N	136	78	−73	−12	7	159	56	−26	−30	146	69	−53	−20	4
20–30°N	151	100	−85	−7	−8	184	64	−23	−41	163	86	−60	−20	−6
10–20°N	156	110	−89	−5	−16	182	74	−36	−38	163	101	−75	−14	−12
0–10°N	149	107	−76	−5	−26	149	79	−58	−21	149	101	−72	−9	−20
0–10°S	152	107	−81	−7	−19	143	75	−59	−16	150	99	−76	−9	−14
10–20°S	155	107	−97	−9	−1	161	69	−44	−25	156	99	−85	−13	−1
20–30°S	147	94	−87	−10	3	169	62	−29	−33	152	87	−74	−15	2
30–40°S	128	73	−77	−12	16	149	55	−29	−26	130	71	−72	−14	15
40–50°S	104	53	−57	−5	9	112	39	−24	−15	104	53	−56	−5	8
50–60°S	84	31	−37	−12	18	80	26	−18	−8	83	31	−37	−12	18
Earth as a whole	128	77	−68	−9	0	132	46	−27	−19	129	68	−56	−12	0

[a] Q = total radiation; R = radiation balance; LE = loss by evaporation; T = turbulent transport; C = transport by ocean currents.
The figures in the body of the table are mean latitudinal values in kcal per cm² per year.

For practical evaluation, the components of the heat-balance equation must be further broken down. The component P is rather complex and turbulence theory is involved in its analysis. The other components are easier to determine, from the following expressions:

$$R = (Q + q)(1 - \alpha) - I; \tag{5.2}$$

$$A = B + F. \tag{5.3}$$

In Equation 5.2, Q is the sum of direct radiation, q the sum total of diffuse radiation, α the albedo of the Earth's surface, and I the difference between incoming and out-going heat amounts at this surface, i.e., the effective outgoing radiation. In Equation 5.3, B is the change in heat amount inside the Earth-atmosphere column during the given period of time, and F is the heat exchange between the column and the ambient space of lithosphere or hydrosphere in a horizontal direction. Heat flow from the depth of the Earth into the base of the column can be taken as negligible.

In the lithosphere, F is usually very small; hence for the land $A = B$, and for long-term considerations, $A = B = 0$. For uniform water bodies, A is approximately equal to B, except where there are currents. Using mean annual values, $R = LE + P$ for the land surfaces, and $R = LE + P + F$ for individual oceans, or $R = LE + P$ for the oceans as a whole. For deserts, $R = P$.

Table 5.1 presents some components of the heat balance in climatological terms, as annual totals, while Figures 5.2 and 5.3 show the distribution of various components in space and time. World maps show that, on a yearly average, almost the entire land surface of the globe transmits heat to the atmosphere (via turbulent conduction), and does not receive heat from it. The quantity of heat transported from low to high latitudes in the atmosphere is twice that transferred in the hydrosphere.

If the entire Earth-atmosphere system is considered, not just the Earth's surface, different equations must be used. The general equation is

$$R_s = C + F + L(E - r) + B_s. \tag{5.4}$$

Here R_s is the heat exchange between the air column and outer space, i.e., the solar radiation absorbed by the column minus the total outgoing long-wave radiation from the column during a given period; C is the difference between heat gain and heat loss due to horizontal transfer in the atmosphere; and F is this difference for the hydrosphere. LE and Lr represent heat sources (positive and negative) within the air column: LE equals the expenditure of heat on evaporation from soil, vegetation, lakes, reservoirs, etc., minus the gain of heat from condensation on these objects; Lr is the gain of heat due to condensation of water vapor within the air column minus the loss of heat for evaporation of water droplets.

For annual averages, B_s is zero, so that, for a mainland area,

$$R_s = C + L(E - r). \tag{5.5}$$

For the Earth as a whole, $E = r$, and $C = F = 0$. The heat balance of the atmosphere alone is given by

$$R_a = R_s - R, \tag{5.6}$$

in which

$$R_s = Q_s(1 - \alpha_s) - l_s, \tag{5.7}$$

where Q_s is the short-wave solar radiation received on the outer boundary of the atmosphere, α_s is the albedo of the Earth-atmosphere system, and l_s is the total long-wave radiation going into space. The change in heat content of the atmosphere, B_a, is given by $B_s - B$; therefore

$$R_a = C - Lr - P + B_a, \qquad [5.8]$$

or, for annual averages;

$$R_a = C - Lr - P. \qquad [5.9]$$

The water-balance equations express the conditions in which the algebraic sum of all gain or loss of liquid, solid, or gaseous water by a horizontal surface from or to its surroundings, during a certain time interval, equals zero. For the land surface,

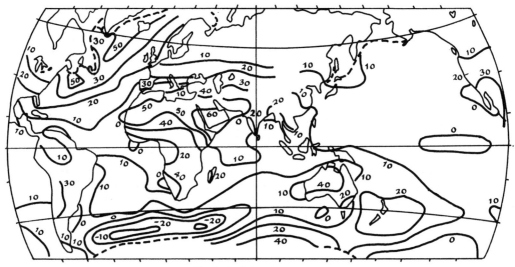

Turbulent heat exchange.

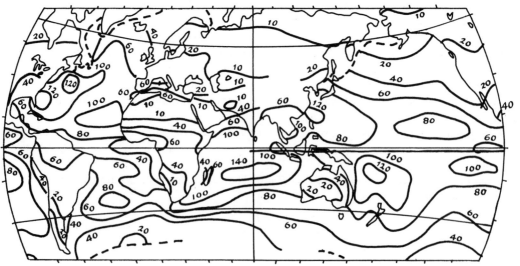

Expenditures of heat for evaporation.

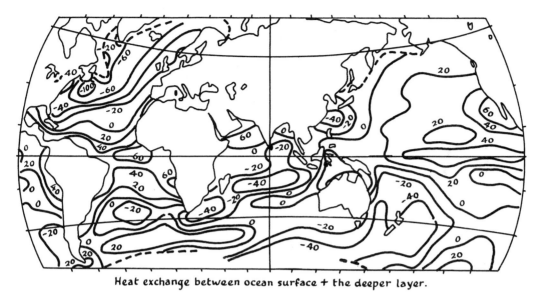

Heat exchange between ocean surface + the deeper layer.

FIGURE 5.2.
Heat exchange and evaporation. Annual values in kg-cal per cm² per year. Highland areas omitted.
(After M. I. Budyko.)

$$r = E + f_w + m, \qquad [5.10]$$

where r is precipitation, E is the difference between evaporation and condensation on the Earth's surface, f_w is the surface runoff, and m is the moisture exchange between the Earth's surface and the underlying layers, i.e., the algebraic sum of the gravitational flux of fluid moisture from the soil surface into the deeper layers, the vertical flux of water-film moisture between various soil layers, the vertical flux of water vapor, the flux of water that is raised by the roots of plants, etc., for a given period. The surface runoff plus the underground runoff equals the full runoff, f. Therefore

$$r = E + f + b. \qquad [5.11]$$

For reservoirs and lakes, b can be found from the change of water level. For annual averages,

$$r = E + f. \qquad [5.12]$$

For the whole globe, $r = E$, since the net horizontal redistribution of moisture is zero.
The water balance of the atmosphere is given by

$$E = r + C + b_a, \qquad [5.13]$$

where C is the moisture gained or lost by the vertical air column, as effected by air currents and by horizontal turbulent exchange, and b_a is the change in water content of the column during a given period; b_a is usually very small, and equals zero for average annual values.

Some figures for water-balance components are given in Table 5.2; Figure 5.3 summarizes climatological aspects of heat balance. For the relationship between heat and water balance, see Appendix 5.8.

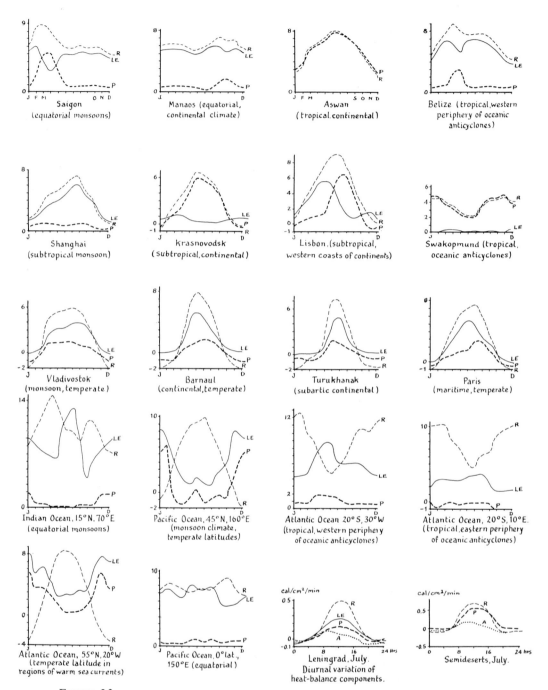

FIGURE 5.3.
Climatological aspects of heat balance. (after M. I. Budyko). All horizontal scales are graduated in months unless otherwise indicated. Vertical scale units are kg-cal per cm² per month.

TABLE 5.2
Water balance of the Earth's surface

Area	Radiation Balance[a]	Heat expended on evaporation[a]	Turbulent heat exchange[a]	Precipitation[b]	Evaporation[b]	Runoff[b]
Europe	33	22	11	60	36	24
Asia	41	23	18	61	39	22
North America	38	24	14	67	40	27
South America	71	52	19	135	86	49
Africa	69	31	38	67	51	16
Australia	66	25	41	47	41	6
Atlantic Ocean (−6)[c]	75	63	12	78	104	−20[d]
Indian Ocean (−30)[c]	78	83	8	101	138	−7[d]
Pacific Ocean (13)[c]	82	68	7	121	114	−6[d]
Arctic Ocean (35)[c]	—	7	—	24	12	−23[d]

[a] Annual totals in kcal per cm^2.
[b] Annual totals in cm.
[c] Figures in parentheses indicate water exchange with other oceans.
[d] Figures refer to runoff from land.

Some examples of heat-balance considerations illustrate the value of this approach. Table 5.3 presents observations of a level surface of short grass at Kew, meaned over periods of 30 minutes (the maximum allowable time if diurnal effects are to be excluded), that indicate the order of magnitude of the heat fluxes during short periods. The size of these fluxes indicate the minute-by-minute rates at which the atmospheric heat engine is being supplied with energy. The algebraic sum of various heat-balance components for Nashville, Tenn., using monthly means, shows that a positive heat balance of 36 langleys in the lower 5,000 feet of the atmosphere produces a rise in surface temperature of 1°C on clear summer days.[12]

From the heat-balance equation, a climatological parameter, H, has been devised that is useful in maximum temperature prediction (see Appendix 5.9). The fact that a climatological parameter can be produced suggests that surface temperatures cannot rise indefinitely. As the surface layers become hotter and hotter, the heat is convected to higher and higher elevations, either by the development of winds in desert regions, or by the growth of cumulonimbus in humid regions. In the latter, air in the upper part of the cloud is warmed by release of latent heat of condensation, while air in its lower part is cooled by the evaporation of rain.

TABLE 5.3
Energy Balance Components Observed on a Level Grass Surface, 1-2 cm high, at Kew.

Date	Time	Downward flux of short-wave radiation[a]	Downward flux of long-wave radiation[a]	Upward flux of long-wave radiation[a]	Total vertical heat flux at 100 cm[b]	Heat flux at ground surface[a]
June 19, 1948	2.15 A.M.	177	80	135	−10	+5(±3)
Oct. 2, 1948	11.10 A.M.	135	72	111	−1	+10(±3)
June 20, 1949	11.21 A.M.	178	79	134	−13	+25(±10)

[a] The units are 10^4 cal per cm^2 per sec.
[b] The units are 10^6 cal per cm^2 per sec.

A large-scale difference in heat balance is that the annual temperature near the North Pole is 30°C higher than that at the South Pole, even though the elliptical orbit of the Earth carries it three million miles farther from the sun at aphelion than at perihelion, so that during midsummer, 7 per cent less solar radiation impinges on the top of the Arctic atmosphere than on the top of the Antarctic atmosphere. Heat from the Arctic Ocean is conducted upward through the thin pack ice to a relatively cold surface, where temperatures average -20°C, but the flux of heat upward through the ice of central Antarctica is quite small, so that surface temperatures remain low. A poleward movement of heat is necessary in both hemispheres to make good the net loss of heat in polar regions. That this flux is slightly higher in the southern hemisphere than in the northern—for latitude 55°N, the meridional rate of heat transport is a maximum of 8×10^{14} cal per sec, as against 9.6×10^{14} cal per sec for 55°S— explains to some extent the storminess of the southern oceans.[13]

Variations in the physical state of the Earth's surface have considerable influence on the heat energy supplied to the atmosphere, particularly fluctuations in the extent of frozen land or sea, the change from a land surface covered with moist green vegetation to a desert expanse, anomalies in sea-surface temperatures, and early or late snowfall in middle latitudes. Albedo causes a difference of 50 to 65 per cent in the heat supplied to the Earth's surface from the sun between highly reflective (snow-covered) and relatively absorbent (vegetated) surfaces.

The continents are poor heat-storage reservoirs compared with the oceans; very little of the heat received from the sun is stored in the ground for more than 24 hours. The ways in which the variegated surfaces of the continents affect the amount of heat supplied to the atmosphere by the ground are summarized in Table 5.4.[14]

Local factors affect heat balance greatly. Variations in the state of the ground can cause changes of more than 4°F in maximum temperatures reached. More important is the effect of the thermal behavior of soils on heat balance in the lowest layer of the atmosphere. In humid regions, the heat flux into the soil is small in comparison with net radiation, but in arid regions, most of the thermal energy in the surface soil can be used for heating the overlying air. Annual amplitudes of heat flux at the surface in Australia average 0.12×10^{-3} cal per cm^2 per sec. In comparison, the amplitude of net radiation flux in arid and semiarid Australia is 1.5×10^{-3} per cal cm^2 per sec. Theoretical considerations indicate that annual temperature variations penetrate into the soil to a depth of about 10 m.[15]

TABLE 5.4
Effects of surface conditions on ground-to-atmosphere heat supply.

Surface condition	Maximum effect on heat supply[a]
Changes that affect conduction	± 28
Freezing of surface water	± 80
Thawing ground in sunlight	-150
Thawing ground under cloud cover	-80
Thawing snow	-40
Changes in evaporation caused by changes in transpiration by vegetation	± 200
Changes in albedo caused by snowfall in middle latitudes	-80
Changes in albedo of drought-stricken vegetation	-50

[a] Units are cal per cm per day.

Observations at Rothamsted, England, of soil-heat balance under crops show that the diurnal variations of net radiation and soil-heat flux are similar, but the soil-heat storage "day" is shorter than the radiation "day." Soil-heat storage during the day amounts to only about 20 per cent of the net incoming radiation, but soil-heat loss at night approximately equals the net outgoing radiation. The heat balance is influenced by the type of crop: heat storage under grass is greater than under potatoes or wheat.[16]

The first factor deciding soil-heat balance is the color of the soil, since this decides how much of the incoming radiation is absorbed by the soil. Measurements at Poona, India, indicate that grey alluvial soil absorbs 50 per cent of the radiation, (green) grass-covered soil absorbs 68 per cent, black soil 84 per cent, but a white chalk soil absorbs no radiation at all.[17]

Once the radiation has been absorbed by the soil, the latter's thermal conductivity and moisture content become the critical factors. The change in heat content of the soil depends largely on its moisture content. The thermal conductivity of a soil increases rapidly up to a certain wetness value, then shows little further increase. Heat capacity increases as moisture content increases; the thermal diffusivity of a soil rapidly increases for a time, then decreases as soil moisture increases. Water content in a field soil varies with depth. Classic heat-conduction theory therefore cannot be applied directly, since a natural soil has different values of thermal diffusivity at different depths (see Appendix 5.10).

Table 5.5 demonstrates the order of magnitude of soil moisture effects on heat balance: the soil was dry, containing only hygroscopic moisture, and balance was not perfect, so that there was a carryover of 11 calories. Thus although the quantities contributed by soil heat to total energy balance are small, they are large enough to determine whether fog or dew will form.

Invisible condensation is important, particularly in dry regions. The soil surface absorbs water vapor from the air layers near the ground, once it attains the state of hygroscopic moisture content early in the dry season. When temperatures reach their minimum values, the soil attains its maximum weight, owing to its relatively high moisture content. When temperatures rise, the evaporating period begins, in which

TABLE 5.5
*Thermal balance of ground surface at Poona
on a clear day in April.*

Heat	g cal per cm²
Gain	
Visible radiation (sun and sky)	655
Radiation from atmosphere	691
Absorption of water by soil surface	20
	1,366
Loss	
Nocturnal earth radiation	950
Convection loss from ground	350
Transfer by conduction	35
Used for evaporation	20
	1,355
Surplus	11

vapor pressure decreases with height above the ground. When temperatures fall, the invisible condensation period commences, when moisture flows downward in the lower layers of the atmosphere, and vapor pressure increases with height. If there is intense radiative cooling of the ground surface and enough moisture, the water will be deposited as visible dew. The intensity of dew deposition increases with height up to a few meters above the ground, then gradually decreases.

The heat stored up in a bare soil during the day is often enough to prevent nocturnal inversions, although they may occur within the lowest meter of air over adjacent fields of grass. If soil and air are originally at different temperatures, the rate of heat exchange falls off rapidly with increasing time of contact. For a difference of 1°C, heat exchange is still appreciable after one hour, and for a difference of 10°C, after one day.

The diurnal amplitude of the heat flux from the soil to the air in the Caspian region of the U.S.S.R. is about 10^{-3} to 10^{-2} cal per cm^2 per sec. This heat flux increases the rate of evaporation by 3.5 mm per day. Air temperatures above irrigated fields at 1 m differ by 4 or 5°C from temperatures over adjacent nonirrigated fields, owing to the effect of the soil moisture on heat balance.[18]

The thermal balance at the soil surface profoundly affects vertical temperature gradients. On clear days, the bare soil gives rise to a shimmering layer with rising filaments of heat air next to downward filaments of cold air from above. This produces lapse-rates 200,000 times greater than the dry-adiabatic rate near the ground, 50 times at 2 m, and 5 times at 6 m, at the time of maximum temperatures.

During clear nights, the ground surface rapidly cools, and the lowest air layers in turn cool by radiative heat exchange with the cooler layers above, because outgoing radiation is greater than downward radiation from the water vapor, carbon dioxide, and ozone in the atmosphere. Hence the lowest air temperatures are found 5 to 10 cm above the ground surface, with an inversion above. The daytime shimmering layer persists on clear calm nights, and the layer of coldest air is at the upper boundary of this layer. A regular winter phenomenon in India (also found to some extent in Britain and the United States) is for a layer of zero lapse-rate (initially at the top of the shimmering layer just before sunset) to rapidly separate into two zero lapse-rate layers, with a developing inversion layer in between.

Largely owing to the rapid changes in soil type, vegetation, and land use on land surfaces, it is very difficult to prepare detailed maps of heat balance at the Earth's surface for the continents, except in highly generalized form. The oceans, however, present much more uniform surfaces, so that reasonably detailed maps are quite feasible for average conditions. Such maps indicate the importance of different oceanic areas as heat and moisture sources for atmospheric phenomena.[19]

The oceans constitute over 70 per cent of the Earth's surface, and all but a small fraction of the total heat energy of the atmosphere is received by conduction and radiation from their surface. When the sea surface is warmer than the air above it, heat is transported upward from sea to air by convection and long-wave radiation. When the sea surface is colder than the overlying air, heat is transferred downward by turbulent diffusion and long-wave radiation.

In general, heat exchange between sea and atmosphere is negligible when the sea is frozen. The rate of heat movement by conduction through ice 1 m thick is only 20 cal per cm^2 per day for a 10°C temperature difference across the ice, and the air layer immediately above an ice surface loses heat by long-wave radiation at the rate of

380 cal per cm² per day. When the sea surface is colder than the overlying air, heat exchange is small (around 20 cal per cm² per day); under shallow convection it is appreciable (50 cal per cm² per day); in strongly convective conditions it can be very large (up to 1,200 cal per cm² per day). A 1°C anomaly in sea-surface temperatures results in a maximum change of ± 30 cal per cm² per day in sensible heat transfer, and one of ± 60 cal per cm² per day in latent heat transfer, from sea to air.[20]

For the oceans as a whole, the annual heat balance is given by

$$Q_r = Q_a + Q_s + Q_v,\qquad\qquad [5.14]$$

where Q_r represents the net radiation received by the sea surface, Q_a the total heat loss by evaporation and sensible heat exchange, Q_s the heat used up in changing the temperature of the water column considered, and Q_v the advective heat flux out of this column due to currents and mixing processes. The storage term, Q_s, is zero if the mean annual temperature of the water column does not vary from year to year.[21]

In Figure 5.4, the rate of exchange of sensible heat between sea surface and atmosphere, Q_h, is based on observations of vapor pressure, atmospheric pressure and temperature, and sea-surface temperature, combined by means of the Bowen ratio; the rate of evaporation, E, is also based on these observations, combined according to turbulence theory. Of the total solar energy absorbed at the sea surface during the course of a year, 50 per cent is used for evaporating sea water, hence is made available to the atmosphere in latent form as water vapor. This latent energy, Q_e, may be estimated from evaporation computations. The sum of Q_h plus Q_e then represents the rate of total heat loss from the oceans through convection, Q_a, which is purely in the form of sensible heat with reference to the sea surface, but with reference to the atmosphere only a fraction of it, Q_h, is immediately available for heating the air; the remainder is in latent form and water vapor. The rate of total heat gain by the atmosphere through convection, Q_{ph}, is then given by the sum of Q_h plus Q_p, where Q_p is the heat supplied to the atmosphere by condensation of water vapor. The difference between Q_a and Q_{ph} then represents the availability of surplus energy, i.e., the amount of latent energy that is locally surplus and is available for transport to other oceanic areas or to the continents, where it is made available as real heat through condensation. (On all of this paragraph, see Appendix 5.11.)

Assuming Q_s in Equation 5.14 to be zero, the total heat acquired by the atmosphere through conduction, condensation, and radiation must exactly balance the radiative loss. Figure 5.5 shows the net transport of water vapor across latitude circles required to balance evaporation and precipitation for the Earth as a whole.

If actual figures are studied instead of mean values, the geographical pattern of heat balance for the oceans is more complicated than Figure 5.4 suggests. For example, the regions of greatest interdiurnal temperature variability for the North Atlantic are different for air and sea temperatures. The greatest surface air temperature variabilities occur off the east coast of North America, but the greatest sea-surface temperature variabilities follow the core of the Gulf Stream. The latter reflects a dynamic balance: eastward-moving meanders, 100 to 400 km long in summer and 300 to 700 km in winter, are found superimposed on the Gulf Stream. These oceanic eddies are smaller than atmospheric cyclones and anticyclones. If actual daily figures are mapped, rather than annual or seasonal means, variations in sea-surface temperature are not necessarily controlled by local variations in the rate of eddy-heat exchange across the air-ocean interface, as they appear to be in the average patterns.[22]

A

Annual exchange of sensible heat from sea to air

FIGURE 5.4.
Mean annual sea to air heat
exchange in the southern
hemisphere to 60°S (after
D. W. Privett). A, C, and D are
in cal per cm² per day; B is in
grams per cm² per day. Shaded
areas indicate a zero value.

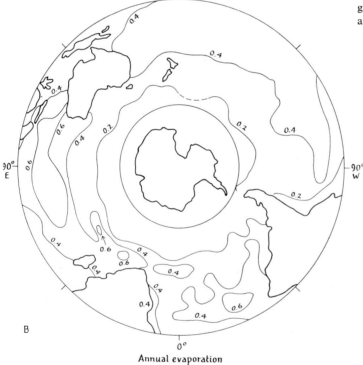

B

Annual evaporation

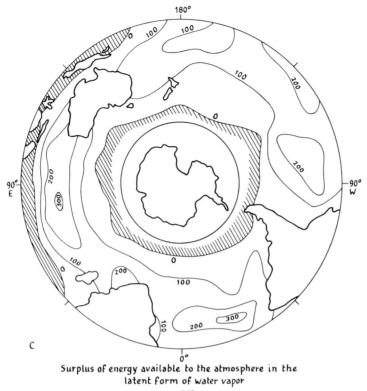

C

Surplus of energy available to the atmosphere in the
latent form of water vapor

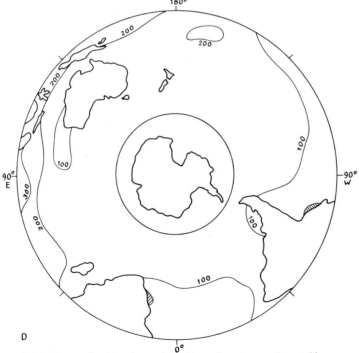

D

Total energy gained by atmosphere through exchange of sensible
heat at sea surface and condensation of water vapor.

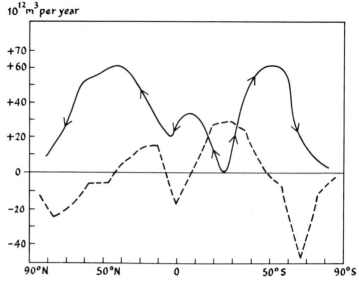

$10^{12} m^3$ per year

FIGURE 5.5.
Annual net transport of water vapor across latitude circles (after D. W. Privett). Solid line arrow indicates net transport required for balance. Broken line indicates total surplus or deficit in each 10° zone of world.

The heat balance of the air layers immediately above a sea surface is often critical in determining the stability characteristics of air masses moving across the surface. A net upward flux of sensible heat from the surface of the North Atlantic Ocean, averaging 2.1 mcal per cm^2 per sec in middle latitudes and 0.1 mcal per cm^2 per sec in tropical latitudes, is enough to produce instability. Such a flux allows a heat accumulation of more than 1,000 cal per cm^2 in a unit vertical air column of southward-moving polar air by the time it reaches latitude 20°N.[23]

Variations in the heat content of the upper layers of the oceans are very significant in deciding the rate at which the lower layers of the overlying atmosphere become heated. For example, the upper 200 m of the North Atlantic between 20°N and 65°N show an annual range in heat content of 36 kg-cal per cm^2 in a unit vertical water column. Maximum heat storage is in middle latitudes, just north of the mean location of the Gulf Stream. The effect of ocean currents and seasonal variations in heat exchange with the deeper layers of the ocean is to increase the amount of heat stored near the surface in middle latitudes, and to reduce it north of 50°N.[24]

In the North Pacific Ocean, the heat gained by the northeast trades during their passage over the sea surface is necessary to maintain the wind system: without this energy input, they would die out. The trades also export a heat amount equivalent to 20 per cent of the total heat transferred to them from the sea surface. This represents a latent heat surplus, which is used to balance heat losses by radiation elsewhere.[25]

The heat balance of the atmosphere alone is a most complex study. The tropospheric and stratospheric heat balances present different problems. The normal regions of heating and cooling in the troposphere were first computed from aerological data in 1944, on the assumption that the mean isotherms for the layer between mean sea level and 10,000 feet are stationary, the air moving across them (see Appendix 5.12). The regions of greatest heat accumulation in February were then found to be situated off the east coast of Asia and North America. Heating above the eastern part of North America was dynamic, caused by subsidence of polar air masses, and that off the Atlantic coast of the continent was due to heat flux from the sea surface.

A pattern of alternating regions of heat gain and heat loss was discovered in the lower troposphere between Europe and eastern Asia.

A more detailed investigation based on energy considerations as well as heat balance shows (see Figure 5.6) that the horizontal scale of atmospheric heating in the layer between sea level and 5.5 km is the same as that of its average temperature field in winter, but the heating and temperature fields are 90° out of phase (see Appendix 5.13). Maximum heating in the troposphere is near the equator, with other maxima in middle latitudes, east of the thermal troughs along the coasts of Asia and North America, and in the central Mediterranean. Tropospheric cooling is found over Eurasia, and both in and eastward of the long-wave ridges situated over the eastern Atlantic and western North America. Heating is transformed into potential energy in the mean long-wave pattern, maintaining the latter against friction with the Earth's surface.

In the troposphere above the North Atlantic, net heating occurs over the northern and western parts of the ocean during the winter half-year. Winter heating and summer cooling are both on the order of 10^{-3} cal per cm^2 per sec, with year-to-year variations being largely associated with rainfall fluctuations.[26]

Tropospheric heating or cooling due to radiation is very important. On the local scale, radiative transfer of heat at the ground surface is of equal importance with convection in deciding the rate of heating or cooling of the lowest few feet of the atmosphere, and hence determining the magnitude of the diurnal temperature cycle. On cloudless days, radiative heating rapidly decreases with increasing height, changing to radiative cooling at approximately 850 mb. In cloudless summer anticylones over southern England, radiative cooling is almost balanced by convective heating in the layer between ground level and 500 mb. The existence of a cloud cover destroys this balance: convection is reduced, radiation is increased, since the cloud cover provides a new radiating surface, and heat is redistributed throughout the layer by release of latent heat during condensation.[27]

Heating or cooling due to radiation absorption or emission by water vapor at 700 mb over the eastern North Pacific forms geographical patterns that vary from day to day as the airflow pattern changes. However, the patterns do not move uniformly with the 700-mb winds.[28]

The troposphere over the entire northern hemisphere loses heat by radiation, the loss amounting to 200 cal per cm^2 per day in March, with maximum losses in the middle troposphere for all latitudes. This loss is balanced by heat gains from sensible heat transferred by eddy conduction from the Earth's surface to the atmosphere (60 cal per cm^2 per day), plus latent heat release through condensation (145 cal per cm^2 per day). Although radiational balance is perfect on the average, it may not be at a particular place or time. Inequalities in the amount of incoming and outgoing radiation within the troposphere transform radiative into kinetic energy, i.e., generate winds.[29]

It has been known for many years that distinctive heat-balance effects must obtain in the upper atmosphere. The existence of a warm layer at a height of 55 km was deduced by Lindeman and Dobson in 1922; in this layer absorption of short-wave radiation from the sun by ozone resulted in temperatures of $+50$ to $+80°C$. This high-altitude heat source was proved to be capable of damping wind systems near the ground. It also provided a second "troposphere" above 60 km in which noctilucent clouds had been observed. The occurrence of mother-of-pearl clouds at 22 to 30 km implied the existence of a layer of cooling at the top of the stratosphere.[30]

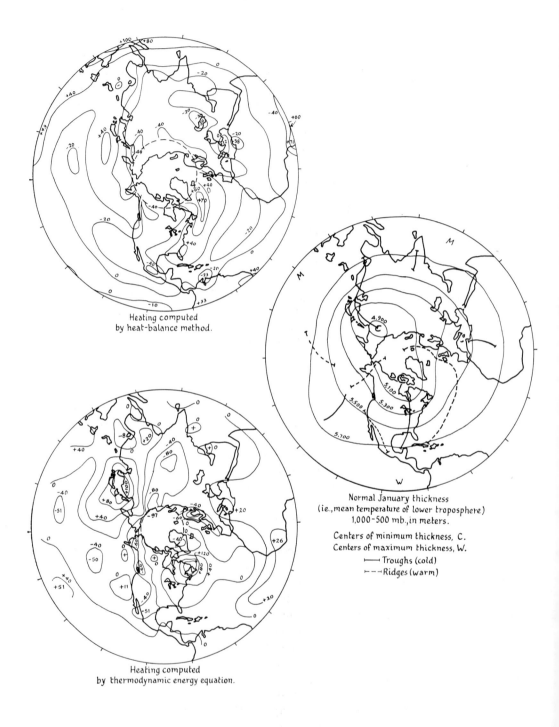

Heating computed
by heat-balance method.

Heating computed
by thermodynamic energy equation.

Normal January thickness
(ie., mean temperature of lower troposphere)
1,000–500 mb., in meters.

Centers of minimum thickness, C.
Centers of maximum thickness, W.
⊢——— Troughs (cold)
⊢- - -Ridges (warm)

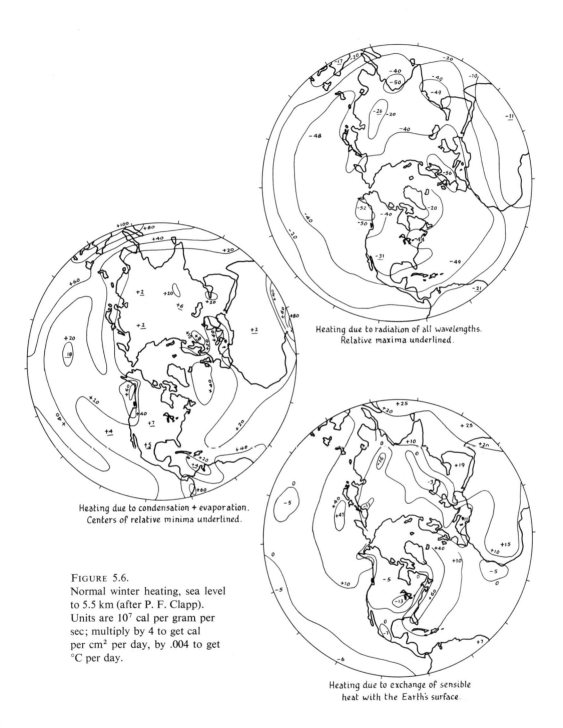

Heating due to radiation of all wavelengths.
Relative maxima underlined.

Heating due to condensation + evaporation.
Centers of relative minima underlined.

FIGURE 5.6.
Normal winter heating, sea level
to 5.5 km (after P. F. Clapp).
Units are 10^7 cal per gram per
sec; multiply by 4 to get cal
per cm² per day, by .004 to get
°C per day.

Heating due to exchange of sensible
heat with the Earth's surface.

Early rocket ascents over New Mexico showed that two temperature maxima must exist between 40 and 70 km: the lower one was attributed to ozone absorption of solar energy, and the upper to absorption of solar ultraviolet radiation by oxygen. Temperatures were found to reach a minimum at 80 km, then to continually increase with height above this. Rocket observations indicate that the upper atmosphere between 70 and 100 km is warmer in winter than in summer. The heat loss measured in the middle of this layer (10°C per day) is probably compensated for by large-scale subsidence at a rate of 42 m per day; the heating that results is not due to adiabatic compression, however, but is of a chemical nature, due to release of energy by the recombination of atomic oxygen.[31]

For the heat balance of the stratosphere, the presence of the ozone layer, above the tropopause and below 60 km, is of crucial importance. In the lower isothermal region of the stratosphere, temperatures are lowest at the equator and highest at the poles in summer; in winter, maximum temperatures occur in middle latitudes, with minima in tropical and polar regions. The total ozone content of the atmosphere is at a minimum at the equator, increases to a maximum at 60°N in the northern hemisphere, then decreases toward the pole. In the upper troposphere, temperature extremes follow the solstices. In the lower stratosphere, however, they occur at intermediate times, because the flux of radiation reaches its maximum and minimum after the solstices, but the ozone extremes are earlier: maximum ozone content in early spring, and minimum in late autumn.

The annual variation of temperature in the troposphere is small (10°C) in comparison with that at the Earth's surface (20 to 40°C), but both are relatively insignificant compared with that in the stratosphere (50 to 60°C). Differential cooling of the atmosphere during the polar night is especially marked over Antarctica. Tropospheric temperatures in the south polar regions decrease only slightly as winter approaches, but stratospheric temperatures continue to drop steadily at 0.25°C per day. This effect, due both to the intense advection of maritime air into the Antarctic troposphere and to the strong stratospheric jetstream that encircles the Antarctic continent and prevents similar advection into the stratosphere, sometimes results in the complete disappearance of the south polar tropopause. Despite these great temperature changes in the Antarctic stratosphere during the winter half-year, a surface inversion persists even on mean cross sections.[32]

Stratospheric temperature gradients are comparable in both northern and southern hemispheres in corresponding months of the year, but gradients are generally stronger in the south polar regions during winter. After the winter solstice, the Antarctic stratosphere continues to cool until the sun reappears. The return of the sun to the polar regions in spring results in a pronounced warming in the stratospheres of both Arctic and Antarctic, but the south polar warming is very pronounced indeed: 1°C per day at 50 mb, or up to 50°C per month. The Arctic stratosphere warms a total of only 30 or 40°C, but begins to heat up as much as six weeks before the sun returns.[33]

The temperature regime of the tropical stratosphere shows a pronounced semi-annual variation above 30 mb (24 km), especially near the equator, where it averages 2°C. At Balboa, Panama, and at Clark Field, in the Philippines, for example, the temperature variation is purely annual at 50 mb and lower stratospheric levels, but has two maxima and minima at 10 mb. Thus the tropical atmosphere shows unusual heat-balance features: the annual variation of temperature has double maxima both at the ground and in the upper stratosphere, but the middle and lower stratosphere has a single maximum. The double maxima at 10 mb is associated with ozone absorption, but downward diffusion of the semiannual wave by eddy conduction is a slow process, so that the double maxima are not evident in the lower stratosphere.[34]

A very distinctive feature of the stratosphere is its proneness to sudden, almost "explosive," warming episodes. The spring warming of the Antarctic stratosphere is, indeed, often very rapid: for example, an increase of 60°C (140°F) in less than three months in 1957, 80 per cent of this rise taking place in October.[35] More noteworthy, however, are the irregular warmings.

The first observed case of an explosive stratospheric warming was recorded in Britain in February 1951. At its peak on February 12, temperatures at 18, 24, and 30 km were 20, 25, and 40°F respectively higher than those outside the warming area. The temperature increases took place almost simultaneously at all heights in the stratosphere, but no similar rises were recorded in the troposphere. The episode was not associated with sunspot or geomagnetic activity.[36]

An explosive warming in January 1958 was followed by distinct developments in surface weather. The warming began at 25 mb and above southeastern Europe, reaching its greatest intensity above the North Sea on January 25. The highest temperatures were recorded above Iceland by January 30, after which the warming spread upstream, crossing northern North America and ultimately disappearing over the North Pacific. During the same period, a second explosive warming moved from the Aleutians across Siberia to Greenland. Both these warmings were associated with solar disturbances; they appeared a few days after heated air (due to a great increase in the density of the ionosphere) had moved from east to west around a large part of the northern hemisphere. They took the form of anticyclones, and were followed by large surface anticyclones whose centers moved west from Siberia to the Azores a few days later.[37]

A remarkable correlation between stratospheric warming and general weather conditions was recorded in New Jersey between January 8 and 14, 1960. Temperatures at 35 km and 26 km remained constant at −30° and −55°C, respectively, but temperatures at 30.5 km increased from −50 to −25°C. The general zonal wind at Belmar during the period changed from 200 knots westerly to light easterly, then back to 100 knots westerly.[38]

The Geographical Pattern of Temperature

The broad facts about the geographical distribution of temperature for the world are widely publicised in atlas maps, but many important features of geographical pattern cannot be derived from these maps, for example, temperature anomalies, and the likelihood of very high or very low surface temperatures. They have, too, the defects of all maps that depict average conditions, which defects may be serious when information is required on actual, not mean, conditions. For example, the isotherms of upper-air temperature (Figures 1.5–1.9 in *Foundations*) are very smooth, and give no idea of the intensity of horizontal temperature gradients which occur fairly frequently. Gradients of 7°C per degree of latitude are common at 200 mb and (in the opposite direction) at 500 mb; more than 10°C per 30 minutes of latitude—equivalent to a vertical wind shear of 60 knots per 1,000 feet—has been reported by aircraft.[39]

Temperature anomalies, i.e., the difference between actual and average temperatures for a given place, form distinctive geographical patterns that may conveniently be studied synoptically by means of thickness charts. If the anomaly for one month is compared with that for the previous, or earlier, months, a geographical pattern is also discovered. In some areas, for example, southern Scandinavia, western Greenland, the Azores, and Madeira, monthly mean temperature anomalies show strong persistence from one month to the next. In the Moscow-Leningrad area, by contrast, there is no persistence at all. In a few areas the anomalies are associated with those occurring two or three months previously. Areas with high persistence generally have water surfaces, a variable snow cover, or breaking or freezing sea ice.[40]

The maximum temperature on any one day is a unique, instantaneous value. In routine practice, temperatures are not measured with enough precision to pinpoint this value, and therefore it is impossible to obtain it from the resulting temperature maps. In general, maximum temperatures of the hottest days in hot deserts occur about three hours after local noon, and persist for two hours or so. Temperatures remain within 4°F of the maximum value for approximately six hours.[41]

Very high temperatures recorded in hot deserts are caused by specific combinations of synoptic and local geographical conditions, not, apparently, by fortuitous events, such as bubbles of hot air being carried up from the hot sand surface into the thermometer screen by gusts of wind. Apart from the conditions favoring strong insolation (i.e., a clear atmosphere, and a receiving surface of bare, dark rock or dry sand), the following combination of conditions is particularly conducive to extremely high temperatures: (1) advection of air from already hot areas; (2) a trajectory providing a long passage of air over a hot ground surface; (3) subsiding motion; (4) a föhn effect. The latter is a threefold effect, including: adiabatic heating; increased temperatures because latent heat, released from cloud and precipitation development on the windward side of the range, is being stirred into the air during overturning after the crest has been passed; and intensification of insolation to leeward of the range because the suspended impurities have been washed out of the airstream by the windward precipitation.

The area just ahead of a cold front is one in which warm air advection is speeded up, and has an extensive fetch. Many peak temperatures have been recorded in such a synoptic situation.[42] Very high temperatures are probably restricted to very small geographical areas. Local geographical effects, for example, the movement of an airstream across a built-up area or down a slope, result in adiabatic temperature rises of 1 or 2°C, which, if taking place in already hot air, can cause exceptionally high temperatures locally.

The world-record high temperatures to date are 57.7°C (136°F) at Azizia, Tripolitania, on September 13, 1922, and 56.6°C (134°F) in Death Valley, Calif., on July 10, 1913. It is difficult if not impossible to estimate the maximum possible temperature that can occur at a given place, because there is no obvious limit to the amount of radiation that may be provided by the sun. For very low temperatures, the situation is different, because every point on the Earth's surface receives nil radiation from the sun for some time during the night.

Minimum insolation is received during the six-month polar night. Assuming that only radiation-exchange processes are in operation, i.e., assuming that the combination of snow-covered ground plus troposphere constitutes a partially closed system with no external energy sources such as advection or radiation, the minimum possible surface temperature at the South Pole is $-200°C$. Record low temperatures in Antarctica are: $-88.3°C$ ($-126.9°F$) at Vostok IGY station on August 24, 1960; $-87.4°C$ ($-125.3°F$) at the same station on August 25, 1958; $-86.7°C$ ($124.1°F$) at Sovietskaya, also in August 1958; and $-74.5°C$ ($-102.1°F$) at Amundsen-Scott base on September 17, 1957.[43]

Unlike very high temperatures, which are geographically local features, very low temperatures often cover extensive areas, and the atmospheric general circulation (particularly in the upper air) is of critical importance in their determination. The exception to this rule is frost hollows, which are very localized products of air drainage that occur only under weak surface circulation regimes.

Some examples will make clear the close connection between the upper-air circulation and widespread low surface temperatures. On the night of June 12–13, 1952, a shift in the orientation of an upper trough permitted cold surface air to move inland, resulting in extensive frosts in the far western United States as far south as northern California. A very intense cold spell occurred in the east central United States between October 20 and 22, 1952, because of an unusually strong southward surge of air at 500 mb into the eastern United States. This caused the surface anticyclone typical of this time of year to move southward, rather than in its more normal easterly direction, resulting in a deep northerly current of cold air over the eastern states.[44]

Even in summer, intensification of the coldness of an airstream by nighttime radiation may result in extensive areas of very low temperature. This happened on the night of July 13–14, 1950, in the north central United States, when the coldness of southward-flowing air behind a depression moving across southern Canada was intensified by nocturnal long-wave radiation in the cloud-free anticyclone building up over the north central states and the western plains.[45]

The combination of cold airstream and nocturnal earth radiation, with a topographic situation favoring local air drainage, can produce extremely low temperatures over quite large areas. For example, on the night of April 4–5, 1954, a minimum temperature of $-17°F$ was recorded in an area 35 miles in diameter centered on First Connecticut Lake, New Hampshire. Temperatures dropped $30°F$ between sunset and midnight, and then rose $25°F$ between midnight and 7:00 A.M. The synoptic situation favored very cold polar continental air, with clear skies and calm air. The ground surface in the area is gently sloping, and there was a deep snow mantle over the open grass, covering the greater part of it.[46]

Maps of normal surface temperatures are usually of two types. World maps, except on large scales, commonly show temperatures reduced to mean sea level. Their interpretation is complicated by the fact that the rate of change of temperature with height is a function of temperature as well as of difference in elevation (see Appendix 5.14). These maps are useful for meteorological studies of the atmospheric circulation, but are of little value for geographical purposes. Continental maps, and maps covering smaller areas, usually show the distribution of actual temperatures. Owing to the great influence of orography, these latter maps often show very complex patterns, and, although valuable to the geographer, are too confusing for general meteorological work.

The local pattern of temperature distribution is a very complex subject to study; such study depends very much on microclimatology, which involves turbulence theory. In general, the important fact to be born in mind is that most temperature maps are based on measurements under standard conditions: i.e., the thermometer bulbs are at a height of 4 feet or so above a flat surface of short grass in an exposed area. Any other height for the measurement, above ground or below, can be expected to result in different temperatures, which can be estimated from the laws of microclimatology.

For example, daily weather reports frequently indicate that grass minimum temperatures are much below the screen minimum temperatures, although if there is fog, the reverse may be true. On cool nights, temperatures one inch above soil level are much lower in short grass than in long grass: at Hurley, Bucks., one-third more frosts were recorded in short grass than in long. On warm nights, the short-grass temperatures exceed those in the long grass.[47]

In built-up areas, temperatures at the standard heights and exposures are influenced by the local increase in the heat capacity of the Earth's surface brought about by the existence of the urban area. The creation of this local heat island may also affect atmospheric stability. The normal range of stability conditions in the layer 60 to 524 feet above Louisville, Kentucky, is much smaller than in the surrounding open country; surface inversions are comparatively rare above the urban area, but occur regularly above the nonurban fringe.[48]

In comparison with the geographical pattern of temperatures on the surface of the continents, the pattern shown by sea-surface temperatures is much less well-charted, despite the fact that the oceans cover a much larger area than the land surfaces of the globe. There are several reasons for this, other than the obvious one of the expense of maintaining permanent ocean stations.

It is rarely possible to obtain the "skin" surface temperature of the sea. The standard techniques measure the mean temperature of the uppermost layer of the ocean masses. The bucket method measures the mean temperature of a layer one foot deep; the intake method measures the temperature of the water flowing through an intake 10 to 30 feet below the surface of the sea. The movement of the ship through the sea stirs up the water to an appreciable depth, and both bucket and intake techniques merely sample this disturbed water.[49]

The temperature gradient in the uppermost layer of the sea depends on weather conditions. On rough days, there is almost no gradient in the upper one or two feet, because of turbulent mixing, but marked gradients may exist on calm sunny days; a daytime temperature decrease with depth of $1°F$ in the first ten feet appears to be typical.[50]

Since sea-surface temperatures represent area-mean values for a layer, and not point values as for land-surface temperatures, they are much less variable than the latter. Typical variability is less than $1.5°F$ in 5 nautical miles, and typical diurnal variation is just over $1°F$. Despite this, the variability of surface air temperatures at ocean stations is greater than that at land stations. Thus the ratio of winter to summer values of monthly interdiurnal variation of surface air temperature (in $°F$) is between 3 to 1 and 6 to 1 for North Atlantic Ocean weather-vessel stations, as against 2 to 1 for land stations.[51]

Sea-surface temperatures have different geographical patterns from day to day, and although it is not routine practice to construct daily synoptic charts for them as for the surface air, five-day synoptic maps are both feasible and useful (see Figure 5.7). These show that the coastal waters around the British Isles have two basic patterns of mean sea-surface temperature. The winter pattern is typical of November to March, and the summer pattern of May to September. April is a transitional month, and generally exhibits weak surface-temperature gradients.[52]

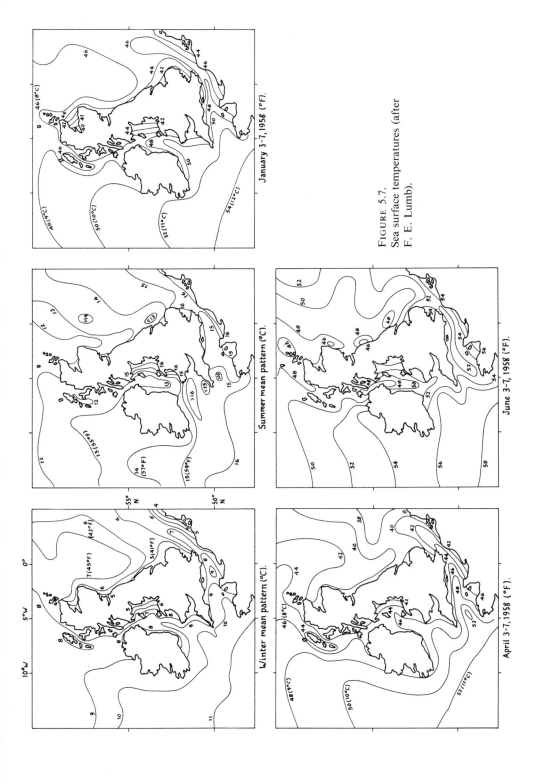

January 3–7, 1958 (°F).

Summer mean pattern (°C).

June 3–7, 1958 (°F).

Winter mean pattern (°C).

April 3–7, 1958 (°F).

FIGURE 5.7.
Sea surface temperatures (after
F. E. Lumb).

Compared to the complexities revealed by maps showing the horizontal distribution of temperature, the vertical distribution of temperature in the atmosphere is roughly the same over both land and sea. Vertical soundings above the surface layer, in which an inversion may or may not be present, show a general decrease in temperature up to the tropopause, and then either an isothermal layer or one in which temperature increases with height in the lower stratosphere.

The variation of temperature with height in and above the stratosphere could not be studied in detail until the advent of rockets. Early rocket data, for example, those provided by soundings from modified V-2's over New Mexico, indicate profiles that are probably of general application. These revealed an increase in temperature from 210°K at 15 km to 308°K at 55 km (i.e., in the layers comprising both stratosphere and stratopause), a decrease to −180°K at 85 km (covering both mesosphere and mesopause), and then a further increase above 85 km in the thermosphere to 266°K at 120 km. A notable feature was a temperature decrease at 100 km in the transition layer where molecular oxygen dissociates into atomic oxygen.[53]

The first seven years of data from Cape Canaveral show quite clearly the "sandwich" structure of the atmosphere, with the tropopause forming a layer of cold air between two warmer layers. On a global scale, the minimum temperature layer forms a wedge of almost isothermal temperatures, which thins towards the equator and thickens toward the pole. The wedge is thicker in the northern than in the southern hemisphere.[54]

Everyday observation of the vertical growth of cumulus clouds suggests that the thermal influence of the Earth's surface on the atmosphere extends vertically upwards a few tens of thousands of feet at the most. However, seasonal trends in temperature are recorded at very high levels. For example, they have been noted over Belmar, New Jersey, at all heights up to 40 km (150,000 feet).[55]

Despite apparent worldwide similarity in broad-scale vertical temperature profiles, numerous local differences are found if the fine detail of individual soundings is compared. The question arises whether the irregularities represent real geographical features in the atmosphere, or are the results of instrumental inaccuracies. In general, irregularities in a temperature sounding that do not depart from the smoothed curve by more than 1.5°F cannot confidently be considered to be real features of the atmosphere. Instrumental error is on the order of 0.5°F, and temperature fluctuations of up to 2°F are quite common over horizontal distances of 50 km, which is a likely distance to be traversed by a sounding balloon during its ascent.[56]

The Statistical Pattern of Temperature

Before subjecting a long series of temperature records to any geographical or historical analysis, one must investigate the statistical properties of the data. Once the requirement of homogeneity has been satisfied, the first problem is, what is the most satisfactory time unit to use in the design of the basic parameter? Apart from the obvious hour, day, and year units, which clearly have justification from the movements of the Earth on its axis and around the sun, the most convenient units appear to be the month and the five-day mean.

The most detailed temperature records published are usually in hourly values. Frequency graphs of hourly temperatures for individual stations often show considerable deviations from the normal probability curve that one might expect. This fact indicates that the data for a single station represent a mixture of several "populations," formed by the temperature variations due to the diurnal, interdiurnal, and interannual cycles.

In the United States, different geographical areas have different types of frequency distribution. Type I is common on the California coast and in the south and northeast in July; its main features are a very small range of temperature, with many hours at or close to the modal temperature. Type II approximates to the normal distribution: it predominates over the whole country (except for California and the southern Great Plains) in January, and is found from Nebraska to upper New York in July. Type III covers a wide range of temperatures with no pronounced peak, and is found in the southern Great Plains in January and throughout the area from Oregon to Oklahoma in July. In general, positive skewness predominates in summer and negative skewness in winter. Differences between stations in the same area for a type can mainly be explained by the differing amounts of water present in the form of clouds, bodies of water, or soil moisture.[57]

The monthly mean is the most popular of all climatological parameters, even though it can be misleading in the absence of other information.[58] Thus Jacksonville, Florida, and Williams, Calif., both have July mean temperatures of 80.1°F (for some years), but their frequency distributions differ considerably: their mean maxima differ by 6°F, their mean minima by 13°F, their absolute maxima by 16°F, and their absolute minima by 14°F.

The temperature unit to be adopted as standard must be both stable and representative of the instrumental record. In general, mean monthly temperatures are not normally distributed, but in many areas (for example, the whole of the United States except for the eastern states in summer) January and July mean temperatures constitute a random sample, so that standard methods of probability and statistical analysis may be applied to them with confidence.[59] One advantage that results is that confidence limits can be provided. For example, the normal temperature for Jacksonville in January (based on 25 years' data) may be stated as 57.8 ± 1.0°F with a confidence of 0.68. Here 1.0°F is the standard deviation of the mean (57.8°), and the confidence figure indicates that if 100 samples of January data were selected at random, then on the average 68 of them would include the true mean in the range of 57.8 ± 1.0°F.

How many years of recorded data are needed to establish a stable mean value is debatable. Much depends on the homogeneity of the data: with nonhomogeneous records, averages over 10 years will give more reliable results than averages over 30 or more years. Increasing the number of years beyond a certain point merely increases the mean-square error (i.e., the "random" element in the data), which decreases the stability of the figures.[60]

Monthly mean temperatures have certain disadvantages. For example, the standard deviation, σ, of monthly mean temperature is widely used as a measure of temperature variability within an area. More than 95 per cent of months of a given name will have mean temperatures within $\pm 2\sigma°$ of the average. The probability that given departures from average will be equaled or exceeded in 100 years may then be deduced from the value of σ.[61] However, this is only true if the monthly mean temperatures are normally distributed, which is often not the case.

Monthly mean temperatures are not really satisfactory for the calculation of accumulated temperatures, i.e., the integrated excess or deficiency of temperature with

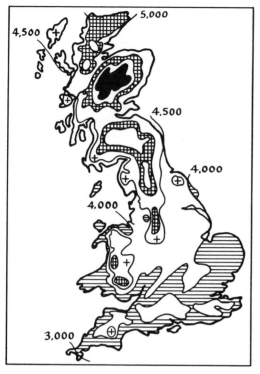
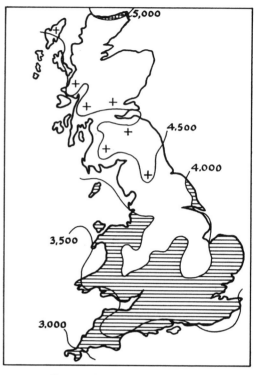

At screen level   Reduced to mean sea level

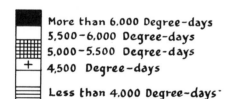

More than 6,000 Degree-days
5,500 – 6,000 Degree-days
5,000 – 5,500 Degree-days
4,500 Degree-days

Less than 4,000 Degree-days

FIGURE 5.8.
Average annual accumulated temperature below
60°F in Britain (after H. C. Shellard).

reference to a fixed datum or base value over an extended period of time. If the mean
temperature of the month is close to the base temperature, then the accumulated
amount will be close to or even equal to zero, despite the fact that on certain days
temperatures may have been much above or much below the base value. Monthly
means of the mean daily maxima and mean daily minima give more reliable results,
and, in general, averages of the latter enable finer distinctions to be drawn between
stations than does the simple monthly mean.[62]

Accumulated temperatures are an example of a common tool in climatology: a
statistical entity, of undoubted practical utility, that does not represent any real
feature of nature. The standard formula for computing monthly accumulated temperatures uses both the long-period average and the standard deviation of the monthly
means (see Appendix 5.15). In Great Britain (see Figure 5.8), average annual accumulated temperature increases rapidly with altitude, so that Buxton, Derbyshire
(1007 feet), has an average that is 15 per cent greater than that at Macclesfield (500
feet), about ten miles to the west. Figure 5.8 does not give local values correctly in
areas away from the standard stations. For example, it indicates 4,400 for Sheffield,

whereas the actual value of 4,170 reflects the local warming effect of the built-up area.

The frequency distribution of hourly temperatures for a given month may be found empirically from the monthly mean, mean maximum, and mean minimum temperatures. The relationship is of general application, which indicates the existence of some underlying unity between climates as diverse as those of the Arctic and the tropics. It would appear to be reasonable to expect that the annual temperature curve should be derivable from the twelve monthly means, but this does not necessarily apply, particularly in Arctic regions. At Resolute, Northwest Territory, for example, minimum temperatures occur in early March. The sun first appears on February 5, but insolation is negligible for two or three weeks. The seven-day means of temperature show this correctly, but a smooth annual variation curve derived from the monthly means incorrectly gives February as the month of minimum temperature.[63]

If information is required in more detail than is available from monthly means, the five-day mean is perhaps the most likely parameter to be chosen. The variations shown by hourly temperature readings over a period of years are very irregular, and the daily and annual cycles swamp all other possible regularities or periodicities. Use of daily mean temperatures eliminates the diurnal cycle, but the variation is still very complex. Monthly means are really only a way to represent the annual cycle as twelve equidistant values. However, five-day or seven-day means provide a compromise averaging period that is short enough to avoid being merely a reflection of the annual cycle, and long enough to eliminate the daily cycle. In effect, they eliminate the "random" temperature variation due to such factors as cloudiness or local wind changes, but the range of both maximum and minimum temperatures likely to occur at a station is still so large that only 80 per cent of the observations can be pinpointed to within even 10 or 15°F.[64]

Weekly averages of daily mean, daily maximum, and daily minimum temperatures give nonnormal distributions, but the errors involved in applying the normal theory still leave the probabilities accurate enough for most purposes. Harmonic regression further enables probabilities for any week to be obtained when the temperatures for only five days out of the seven are known.[65] Indeed, harmonic analysis has proved very valuable in the analysis of mean temperatures generally.

The distributions of harmonic parameters for surface temperatures of the North Atlantic Ocean provide detailed information concerning seasonal changes that is not apparent from simple inspection of the original data. The rate of change of sea-surface temperature is found by differentiating the sum of the first three harmonics (see Appendix 5.16). The annual component in the North Atlantic is usually much greater than the semiannual, whose maximum value is around 3°F. The greatest variability of sea-surface temperature is found in the North Sea and in British coastal waters. The times at which the sea temperature reaches its maximum and minimum values are determined by the phase angle of the annual component, which varies from 220°, giving a maximum on September 7 and a minimum on March 7, to 240°, giving a maximum on August 16 and a minimum on February 14. Separation of the harmonic components also makes it possible to distinguish between areas where temperatures increase more rapidly in spring and summer than they decrease in autumn and winter, and vice versa.[66]

The annual variation shown by five-day mean temperatures for many European stations can be almost completely represented by a two-term harmonic curve. The semiamplitude of the first harmonic increases from 5°C in Ireland, Wales, and western

Scotland to 12°C along the Norway-Sweden border, and to 10 or 11°C on the North German Plain. Deviation of the actual values for five-day mean temperatures from the harmonic values may be taken to indicate the presence of singularities: the most evident singularities are positive temperature anomalies in late May and early June, and negative anomalies in late June. The agreement found between distant stations indicates that the anomaly-producing processes work on a large scale, approaching the size of half-waves in the general westerly circulation. The magnitude of the anomalies is greatest (nearly 1°C) in Germany and least in Scotland, and hence the May-June singularity is tied mainly to central Europe. This indicates that large-scale circulation features may tend to persist in fixed positions for several months; the central European temperature pattern is probably associated with the monsoonal effect due to solar heating in the continental interior.[67]

The sequence of five-day mean temperature anomalies at Kew is vaguely oscillatory in character. Persistently positive or negative regimes tend to develop at specific times of year, and especially between August 14 and September 7. In winter, the temperature regime tends to perpetuate itself, whereas in summer there are two periods of self-perpetuation with a marked break in between. In spring and autumn, the regime tends to revert to normal in more than ten but less than twenty days.[68]

The Historical Pattern of Temperature

Historical variations in temperature may be divided into (a) short-period and (b) long-period fluctuations, each of which may be considered in terms of trends, regular and aperiodic oscillations, and random variations. From the account of time-series analysis in Chapter 2 it will be evident that a full analysis of the pattern shown by any lengthy series of temperature observations is a very technical matter, and that little reliable information can be expected from simple examination of the data.

Short-period fluctuations are defined here to be those variations shown by averages during less than one year. These include microoscillations of temperature, which are "random" in the sense that they do not appear in the standard climatic tables, and irregular variations, such as sudden drops of temperature, as well as periodic or aperiodic fluctuations shown by monthly, weekly, or daily means.

That microoscillations of temperature, recorded by sensitive rapid-response resistance thermometers, show consistent patterns indicates that some underlying mechanism is controlling them. That mechanism is atmospheric turbulence. Individual variations shown by, e.g., five-second average temperatures plotted on height-time coordinates, fit into an over-all turbulence pattern, in which fluctuations at a given height above the ground are highly correlated with those at all other heights at the same time. The effect of increasing instability is to reduce the "wave length" of the turbulence pattern; when a superadiabatic lapse-rate prevails, two types of turbulence, giving rise to local horizontal divergence and local horizontal convergence, interfere to produce a distinctive pattern.[69]

On a longer time-scale, day-to-day changes in the general synoptic situation sometimes result in sudden rises or falls in daily mean temperature that, although apparently fortuitous at the time, show a certain pattern when several years' worth are examined. Numerous cases of abrupt drops of 5°C or more in daily mean temperature have been recorded at Athens between 1901 and 1950. They may be separated into four categories, each of which has a preferred season. Category A temperature drops (which make up 40.8 per cent of all drops in one year) are associated with anticyclones covering northern Europe, and are most frequent in February. Category B drops (37.5 per cent) are associated with an extension of the Azores anticyclone over Europe, and are most frequent in March. Category C drops (6.0 per cent), most common in autumn and winter, are associated with a westward extension of Russian anticyclones. Category D drops (15.7 per cent) take place mainly in February and November, and are associated with the cold fronts of eastward-moving depressions that pass over Athens.[70]

Isolated, sudden temperature changes, without any apparent historical pattern, may be brought about by a sharper-than-average boundary between adjacent air masses. An increase of 11°F in 72 minutes on May 10, 1960, at Edinburgh, followed by a 16°F drop in 15 minutes, indicates how sharp the boundary may be between cool moist air over the North Sea coast on the east and adiabatically-heated air descending from the hills on the west. A southeasterly airstream prevailed during the temperature changes, the wind remaining steady the whole time.[71]

Abrupt temperature falls due to advection—often intensified by radiation cooling—are fairly common in many areas. In some places they are very pronounced; New York City, for example, has an average of 13 per year, mainly between October and May. They usually last one or two days, and falls of 15°F or more are common. The greatest drop in one hour (between 1891 and 1940) was one of 29°F on March 28, 1921; the longest period with a continually falling temperature was a 21-hour interval on November 27, 1932, with a fall of 34°F.[72]

Even soil temperatures, which normally show very regular and reliable variations, very occasionally are subject to sudden changes by exceptional weather conditions. The diurnal variation of soil temperatures at 12 inches depth rarely exceeds 5°F in Great Britain, but at West Linton, Peeblesshire, in March 1940, percolating rainwater caused an increase of 7°F in a very few hours.[73]

In the search for regularities in short-period fluctuations, two main techniques have been used. Power-spectrum analysis indicates the importance of fronts in accounting for fluctuations in daily mean temperature, and harmonic analysis provides a smooth curve that can be regarded as a model, any calendar-bound deviations from which may be evidence of spells or singularities. Harmonic analysis of daily mean maximum temperatures at Bidston confirms the existence of three of Buchan's cold spells (February 7–14, August 6–11, November 6–13) and one of his warm spells (July 12–15), but denies that of his warm spell of April 11–14. In contrast to these results, harmonic analysis of Edinburgh temperatures for the period 1901–1950 shows that the recurrences on days of Buchan spells do not depart significantly from the chance expectation. The only statistically significant cold spells are on November 16 and 18. Singularities that seem well-established for central and northern Europe by harmonic analysis are: warm singularities, January 31 to February 4, April 1–5, May 21–25, and 26–30, May 31 to June 4, and September 28 to October 2; cold singularities, March 12–16, June 20–24, and November 12–16 and 17–21.[74]

An unusual correlation between January temperature recurrences and the general circulation occurs at Victoria, British Columbia. Warm periods tend to recur around January 14 and 24–26, the former being a high-index singularity, and the latter a low-index period. In the northeastern United States, warm spells often occur in January, usually during low-index periods. A notable warm singularity in early

January—whose actual date during several decades may vary over a two-week period —is connected with a change in the general planetary circulation that displaces Gulf Coast cyclones into a track along the Appalachians, thereby bringing warm air to the northeastern states.[75]

Sampling difficulties mean that the delimitation of singularities is very difficult. The annual curve of daily mean temperatures is always irregular, whatever the length of the instrumental record, and the departures of the actual means from the smooth harmonic curve fitted to them may be simply statistical fluctuations caused by the fact that the figures are computed from a finite sample, rather than true singularities, but even when temperature recurrences are largely casual, there is only a very small probability that they are completely random.[76]

Long-period fluctuations in climate can only be studied accurately when allowances for changes in station location, hours of observation, type of instrumentation, and so on, have been allowed for as far as possible. Examples of such homogenized tempera- ture series are those of monthly mean temperatures for central England, Utrecht, London, Edinburgh, and New Haven, dating from 1698, 1706, 1763, 1764, and 1780, respectively. In countries of uniform relief, it is possible to build up such tables if overlapping records are available from different locations or different observers. In areas of more complicated topography, the matter is more difficult. An increase in the extent of built-up areas around the station, or the change in exposure when a station is moved from within the city to an exposed location at an aerodrome outside the urban area, are important influences that may have to be filtered out.[77]

Detailed investigations of temperature trends for the whole world were published in 1950, 1961, and 1963. The earliest investigation used 129 stations with 50 years or more of records, plus 54 other stations with shorter records. The results indicated an upward trend since 1885, which was not uniform throughout the world. The positive trend was particularly pronounced in higher middle latitudes and polar regions of the northern hemisphere, but a negative trend was found in south polar latitudes. A general increase since 1885 of 1.0°F in annual mean temperature and 2.2°F in winter mean temperature was deduced. There was no evidence for any influence of popula- tion increase or industrial growth on temperature.[78]

The 1961 study examined the deviations from average in data from over 400 stations, weighted according to area to give the zonal or worldwide fluctuations. These show a gradual rise in temperature for several decades, with maximum rise in the higher latitudes of the northern hemisphere. The main fluctuations between 1875 and 1915 were simultaneous in tropical and middle latitudes, but since 1915 the pattern has become more confused: a change appears to have taken place in the zonal fluctuations since 1920. Temperatures in a few areas, including most of Australia, the extreme South Atlantic, and the Black Sea—Caspian region, have remained stable. The average increase in temperature at town stations in comparison with rural sites, over a 30-year period, is between 0.03 and 0.1°C.[79]

In 1963, the global pattern of change as deduced from data for 1880–1960 was published. Differences between consecutive pentad (five-year) averages at each of a large number of uniformly distributed stations—not more than one station per 10° latitude-longitude square—were averaged for all stations within 10° bands of latitude. The cumulative sums of these averaged differences were then used as a measure of interpentadal temperature changes in each band, a procedure that minimizes irrelevant influences. Weighted averages of the 10°-band indices were then formed to indicate the

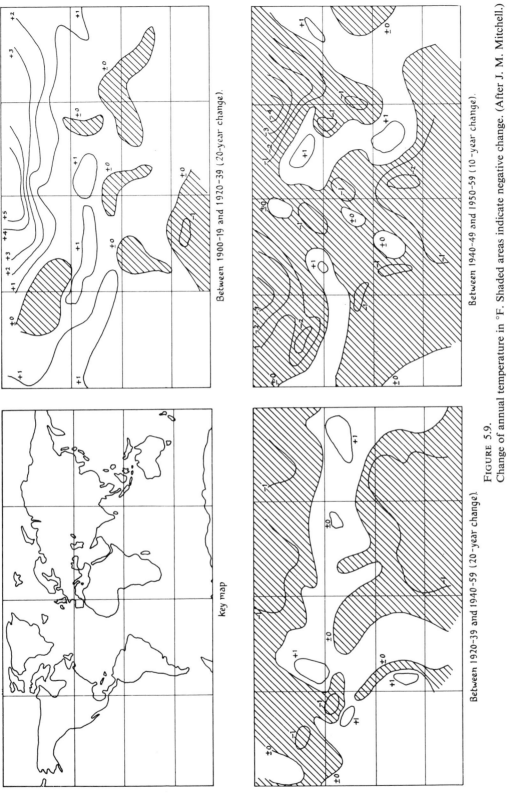

key map

Between 1900–19 and 1920–39 (20-year change).

Between 1920–39 and 1940–59 (20-year change).

Between 1940–49 and 1950–59 (10-year change).

FIGURE 5.9.
Change of annual temperature in °F. Shaded areas indicate negative change. (After J. M. Mitchell.)

hemispheric and global average trends. The results (see Figure 5.9) indicate that changes of average global temperature are not necessarily associated with particular geographical modes of change, or even with those of atmospheric circulation change. Since the secular trends are very irregular, maps representing mean temperature changes from one time period to another may be radically altered in detail by relatively small changes in the length or dates of the time periods employed.[80]

Different investigators have chosen different methods to demonstrate long-period temperature changes, but the broad conclusions remain the same. European data (1780–1930), arranged in the form of 30-year means at 15-year intervals, and expressed as departures from the 1901–1930 normal, show that winter warming began in 1840 at Copenhagen and 1885 at Vienna, and that annual temperatures have increased since 1825 at Edinburgh and 1810 at Berlin. The ten-year averages in Britain and eastern Canada show very similar fluctuations, reflecting the influence of the long waves in the westerlies. The ten-year area means showed rises from 49.5 to 50.2°F for England and Wales, and from 47.0 to 47.7°F for Scotland, between 1922–31 and 1929–38. The increases were mainly due to higher summer temperatures, because winters became rather colder in the latter period. Linear regressions for stations in South Africa indicate a significant increase in temperature between 1901 and 1930.[81]

In general, all the evidence points to the conclusion that temperatures have increased in most parts of the world from 1850 to at least 1940, since when they have leveled off or begun to fall. The warming was slow and irregular at first, but gradually increased its rate to reach a peak in the 1930's and early 1940's. It was neither symmetric nor uniform with respect to the pole, being most marked in the higher latitudes of the northern hemisphere, particularly in areas bordering on the Atlantic Ocean, and least marked (or even nonexistent) in middle latitudes of the southern hemisphere. Furthermore, the global trend is not representative of most geographical locations. The temperature rise in some areas has lagged behind that in others, and the fluctuations have been opposite for a time in adjacent regions. The areas of most rapid warming before 1940 tend to coincide with the areas of most rapid cooling since 1940.[82]

An interesting question is the statistical significance of the recorded temperature changes. Some areas, especially India, show no significant departure from randomness in their variations, but instead appear to have undergone oscillations with periods of 30 to 40 years. The 30-year overlapping means, undoubtedly the most popular method of representation of temperature variations, would filter out such 30-year oscillations. Short-period fluctuations, which are not significant within a long record, may nevertheless have very important effects on agriculture and other activities.[83]

There is little doubt that significant temperature oscillations have been superimposed on the general trends of the past 100 years. In particular, a periodicity of 2.1 years, the surface expression of a biennial atmospheric pulse, seems well-established. Brought into prominence by the use of a narrow-bandpass filter, the pulse has greatest amplitude when its extremes coincide with the winter months. The pulses in intertropical regions are mutually inphase, but those at higher-latitude stations are out of phase with those at tropical stations. Half the variance of surface temperatures in this particular frequency range is associated with background "red noise," so that the pulse is generally deeply embedded in atmospheric noise and hence is distorted. Amplitude modulation occurs in places, the pulse tending to become locked in phase with the season of the year; this results in a seven- to nine-year oscillation in pulse amplitude at many stations.[84]

Notes to Chapter 5

1. O. M. Ashford, *QJRMS*, 74 (1948), 99. C. E. N. Frankcom, *loc. cit.*, p. 403.

2. C. G. Webb, *QJRMS*, 82 (1956), 518.

3. A. P. Crary, *JM*, 7 (1950), 223. A. P. Crary and V. C. Bushnell, *JM*, 12 (1955), 463. W. B. Kennedy, L. Brogan, and N. J. Sible, *JM*, 12 (1955), 519.

4. R. G. Fleagle, *JM*, 13 (1956), 160.

5. N. E. Rider, *MM*, 82 (1953), 329.

6. W. Harrell and E. A. Richardson, *MWR*, 88 (1960), 269.

7. For details, see *EAFM*, p. 17. For a history of thermometer scales, see H. E. Landsberg, *W*, 19 (1964), 2. On the standard error of sea-surface temperatures obtained by means of a canvas bucket, see M. W. Stubbs, *MM*, 94 (1965), 66. For details of instrumentation, see: M. Fuchs and C. B. Tanner, *JAM*, 4 (1965), 544, on radiation shields for air-temperature thermometers; P. Hyson, *JAM*, 7 (1968), 908, on thermistor mountings; D. W. Stevens, *JAM*, 6 (1967), 179, on micrometeorological measurements of temperature to within 0.05°C and temperature-differences to within 0.01°C; G. Q. Clark and J. G. McCoy, *JAM*, 4 (1965), 365, and H. N. Ballard, *JAM*, 6 (1967), 150, on measurement of stratospheric temperatures. See P. G. Hookey, *W*, 20 (1965), 259, for an inexpensive sun-maximum thermometer. See also D. H. Sargeant, *JAM*, 4 (1965), 644, on the theory of junction diodes as temperature sensors. See G. E. Daniels, *JAM*, 7 (1968), 1026, for air-temperature measurement with a radiation compensating thermocouple. For details of a balloon-borne temperature sensor, with an error of 0.4°C at ground level, rising to 0.9°C at 18 km, see D. W. Camp and F. E. Caplan, *JAM*, 8 (1969), 159. On the accuracy of temperatures measured by rocketsonde thermistors, see A. J. Miller, *JAM*, 8 (1969), 172. On the indirect measurement of atmospheric temperature profiles from satellites, see D. Q. Wark and H. E. Fleming, *MWR*, 94 (1966), 351, and D. T. Hilleary *et al.*, *loc. cit.*, p. 367.

8. R. W. Gloyne and L. P. Smith, *MM*, 80 (1951), 203.

9. E. P. Ney, R. W. Maas, and W. F. Huch, *JM*, 18 (1961), 60.

10. For further information, see: D. Bryant, *MM*, 97 (1968), 183, on the effect of windspeed on the lag of a screen-mounted thermometer; T. J. Chandler, *QJRMS*, 90 (1965), 333, on north-wall vs. screen temperatures at Kew; D. M. Love, *MM*, 96 (1967), 353, on the differences between temperatures measured at ground-level and roof-top sites at Southampton; L. Baer and D. Hamm, *JGR*, 70 (1965), 4579, on the representativeness of ocean temperature measurements; R. S. Lindzen, *JAS*, 24 (1967), 317, on the theory of upper-air temperatures measured by thermistors; and C. E. Duchon, *JAM*, 3 (1964), 327, on the theory for correcting thermocouple readings for long-wave radiation effects. For errors in temperature soundings, see H. N. Ballard and R. Rubio, *JAM*, 7 (1968), 919, on rocketsondes and balloonsondes. For comparisons of temperature measurements obtained in different ways, see: B. Ingham, *MM*, 96 (1967), 363, on moorland stations in Britain vs. radiosondes; J. D. McFadden and J. W. Wilkinson, *MWR*, 95 (1967), 936, on aircraft vs. shipborne observations; O. W. Thiele and N. J. Beyers, *JGR*, 72 (1967), 2485, on rocketsondes vs. radiosondes; and J. E. Masterton *et al.*, *JAM*, 5 (1966), 182, on a comparison of different types of temperature soundings over the tropical Pacific to 100 km. For more details on the thermometers discussed in the text, see *MOHSI*, pp. 82–153.

11. For full details, see *HBES*. See also M. J. Budyko, *BWMO*, 7 (1958), 166.

12. N. E. Rider and G. D. Robinson, *QJRMS*, 77 (1951), 375. V. A. Myers, *MWR*, 86 (1958), 149.

13. K. J. Hanson, *MWR*, 89 (1961), 173. J. F. Gabites, *AM*, p. 370.

14. The data in Table 5.4 is from D. M. Houghton, *MM*, 87 (1958), 132.

15. G. D. Hembree, *MWR*, 86 (1958), 171. D. A. de Vries, *UCSCM*, p. 109.

16. J. L Monteith, *UCSCM*, p. 123

17. L. A. Ramdas, *UCSCM*, p. 129.

18. D. A. de Vries, *UCSCM*, p. 114.

19. W. C. Jacobs, *CM*, pp. 1059–68.

20. D. M. Houghton, *MM*, 87 (1958), 132.

21. D. W. Privett, *MOGM*, vol. 13, no. 104 (1960).

22. D. L. Bradbury, *JAM*, 1 (1962), 421.

23. For details of the computations involved, see H. Riehl *et al.*, *QJRMS*, 77 (1951), 598, and H. Riehl and J. S. Malkus, *QJRMS*, 83 (1957), 21.

24. K. Bryan and E. Schroeder, *JM*, 17 (1960), 670.

25. Riehl, *op. cit.* (1951); Riehl and Malkus, *op. cit.* (1957).

26. G. B. Tucker, *MM*, 91 (1962), 184.

27. G. D. Robinson, *CPRMS*, p. 26.

28. A. H. Thompson and M. Neiburger, *JM*, 10 (1953), 167.

29. J. London, *JM*, 9 (1952), 145.

30. G. M. B. Dobson *et al.*, *Proc. Roy. Soc.*, 185, series A (1946), 144.

31. W. W. Kellogg, *JM*, 18 (1961), 373.

32. H. Wexler, *QJRMS*, 85 (1959), 196. E. Flowers, *AM*, p. 439.

33. R. C. Taylor, *AM*, p. 439.

34. R. J. Reed, *MWR*, 90 (1962), 211.

35. *TAGU*, 41 (1960), 384.

36. F. J. Scrase, *MM*, 82 (1953), 15.

37. R. Scherhag, *JM*, 17 (1960), 575.

38. W. C. Conover, *JM*, 18 (1961), 410. For a theory of the mean temperature of the Earth's surface, based on surface heat balance considerations, see B. Saltzman, *T*, 19 (1967), 219. For a synoptic study of the thermal nature of the Earth's surface, see R. R. Dickson, *MWR*, 92 (1964), 195. For Antarctic micro-meteorological observations of the exchange of sensible heat between a snow surface and the air above, see T. E. Vinje, *AMGB*, 16, series A (1967), 31. On soil temperatures, see: W. R. van Wijk, *Meteorol. Monogr.*, 6, no. 28 (1965), 59, for the theory of heat flow in soil; P. B. Sarson, *MOSP*, no. 14 (1962), for a theory for Britain; T. R. Oke and F. G. Hannell, *W*, 21 (1966), 21, for an empirical study in Hamilton, Ontario; and C. R. Stearns, *JGR*, 74 (1969), 532, for a theory for Peru. On oceanic heat balance, see: W. L. Donn and D. M. Shaw, *JGR*, 71 (1966), 1087, for the Arctic Ocean, ice-free and ice-covered; Z. W. Zillman and J. A. Bell, *JGR*, 73 (1968), 7057, for sea-air heat flux in the Indian Ocean; M. Garstang, *T*, 19 (1967), 492, for heat exchange on a synoptic scale in the tropical Atlantic; and H. C. Shellard, *MOSP*, no. 11 (1962), for the climatology of North Atlantic Ocean weather stations. For a theory of temperatures at the ocean-air interface, see P. M. Saunders, *JAS*, 24 (1967), 269. On the numerical prediction of monthly mean temperatures in the atmosphere-ocean-continent system, see J. Adem and W. J. Jacob, *MWR*, 96 (1968), 714. On horizontal heat flux, see P. Welander, *T*, 18 (1966), 63, whose mathematical theory indicates that there ought to be a general large-scale eastward eddy diffusion of heat in the atmosphere, driven by the Coriolis force. On vertical heat flux in the northern hemisphere, see: E. Palmén, *T*, 18 (1966), 838, on kinetic energy correlation; R. Berggren and A. Nyberg, *T*, 19 (1967), 18, on eddy transports; and W. P. Elliott, *QJRMS*, 90 (1964), 260, on height variation of vertical heat flux near the ground over flat prairie in Nebraska. On atmospheric heat balance, see D. B. Shaw, *QJRMS*,

92 (1966), 55, on heat sources and sinks in the northern hemisphere, and S. L. Hastenrath, *AMGB*, 17, series A (1968), 114, on meridional heat transport associated with standing eddies. On the climatology of the local rate of heating in the northern hemisphere troposphere, see G. B. Tucker, *MM*, 94 (1965), 205. On the stratosphere, see: D. A. Stewart, *AMGB*, 17, series A (1968), 101, for the climatology of eddy heat flux; A. E. Cole, A. J. Kantor, and P. Nee, *JGR*, 70 (1965), 5001, on temperature variations at latitude 15°N; D. M. Hunten and W. L. Godson, *JAS*, 24 (1967), 80, on the correlation between upper-atmosphere sodium and stratospheric warming; K. Labitzke, *JAM*, 4 (1965), 91, on the mutual relations between stratosphere and troposphere during winter stratospheric warmings; K. Labitzke and H. van Loon, *JAM*, 4 (1965), 292, and P. R. Julian, *JAM*, 6 (1967), 557, on midwinter stratospheric warmings in the southern hemisphere. For examples of sudden stratospheric warmings, see: K. Labitzke, *QJRMS*, 94 (1968), 279; R. S. Quiroz, *JAM*, 5 (1966), 126, in the midwinter Antarctic; and F. A. Berson, *T*, 19 (1967), 161, in the spring Antarctic. On the thermal properties of a soil near the surface, see W. R. van Wijk and W. J. Derkren, *Agric. Meteorol.*, 3 (1966), 333. On the geographical distribution of the coefficient of sensible heat transfer at the Earth's surface in Japan, see T. Nishizawa, *JPC*, 3 (1966), 79. For observations of microfluxes of sensible and latent heat on the southeast coast of India, see R. J. Polavarapu, *Agric. Meteorol.*, 5 (1968), 225. For an estimate of the ratio of artificial heat generation to natural heat radiation at Sheffield, England, see A. Garnett and W. Bach, *MWR*, 93 (1965), 383.

39. A. Kochanski, *BAMS*, 37 (1956), 47.

40. J. M. Craddock and R. Ward, *MOSP*, no. 12 (1962).

41. A. Court, *BAMS*, 33 (1952), 140.

42. H. H. Lamb, *MM*, 87 (1958), 39.

43. For the theory, see R. A. McCormick, *MWR*, 86 (1958), 1. The record low temperatures are noted by N. A. Stepanova, *MWR*, 87 (1959), 145, and by D. J. George, *W*, 16, (1961), 144.

44. G. D. Hughes and R. B. Ross, *MWR*, 80 (1952), 105. H. D. Parry and C. Roe, *loc. cit.*, p. 195.

45. C. M. Lennahan and L. C. Norton, *MWR*, 78 (1950), 128.

46. J. K. McGuire and E. Sable, *MWR*, 82 (1954), 281.

47. R. M. Poulter, *W*, 9 (1954), 144. M. J. T. Norman, A. W. Kemp, and J. E. Taylor, *MM*, 86 (1957), 148.

48. G. A. DeMarrais, *BAMS*, 42 (1961), 548.

49. C. E. N. Frankcom, *MM*, 77 (1948), 281. K. Stormonth, *loc. cit.*, p. 172.

50. D. M. Houghton, *QJRMS*, 82 (1956), 515.

51. P. M. Wolff, *JAM*, 2 (1963), 430. S. L. Rosenthal, *JM*, 17 (1960), 1 and 78.

52. For maps depicting these patterns, see F. E. Lumb, *MOSP*, no. 6 (1961).

53. A. Nazarck, *BAMS*, 31 (1950), 44.

54. J. W. Smith, *JAM*, 2 (1963), 655.

55. W. C. Conover and C. J. Wentzien, *JM*, 12 (1955), 160.

56. J. S. Sawyer, *MM*, 82 (1953), 257. For a theory explaining the amplitude of the temperature variation along a latitude circle, see A. Wiin-Nielsen, L. Steinberg, and M. Drake, *JGR*, 72 (1967), 461. According to H. Geise, *BAMS*, 47 (1966), 405, horizontal fields of surface temperature can be extrapolated like surface pressure fields, as a basis for hour-by-hour local temperature forecasts. For statistical relationships between temperature anomaly patterns in Europe and western Siberia, see J. M. Craddock and R. Ward, *MOSP*, no. 12 (1962). On maximum temperatures, see L. J. Ash, *W*, 21 (1966), 459, who contests Azizia as the holder of the world-record screen temperature. At Dallol, in northeastern Ethiopia, the mean *monthly* maximum temperature exceeds 100°F in all months; see D. E. Pedgley, *MM*, 96 (1967), 265. On minimum temperatures, see: G. W. Hurst, *MM*, 95 (1966), 273, 96 (1967), 135, and 98 (1969), 39, on Thetford Chase forest, Suffolk; A. A. Harrison, *MM*, 96 (1967), 257, on a mid-Kent valley; J. Oliver, *MM*, 95 (1966), 13, on Norfolk; and J. S. Hay, *MM*, 98 (1969), 55, on night-minimum road temperatures in Britain. On frost hollows, see: D. E. Waco, *W*, 23 (1968), 456, on the Santa Monica mountains, California; and K. Smith, *MM*, 96 (1967), 300, on Houghall, Co. Durham. On urban influence, see E. N. Lawrence, *MM*, 97 (1968), 43, on Manchester, England, and *MM*, 98 (1969), 1, on London, England. On urban heat islands, see: T. J. Chandler, *MM*, 96 (1967), 244, on Leicester; C. A. Woolum, *WW*, 17 (1964), 263, on Washington, D.C.; and R. J. Hutcheon *et al.*, *BAMS*, 48 (1967), 7, on the urban heat island in Corvallis, Oregon, a small city of only 20,000 population. On sea-surface temperatures, see: F. Stearns, *JGR*, 70 (1965), 283, on the North Atlantic; J. D. Perry, *MM*, 97 (1968), 33, on O.W.S. "India" in the North Atlantic; G. A. Tunnell, *Marine Observer* (Oct. 1963), p. 192, on the Red Sea, Gulf of Aden, and Arabian Sea; C. L. Jordan, *T*, 19 (1967), 107, on the tropical Atlantic; P. Fergusson, *W*, 20 (1965), 294, on their influence on seasonal temperature change in Britain; R. C. Landis and D. F. Leipper, *JAM*, 7 (1968), 554, and J. B. Hazelworth, *JGR*, 73 (1968), 5105, on the effect of hurricanes; and I. Perlroth, *T*, 21 (1969), 230, on the effects of sea-surface temperatures on equatorial Atlantic hurricanes. See W. J. Jacob, *JGR*, 72 (1967), 1681, on numerical prediction of monthly mean sea-surface temperatures. On temporal charges in vertical temperature profiles of the lowest 300 feet after sunrise in North America, see G. A. De Marrais, *JAM*, 4 (1965), 533. For vertical lapse-rates of temperature

in the mountains of central Japan, see T. Kawamura, *JPC*, 3 (1966), 992.

57. For discussions of each frequency-distribution type, see A. Court, *JM*, 8 (1951), 367.

58. See W. A. L. Marshall, *MM*, 83 (1954), 100, for definitions.

59. For mathematical details, see R. W. Lenhard and W. A. Baum, *JM*, 11 (1954), 392.

60. J. M. Craddock and M. Grimmer, *W*, 15 (1960), 340.

61. For details, see H. C. Shellard, *MOPN*, no. 125 (1959).

62. W. A. L. Marshall, *MM*, 83 (1954), 100. For the derivation of accumulated temperatures, see H. S. Thom, *MWR*, 82 (1954), 111.

63. W. C. Spreen, *JM*, 13 (1956), 351. R. W. Longley, *QJRMS*, 84 (1958), 459.

64. For examples from temperature data for London and Glasgow, see M. H. Freeman, *MM*, 91 (1962), 227.

65. C. Bingham, *MWR*, 89 (1961), 357.

66. See T. H. Kirk, *MOGM*, no. 90 (1953), for details.

67. J. M. Craddock, *QJRMS*, 82 (1956), 275. See also K. Bender *et al.*, *GPA*, 38, no. 3 (1957), 265.

68. J. M. Craddock, *QJRMS*, 89 (1963), 461. For a spectral analysis of temperature data for North American stations, see V. M. Polowchak and H. A. Panofsky, *MWR*, 96 (1968), 596. On the geographical distribution of persistence as revealed by monthly mean data for the United States, see R. R. Dickson, *JAM*, 6 (1967), 31. For maps of standard deviation of monthly mean temperatures in the northern hemisphere, see J. M. Craddock, *MOSP*, no. 20 (1964). On the estimation of numbers of hours per day above certain temperature thresholds, see R. E. Neild, *MWR*, 95 (1967), 583. On the relation between non-seasonal temperature oscillations on the west coast of North America above and below arbitrary levels and the Gaussian normal distribution, see G. J. Roden, *JAS*, 21 (1964), 520. On the variability of accumulated temperatures in Britain, see G. W. Hurst, *MM*, 98 (1969), 78. For simple indices of winter conditions, see: K. J. Hosking, *W*, 23 (1968), 80, on mildness in the Isle of Wight; R. E. Booth, *W*, 23 (1968), 477, on severity in Britain; and C. G. Smith, *W*, 24 (1969), 23, on severity at Oxford. For the theory of normal degree days above any base, see H. C. S. Thom, *MWR*, 94 (1966), 461.

69. For cross-sections illustrating this pattern, see J. R. Gerhardt and W. E. Gordon, *JM*, 5 (1948), 197.

70. L. N. Carapiperis, *GPA*, 37, no. 3 (1956), 143.

71. D. Girdwood, *MM*, 89 (1960), 336.

72. Pao-kun Chang, *BAMS*, 30 (1949), 107.

73. P. B. Sarson, *MM*, 89 (1960), 201.

74. H. L. Griffiths, H. A. Panofsky, and I. Van der Hoven, *JM*, 13 (1956), 279. G. Reynolds, *QJRMS*,

81 (1955), 613. D. H. McIntosh, *QJRMS*, 79 (1953), 262. J. M. Craddock, *QJRMS*, 82 (1956), 275.

75. E. J. Rebman, *W*, 9 (1954), 131. R. T. Duquet, *MWR*, 91 (1963), 47.

76. I. Enger, *JM*, 16 (1959), 238.

77. G. Manley, *QJRMS*, 79 (1953), 242.

78. H. C. Willett, *CPRMS*, p. 195.

79. G. S. Callendar, *QJRMS*, 87 (1961), 1.

80. J. M. Mitchell, *USCC*, p. 161.

81. R. Fay, *JM*, 15 (1958), 467. G. S. Callendar, *QJRMS*, 81 (1955), 98. J. Glasspoole, *MM*, 84 (1955), 33. W. L. Hofmeyr and B. R. Schulze, *USCC*, p. 81.

82. R. G. Veryard, *USCC*, p. 3.

83. S. K. Pramanik and P. Jagannathan, *SPIAM*, p. 106. L. Lysgaard, *USCC*, p. 151. C. C. Wallén, *USCC*, p. 467.

84. H. E. Landsberg *et al.*, *MWR*, 91 (1963), 549. On very short-period fluctuations in temperature ("thermograph antics"), see R. E. Lautzenheiser and R. Fay, *MWR*, 92 (1964), 33, for details of jumps of up to 21°F in 2 hours at Keene, New Hampshire, and G. Froude and J. Simmonds, *MM*, 94 (1965), 184, for oscillations of 20°F in 5 minutes at Aden. On persistence of monthly mean temperatures in central England, see R. Murray, *MM*, 96 (1967), 356. For temperature trends, see: P. C. Clarke, *W*, 21 (1966), 364, on the last 90 years in Britain; S. A. Changnon, Jr., *MWR*, 92 (1964), 471, on air and soil temperatures, 1901–50, in Urbana, Illinois; E. W. Wahl, *MWR*, 96 (1968), 73, on the eastern United States, comparing the 1830's with today; and P. Jagannathan and A. A. Ramasastry, *JGR*, 69 (1964), 215, on sea-surface temperatures since 1900 in the Arabian Sea and Bay of Bengal. For singularities, see E. Newman, *JAM*, 4 (1965), 706, on Boston, Mass., and R. H. Frederick, *W*, 21 (1966), 9, who regards the January thaw in North America as an eastward-moving phenomenon. See F. Loewe, *W*, 21 (1966), 241, for evidence of a temperature seesaw between Greenland and Europe. On the theory of the diurnal variation in temperature, see: O. Lönnqvist, *T*, 14 (1962), 96, for global radiation correlation; P. J. Gierasch, *JAS*, 26 (1969), 65, and H. L. Kuo, *JAS*, 25 (1968), 682, on Earth-atmosphere thermal interaction. On diurnal temperature variation in Scottish glens, see F. H. Dight, *MM*, 96 (1967), 327. For diurnal variation in the stratosphere, see J. G. Sparrow, *JAM*, 6 (1967), 441, on the equatorial stratosphere, and F. G. Finger and R. M. McInturff, *JAS*, 25 (1968), 1116. See also N. J. Beyers and B. T. Miers, *JAS*, 22 (1965), 262, on the diurnal cycle at 30 to 60 km over White Sands, New Mexico. On seasonal temperature change, see D. C. Thompson, *QJRMS*, 95 (1969), 404, on the temporary reversal of the downward trend in monthly mean surface air temperatures in early winter at Scott Base, Antarctica, and H. C. Willett, *JAS*, 25 (1968), 341, on polar stratospheres. On annual temperature variation in the Arctic stratosphere, see A. J. Kantor, *JGR*, 71 (1966), 2445. On biennial variations in springtime temperature, see J. K. Angell and J. Korshover, *MWR*, 95 (1967), 757, on extratropical latitudes. For the synoptics of a rapid temperature change at Suffield, Alberta, see E. R. Walker, *W*, 21 (1966), 410. On mesoscale variability in temperature below 60,000 feet, see P. F. Nee, *JAM*, 5 (1966), 847. For a synoptic climatology of interdiurnal change in air temperature in Japan, see T. Asai and T. Nishizawa, *JPC*, 1 (1964), 979. On the annual temperature variation in the lower tropical stratosphere, and its relation to the northern Hadley cell, see R. J. Reed and C. L. Vlcek, *JAS*, 26 (1969), 163.

CHAPTER 6

Clouds and Climate

Given their familiarity, variety, and ubiquity, clouds are described in geographical terms with amazing infrequency. Although numerous empirical descriptions of local clouds may be found in meteorological literature,[1] almost all the geographical portrayals of cloud distribution are maps showing the frequency of different amounts of cloud (see Figure 6.1). Information concerning other aspects of cloud distribution, for example, the frequency of different types of cloud or the typical cloud cover to be expected with a given airstream, is not easy to find. In particular, there is a dearth of geographical descriptions of the cloudscapes associated with meteorological conditions in specific localities. Despite this lack of geographical work, enormous progress has been made in recent years in the physics of clouds. Many of these advances are of direct relevance to geographical study.

Topics that the geographical student of clouds needs to study are: the problems involved in cloud observation, which is often highly subjective; the taxonomy of clouds, a detailed system of classification being very necessary in view of the complexity of cloud forms; the mechanics of cloud formation, an appreciation of which facilitates cloud classification; and finally, the problem of describing the appearance of the whole sky, as distinct from individual clouds. This latter problem involves not only the question of how best to describe the dome-shaped skyscape visible from a high building or a hilltop, but also that of how to describe cloud systems covering areas of thousands of square miles, as visible from an artificial Earth satellite, for example.

January

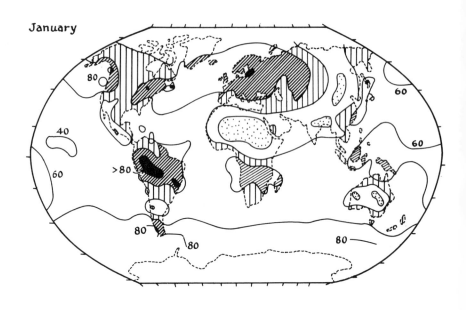

July

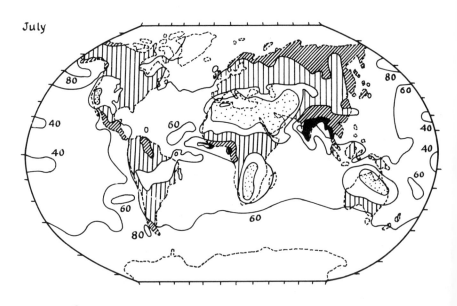

● More than 80%
◒ 60 - 80 %
◐ 40 - 60 %
○ 20 - 40 %
◉ Less than 20 %

FIGURE 6.1.
Mean cloudiness in percentage of the sky covered by cloud (after H. E. Landsberg).

A first distinction must be made between individual clouds, cloudscapes, and cloud systems. *Individual clouds* are well catered for by international classification; much is known concerning the physics of their formation and disappearance; and several standard instrumental techniques are available for their study. Although *cloudscapes* are to some extent taken into account in the international "states of the sky" classification, they have been much less studied, and the instrumental means of observing them are largely confined to experimental whole-sky photography. The geography of *cloud systems* has, of course, been inferred indirectly from synoptic observations for many years, but it is only since the advent of the American TIROS satellites that direct charting of cloud systems covering extensive areas of the Earth's surface has become possible.

Figure 6.1 presents world maps of the distribution of cloud amount, based on subjective estimates by eye of the proportion of the sky covered by cloud at a specified time. Cloud amount is now estimated in terms of eighths (*oktas*) of the sky covered; formerly, it was estimated in tenths. There appears to be a bias in reporting cloud amount.[2] Most climatic elements give a frequency distribution that is bell-shaped, so that the mean value coincides with the maximum number of observations. For cloudiness, however, the frequency histogram is U-shaped, with the extremes (clear skies, overcast skies) more frequent than the intermediate values, except in the tropics, where middle values of cloudiness are most common. The explanation is that it is easier to distinguish between one-tenth and two-tenths of the sky being covered than between four-tenths and five-tenths, because there is a greater relative difference in sky appearance between the former than between the latter. It is also easier to distinguish between eight-tenths and nine-tenths than between five-tenths and six-tenths, because the relative amount of blue sky is rapidly assessed. Therefore Figure 6.1 is probably more reliable for low and high cloud amounts than for medium cloud covers. Analysis of over 1,700 observations of cloud amount made at Greenwich many years ago showed that overcast or almost clear skies were far more frequent than half-clouded skies, and the climatological significance of the fact was regarded as doubtful.

Maps showing cloud amount may, in certain areas, give quite a good idea of the geographical variation of cloud height. For the eastern United States, this correlation appears to work well for cloud amounts of six-tenths or more, for cloud heights between 1,000 and 35,000 feet above the ground. If two parameters are used, one representing local topographical effects and the other large-scale geographical effects, the relation between cloud height (above station level) and frequency of ceilings at or below this height is linear. In general, for all regions poleward of 20°N and 20°S, mean cloudiness is a linear function of the difference between the number of cloudy and clear days in a month. Thus a mean cloudiness of 20 per cent corresponds to 22 clear days per month in January and 23 in July, and a mean cloudiness of 70 per cent to 6 clear days per month in January and 4 in July.[3]

The Study of Individual Clouds

A *cloud* is defined as a visible aggregate of minute particles of water or ice, or both, in the free air. This aggregate may also include larger particles of ice or water, and nonaqueous liquid or solid particles, for example, those present in fumes, smoke, or dust. Definition is necessary, because visible phenomena other than clouds appear suspended in the atmosphere: these are distinguished as *meteors* (see Appendix 6.1).

Description of clouds must obviously be based, initially, on their visual appearance, which is determined by three factors: first, the nature, sizes, number, and distribution in space of the particles constituting the cloud; second, the color and intensity of the light received by the cloud; third, the relative positions of the observer and the source of illumination with respect to the cloud. The most obvious visual attributes of a cloud are its brightness, color, and shape.

The *luminance* (the term adopted at the 1948 Paris meeting of the International Lighting Commission to supersede the term "brightness") of a cloud depends on the light reflected, scattered, and transmitted by its constituent particles. This light comes not only from the luminary source (the sun or the moon), but also from the sky and the solid surface of the Earth. It may be considerably modified by the time it reaches the eye if it passes through haze or various optical phenomena between the cloud and the observer. The degree to which the cloud prevents light from passing through it is the *optical thickness* of the cloud: this depends on its dimensions and physical constitution.

Clouds diffuse light of all wavelengths more or less equally, and therefore the *color* of a cloud varies with the time of day, the height of the cloud, its position relative to the observer and to the sun, and the color of the incident light. When the sun is high in the sky, clouds receiving light from the sun appear white or grey, while those receiving light mainly from the blue sky appear bluish-grey. When the sun is near the horizon, differences in cloud color indicate the relative heights of the clouds: high clouds appear almost white, and low clouds orange or red. When the sun is just below the horizon, high clouds still remain white, but middle clouds become rose-tinted and low clouds (in the shadow of the Earth) grey.

During the hours of darkness, cloud luminance is too weak to result in color vision, and so the clouds appear grey to black, except for those illuminated by the moon, which appear whitish. During weak daylight, clouds assume the color of the surface beneath them. The effect of haze between the observer and the cloud, under daylight conditions, is to modify cloud coloring: distant clouds become yellow, orange, or red.

Clearly, therefore, color cannot be used to classify clouds. *Shape* is a much more useful attribute, and the first widely known classification, published by Luke Howard in England in 1803, distinguished four basic types of cloud from their characteristic form. These were thread clouds or *cirrus*, heap clouds or *cumulus*, flat (or level) sheet clouds, or *stratus*, and rain clouds or *nimbus*. The previous year, Lamarck in France had produced a less useful classification, which did not attract much attention. Both Lamarck and Howard believed that cloud forms were not the same all over the world, but Abercromby's journeys around the world, as reported in 1887, disproved this belief.[4]

Howard believed that his four types were fundamental, and that all other observed forms of cloud could be derived from them by transition or association, but distinct new types were recognized later: *stratocumulus* by Kaemtz in 1840, medium-height clouds (*altocumulus* and *altostratus*) and the distinction between detached and continuous clouds by Renou in 1877, and thundercloud (*cumulonimbus*) by Weilbach in 1880.

In 1879, Hildebrandsson published the first cloud atlas, which included 16 photographs of clouds. Hildebrandsson, a Swede, and Abercromby, an Englishman, together collected all the available data, and published in 1887 a detailed classification of cloud forms on the basis of Howard's scheme. They attached great importance to height as a criterion in this classification: clouds were grouped into four height ranges, the mean heights being fixed from measurements made in Sweden. They believed the main value of cloud observation to be the information it provides on wind conditions at different altitudes; today, of course, it is recognized as having many other uses, as well as being valuable in itself.

The first international conference to deal with the question of cloud classification, that of the IMC at Munich in 1891, used the Hildebrandsson-Abercromby scheme, together with the atlas published as a complement to it in 1890, as the basis for an official classification. The first atlas illustrating this international classification appeared in 1895; it contained 27 sketches and photographs, including colored plates. Cloud reporting all over the world was standardized on the basis of its conventions.

New cloud types had to be introduced later, when the basic ones given in the atlas were found not to be comprehensive. The atlas was reprinted, with these modifications, in 1905 and in 1910. The International Commission for the Study of Clouds was formed in London in 1921, and, after considering the new knowledge resulting from observations of the upper surfaces of cloud from the air, and the advances in what has become known as cloud physics, decided to produce a completely new atlas of cloud types. This atlas, published in 1932, contained a more comprehensive classification of clouds, which were divided into four families, each subdividied into genera. Ten genera were recognized; twenty species of cloud forms were then described explicitly for seven of these genera.

A further edition of the 1932 International Atlas was published in 1939, but wartime observations soon revealed that it still suffered from many deficiencies and ambiguities, principally because when it was written, very few meteorologists had ever viewed clouds from above.[5] Hence in 1947, the Washington Conference of the IMO set up a small group under Viaut to prepare a new *International Atlas of Clouds and Types of Sky*. This group extended and modified the definitions of species and varieties, redefined the so-called " accidental details " described in the old atlas as supplementary features and accessory clouds, and introduced the concept of " mother clouds " to indicate the transformation processes taking place in various genera. The new atlas, in two volumes, one of which is a magnificent collection of over 200 plates, was published in 1956 by the WMO in Geneva.

Throughout the history of the observation and description of individual clouds, there have been two dominant approaches to classification: one based on the appearance of the cloud from the ground, the other on the presumed process(es) that formed the cloud. The first is still the most popular approach, for agreement on processes is by no means universal.

The current system of classification, set out in detail in volume I of the WMO *Atlas*, follows the Linnaean binomial system for classifying natural phenomena: the various cloud forms are grouped into *genera*, while *species* and *varieties* also occur. The international atlas of 1895 distinguished ten principal cloud types: cirrus, cirrocumulus, cirrostratus, altocumulus, altostratus, stratus, cumulus, stratocumulus, nimbus, and cumulonimbus. The 1956 atlas recognizes these ten types as genera, except for nimbus, which is replaced by cumulonimbus. The cloud types in the former " nimbus " class have, since 1939, been included as species within other genera.

Observation indicates that there are three basic forms of cloud everywhere on

Earth, and at all altitudes. These are (a) isolated heap clouds, which tend to develop rapidly and vertically when forming and to spread out when dissolving; (b) sheet clouds, in the form of filaments, rounded masses, or scales, usually unstable or in process of disintegrating; and (c) more or less continuous cloud sheets, which are often in process of forming or growing. Clouds are always either evolving or devolving, and appear in an infinite variety of forms. There are intermediate and transitional forms besides the basic genera and species, but they are generally recognized as being less stable than the basic types, from which they do not radically differ in appearance.

Cloud genera are defined as mutually exclusive groups. A cloud belonging to a certain genus must belong to only one species within that genus, but a certain variety of cloud may exist in several genera: in fact, an individual cloud may show characteristics of more than one variety. *Supplementary features* and *accessory clouds* are also distinguished, to make the system more complete.

A valuable feature of the 1956 international classification is that it takes into account the fact that individual clouds not only form in clear air, but may also grow out of preexisting "mother" clouds. The new extension developing from the mother cloud may or may not be of the same genus as the parent. Therefore the suffix *genitus* is added to the name of the cloud if only a part of the mother cloud changes genus, but the suffix *mutatus* is used if either the whole or the greater part of the mother cloud is involved in the transformation. (The classification system is summarized in Tables 6.1 and 6.2.)

Even this detailed system does not cover all observable types of cloud. The categories are based on the shape of the cloud, which depends on its constituent particles or droplets (water, ice, or both), its color and luminosity, and the relative positions of the observer and the source of light illuminating the cloud. The luminance of the cloud is a function of, among other features, its thickness and density, which are revealed by shading of the underside of the cloud; the darker the underside of the cloud, the thicker and denser it is. Occasional difficulties result when this relationship is used as a criterion for classification.

Cirrocumulus, for example, is defined as "thin, white patch, sheet, or layer clouds without shading, composed of very small elements . . . most of the elements having

TABLE 6.1
The international cloud classification: cloud genera.

Genus	Mother Clouds	
	Genitus	*Mutatus*
Cirrus (Ci)	Cc, Ac, Cb	Cs
Cirrocumulus (Cc)		Ci, Cs, Ac
Cirrostratus (Cs)	Cc, Cb	Ci, Cc, As
Altocumulus (Ac)	Cu, Cb	Cc, As, Ns, Sc
Altostratus (As)	Ac, Cb	Cs, Ns
Nimbostratus (Ns)	Cu, Cb	Ac, As, Sc
Stratocumulus (Sc)	As, Ns, Cu, Cb	Ac, Ns, St
Stratus (St)	Ns, Cu, Cb	Sc
Cumulus (Cu)	Ac, Sc	Sc, St
Cumulonimbus (Cb)	Ac, As, Ns, Sc, Cu	Cu

TABLE 6.2
Subtypes of cloud.

Subtype	Can belong to genus
Species	
Fibratus: fibrous	Ci, Cs
Uncinus: hooked	Ci
Spissatus: thick	Ci
Castellanus: turret-like	Ci, Cc, Ac, Sc
Floccus: woolly	Ci, Cc, Ac
Stratiformis: flattened	Cc, Ac, Sc
Nebulosus: misty, nebulous	
Lenticularis: lens-shaped	Cc, Ac, Sc
Fractus: broken	St, Cu
Humilis: small-sized	Cu
Mediocris: medium-sized	Cu
Congestus: heaped up	Cu
Calvus: bald, bared	Cb
Capillatus: hair-like	Cb
Varieties	
Intortus: twisted	Ci
Duplicatus: repeated	Ci, Cs, Ac, As, Sc
Vertebratus: vertebrae-like	Ci
Translucidus: transparent	Ac, As, Sc, St
Undulatus: wave-like	Cc, Cs, Ac, As, Sc, St
Perlucidus: opaque	Ac, Sc
Radiatus: ray-like	Ci, Ac, As, Sc, Cu
Opacus: shadowy or thick	Ac, As, Sc, St
Lacunosus: with holes or furrows	Cc, Ac, Sc
Supplementary features and accessory clouds	
Incus: anvil-like	Cb
Mamma: udder-like	Ci, Cc, Ac, As, Sc, Cb
Virga: rod or branch-like	Cc, Ac, As, Ns, Sc, Cu, Cb
Praecipitatio: precipice-like	As, Ns, Sc, St, Cu, Cb
Arcus: arch-like	Cu, Cb
Tuba: tube-like	Cu, Cb
Pileus: cap-like	Cu, Cb
Velum: sail-like	Cu, Cb
Pannus: ragged or tattered	As, Ns, Cu, Cb

an apparent width of less than one degree." Altocumulus is defined as "white or grey patch, sheet, or layer clouds generally with shading, . . . most of the small elements usually having an apparent width of between one and three degrees." The critical difference between cirrocumulus and altocumulus, apart from the question of height, is thus that of shading. If the cloud has shading on its underside, the convention is that it should be classified as altocumulus, even if the small elements composing it have an apparent width of less than one degree. The definition may be qualified by "most of," "generally," and "usually," but conventions still have to be adopted.

How, for example, should a high cloud be classified if it has elements of cirrocumulus form, and is thick enough to show shading (i.e., might be termed altocumulus), but its smallest elements change in size from an angular width of less than one degree to the north of the observer to a width of three or four degrees to the south? Ambiguities such as this are not infrequent in practical observation of clouds, and the WMO atlas describes conventions that indicate into which category the awkward cloud should be placed. Of course, no classification is wholly satisfactory if it still needs such conventions.

Cloud reports demand much practical experience with the international classification if the observer is to produce meaningful results. Even experience is not the last word, unfortunately, as was demonstrated in 1942 by an American meteorologist who showed a set of colored cloud pictures to a group of his professional colleagues, each of whom had at least three years' experience in observation. Only just over half of them agreed on the genus of each cloud, and the opinions of one-fifth of these experienced observers ranged over six additional genera for each cloud. Even the reporting of cloud height shows considerable variation: a range of estimates for a given low cloud from 3,500 to 7,000 feet has been reported, a range of 5,500 to 15,000 feet for a given middle cloud, and a range of 7,000 to 12,000 feet for a given high cloud. The bias that enters into the estimation of cloud amount is partly due to the effect of perspective. It is very difficult to see what is happening near the horizon: gaps may exist between clouds, but they will be invisible to the observer. The convention is that only visible patches of blue sky should be taken into account in estimating the fraction of the celestial dome covered by clouds, so that quite often an underestimate will result. When it is obvious that the sky is "moderately" cloudy, it seems that (unconsciously, no doubt) most observers adopt a convention of their own on whether they will report it as 3, 4, or 5 oktas.[6]

Although there is not, as yet, international agreement on a classification of clouds based on their mode of origin, consideration of such a classification is very relevant geographically, because the cloud-forming processes vary widely, not only from time

TABLE 6.3
Petterssen's cloud classification

Conditions of formation	Type of cloud
In unstable air	Cu humilis Cu congestus Cb calvus Cb capillatus Ci from Cb incus
In stable air	Fog St
In inversions with stable air	Sc Ac
In inversions with unstable air	Sc castellanus Sc cumulogenitus Cu undulatus Ac castellanus
In fronts	As (if upglide is stable) Cb arcus (if upglide is unstable)

TABLE 6.4
Ludlam's cloud classification

Type of vertical motion	Type of cloud
Orographic disturbances	Lenticularis Undulatus
Widespread irregular stirring	Fog St Sc
Widespread regular ascent	Ci Cs As Ac Ns
Local penetrative convection	Cu Cb

to time but also from place to place. Two useful genetic classifications of clouds have been developed by Petterssen and Ludlam. Both recognize four main types of cloud.[7] Petterssen distinguishes between clouds forming in unstable and stable air masses, clouds forming in inversions, and frontal clouds (see Table 6.3). Ludlam classifies clouds according to the kind of air movement producing them, this method resulting in three types of layer cloud and clouds formed by convection (see Table 6.4). Familiarity with the WMO system is still an essential preliminary before their schemes can be usefully applied.

The Study of the State of the Whole Sky

The detailed analysis of individual clouds should follow, not precede, observation of the state of the sky as a whole. We discussed individual clouds first in order to be able to describe the state of the sky with economy. The appearance of the sky is continually changing, and many transitional types of cloud appear. One must have a classification of sky types, because whereas individual clouds are usually migrant forms, soon passing out of sight of a fixed observer, the evolution of the state of the whole sky can readily be followed at a single station.

The 1932 atlas contained an international classification of states of the sky, based on both the physical processes involved in cloud formation and synoptic study of atmospheric disturbances. The observation of states of the sky and their evolution provides a key to future weather developments as well as to the existing distribution of weather in an extensive area. By observing the state of the sky, for example, one can determine in what sector of a depression the station is situated, which not only gives the observer an idea of the instantaneous geographical pattern of weather in adjacent regions, but also enables him to practise "single-station forecasting."[8]

Upper, middle, and lower skies are recognized. The state of the upper and middle skies is determined by thermodynamic processes dominant throughout a large section of the atmosphere. The state of the lower skies is determined mainly by conditions

Classification of Skies for Synoptic Purposes

A. *Upper and middle skies*

    1. Emissary sky
    2. Front sky, typical
    3. Front sky, weak
    4. Lateral sky
    5. Central sky, typical
    6. Central sky, weak
    7. Rear sky
    8. Prethunderstorm sky
    9. Thundery sky

B. *Lower skies*

  a. Simple skies
    i. Convection sky. 1. Fine-weather cumuliform sky
                        2. Disturbed cumuliform sky (without Cb)
                        3. Disturbed cumuliform sky (with Cb)
    ii. Turbulence sky. 4. Stratiform sky (St or Sc)
                        5. Amorphous sky (Fracto St or Fracto Cu)
  b. Mixed skies
    i. Fine-weather cumuliform sky combined with stratiform sky
    ii. Disturbed cumuliform sky combined with stratiform sky
    iii. Disturbed cumuliform sky combined with amorphous sky

in the lowest layer of the atmosphere, which are directly influenced by the surface of the Earth: thus the clouds characterizing lower skies are formed by convection, radiational cooling, or turbulence.

The upper and middle skies characterize major synoptic disturbances, and are divided into zones: front, central, rear, and lateral. The "emissary sky" forms the first indication of the presence of a disturbance at a distance. The "rear sky" is one of unstable weather, in which short sunny periods of exceptional visibility alternate with threatening, disturbed skies accompanied by showers and squalls; cloud patterns are often complicated, because of the remains of dense, relatively low, high or middle clouds.

For tropical skies, the above classification is unsuitable. In these areas, a new classification suggested by Alaka employs 17 specifications. The first five categories describe types of undisturbed trade-wind skies; the next five describe trade-wind disturbances; and the final seven groups are reserved for strong disturbances. Further refinements, based on wind shear (as revealed by the trade cumuli), marked organization of the clouds into rows, the presence of distant disturbances, and the degree of intensity of cumulus activity, may be denoted by subscripts to the code figures used to indicate the specifications.[9]

Synoptic Cloud Reporting

The international codes employed for synoptic and climatological cloud reporting represent a compromise between the recognition of individual clouds and the description of states of the sky. The 1956 WMO classification recognizes more than 120 distinct species or varieties of individual clouds; the synoptic codes allows the reporting of clouds in only 24 classes. The original system of radio transmission adopted for the codes allowed each element to be assigned one of the figures 0 to 9. Clouds were recognized to be low, middle, or high; since the code figure 0 had obviously to be reserved for reporting "no cloud" at that altitude, only nine categories remained for each element.

Since cloud observations every six hours are published in *Daily Weather Reports* in these codes, a vast amount of material is available for the geographical study of clouds. Very little of this material has been mapped. For example, the Meteorological Office's *Climatological Atlas of the British Isles* does not give maps of cloud distribution, stating that it is not feasible to do so. Instead, it gives cloud data for 44 stations in tabular form. The data is from instrumental observations of the height of cloud bases made every hour for a period of several years during the decade 1940 to 1950. The data show that low cloud was more frequent in winter than in summer, and more common inland in the early hours of the morning than in the early afternoon, except in winter.

Although climatological maps of synoptic cloud data are not very common, the figures are often used for a particular type of map in *nephanalysis* (see Figure 6.2). This type of analysis is very useful in polar and oceanic regions, or anywhere dense networks of stations do not exist. It involves plotting clouds and hydrometers on a map, on which synoptic entities termed nephsystems are then identified. *Nephsystems* are large areas of cloud such that, if an itinerary is traced within one system, all the cloud forms successively encountered constitute a continuous succession; i.e., the species and genera encountered are related by a series of intermediate forms. The nephsystem thus forms an organized society of clouds: a synoptic continuum.

Nephsystems can be divided into sectors characterized by different kinds of weather. A nephsystem may be surrounded by clear skies, by cumulus humulis, or by cirriform clouds. The basic cloud of the nephsystem is altostratus, to which almost all the other cloud forms found in it can be related via a continuous succession of intermediate forms. Altostratus is usually associated with warm fronts, but not necessarily in a nephsystem.

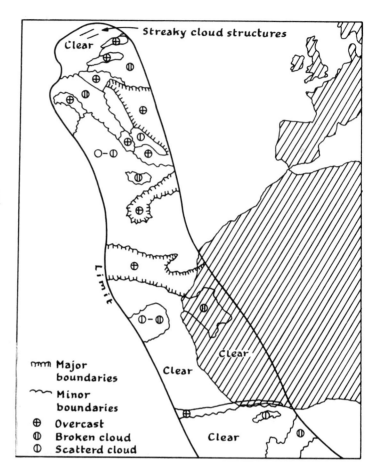

FIGURE 6.2.
Nephanalysis based on
photograph taken on the 118th
orbit of TIROS on December, 1
1960.

Various isopleths may be used to define nephsystems; for example, isolines of a particular ceiling amount, or isolines separating zero cloudiness from one okta, or nil precipitation from some precipitation, or isolines separating stable from unstable areas. Borderline curves separating stations reporting altocumulus from those reporting no altocumulus are also plotted, as are weakness curves, which mark the lines of weakness that separate two nephsystems not otherwise separated by an area of fair weather. A nephsystem usually splits into two or more adjacent systems along a line of weakness. Such lines are marked by the total cloud cover decreasing, or by the sky remaining overcast but the ceiling rising along a linear zone, or by the distance between neighboring borderlines decreasing. The presence of stratocumulus is often considered to indicate the boundary between adjacent nephsystems.

One reason there has been little climatological treatment of synoptic cloud observations is the subjectivity of such observations. The same cloud is often classified differently by adjacent stations, and often the time available for studying the evolution of the sky is very limited. Objective, instrumental records are much to be preferred. Instruments are available that estimate cloud heights and speeds of movement moderately accurately, and that measure cloud ceilings with good accuracy, given certain assumptions. For recording individual clouds and states of the sky, photographic techniques are valuable.

Cloud height is measured above ground level in the vicinity of the observer. For middle and high clouds, correct visual estimating of height can only be learned by experience: by checking one's own estimates against reports from aircraft that have flown through the cloud in question, or by using mountains or hills to provide a convenient vertical scale. For low clouds, searchlights and pilot balloons provide more accurate measures than visual estimation.

The horizontal extent of cloud cover is usually estimated using a conical cardboard tube whose length is 1.07 times the diameter of its large end. The small end of the tube is covered by a disc, which has a peephole in its center. This aperture subtends a solid angle at the eye equal to one-eighth of a hemisphere.

The determination of the exact location of the base of a low cloud is not easy when the cloud is vertically overhead. According to an IMO convention, when visibility is poor, cloud base is to be taken as the lowest zone in which the type of obscuration perceptibly changes from that corresponding to clear air or haze to that corresponding to water droplets or ice crystals. This zone may be determined optically, because in the air below the cloud the obscuring particles show spectral selectivity, whereas in the cloud, owing to the different droplet size, there is virtually no spectral selectivity.

The height of low clouds may be determined approximately by using *ceiling balloons*, i.e., small pilot balloons that rise at a constant rate. The balloons are followed with binoculars, telescope, or theodolite, and timed with a stopwatch; the cloud base is assumed to be at the height where a balloon appears to enter a misty layer just before finally disappearing. A 4-gram balloon, which rises at the rate of 300 feet per min if filled with hydrogen to a diameter of 10 inches, is used for daylight observation of cloud heights up to 1,000 feet in light winds. A 10-gram balloon, rising at 400 feet per min, is used for daylight observations at other times, and a 20-gram balloon, rising at 500 feet per min, to which lantern and candles are attached, is used for night observations. If the balloon develops a leak, or if there are vertical air currents, the estimation of cloud height may be in error by as much as 20 per cent.

Searchlights enable cloud height to be estimated with more accuracy. At night, an ordinary searchlight with a 250- or 500-watt lamp is used in conjunction with a clinometer or alidade. A 36-watt car headlamp is suitable for measurements of cloud height up to 3,000 feet above the ground. The searchlight projects a narrow beam of parallel light in a known direction to form a spot on the underside of the cloud. Since at night the spot is visible, its angle of elevation is easily determined using the clinometer or alidade from the other end of a baseline (usually 1,000 feet long) measured from the searchlight. During daylight, a modulated searchlight beam is used, in conjunction with a photoelectric cell, at the focus of a lens or mirror, which acts as a detector telescope. The instrument is sometimes known as a *nephohypsometer*. The photocell output is fed to an amplifier tuned to the same frequency as that of the modulated beam. The output of the amplifier reaches a maximum when the telescope is pointed at the spot. The accuracy of the measurement of cloud height is limited by the fact that the larger the angle of elevation is, the less accurately it can be measured. The error in height for an error of 1° in the determination of this angle varies from

Synoptic Categories for Reporting Clouds

1. Low Clouds

CODE FIGURE

1	Small, flattened Cu
2	Strongly developed Cu
3	Cb
4	Sc formed by spreading out of Cu
5	Sc not formed by spreading out of Cu
6	St in continuous sheet
7	Fractostratus—bad weather, below As or Ns
8	Cu and St not formed by spreading out of Cu
9	Cb with anvil

2. Medium Clouds

CODE FIGURE

1	As, semitransparent, sun or moon weakly visible
2	Thick As, sun or moon invisible
3	Ac at single level, semitransparent
4	Ac at several levels
5	Ac in bands or continuous sheet(s), thick and opaque
6	Ac formed by spreading out of Cu
7	Ac not progrssively invading sky
8	Ac with tufts or "battlements"
9	Ac in several layers in chaotic sky, dense Ci usually present

3. High Clouds

CODE FIGURE

1	Ci not progressively invading sky (mare's tails, tufts, hooks)
2	Dense Ci
3	Ci in form of anvil, remains of upper part of Cb, or part of distant Cb whose cumuliform portions are invisible
4	Ci progressively invading sky (filaments, hooks, etc.)
5	Ci or Cs in converging bands, increasing and growing denser, but continuous veil does not reach 45° above horizon
6	Same, but with veil, more than 45° above horizon
7	Veil of Cs completely covering sky
8	Cs not completely covering sky, and not progressively invading sky
9	Cc alone, or Cc with Ci or Cs

Synoptic Categories for Reporting Clouds (Continued)

4. *Cloud Height*

CODE FIGURE	*Height of base cloud in meters*
0	0–50
1	50–100
2	100–200
3	200–300
4	300–600
5	600–1,000
6	1,000–1,500
7	1,500–2,000
8	2,000–2,500
9	Above 2,500
/	Height of base of cloud not known, or base of clouds lower and tops higher than station

5. *Cloud Amount*

CODE FIGURE	
0	None
1	One-eighth of sky covered or less, but not zero
2	Two-eighths of sky covered
3	Three-eighths of sky covered
4	Four-eighths of sky covered
5	Five-eighths of sky covered
6	Six-eighths of sky covered
7	Seven-eighths of sky covered or more, but not eight-eighths
8	Eight-eighths (sky completely covered)
9	Sky obscured or cloud amount cannot be estimated

35 feet for a ceiling of 1,000 feet to 450 feet for a ceiling of 5,000 feet, assuming a baseline 1,000 feet long. (On this paragraph, see Appendix 6.2.)

Simultaneous observations from two or more points by means of *theodolites* enable the absolute speed and direction of movement of clouds to be measured. If, as is usually the case, observations must be made from a single point, then the height of the cloud must be known before its true rate of movement can be found. Single-station observations are based on the principle that the angular velocity of a cloud about a point on the Earth's surface vertically beneath the cloud can be found from the velocity-height ratio (see Appendix 6.3), even if the observer is not vertically beneath the cloud. The observations are usually made with nephoscopes, of which there are two types.

The *comb nephoscope*, introduced by Besson in 1897, is a direct-vision fixed instrument. It consists of a vertical brass spindle seven to eight feet long, with a 3-foot long crosspiece to which are attached seven equidistant vertical prongs, fixed to its upper end. The spindle is mounted on a vertical post at least five feet high, so that it can be rotated to bring any particular cloud into line with the prongs. The instrument is mounted in an open position with as complete a view of the sky as possible.

In use, the instrument is first adjusted to the observer's eye level, and then a 12-foot loop of cord attached to the spindle is used to rotate it until the selected cloud appears to move along the tips of the prongs in the direction of a metal arrow. The initial sighting must be on the central prong. A compass plate attached to the spindle then indicates the direction of cloud movement, and the speed of movement can be found by timing the cloud from one prong to the next (see Appendix 6.4). The instrument is a little tricky to use, since the observer's head must be kept perfectly still during an observation.

The *reflecting nephoscope*, devised by Fineman, is a singular, portable instrument. It consists of a black glass disc, forming the reflector, on which are engraved two concentric circles. The disc is mounted horizontally on a tripod fitted with leveling screws, and is surrounded by a circular brass ring engraved with a 360° scale. A compass needle is mounted beneath the glass disc, and a movable vertical pointer, carrying a scale of millimeters, is attached to the perimeter of the mirror. The disc assembly can be rotated about a vertical axis through its center.

The image of a prominent part of the cloud is made to appear in the center of the mirror by the observer's placing himself accordingly. The pointer is adjusted so that its tip also appears in the center of the mirror, and the observer then keeps the images of cloud and pointer tip in coincidence by moving his head. The point on the circumference of the disc at which the cloud appears to leave the mirror then gives the direction from which the cloud is moving. The velocity-height ratio is found from the time required for the cloud image to travel from the inner engraved circle to the outer circle, and the height of the pointer tip above the mirror (see Appendix 6.5).

Nephoscopes yield only the relative speed of cloud movement, and their use is restricted to high, and occasionally medium, clouds. If the actual height of the cloud is known, then its actual rate of movement can be calculated.

Pilot-balloon theodolites may also be used as nephoscopes. A sharply defined part of a cloud is held in view in the telescope for one minute, and its azimuth and elevation found for several instants during this time. Assuming a constant cloud-base height of one nautical mile (6,080 feet), the computed speed of movement in knots will then numerically equal the angular velocity of the cloud in radians per hour (see Appendix 6.6).[10]

Cloud Photography

Although cloud photography obviously has an artistic appeal, it can be an important scientific technique, particularly whole-sky photography. Photographs taken at or near the main synoptic hours provide a natural history of the sky that may reveal new synoptic associations.*

Two photographs should be taken at each observing hour, in opposite directions, with the sun to one side of the observer; the horizon should be included, also a tree, building, pole or chimney to indicate the size of the clouds. It is usual to emphasize high and medium clouds, which are of general meteorological importance, rather than low clouds, which are often of only local significance. If there are several cloud layers, photograph that part of the sky in which clouds are visible at several levels. Given cloud sheets, photographs of a break in the sheet will reveal details of its structure.

The films employed in cloud photography should have low sensitivity but high contrast, the effective sensitivity being increased by use of a fine-grain developer. Fine-grain panchromatic film is usually used, the speed selected depending on the conditions under which the observer is working. Miniature cameras with fast lenses are convenient, particularly for color transparencies, and interchangeable lenses enable cloud details to be emphasized.

Photographic films are all somewhat sensitive to blue light; unless a filter is used to absorb this light, the blue sky will appear white in a black-and-white print, and the normal contrast of cloud against sky will be lost. Red filters are useful for grey skies or for bright white clouds against a dark background; yellow filters for blue or broken skies, or for dark clouds against a bright background; and dark yellow or orange filters for intermediate skies, or for photography above 6,000 feet. Exposures are normally only one-tenth or so of those required for distant landscapes under the same illumination. Small apertures are required for cloud photographs from the ground, because of the relatively high brightness of clouds, but larger apertures may be necessary for shots from aircraft.

The development of *cloud photogrammetry* has opened up a whole new era of cloud study. If a single camera is used, the dimensions of a cloud can be calculated from its photograph, knowing the distance of the cloud and the focal length of the lens (see Appendix 6.7). If two cameras are used simultaneously at the two ends of an accurately measured baseline, the resulting stereoscopic pair of photographs enables the distance of the cloud to be calculated and its contours to be traced. Stereoscopic pairs taken at known intervals from the same baseline enable the velocity and direction of movement of the cloud to be found.

Two photographs of the same cloud taken from stations one mile apart make it possible to measure the height of the cloud. Two successive photos taken from the same aircraft, the camera pointing laterally at the same cloud formation for both, make fairly accurate quantitative estimates feasible if the time between exposures is carefully assessed, since the speed of the aircraft is as important as the distance of the cloud. For low clouds, photos taken one mile apart should be suitable, but for medium or high clouds much greater distances are necessary.

---

* For general introductions to the methods of cloud photography, see: *HM*, p. 912; *MOHSI*, p. 398; H. Baines and S. C. Goddard, *W*, 9 (1954), 37; M. Koldovsky, *W*, 13 (1958), 268. For examples of whole-sky photograms at Columbia, Missouri, see L. O. Pochop and M. D. Shanklin, *WW*, 19 (1966), 198. For examples of a continuous photographic record of daytime clouds in Hawaii, see L. E. Eber, *T*, 9 (1957), 581.

Terrestrial photogrammetry, employing horizontally mounted K-17 aerial cameras with 74° fields of vision and taking stereopairs at regular intervals, enables individual cloud features to be determined with an accuracy of between 100 and 300 feet in height and 0.1 to 0.7 miles in range at a distance of 13 miles. By this means, cloud heights can be measured up to 40,000 feet at 10 miles, and up to above cumulo-nimbus height at 30 miles.[11] Two plotting methods have been employed. A graphic method of radial triangulation requires the measurement, from each photograph in the stereopair, of two distances, or one distance and one angle, or two angles; from this the height and range of individual parts of a cloud can be determined. A method that requires much practice allows the outlines of a cloud to be traced using a radial plotter: stereopairs taken every few seconds then indicate the effect of vertical veloci-ties and accelerations in the cloud as its size changes. (See Appendix 6.8.)

Time-lapse cinephotography provides really valuable information on the mode of formation and dissipation of clouds, as well as a means by which the sequence of sky changes at one station during, say, one day, can be demonstrated in a few minutes. The critical question is the selection of a suitable exposure time. One exposure per minute is quite adequate to show the life cycle of clouds. One exposure every 10 seconds enables the complete sequence of cloud changes during the passage of a trough, a ridge, or a small depression to be viewed in detail if projected in 3 minutes at 24 frames per sec. One exposure every 15 seconds, projected at 16 frames per sec, will record in detail the motion of all but very low clouds. (See Appendix 6.9.)

The disadvantage with normal still or movie cameras for cloud photography is that their field of view is comparatively small. Commercially made wide-angle pan-oramic cameras are available, but actually it is photographs of the whole sky, i.e., covering the complete sky hemisphere, that are required, and much experimental work has been devoted to this problem (see Appendix 6.10).

Ordinary cameras in effect employ the *gnomonic projection* to form an image. For photography of the hemispheric sky, the *orthographic projection* is more suitable. In the gnomonic projection, if a light source is imagined at the center of a transparent sphere (one half of which represents the sky), projecting on to a plane tangent to the sphere, all "great circles" plotted on the sphere will be represented on the plane sur-face as straight lines, but "small circles" will be represented as hyperbolas. The radial scale rapidly increases away from the point of contact of sphere and plane surface, and thus a photograph of the projected image will show rapidly increasing distortion with distance from its center, assuming this is the point of contact. The horizon, assuming an 180° vertical field of view, cannot be projected onto the plane sur-face.

With an orthographic image, small circles are represented as straight lines and great circles as ellipses of different eccentricity. The radial scale diminishes with in-creasing angular distance from the point of contact of sphere and plane surface, and becomes zero when the angle reaches 90°. If an orthographic image of the whole sky is projected onto a flat screen, it will give the impression of a sphere, if viewed from a distance. Such an image can be projected by a standard movie projector onto the exterior of a small spherical screen; a miniature atmospherium then results (see Appendix 6.11). If a small theodolite is also mounted at the center of curvature of the screen, the angles and shapes of the clouds in the projected photograph can then be correctly measured. If the camera is rotated about a vertical axis, enabling it to completely cover three-dimensional space, the vertical dimensions of the objects represented do not change as they traverse the field of view, and the horizon cuts through the center of the field.

Although what is desirable can thus be easily established in theory, the practical problems are difficult; in particular, how does one arrange the optical system of the camera so that light can be collected over a solid angle of one hemisphere? One of the first attempts, Hill's cloud camera of 1923, used a specially constructed lens in a camera pointing vertically upward.

Georgi in 1933 employed a downward-pointing camera, supported 50 cm above a glass hemisphere silvered inside. In the resulting photographs of the image seen on the hemisphere, the altitude-angle is transformed almost perfectly, at least from the horizon to 60°, but objects near the horizon are extended horizontally. To determine how angles of equidistant elevation in the sky will appear on a photograph, the apparatus is mounted horizontally and the horizon photographed. Details of the objects appearing on the photo are measured in the field with a theodolite sited at the camera location, and thus an empirical correction diagram is compiled, enabling the angles of elevation and azimuth as measured from future photos to be corrected.[12]

In Georgi's portable sky mirror, the camera is mounted below a large spherical mirror, and light rays from the object are reflected into the lens via a small plane mirror fixed above it (see Figure 6.3). Reflections are doubled, but photos can be taken with the lens fully open, so that skies of low luminosity can be photographed, and short exposures or strong filters used.

Condit reflected the image from a silvered hemispherical bowl, via an inclined plane mirror mounted vertically above the bowl, into a movie camera with a 2.5-inch lens specially treated to reduce flare and ghosting. The camera was operated continuously from sunrise to sunset, at a speed that could vary from 1 to 20 frames per min; at the former setting, skies for a complete day could be viewed in 45 seconds, at the latter in 30 minutes. Clouds near the center of the field of vision appeared to move faster than those near the edge, because the radial scale is greater in the center of the mirror than near its periphery; this effect is difficult to allow for accurately in viewing the films.

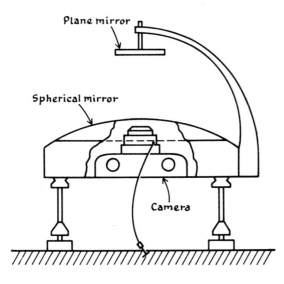

Plane mirror

Spherical mirror

Camera

FIGURE 6.3.
Photographic sky mirror (after J. Georgi).

Von Arx has experimented with both refracting and reflecting systems.[13] The principle of *reflecting systems* is that the field of view of the camera is expanded by the use of reflectors. Although the camera photographs itself, it appears very tiny on the resulting photograph. Convex spherical mirrors with 180° fields of vision have been constructed, as have concave spherical mirrors with 140° fields of vision and paraboloid mirrors with 240° fields of vision. In the latter a field of vision of 300° is theoretically possible, but commercial-grade mirror surfaces for this type of paraboloid are very expensive. One parabolic reflector used by von Arx covers a field of vision of 270° in nearly equidistant projection. The system consists of a glass searchlight reflector 16 inches in diameter, with a focal length of 3.75 inches, silvered and coated on its concave side to form a second surface mirror. This is mounted 30 inches (i.e., eight focal lengths) beneath a movie camera fitted with a 15-mm lens. Recording at 4 frames per minute and projecting at 16 frames per sec enables cloud sequences for one whole day to be photographed with 100 feet of film and viewed in just over 4 minutes. Cloud sequences may also be photographed at night, using an exposure of one frame per minute and setting the shutter so that it remains open between film shift cycles.

The strong curvature of the field of view that would otherwise result can be eliminated by the use of compound reflectors placed ahead of the camera. These consist of a convex paraboloidal primary reflector, 16 inches in diameter and with a 3.75 inch focal length, and a convex spheroidal secondary mirror, 4 inches in diameter and with a 14-inch radius of curvature, mounted 7 inches above the vertex of the primary. The latter is perforated to allow the camera to be fixed below the primary reflector, with a full view of the secondary. The obstruction due to the latter covers less than 1 per cent of the total field of vision. The curvature of the visual field produced by the primary mirror is nearly canceled by the effect of the secondary.

The principle of the *refracting system* involves photographing of the small, uniformly bright image that is formed by viewing the sky backwards through a binocular or a wide field telescope. This image is formed at infinity, because the objective lens acts as a collimator; thus it can be photographed by an ordinary camera with lens set at infinity, using the same exposure as if the telescope were absent. The field of view of the telescope or binocular is enlarged by placing the curved surface of a short-focus planoconvex lens against its eyepiece, which of course faces the sky.

In one refracting system developed by von Arx, a M-71H gunsight-telescope is used. The eyepiece has a 65° field of vision, which is expanded by placing a 65-mm diameter block of low dispersion glass some 60 mm ahead of it. A movie camera with a 50-mm lens placed at the objective lens of the telescope then gives a field of vision of almost 180° with very little internal reflection or lateral chromatic aberration. This system is particularly suitable for cloud photography from aircraft.

If the camera is pointed directly ahead from an aircraft flying above 2,000 feet with an airspeed of 150 knots, a speed of one frame per second results in smooth relative motion in the foreground on projection. If the camera is directed at right angles to the path of the aircraft, time-lapse photography results in a succession of stereoscopic pairs of photos. If the time-lapse is one second or more, so that the resulting stereo baseline is longer than the aircraft, the three-dimensional forms of clouds from 5 to 15 miles distant can be recognized. For viewing such movie shots, either a rotating polarized disc is placed just in front of the projector lens, or a special shutter is fitted to the projector; this shutter is a rotary type with a single clear sector 30° wide, coupled in correct phase to the shutter shaft of the projector by means of a flexible shaft.

In Australia, a cloud-photography device has been developed which can be carried by standard constant-level balloons. The device consists of an inverted whole-sky camera with a focal length of 7.7 cm, mounted 25 cm below the center of a hemispheric mirror. The camera shutter is mechanically operated by an alarm clock, and the camera photographs the face of the clock and the reading of a magnetic compass as well as the required cloudscape. A sheet of cardboard fixed behind the mirror shields the camera lens from both direct sunlight and scattered light from the balloon. A valve keeps the balloon at a constant height (95,000 feet for a 19-foot balloon). After the photographs have been taken, the camera is dropped by parachute, displaying a target of metalized paper for radar tracking. This technique produces photos of the clouds from above, showing cloud cover and the exposed terrain within a solid angle of about 2π steradians. A small area vertically below the camera is obscured by it. Radial distances on the photos can be measured correctly from a knowledge of the geometry of the mirror (see Appendix 6.12).

A logical extension of this technique is the photography of the cloud cover over much of the Earth's surface from an artificial satellite. Since the advent of the TIROS and later satellites, many thousands of photographs of cloud patterns have been made available for climatological studies, quite apart from the routine transmission of satellite nephanalyses for synoptic forecasting. These photographs are of obvious value for those parts of the world, such as the tropics and the oceans, in which no observations are made regularly, and numerous published studies describe cloud patterns in these regions that otherwise would be completely unknown. But satellite photographs of clouds have many other uses. For example, TIROS nephanalyses show that characteristically the Earth's cloud cover has a "streaming" appearance that enables the broad features of the airflow at 500 mb to be determined by kinematic analysis. The cloud images on the photograph may be used as indicators of the general pattern of vertical motion, and thus the state of the atmosphere at a given time may be known in three dimensions; this is particularly valuable for providing a realistic picture of the "initial state" when a numerical prediction model is to be applied.[14]

Ordinary photographs of the Earth's cloud cover can be taken from an artificial satellite only during the daytime, but nighttime "pictures" may be obtained indirectly from radiation measurements. Measurements of infrared radiation in the 10-micron water-vapor "window," made from TIROS II, enabled the distribution of large-scale systems of middle and dense high cloud overcasts to be mapped during the night, and the height of the cloud tops to be estimated (see Figure 6.4). Preliminary examination of the observations revealed a marked contrast between the northern and southern hemispheres: in the latter, the percentage of cloudiness at night appears to be considerably higher than that during the day, whereas in the northern hemisphere the sky is cloudier during the day than during the night. Earlier analysis of conventional cloud data had indicated that there is no diurnal variation in cloudiness over the open sea, and that any apparent difference between daytime and nighttime cloud covers must be caused by adjacent land masses.[15]

An early purpose of TIROS photographs was to see if the actual distribution of clouds in an extratropical depression corroborates the Bjerknes-Bergeron models. As a result of the TIROS observations, a new series of model cloud distributions has

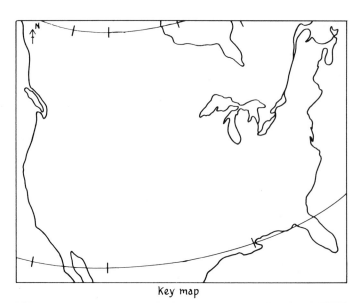

Key map

FIGURE 6.4.
Heights of tops of highest cloud layers over the United States, at 6:00 P.M. G.M.T., November 23, 1960. Units are in thousands of feet. (After S. Fritz and J. S. Winston.)

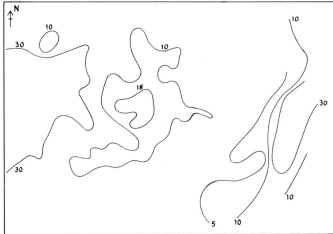

Heights based on conventional pilot reports.

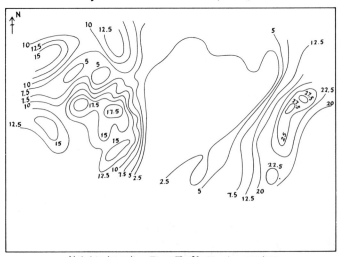

Heights based on Tiros II effective temperature (i.e., radiation) data.

been developed (see Figure 6.5). By applying this sequence of visible cloud patterns, much additional information may be obtained, for example, the location of the low-pressure center, the positions of the fronts, the precipitation distribution, and estimates of surface and upper winds and airmass conditions. A combination of satellite photographs and radar precipitation echoes is particularly useful in the study of depression precipitation. In a large extratropical maritime cyclone associated with a strong blocking pattern, very few of the clouds in the system were precipitating at any one time; the precipitation was mainly showers, and mainly within cellular-type convection clouds. The showers developed in the same fashion as the clouds, and on the average persisted much longer (two hours, at least) than ordinary airmass showers over a continent.[16]

The dominant feature of cyclone clouds as observed from satellites is their spiral patterns. Comparisons between the nephanalyses and the appropriate synoptic charts show that advection can be a dominant mechanism in the formation of these spiral patterns. An unexpected feature is that cloud patterns indicative of cellular convection are frequently observed to be associated with cyclones. In fact, cellular convection is found on a larger scale and over much larger areas of the Earth's surface than had been realized previously. Cumuliform clouds over the open oceans are observed in complex patterns that may be resolved by power-spectrum analysis into lines and cells whose horizontal dimensions are on the order of 20 to 100 miles. Cumulonimbus clouds may usually be distinguished from other cumuliform clouds on the satellite photographs. They are much larger, being 100 to 200 miles in extent, and present uniform, unbroken blobs of brightness that usually have sharply defined scalloped edges. The cirrus anvils in thunderhead cumulonimbi may be used as approximate indicators of wind shear: the longer the anvil, as measured on the photograph, the stronger the vertical shear.[17]

Shadows cast by the clouds are visible in satellite photographs, the apparent size of the shadows being a function of the attitude of the satellite, the sun angle, and the height of the cloud above the surface on which the shadow is cast. These relationships allow jetstreams to be detected under certain conditions, e.g., when the northern edge of the cirrus sheet (which is usually situated immediately south of the jet axis) may be identified by the shadow it casts on the underlying surface. In fact, TIROS photographs have proved to be very useful in sensing anomalies in the general atmospheric circulation.[18]

Satellite photographs are a convenient way to detect the presence of lee waves. For example, parallel cloud bands observed in the lee of the Andes proved to be compatible with theoretically predicted lee waves. Other photographs have shown the presence of standing long waves due to gravity in the wake of small islands in the eastern Pacific; these waves gave rise to low clouds, in particular, to a 60-km wide band of disturbed, diffuse stratus that extended at least 385 km downwind from an island 1,600 m above sea level.[19]

Cloud streets, i.e., long lines of convective clouds of a single type, are very often revealed in satellite and aircraft photographs. These may be lines of cumulus elements that extend between 10–100 miles, particularly in the trade-wind airstreams or in polar outbreaks, or they may be cirrus bands 400 miles or more in length, which radiate from one point on the horizon, stretch across the entire sky, and converge toward a point on the opposite horizon. These cirrus bands were termed *polar bands* by Humboldt, who believed they were oriented in the direction of the Earth's magnetic field, although in fact they are associated with jetstreams and run in the

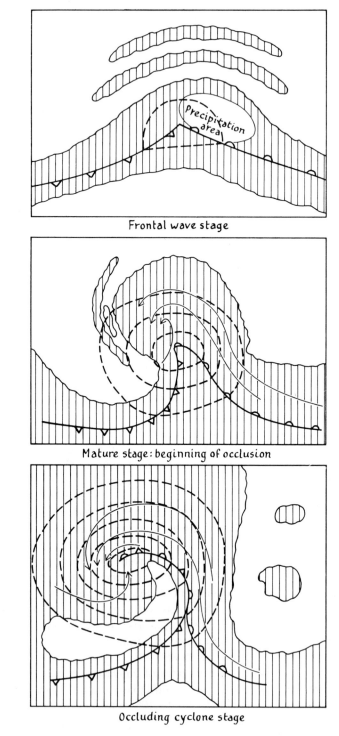

Frontal wave stage

Mature stage: beginning of occlusion

Occluding cyclone stage

FIGURE 6.5.
Model of cloud patterns associated with an extra tropical frontal depression (satellite photos; after R. J. Boucher and R. J. Newcomb). Clouded zone indicated by shading.

direction of maximum wind. Satellite photographs reveal numerous varieties of cloud street, and emphasize the typically "banded" structure of many atmospheric phenomena that is caused by the characteristic "streakiness" of the atmosphere. Cloud streets show definite geographical preferences—for example, they are the normal form of convection in Finland and northern Scandinavia generally—but they are dynamic features, and may form quite independently of orography or of the thermal properties of the ground surface, even though they are usually moving downwind from particularly effective thermal sources. According to Kuettner, cloud streets are associated with a "nose" profile of windspeed—i.e., a vertical profile in which the shear of the windspeed changes sign with height—and should be parallel to the wind direction. According to Haurwitz, cloud streets (and billow clouds) are associated with internal waves within the atmosphere: they are composed of polygonal cells, form at an internal surface that bounds a density discontinuity, and should form at right angles to the wind shear across the boundary. Satellite photographs show both types of cloud street. A further type of cloud street may be caused by stationary ridges on an inversion over a mass of moist air. An inversion or frontal surface over moist air in which there is a geostrophic wind balance must be curved, so that a steady-state distribution is possible in which a discontinuity takes the shape of the surface.[20]

Cloud lines, of course, may often be observed at the Earth's surface without recourse to satellite photographs, and often prove to be associated with specific localities or geographical structures. The Hawaiian Islands provide a good example. Here land-breeze and sea-breeze circulations meet and interact with the prevailing trade-wind regimes to produce very distinctive types of cloud line, which not only provide very localized precipitation but also form often enough to give rise to distinct local climates. Four types of interaction between land and sea breezes and the trades are possible, depending on the height and size of the orographic barrier, and each produces a special type of local weather. The *Lanai type*, typical of the low-lying islands, occurs when the trades meet the sea breeze in a line perpendicular to the direction of the trades, the line being marked by cumulus. The *Maui type* is characteristic of islands with a mountain mass large enough to split the trade wind, but whose lee area is not large enough for convective showers to develop in it from normal trade-wind cumuli, i.e., independently of the interaction between sea breeze and trade wind. In the *Kona type*, by contrast, the mountain has a large lee area where normal tropical convective showers develop, and the cumuli usually move inland in a deep sea-breeze current. Finally, the *Mauna Kea type* is a windward phenomenon, in which the direction of the sea breeze is the same as that of the trade wind, so that the two currents reinforce each other. The windspeed in this type of regime is moderated at night, because the land breeze at night is opposed to and stronger than the trade wind; hence the cumuli disperse and clear skies result.[21]

Despite the predominance of linear cloud patterns in the atmosphere, passengers in jet aircraft will often observe spiral, circular, and concentric-ring patterns of cloudlets, which probably indicate the presence of eddies in unstable air. Brief notes of such observations may often be found in the literature.[22] Since these phenomena are rarely recorded by the synoptic stations, and may appear only as insignificant detail in the satellite photographs, there is much scope for the amateur in photographing these features, but his geographical position must be known accurately if the resulting photographs are to have any scientific value.

The Climatological Importance of Cloud Physics

Clouds are very complex phenomena, and are very small compared with the distances between observing stations. Also, the synoptic codes for cloud reporting are by no means perfect. One can estimate *cloudiness* for an unobserved area by making inferences from the statistics for surrounding stations, but to do the same for *cloud types* is of very doubtful significance. Even the published synoptic data on cloudiness is open to question; for example, a comparison of simultaneous aircraft and surface observations of clouds indicates that the published data may greatly overestimate the amount of low clouds. But if one can accept the synoptic data as accurate enough for the purpose in hand, much useful information may be derived indirectly from the figures; for example, frequencies of the height of cloud ceilings can be estimated from frequencies of cloud cover, and the frequencies of cloud ceilings below various altitudes and within various altitude ranges may be obtained.[23]

If interpolations between synoptic stations are to be meaningful, some knowledge of the norms of cloud behavior is essential. Much knowledge may be obtained by field observation, but it must be supplemented by an understanding of the basic principles of cloud physics. Without such understanding, very serious misinterpretations of observed cloud behavior may be made.

A good example of the importance of cloud physics for the climatologist is orographic rainfall. Cloud physics and precipitation physics obviously are very closely related; both deal with the processes involved in phase changes of water, and with " particles " that differ basically only in size. Cloud particles range in diameter from about 1 micron to between 100 and 500 μ; drizzle droplets are 100 μ in diameter or more; raindrops are on the order of 1,000 μ in diameter (i.e., about 1 mm); and hailstones are on the order of 1 cm in diameter. The normal processes by means of which clouds grow and produce precipitation include: (a) the *Bergeron–Findeisen process*, in which ice crystals grow at the expense of water drops in a mixture of coexisting ice crystals and supercooled water droplets, water vapor from the droplets freezing directly onto the ice crystals; and (b) the *coalescence process*, in which large cloud droplets grow faster than small droplets—i.e., water droplets of different sizes, which may be assumed to exist in a given cloud, will be subjected to upward or downward motions by updrafts and downdrafts, and the larger the drop, the greater its chance of collision with other droplets. The coalescence process takes an hour, and the Bergeron–Findeisen process half an hour, before it can produce effective precipitation. Thus the precipitation mechanism sets a time limit for the development of orographic rainfall. If a cloud moves completely over an orographic barrier before this mechanism can operate, no precipitation can result.[24]

In the hills of Wales, clouds in a 40-knot westerly airstream may be carried from the western Welsh coast to the main watershed in 30 minutes, so that orographic precipitation cannot fall unless there is a cloud containing large droplets or ice crystals (preferably the latter) over the coast before orographic ascent begins. The time available during its passage over the Welsh hills is thus only just adequate to enable a cloud to produce orographic rainfall if it is very thick vertically before it begins its journey. In general, a vertical disturbance of the atmosphere that persists horizontally for 10 miles or less is not enough to cause orographic rainfall. If the spectrum of cloud-drop sizes within an airstream is known, and if this minimum time criterion is satisfied, one can predict how much orographic precipitation there will be in a given airstream. For orographic precipitation over a large area, synoptic considerations are important as well. For example, in the intermountain western United States, the orographic component in the rainfall in upper cold lows is relatively small compared with frontal precipitation, although the orographically induced vertical motion is the same in all situations.[25]

Orographic rainfall is very complicated to study, because it involves not only cloud physics but also atmospheric dynamics, as was noted in Chapter 2 of *Foundations*. Early empirical investigations of the relation between rainfall and topography in the British Isles suggested that if Britain were completely flat, it would experience an annual rainfall varying from 20 inches in the southeast to 40 inches in western Ireland and the Outer Hebrides, but that superimposed on this are varying amounts of " orographic increment." Simple rules like this are of doubtful validity for small areas and short times. The tracking of nuclear-bomb debris—much of which is captured by orographic precipitation—proves that the motions within orographic clouds are very complex, and that each case of orographic precipitation must be considered on its own merits.[26]

Turning to the general problems of cloud physics, and to precipitation within a cloud that is not necessarily orographic, the concept of *warm clouds* is of distinct climatological significance. These are clouds that do not reach up to the freezing level, and that consequently must form precipitation by the coalescence process. Warm-sector rainfall in the British Isles and the middle latitudes generally comes from warm clouds, and they are very important indeed in the tropics; for example, most of the rainfall in Puerto Rico comes from warm cumulus. An investigation of radiosonde data for the northern hemisphere showed that rainfall from nonfreezing clouds is quite frequent in maritime regions, for example, in airstreams from off the sea in summer in middle and high latitudes (even in Alaska), usually from layer clouds, and from convective clouds all year round in tropical latitudes. Warm clouds appear to be rarely reported inland, but warm " rain " may occur (in the form of drizzle) in the coldest parts of the Earth.[27]

Before a cloud droplet can grow into a precipitation droplet, it must be produced out of apparently clear air. The importance of the presence of a condensation nucleus in the air before a cloud droplet can form has been touched on in an earlier chapter. Of equal importance is the existence of *supercooled water* within the atmosphere. Laboratory experiments for many years have indicated a critical temperature of $-40°C$, at which homogeneous nucleation transforms all supercooled water—i.e., all water that is still liquid at temperatures below $0°C$—into ice particles. These ice particles themselves form condensation nuclei, and their formation is facilitated in turn if there are impurities, known as *freezing nuclei*, in the air; thus, even in air that is deficient in condensation nuclei, clouds may form if temperatures drop below $-40°C$, as has been demonstrated spectacularly by the hot springs and geysers of Yellowstone Park, where there are atmospheric temperatures both above and below $-40°C$ during the winter. The geysers provide large amounts of moisture, which condenses in the cold winter air to form dense fog (i.e., a cloud touching the ground) even though condensation nuclei are very rare in the area. Many parts of the Earth's surface in winter may have air temperatures of less than $-40°C$, and under such conditions clouds may appear from nowhere within a few minutes. Their disappearance may be equally rapid.[28]

In order to outline briefly the climatological aspects of the physics of different types of cloud, it is convenient to distinguish between dust clouds, ice clouds, and water clouds. Individual clouds may consist of one, two, or all three forms of matter.

Stratospheric clouds usually seem to be made of dust, and are infrequent, being usually caused by volcanic eruptions. Examples seen over England include a thin dust cloud, with base between 49,000 and 49,250 feet and top between 49,500 and 49,650 feet, reported between July 27 and 30, 1953, and another at 55,000 feet that covered most of southwest England on April 3–4, 1956. The former could be distinguished from ordinary cirrus by its noncrystalline and hazy appearance; it was probably a wavy, continuous sheet of dust from a volcanic eruption in Alaska; it persisted until August 4, according to an observation from Lichfield. The latter was due to the eruption of a volcano in Kamchatka on March 30; it projected ashes to a height of 67,000 feet, from where they were advected by the 100-mb winds. This particular cloud was off-white in color, transparent when viewed directly from below, and in the form of north–south trending undulations with minor transverse ripples on their underside.[29]

Measurements of twilight intensity supplemented by model computations show significant quantities of dust in the atmosphere all the way up to 65 km, most of it being near 20 km. The dust at some heights may sometimes become dense enough to be visible from the ground. Its origin is uncertain—it does not appear to be correlated with meteor activity.[30]

Although frost points are very low in the stratosphere, it appears that ice clouds can exist in it. For example, a very tenuous layer of high cloud covered the area from north Wales to the southeast coast of England on August 10, 1951. This cloud was between 46,500 and 47,500 feet, where frost points were about -50 to $-58°F$. Assuming that a saturated layer of air between 100 mb and 150 mb is a semi-permanent feature of middle latitudes, dynamic cooling of such a layer in the area of intense divergence ahead of a vigorous depression could produce an ice-crystal cloud at that height.[31]

Noctilucent clouds may consist of dust, ice crystals, or both. The dust probably includes minute solid particles produced in the stratosphere by the condensation of gases in the wakes of meteors, as well as volcanic dust. These clouds are only observed in summertime, and only in high latitudes, because certain geometric conditions must be fulfilled before they can become visible. Seen after sunset, they are spectacular and colorful phenomena, which remain in place for a long time; for example, in latitude 63° 15′, they may be observed for nearly an hour for part of the year, and for nearly three hours given a declination of 15° (but not at all if the declination is greater than 21°). The absence of emission lines from the spectrum of noctilucent clouds, and the increase in the intensity of this spectrum toward the shorter wavelengths, indicates that they are seen by reflected sunlight. This light is partly polarized, indicating that the sunlight has been scattered by particles whose diameter is of the same order of magnitude as (or is smaller than) the wavelength of visible light. These particles may be cosmic dust, but an ice-cloud model can equally well explain the observed features of the clouds. On the average, particles have a radius of 1.4×10^{-5} cm, and their density is 1 to 10^{-2} particles per cubic cm.[32]

Photogrammetric studies of noctilucent clouds in central Sweden during the IGY show that the clouds observed were at heights of 81.5 to 85.5 km. They were in long parallel bands that proved to be part of a system of waves of about 50 km in length and up to 4 km in amplitude. The wave crests were oriented in a direction almost perpendicular to that of the general wind, and were continuous for hundreds of kilometers. The waves were propagated with absolute velocities of about 10 or 20 m per sec, and moved in a direction nearly opposite to that of the cloud system as a whole. Smaller billows, with lengths of about 5 to 10 km, and amplitudes of about 0.5 to 1.0 km, moved through the crests of the waves in the general direction of movement of the main cloud system. The visual brightness of the clouds fluctuated in conformity with changes in the optical thickness of the cloud layer.[33]

Since noctilucent clouds were first reported, in 1885 in Germany and Russia, there have been hundreds of cases reported in the northern hemisphere, in particular from northern Europe, the U.S.S.R., Canada, and Alaska. That very few have been reported from the southern hemisphere is due to a lack of observers rather than to a real difference between the two hemispheres.[34]

Mother-of-pearl clouds have somewhat the same iridescent appearance as nocti-lucent clouds, but are much lower, usually between 18 and 30 km, and especially just above 20 km. They have been studied in Norway in particular, where they occur in northwesterly airstreams around an almost stationary depression. They appear to be stratospheric wave clouds, set up in the crests of stationary waves generated in a deep, unidirectional airstream over the Norwegian mountains. They are rare because unidirectional airstreams of the required vertical thickness are not common. They are interesting physically in that they require the existence of a property of water that is not usually observed in the laboratory. Apparently, the clouds consist of supercooled, spherical water droplets at temperatures nearer to $-80°C$ than to $-40°C$. These droplets have a diameter of less than one-tenth of that of particles in ordinary tropospheric clouds. Since their chance of spontaneous freezing is reduced considerably, many of them may not freeze during their movement through the crest of a wave, and so the iridescent clouds remain stationary for an appreciable length of time after sunset at the Earth's surface, illuminated by the rays of the setting sun. Their climatological interest is that they rarely if ever are reported at heights between 10 and 18 km, although orographic waves of the required amplitude almost certainly extend upward through this layer. It may be that the air between 10 and 18 km is drier than that above, or that the moisture in the upper layer may have been derived from ascending motions in the polar-night region of the atmosphere, in which any wave clouds forming between 10 and 18 km would be almost invisible.[35]

The most obvious form of ice cloud is *cirrus*. Fair-weather cirrus is often orographic in origin, and much of the irregular detail at the edges of sheets of frontal cirrostratus may be caused by hills. Streamers of cirrus may extend several hundred miles down-wind from the hills producing them. The vertical air motions induced by even rela-tively small orographic obstacles may be considerably magnified at cirrus height, so that hill-wave cirrus may be produced by quite low obstacles; for example, in England it is generated at 20,000 to 30,000 feet over the Cotswold and Chiltern Hills. It forms where the air temperature is at or around $-40°C$, and, unlike water cloud formed by orographic uplift, does not evaporate in the subsiding air in the lee of the topographic barrier, but continues to grow, becomes fibrous in appearance, and moves away downwind from the hill. This gives rise to somewhat unusual situ-ations, for example, when the Cotswolds produce orographic cirrus but the much higher and more extensive Welsh mountains do not. At a temperature of $-40°C$, ice crystals will not evaporate unless the relative humidity is less than 68 per cent (corresponding to the saturation point for liquid water). Orographic cirrus typically forms in the morning and evening, and is always a shallow cloud, however large the topographic obstacle producing it.[36]

Globally, cirrus is related to the location of the Polar Front jetstream, as was noted in *Foundations*. Bands of cirrus usually run parallel with the jet and within 200 nautical miles of the jet core, particularly on the warm side of the jet axis and in its entrance or neutral areas. The cirrus sheet usually terminates in a sharp northward-facing edge, 100 km south of the jet core, and when frontal cirrostratus has a well-defined boundary, the axis of the jetstream is usually found close to this boundary. In general, cirrus is much more frequent and extensive to the right of the jet axis than to the left (looking east).[37]

In synoptic terms, cirrus is caused by moist air in the upper troposphere. Although in elementary meteorology, cirrus is taken to be the first sign of an approaching warm front, more than one-quarter of the reported occurrences of extensive cirrus are not associated with fronts. The 1,000-to-500-mb thermal wind is stronger when there is cirrus than when there isn't. Cirrus is more frequent in or near warm ridges than in or near cold troughs, and is particularly associated with anticyclonically curved thickness isopleths. Extensive cirrus sheets tend to form at 300 mb in areas of negative absolute vorticity at 300 mb, thick sheets in areas of positive relative vorticity.[38]

Synoptic reporting of cirrus is difficult during hazy conditions, when high cloud is only visible if the sun is low, and at night, when the amount of cirriform cloud reported varies with the phases of the moon and rapidly increases as dawn approaches. Ice-crystal clouds have been reported from all parts of the world. In Britain and in similar latitudes, cirriform clouds usually form at around 35,000 feet, and usually occupy layers between 3,000 and 5,000 feet thick. They are present over southern England almost 50 per cent of the time, according to the synoptic reports, but since this area is free from extensive low cloud only 25 per cent of the time, the figure for cirrus clouds is probably an underestimate of their frequency. Variations in their frequency in middle latitudes seem to be due to changes in the general synoptic situation rather than seasonal. In polar regions in winter, cirrus clouds form at the ground. In tropical regions, cirrus is usually found at 45 to 50 thousand feet. Tropical cirriform cloud is often associated with thunderstorms or is the residual anvil of a cumulonimbus after convection has died down. Some extensive sheets of cirrus or cirrostratus in the tropics result from condensation in the form of ice needles in rising air; the clouds are invisible to an observer flying through them, but are visible from the ground. Widespread cirrus normally forms under a high tropopause, probably in poleward-drifting stratospheric air.[39]

In general, extensive cirriform cloud may be formed by convection—for example, from cumulonimbi or from instability in shallow layers (for cirrocumulus)—or by widespread slow ascent of air. It can only form where the atmosphere is saturated by water, so that water vapor can turn to ice only by first becoming liquid. As already noted, ice crystals once formed persist until the atmosphere is no longer saturated for ice. Cirrus crystals fall at about three cm per sec, and can fall for three hours (i.e., the cirrus cloud can persist for three hours) before they start to evaporate. During this time, the cloud may be transported horizontally several hundred miles. Therefore, the dynamic conditions under which a cirrus sheet formed may differ from those it is associated with two or three hours later.

Aircraft condensation trails provide good examples of man-made cirrus. Various condensation phenomena are caused by high-speed aircraft—for example, low pressure areas develop on the surfaces of the wings and cause adiabatic cooling of the air below its dewpoint—but *contrails* are specifically caused by engine exhausts. Contrails normally consist of ice crystals; by making certain assumptions, one can compute the critical temperature for contrail formation as a function of the pressure and relative humidity of the environmental air and of the amount of air entrained into the exhaust jet, the latter depending on the ratio of water to heat in the exhaust, which in turn depends on the type of engine in the aircraft.[40]

Contrails are useful as synoptic indicators; for example, they normally form only between the Polar Front and subtropical jetstreams, and rarely form on the cyclonic-shear side of the Polar Front jet or on the anticyclonic-shear side of the subtropical jet. They are also useful as very local indicators of windshear and of the boundary or zone of separation between different airstreams. Kinks in a contrail may indicate dynamic instability in large horizontal shear, the distortions in the condensation trail being set up when the airflow passes over a hill; the kinks should be of the same order of magnitude as the hill.[41]

An interesting type of "surface contrail" forms in very cold regions. Fog is very rare at temperatures between 0°F and −30°F in Alaska and Canada, being rarest between −20° and −30°F. However, in the vicinity of inhabited areas, the frequency of fogs rapidly *increases* at temperatures below −30°F, because of the burning of hydrocarbon fuels in the settlements; supersaturation of the air results, and ice fog (i.e., a condensation trail at ground level) forms. By applying the assumptions mentioned two paragraphs back, one can calculate the combinations of temperature and relative humidity necessary for ice-fog formation (see Figure 6.6). At temperatures of −20° to −40°F (depending on relative humidity), burning of hydrocarbon fuels will cause formation of man-made ice clouds. At higher temperatures, it should be noted, the combustion of these fuels hinders the formation of fog because it reduces the relative humidity of the air.[42]

Distrails are narrow ribbons of clear air produced when an aircraft flies through a sheet of very diffuse cloud and evaporates the latter through its exhaust heat. The distrail is visible as a blue channel, in the midst of an area of very light blue or bluish-white. If the distrail is produced at middle-cloud height, cirrus may form in it.[43]

Water clouds have been the subject of hundreds of careful investigations, for obvious reasons. The literature is extensive, and only a few examples of particular climatological interest will be described here. Attention will be given to certain aspects of the physics of cumulus, cumulonimbus, stratus and stratocumulus clouds.

Cumulus clouds provide visible evidence of the importance of convection in the atmosphere, and certain very broad-scale synoptic correlations may be made as a start. There is a distinct correlation between areas of deep or shallow convection and vorticity distribution. Deep or very deep convection (i.e., cumulus tops reaching to 800 or 400 mb, respectively) is associated with cyclonic vorticity at 750-mb, and shallow convection (i.e., cumulus tops not reaching to 800 mb) is associated with anticyclonic or nil vorticity. Shallow convection is often limited in depth by a subsidence inversion. Beyond these very general correlations, each convection cloud or cloud system must be studied on its own terms, and considerations such as stability characteristics of the air mass, and the possibility of entrainment, become paramount.[44]

Much may be learned by following the history of a particular cumulus cloud. Photogrammetric observation of the growth of small cumulus clouds in Arizona

316

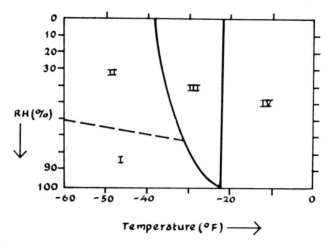

Figure 6.6.
Conditions necessary for formation of ice fog: I, persistent ice fogs; II, nonpersistent ice fogs;
III, no ice fogs—combustion causes drying of atmosphere; IV, no ice fogs—droplets will not freeze.
(After H. Appleman.)

showed that each new cloud formed initially in a small, well defined area. The active cloud elements were similar to laboratory thermals, broadening with height along a cone with an interior angle of 30°; each cloud consisted of comparatively few thermal elements—the larger the cloud, the larger the thermal. Orographic effects prove to be very complex. Mesometeorological study of other cumulus cases in Arizona indicated that morning cumulus convection was caused by a convergence field developing in a very small low-pressure area above the heated part of a mountainside. In a cumulus initiation in Alberta, the cloud formed above a forest fire, reaching a vertical depth of 6,000 feet and a horizontal dimension of several square miles within seven hours. Normally cumuli originate in thermals that rise from specific areas of ground with favorable physical properties, although all thermal source areas do not necessarily always generate thermals on apparently similar days. On the other hand, fair-weather cumuli in Australia do not form because of a specific thermal that originates at a specific spot on the Earth's surface, but instead develop randomly from the top of a deep layer that becomes slightly unstable after the thermals move into it.[45]

Once a cumulus cloud has formed, it is separated from the Earth's surface by a *subcloud layer* that has very distinctive physical properties. For typical large cumuli over England, the lowest thousand feet of the subcloud layer is characterized by a superadiabatic lapse-rate, which decreases with height and which is maintained by a continuous succession of rising bubbles of warm air that have excess temperatures of about 1 or 2°F, excess dew-points of about 5 to 7°F, and an average diameter of about 500 feet. At 1,000 feet above the ground, the lapse-rate changes abruptly to dry-adiabatic, the bubbles average 600 feet in diameter, and the temperature and dew-point excesses decrease to 0.3°F and 4°F, respectively. Finally, the lapse-rate becomes stable some 200 to 400 feet below the cloud base, and rising bubbles become infrequent. Over the sea these three divisions of the subcloud layer are not usually evident, but the stable layer immediately below the cloud base is often present. This stable layer is very important: its existence implies that updrafts within the cloud do not extend down into the dry-adiabatic layer below the cloud. The large cumuli over England usually have a diameter of 10 to 15 miles, and are composed of several distinct cells, each of which is about 3 miles wide and has a lifetime of 20 to 30 minutes. Assuming that all the condensed water within such a cloud is carried upward in the updraft, one can estimate the water content of the cloud. With relatively high temperatures at the cloud base, the cumulus can hold up to 7 grams of water per m^3 of air, but with relatively low temperatures only 2 grams per m^3.[46]

Cumuli in the tropics grow much larger than those in England or in other middle-latitude countries. They also exhibit a stratified subcloud layer —in fact, the subcloud layer was first studied intensively in the Caribbean—but, unlike the temperate variety, do not grow from well-defined "roots," i.e., thermals. By contrast, the trade-wind cumuli are a cluster of formerly independent elements that come together when random eddies in the subcloud layer reach condensation level. The large clouds are produced by coagulation of small cloudlets, which explains the characteristic clear-air gaps in the interior of the lower levels of trade-wind cumuli as distinct from the more solid extratropical cumulus.[47]

The trade-wind cumuli help maintain the general atmospheric circulation in tropical latitudes; 90 per cent of the energy that is transferred from sea surface to atmosphere in the trade-wind zone is in the form of evaporated water, which is transported into the upper troposphere and even the stratosphere by the cumuli. The latent energy transferred from sea surface to upper atmosphere by trade-wind cumuli, assuming that one-half of each square-mile area in all tropical oceans is covered with cumulus, is 40 to 50 times greater than the total rate at which kinetic energy is dissipated by all the wind systems of the world. The relative inefficiency of the atmosphere as a heat engine is largely due to the loss of heat by reradiation to space in tropical and subtropical latitudes.[48]

A noteworthy feature of trade-wind cumuli is their distribution: they tend to develop in clusters a few miles in diameter, separated by clear areas 15 to 20 miles across; even in stormy weather they do not usually merge into cumulus barriers many miles in length, which they do in higher latitudes. Other features are the constant depth of the subcloud layer, which is rarely more than 2,000 feet deep, and the fact that most trade-wind cumuli cease growing vertically well below inversions. The strongest cumulus towers do penetrate the inversion, but their tops very soon become cut off and evaporate, adding moisture to the dry inversion layer, which causes it to weaken as it increases in height downstream. According to the parcel theory of convection, even tiny trade-wind cumuli possess enough energy to enable them to ascend right into the inversion layer. However, entrainment is continuous as the cloud grows vertically, and the development of a trade cumulus may be explained in terms of a steady-state model in which entrainment figures prominently. This model emphasizes that the cloud looks asymmetric because its upper portions move with a different speed from that of the environmental air: the windspeed in the latter increases with increasing altitude, and the upper portion of the cloud moves in the direction of the general wind at its own level, but more slowly.[49]

The rate of growth of trade-wind cumuli over water is not excessive; it is, for example, appreciably less than for cumulus clouds over hot land surfaces in the United States. The distribution of water within tropical cumuli is very uneven in the horizontal plane, usually being asymmetric, with the highest concentration of water being in the up-windshear direction from the center of the cloud. For cumulus clouds generally, when vertical windshear is present in the ambient air, the cloud, if tall enough, tilts away from the vertical and moves, not in the direction of the general wind, but somewhat to the right of the wind within the convective layer.[50]

The point at which a cumulus cloud will begin to precipitate its cloud water as rain is of great climatological interest. Precipitation from tropical cumuli only begins when the upper part of the cloud is positively buoyant, i.e., only when an upward acceleration in the cloud's upper levels exceeds some critical value. For cumuli in general, the larger the cloud, the greater the chance of precipitation, but comparative studies show that the probability of precipitation forming within cumuli of a given size and temperature pattern differs from one climatic region to another. The deciding factor is the relative importance of condensation-coalescence and sublimation-coalescence processes within the cloud. In England, the probability of rainfall from cumulus is 50 per cent with a cloud thickness of 7,000 feet and a cloud-top temperature of 20°F, i.e., warm-type clouds. Continental cumuli are much less efficient than maritime cumuli as producers of warm rain; they contain greater concentrations of large cloud droplets since continents seem to be the main sources of the condensation nuclei that grow into clouds.[51]

The formation of precipitation within a cloud leads to increased storage of water in the cumulus, which decreases the strength and vertical extent of the updraft. The more vigorous the cloud (i.e., the stronger its updraft), the higher is its precipitation potential. The amount of precipitation reaching the ground from a given cumulus cloud does not necessarily depend on the microphysics of the cloud, but is largely controlled by those properties of the air surrounding the cloud that determine the magnitude and duration of the updraft within the latter.[52]

Fair-weather cumuli are usually limited in their vertical growth by a stable layer or inversion, and only produce showers if convection continues so that the cumuli grow laterally into extensive blocks of stratocumulus-like clouds, which must exist for 30 minutes to one hour. Large and medium-size cumuli usually must be at least 2,500 m deep before precipitation is possible, but the average depth that is necessary for showers varies considerably from place to place. For example, the average figure is 2.5 km in much of Europe, but is 3 km in Puerto Rico, 4 km in the central parts of the United States, and 6 km in New Mexico. In general, showers fall from relatively shallow cumuli in maritime areas and from relatively deep cumuli in continental interiors. Normally, the vertical thickness of cloud necessary for shower formation depends on the static stability of the air in the convection layer. If the precipitation is being formed by the Bergeron–Findeisen process, visible evidence of the imminence of showers is provided by the *glaciation* of the cumulus tops; i.e., the rounded tops of castellanus cumulus gradually take on a silky or fibrous appearance, which indicates that ice crystals have grown to precipitation size by sublimation of supercooled water droplets onto the ice nuclei in the upper part of the clouds. Note that glaciation indicates the size rather than the nature of the cloud particles: the ice crystals are not necessarily present in the "anvils" of cumulonimbus, and, particularly in the tropics, well-developed anvils form at the top of cumulonimbi that are essentially warm clouds.[53]

The coalescence process appears to be dominant in shower formation. Cumulus showers may be considered to be synoptically controlled; the presence or absence of particulate matter, such as ice or sea-salt nuclei, does not determine whether precipitation is or is not released. The critical factor is whether or not the cumuli ascend to the *fibrillation level*, i.e., the level above which fibrous streaks make a brief appearance at the summits of the evaporating cumulus towers. In central Sweden, the fibrillation zone is 200 to 400 m deep on summer days, and the mean rate of rise of a thermal, if it is to reach this zone from a cloud base at 900 mb, varies from 2 to 7 m per sec, corresponding to cloud thicknesses of 7,500 and 17,000 feet, respectively. Glaciation proper is a stage beyond fibrillation, and involves the continuation of fibrillation until a persistent anvil results, as the fibers spread out laterally from the cumulus towers. Because of the synoptic correlation, criteria for showers from cumulus may be sought on weather maps. For example, the dew-point temperature at 700 mb is critical for the development of showers in southern Arizona.[54]

Cumulonimbus clouds are extremely tall cumulus clouds that have glaciated and formed great anvils that, even in the tropics, reach almost to the tropopause. These spectacular clouds have been described and sketched for many years, and a number of cumulonimbus models have been produced. Very briefly, the transformation of a cumulus cloud into cumulonimbus proceeds as follows. A large cumulus cloud may be seen to consist of an assemblage of thermals that rise out of the cloud top as a succession of towers. Some of these towers cease rising and evaporate within a few minutes; others continue to rise and begin to develop fibrillation, which is visible to the unaided eye only beyond distances of 40 km. At this stage, the towers do not usually produce a radar echo, showing that solid precipitation particles (i.e., ice crystals) are not present, but the cloud usually begins to precipitate. When the towers rise several hundred meters above the fibrillation level, they become glaciated, producing a radar echo, and billow upward and outward into an anvil or mushroom shape within less than 30 minutes. The cloud has now transformed from a domed or pyramidal outline with a flat base to a mass at least double the original size, characterized by an extensive, flat upper surface; it has now become a cumulonimbus.[55]

Cumulonimbi are important climatologically because they usually form thunderstorms, and for other reasons. For example, whether or not a given cumulonimbus is associated with thunder and lightning, it is characteristically accompanied by a downdraft of cold air caused by the weight of the falling rain inside the cloud. This downdraft tends to spread out beneath the cloud, and moves horizontally across the ground as a shallow mass of cold air behind a sharply defined, mesoscale cold front. This cold front may move very rapidly across land that slopes downward; if cumulonimbi develop above mountains, very strong katabatic winds may develop in the adjacent lowland. Sometimes the cold-air dome may be associated with a small cloud similar to the mountain-top plume of Teneriffe, which forms several hundred feet below the cumulonimbus base. Cumulonimbi are associated with torrential downpours of rain at the Earth's surface. There is some evidence that the local distribution of annual precipitation totals conforms in certain areas to the increase in cumulonimbus activity caused by excess warmth generated by built-up regions. Hail, icing, and severe turbulence also frequently accompany a cumulonimbus.[56]

In tropical latitudes, cumulonimbus towers are related to hurricane activity. Giant "hot towers" are essential to maintain the horizontal pressure gradient needed to produce the high winds. The towers must be little diluted by entrainment of environmental air if the hot air is to be lifted to the high troposphere. Dynamic factors prove to be much more critical for this process than static stability, which is so important in higher latitudes. Cumulonimbus conditions are only marginally unstable in the tropics; a far more important control of cumulonimbus development is synoptic-scale convergence.[57]

The chief problem of penetrative convection in the tropical atmosphere is that in the trade-wind cumuli temperatures and liquid-water contents are much lower than is assumed in adiabatic ascents on the tephigram, and mixing with drier surrounding air reduces their buoyancy so that their ascent terminates well before the appropriate saturated-adiabatic lapse-rate line recrosses the environment sounding, and yet occasional cumulonimbus towers penetrate to, or even above, the tropopause. It appears that towers with such penetrative capabilities require a cloud whose minimum horizontal dimension is about 9 km (in the plane of the wind and the windshear). A cumulonimbus of this size can produce undiluted bubbles 2 km in diameter every 20 minutes or so, which emerge from the top of a 40,000-foot-thick cloud with an upward velocity of 11 m per sec. Applying a mathematical model devised by Levine, we find that there is an inverse relationship between the rate of entrainment and the diameter of the cumulonimbus tower. This relationship shows that thunderstorm-size towers (i.e., 5 to 10 km in diameter) can rapidly penetrate to the high troposphere without experiencing serious dilution, whereas trade-cumuli-size clouds (i.e., with diameters of about 1 km) entrain their own initial mass after moving through a vertical distance of only 1 km.[58]

Unlike middle- and high-latitude cumulonimbi, those in the tropics seem to be entirely confined to synoptic disturbances, particularly tropical cyclones. Even in the latter, the cumulonimbi occur in layers where the static stability is 25 per cent higher than the *average* above the tropical oceans, and penetrative convection is not, of course, a feature of the average tropical atmosphere. The mean lapse-rate within cumulonimbi layers in tropical latitudes is barely as steep as the saturated adiabatic, yet that for the mean tropical atmosphere is 10 per cent steeper than the latter.[59]

Stratocumulus is a very typical cloud type in middle latitudes. It is not normally associated with precipitation, but has an important climatological influence, in that it frequently produces a complete cloud cover that blocks all long-wave radiation from the Earth's surface at night, besides reducing the amount of short-wave radiation reaching the ground. Stratocumulus sheets are in a state of delicate balance with their environment: they may be intensified or dissipated by subsidence or ascent of air at their height but not at the ground; they may be transferred from one area to another by advective processes; and they may be influenced by the terrain. However, the synoptic chart and upper-air soundings often give very little guide to the processes important in maintaining the balance in a given stratocumulus sheet, even when it covers extensive areas.[60]

For stratocumulus over the British Isles, turbulent mixing is an important control of dissipation or intensification, and the lapse-rates of temperature and moisture content of the air directly above the cloud are also very important. Turbulent mixing above the cloud top causes warmer and drier air to enter the cloud, which causes the cloud initially to thin by drying, and then to dissipate by warming. Thin, cold stratocumulus sheets dissipate more readily than do thin, warm clouds. The turbulent heat flux into the top of the cloud is about 1 cal per cm^2 per hour, and both cloud and air lose heat by radiation at the rate of 5 cal per cm^2 per hour. The stratocumulus sheet, as a result, warms 0.4°C per hour during the day, and cools 2.2°C per hour at night. The turbulent transfer of water vapor upward from the cloud top proceeds at a rate of about 2.2×10^{-3} grams of water per cm^2 per hour; combining this with the turbulent heat flux shows that the cloud sheet gains about 3.8×10^{-3}

grams of water vapor per cm^2 per hour during the day, and loses about 0.7×10^{-3} grams per cm^2 per hour during the night. Consequently, stratocumulus sheets tend to develop or intensify during the day and to dissipate, or at least decrease, during the night over Britain. Nocturnal dissipation of stratocumulus is prevented if there is another cloud sheet above it, because the thinning of the cloud by turbulent mixing at the cloud top is not intensified by long-wave radiation loss from the upper surface of the stratocumulus sheet.[61]

The heat and moisture balance needed to keep a stratocumulus sheet in existence can be maintained only by large scale subsidence at the height of the sheet. The subsidence provides heat to replace the deficit of 4 cal per cm^2 per hour between incoming (turbulent) energy and outgoing (radiative) energy mentioned in the last paragraph. Observation shows that the temperature lapse-rate below the base of stratocumulus over Britain is dry-adiabatic almost to the ground, but that there is a sharp temperature inversion, which usually does not appear on the radiosonde sounding, immediately above the cloud top. This inversion has a magnitude of about 2 or 3°C per 100 feet of height, and is present only within a vertical distance of 500 ft of the cloud top; rates of water-vapor increase with height of up to 1.5 grams of water per kilogram of air per 100 feet of height have been measured in the inversion. In a steady-state stratocumulus, subsidence at the rate of 3 mb per hour must be maintained within the inversion layer to offset the turbulence effect. If the subsidence rate is less, the height of the top of the stratocumulus sheet increases and the cloud thickens; if it is greater, the cloud top lowers.

The synoptic parameters that decide whether a given stratocumulus sheet will intensify or dissipate include: (a) the cloud thickness; (b) the average hydro-lapse through the 50-mb layer below the cloud base; (c) the maximum depression of dew-point temperature below the dry-bulb temperature at any pressure surface up to 50 mb above the cloud top. In addition, the heat and water balance of the cloud sheet is appreciably affected by the magnitude of the coefficient of eddy diffusion, which increases if no solar radiation reaches the ground. Geographical factors are also important, as indicated, for example, by occasions on which there is no cloud over England but sheets of stratocumulus cover the North Sea.[62]

In contrast with stratocumulus, *stratus* clouds may be associated with drizzle, rain, or snow. The coalescence of colliding cloud droplets appears to be the main cause of precipitation release from stratus, but the probability of precipitation from a given stratus sheet increases if the latter has a cold top, and particularly if the temperature at the cloud base is above 0°C. Observations of stratiform clouds over northern Ireland show that the cloud layer must be at least 2,000 feet deep, given a cloud-base temperature between $-5°$ and $+16°C$, for the cloud to commence precipitating. If the cloud-base temperature is below 0°C and the cloud-top temperature is above $-12°C$, however, the cloud layer usually must be 7,500 feet deep before rain or snow will fall. Before drizzle may form, large water droplets (radii equal to or greater than 50 microns) must be near the top of the stratus layer; once this condition is satisfied, there is a definite relationship between cloud thickness and drizzle frequency. In northern Ireland, the frequency of drizzle decreases rapidly if the stratus layer is more than 10,000 feet deep and if the cloud base is not more than 2,000 ft above the ground; if the cloud base is between 2,000 and 4,000 feet above the ground, the frequency of drizzle begins to fall off when cloud thickness exceeds 5,000ft.[63]

Stratus usually forms as deep layers of cloud, but occasionally very thin layers are reported. For example, very thin stratus reported over Surrey on April 21, 1954, at a height of 5,000 feet was marked by a well-defined inverted ridge about one mile in length that hung downward from the cloud base for several hundred feet. There were clear lanes on each side of the ridge, with a band of stratus along their center lines, attached to which were clear ribs of cloud. Turbulent cloudlets between the ribs resembled flattened stratocumulus. This feature was remarkable in that it was the only disturbance of an otherwise remarkably even cloud base. The pattern moved, without appreciable change in its appearance, across the sky from east to west, the inverted ridge being at right angles to the wind direction, along the wind shear in the vertical plane. It apparently formed because of a twin-jet aircraft flying just beneath the stratus sheet. Normally the underside of a stratus sheet is very uniform in appearance.[64]

In small and topographically varied countries such as Britain, short-period variations in the low cloud cover may have very important consequences for local climate. Certain parts of England are subject to distinctive types of stratiform cloud. For example, *North Sea stratus* is important in control of local climate in parts of the eastern counties, bringing cool, overcast weather that may persist for days. North Sea stratus is most frequent in the night and early morning; it tends to occur over the sea rather than over the land during spring and early summer, but during other months it often is found over land areas. Two types of situation are quite common. In the first of these, a cloudless air mass moves inland during a period of cooling; stratus then forms over the land, its height and thickness being proportional to the general windspeed. This stratus does not gradually spread downwind, but develops simultaneously over a large area. In the second situation, there are clear skies over the land, but North Sea stratus over the sea in an airflow that is moving inland. The stratus sheet over the North Sea is in a state of dynamic equilibrium: it continually forms over the sea, but continually evaporates along its leading edge whenever the latter advects over the land. With northeasterly winds, the stratus moves inland over East Anglia when diffluence is present in the surface streamlines (see Figure 6.7), and when the surface temperatures at coastal stations fall below the clearance temperature of the stratus sheet; this usually takes place in the evening, and the cloud sheet then moves inland with the speed of the wind at its height. The stratus normally remains at the same height over both land and sea as it progresses inland, but if the ground has been cooling for some hours before the arrival of the cloud, stratus may form over high ground ahead of the main body of the North Sea stratus, in which case the latter appears to advance erratically. Clearance of the stratus usually works in a seaward direction (i.e., backward against the wind), and occurs when the heat balance of the cloud layer over the land favors evaporation of the cloud droplets.[65]

Synoptic studies show that the local development of low stratiform cloud is very much influenced by orography and by slight variations in the direction of the geostrophic wind. For example, Prestwick Airport is sheltered by hills to the south and south-southwest, and very low stratus is associated with surface wind directions of 230° and 250°, which correspond to those air trajectories with the longest possible unbroken track over the sea. A variation in wind direction of 10° beyond this range brings in air whose sea trajectory is at most only one-third as long. Provided the surface dew-point in the airstream is not less than the sea-surface temperature when the air reaches the Firth of Clyde, stratus will then form at Prestwick with a cloud base

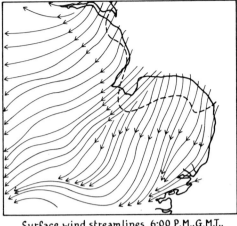

Surface wind streamlines, 6:00 P.M.,G.M.T.,
August 4, 1944.

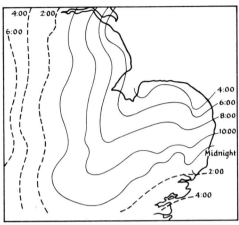

Isochrones of time of arrival of leading
edge of stratus sheet, August 4-5,1944.

- - - - Edge of stratus cloud   ———P.M.,August 4.   - - - -A.M.,August 5. G.M.T.

FIGURE 6.7.
Low North Sea stratus over East Anglia (after W. R. Sparks).

at 600 feet or lower. Low stratus at Prestwick is greatly variable both diurnally and seasonally; this variability is mainly due to fluctuations in the stability and moisture content of the lowest layers of the air. Conditions in stagnant air are particularly complex to predict.[66]

Orographic shelter is also an important influence. Valley, on the Isle of Anglesey, is almost completely free of low cloud below 1,000 feet in easterly or southeasterly airstreams, because of the protection provided by the mountains of North Wales. Cloud bases that are not more than 1,000 feet above the ground are most probable with north-northwesterly winds; cloud bases that are not more than 800 feet up are most probable with southerly to southwesterly winds. The Dale peninsula of south-west Pembrokeshire, a plateau averaging 200 feet in height, is much subjected to low stratus, particularly in winter. The stratus usually forms in warm airstreams with surface directions of 210° to 260°; it may persist for days, but a slight veer in wind direction, from 260° to 270°, causes the cloud base to lift rapidly and visibility to increase to 20 miles. For stratus covering relatively flat areas, the type of airstream and the season are the dominant controls. For example, low stratus over eastern England is more frequent in summer in northeasterly airstreams than in frontal situations, although for the year as a whole, only 25 per cent of the stratus is associated with northeasterly winds, but 40 per cent with fronts. Low cloud in frontal situations tends to be held back by orographic obstacles, and surface effects may influence the prefrontal cloud pattern. For example, in winter very low stratiform cloud develops over frozen ground in the Dartmoor area far ahead of warm fronts, so that Exeter Airport is particularly liable to experience low cloud bases, which are very slow to clear after the passage of the front. The length of time a given station experiences very low stratus (i.e., nimbostratus or fractostratus) during the passage of a warm front or warm occlusion depends on the height of the station above sea level. The

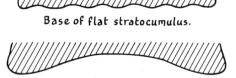

Base of flat stratocumulus.

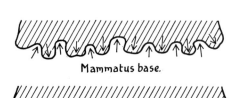

Base of convective cloud.

FIGURE 6.8A.
Cloud base types.

Mammatus base.

Base of normal stratocumulus.

FIGURE 6.8B.
Fallout front. Arrows indicate direction
of vorticity flow.

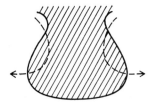

duration of low cloud whose base is within a few hundred feet of the ground varies from 30 minutes or an hour at stations near sea level to several hours at stations only a few hundred feet higher.[67]

The literature of weather forecasting in all parts of the world contains numerous studies of the influence of geographical factors on the formation and dissipation of low cloud that prove both geography and cloud physics must be taken into account if valid climatologies are to be produced.[68]

Our consideration of water clouds must deal, finally, with certain cloud species and accessory clouds which have a distinct climatological interest.

Mamma or mammatus, the rounded protuberances that may be seen at times hanging from the thicker parts of a cumulonimbus anvil, or projecting from the undersides of cirrus sheets, cirrocumulus, altocumulus, altostratus, or stratocumulus, represent negative thermals. Their dynamics are interesting: although they are in effect the visible manifestation of a downward-moving thermal, they do not normally occur in between the usual type of upward-moving thermal.[69] Mamma are fairly common, but only become visible under extensive cloud sheets given certain lighting conditions. Two mechanisms are involved in their formation.

The first mechanism is subsidence of the lower portion of a cloud layer, or of a "shelf" of stratocumulus from the top of a cumulus that has ascended and spread

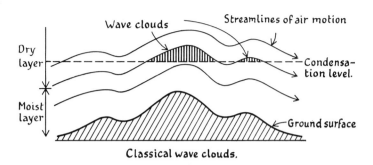

Classical wave clouds.

FIGURE 6.9.
Formation of wave clouds with
arched bases.

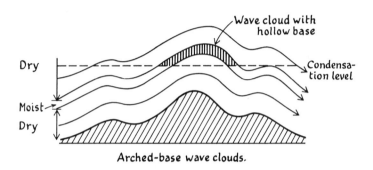

Arched-base wave clouds.

out into a relatively dry stratum of the atmosphere. Subsidence results in the cloud base becoming wrinkled or pitted (see Figure 6.8,A), because air above the cloud base will be heated at the saturated-adiabatic lapse-rate, air below at the dry-adiabatic lapse-rate. Subsidence of 100 to 200 km is enough to result in a temperature variation of 0.5° to 1.0°C along the cloud base. The vertical air displacements necessary to produce subsidence of this magnitude are frequently caused by the airstream's moving over topography with a rather irregular vertical profile.

The second mechanism is the development of a *fallout front* beneath the cloud base. Fallout of liquid or ice particles from a cloud takes place when the particles acquire a terminal fall velocity that exceeds the upward velocity of the air within the cloud, i.e., a velocity of 0.5 to 10 m per sec. The falling particles generate downdrafts by the overlap and interaction of their wakes, and the lower boundary of the air containing the fallout (i.e., the fallout front) widens because vorticity is produced by the greater weight of fallout in the central part of the bulge (see Figure 6.8, B). The vorticity distribution produces horizontal air motions with an outward component around the lower part of the mammatus, and the effectiveness of the downdraft is increased by evaporation of the fallout, which cools the air. This process may produce very extensive downdrafts, which can influence local surface airflow in mountainous areas and the intensity of rain or snow showers, and cause appreciable variations in temperature in the horizontal direction.

Wave clouds—usually the cloud species *altocumulus lenticularis*, but occasionally *cirrocumulus lenticularis* or even *stratocumulus lenticularis*—are always interesting in that they provide visible evidence of the presence of orographic waves. The wave clouds form where the air rises above condensation level, and dissipate where the air descends below this level: consequently, the clouds are almond- or lens-shaped, and remain stationary while the wind blows through them. Numerous examples of wave clouds have been described, ranging from relatively small to relatively large. Sometimes a wave cloud may be extremely persistent; for example, one persisted for at least 11 hours over North Breconshire on April 13, 1958, stretching for about 13 miles in a north-northwest—south-southeast direction. A system of wave clouds may cover an extensive area; e.g., on November 29, 1953, lenticular altocumulus covered northern Ireland. Comparatively small orographic features may produce wave clouds; for example, the Cotswolds give rise to cirrocumulus lenticularis at heights of up to 30,000 feet.[70]

Modifications of classical wave-cloud forms are known. If convection gives rise to thin, moist layers of air with drier air both above and below, the base of the wave cloud may be arched (see Figure 6.9). A wave cloud at cirrus height will not evaporate if the supercooled droplets of which it is composed freeze into ice crystals; the cloud then tends to spead out downwind of the wave, often for several hundred miles. If a cloud layer exists before an atmospheric stratification conducive to orographic waves develops, then the initiation of the wave system may result in the cloud's evaporating in the downward parts of the waves; the holes so formed are "negative" wave clouds. *Pileus* are small clouds formed in the top of a rising thermal that ascends above the condensation level. They have the appearance of wave clouds, and in fact sometimes form under true lenticular clouds, increasing in size so that a double wave cloud appears to exist; but if this double "wave" cloud is not stationary but moves with the wind, it cannot represent an orographic wave. Numerous other cloud species and accessory clouds are climatologically significant—for example, billow clouds and funnel clouds—and the reader is referred to the literature for details.[71]

Notes to Chapter 6

1. See, in particular, numerous articles in *Weather* (monthly, Roy. Meteorol. Soc., London), the *Meteorological Magazine* (monthly, MO), *Weatherwise* (monthly, Amer. Meteorol. Soc., Boston), and the *Monthly Weather Review* (Environmental Science Services Administration, Rockville, Md.).

2. E.g., A. M. Galligan, *AFSG*, no. 33 (1953).

3. G. L. Arnold, *TAGU*, 34 (1953), 189. H. Landsberg, *HM*, pp. 882–926. On the relation between cloud cover and the mean surface temperature of the Earth, see G. Ohring and J. Mariano, *JAS*, 21 (1964), 448. For the diurnal variation of cloud patterns, as obtained from measurements with a sunshine recorder,

see P. A. Huxley, *W*, 20 (1965), 84. On the numerical prediction of total cloud cover, see H. Edson and R. L. Daye, *JAM*, 7 (1968), 759.

4. L. Howard, *Tilloch's Phil. Mag.*, 16 (1803), 97 and 344, and 17 (1803), 5. J. B. Lamarck, *Ann. Meteorol.*, 10 (1802), 149. R. Abercromby, *QJRMS*, 13 (1887), 62. From time to time, however, reports are made of clouds which are unique, or at least unusual: e.g., see R. Tolley, *W*, 20 (1965), 359, for a note concerning a doughnut-shaped cloud due to a local air circulation.

5. W. Bleeker, *BWMO*, 4 (1955), 135.

6. W. E. Howell, *CM*, p. 1164. Galligan, *op. cit.*

(1953). For a historical review of cloud study and observation, see J. A. Kington, *W*, 23 (1968), 348. For a history of cloud classification, see J. A. Kington, *W*, 24 (1969), 84.

7. See *WAF*, II, 72. F. H. Ludlam, in B. J. Mason, *The Physics of Clouds* (Oxford, 1957), p. xv. See also R. S. Scorer, *ISRSM*, p. 393.

8. For full details, see: P. L. Schereschewsky, *HM*, pp. 893–900; V. J. Oliver and M. B. Oliver, *HM*, p. 858.

9. M. A. Alaka, *BWMO*, 9 (1960), 105.

10. For guides to the identification and reporting of clouds, see R. S. Scorer and H. Wexler, *A Color Guide to Clouds* (London, 1964) and *Cloud Studies in Color* (Oxford and London, 1967). For difficulties in reporting cumulonimbus, see B. Vonnegut *et al.*, *BAMS*, 48 (1967), 886. On the reliability and representativeness of synoptic reports of sky cover in Antarctica, see E. S. Merritt, *JAM*, 5 (1966), 369. On estimation of the height of the base of convective cloud from screen temperatures, see D. J. George, *W*, 22 (1967), 147. For tables converting angular velocities of clouds into knots, given absolute altitude and zenith distance, see F. Burdecki, *NOTOS*, 10 (1961), 91. On the use of photovoltaic (or solar) cells to obtain continuous records of daytime cloud types, see R. E. Falconer, *GPA*, 60, no. 1 (1965), 236. For details of a laser cloud-base recorder, see L. G. Bird and N. E. Rider, *MM*, 97 (1968), 107. For details of cloud detection with an 8.6-mm radar, see W. G. Harper, *MM*, 93 (1964), 337, and J. Stewart, *MM*, 95 (1966), 112. For a more detailed account of synoptic cloud reporting, see P. L. Schereschewsky, *HM*, pp. 881–926, and *MOHSI*, pp. 374–98.

11. H. D. Orville and A. R. Kassander, *JM*, 18 (1961), 682. For a review of the principles of stereoscopic cloud photography, see A. B. Fraser, *W*, 23 (1968), 505.

12. J. Georgi, *W*, 12 (1957), 149.

13. W. S. von Arx, *W*, 13 (1958), 179.

14. For instructions concerning analysis (including rectification) of details of satellite photographs, see the *User's Guides* to the various satellites. For bibliography of technical papers and sources of satellite data, see W. K. Widger, Jr., *Meteorological Satellites* (New York, 1966). For examples of studies in the synoptic climatology of satellite nephanalyses, see: M. H. van Dijk and G. T. Rutherford, *ISRSM*, p. 305, for southeast Australia; C. J. van der Harn, *ISRSM*, p. 317, for Europe; and I. Jacobs-Haupt, *ISRSM*, p. 323 for North Africa and the Mediterranean. Other sets of photographs are also available; in particular, those of the NOMS (National Operational Meteorological Satellite) system of the United States, on which see J. R. Blankenship, *JAM*, 1 (1962), 581, and 2 (1963), 808; this system is based on more advanced, Earth-oriented satellites. For cloud patterns in the tropics, see J. C. Sadler, *ISRSM*,

p. 333. On TIROS nephanalyses, see F. R. Miller, *MWR*, 91 (1963), 433. On identifying vertical motion, see C. S. Bristor and M. A. Ruzecki, *MWR*, 88 (1960), 315. For reviews of cloud photography from satellites with examples, see: W. K. Widger, Jr., *WW*, 19 (1966), 100; W. Nordberg and H. Press, *BAMS*, 45 (1964), 684, on Nimbus I; for British examples, R. A. S. Ratcliffe, *MM*, 95 (1966), 257; and E. C. Barrett, *W*, 22 (1967), 151. For problem photographs provided by satellites, see the *Picture of the Month* series in *MWR*, vol. 91 (1964), January onward. For classifications of the features visible on satellite photographs, see C. J. van der Harn, *W*, 19 (1964), 180, and M. M. Hopkins, Jr., *JAM*, 6 (1967), 164. On the processing of satellite photographs, see: T. Fujita, *JGR*, 70 (1965), 5997, on graphic rectification; H. V. Senn and P. J. Davis, *JAM*, 5 (1966), 334, on optical rectification; C. L. Bristor, C. M. Callicott, and R. E. Bradfort, *MWR*, 94 (1966), 415, on computer processing; and R. E. Nagel *et al.*, *MWR*, 95 (1967), 171, and B. F. Watson, *loc. cit.*, p. 645, on objective assembly of photos. On the routine interpretation of satellite photographs of layer-cloud evolution in North America, see R. E. Nagel, J. R. Clark, and M. M. Holl, *MWR*, 94 (1966), 55. For global climatological pictures of cloud cover derived from satellite photographs, see: P. F. Clapp, *MWR*, 92 (1964), 495, for seasons, 60°N–60°S; J. Kornfield *et al.*, *BAMS*, 48 (1967), 878, for ESSA III and V photomosaics; and E. C. Barrett, *W*, 23 (1968), 198, for a review of global photomosaics. On the subjective element in estimating total cloud cover from satellite photographs, see M. J. Young, *JAM*, 6 (1967), 573. On the integration of satellite and conventional data in extratropical regions, see P. E. Shear, *JAM*, 5 (1966), 64. For important satellite photographs, see R. H. McQuain, *BAMS*, 48 (1967), 74, for the first nearly continuous photos of the Earth's cloud cover from space, taken by the ATS I satellite, and G. Warneke and W. S. Sunderlin, *BAMS*, 49 (1968), 75, for the first color photo of the whole Earth from space, taken by the ATS III satellite. See H. D. Ausfresser, A. C. Johnson, and R. A. Kowalski, *BAMS*, 50 (1969), 76, on distortion in the ATS satellite photo series. For cloud photographs from spacecraft, see K. M. Nagler and S. D. Soules, *BAMS*, 46 (1965), 522, from Gemini 4, and *BAMS*, 50 (1969), 58, from Apollo 7.

15. S. Fritz and J. S. Winston, *MWR*, 90 (1962), 1. S. I. Rasool, *JAS*, 21 (1964), 152. J. S. Winston, *JAS*, 22 (1965), 33. F. A. Sharp, *MM*, 87 (1958), 202. On nighttime cloud-cover and cloud-height determination from TIROS radiation data, see S. I. Rasool, *JAS*, 21 (1964), 152, and, for global climatological maps, J. S. Winston, *JAS*, 22 (1965), 333. For cloud heights by stereointerpretation of NIMBUS II photographs, see R. J. Ondrejka and J. H. Conover, *MWR*, 94 (1966), 611. For cloud-top determinations

from Gemini 5 photographs, see F. Saiedy, J. Jacobowitz, and D. Q. Wark, *JAS*, 24 (1967), 63.

16. R. J. Boucher and R. J. Newcomb, *JAM*, 1 (1962), 127. J. W. Deardorff, *JAM*, 2 (1963), 173. W. K. Widger, Jr., *MWR*, 92 (1964), 263 and 598. R. E. Nagle and S. M. Serebreny, *JAM*, 1 (1962), 279. For a satellite photo analysis of an old occluded depression, see I. J. W. Pothecary and R. A. S. Ratcliffe, *MM*, 95 (1966), 332. For satellite photographs of a destructive cyclonic storm on Rameswaram Island, India, see S. M. Kulshrestha and M. G. Gupta, *JAM*, 5 (1966), 373. For maximum windspeeds in tropical disturbances as inferred from cloud patterns on TIROS photographs, see S. Fritz, L. F. Hubert, and A. Timchalk, *MWR*, 94 (1966), 231.

17. A. Leese, *T*, 14 (1962), 409. A. F. Krueger and S. Fritz, *T*, 13 (1961), 1. For details of the analysis of convective nephsystems on the mesoscale, see T. Fujita *et al.*, *ISRSM*, p. 357. For the power-spectrum analysis, see J. A. Leese and E. S. Epstein, *JAM*, 2 (1963), 629. L. F. Whitney, Jr., *JAM*, 2 (1963), 501. C. O. Erickson, *MWR*, 92 (1964), 283. For analyses of vortical cloud patterns visible in satellite photographs, see: W. E. Shenk and E. M. Brooks, *JAM*, 4 (1965), 673, on extratropical vortices; W. E. Shenk, *MWR*, 93 (1965), 613, on a tight, frontless vortex over the Caspian Sea; S. Barr, M. B. Lawrence, and F. Sanders, *MWR*, 94 (1966), 675, on vertical motion in vortices over eastern North America; and N. L. Frank and H. M. Johnson, *MWR*, 97 (1969), 124, on the tropical Atlantic. For satellite photographs of organized patterns of cumulus, see J. P. Kuettner and S. D. Soules, *BAMS*, 47 (1966), 364; on the estimation of low-level wind velocities from such patterns, see C. W. C. Rogers, *JAM*, 4 (1965), 387.

18. V. J. Oliver, R. K. Anderson, and E. W. Ferguson, *MWR*, 92 (1964), 441. P. F. Clapp, *MWR*, 92 (1964), 495. On the formation of shadows of low clouds on clouds at higher levels, see F. Anýž, *W*, 19 (1964), 22. On the relation between the size of satellite-observed cirrus shields and thunderstorm severity, see R. J. Boucher, *JAM*, 6 (1967), 564.

19. B. R. Döös, *T*, 14 (1962), 300. C. J. Bowley *et al.*, *JAS*, 19 (1962), 52. For unique cloud forms revealed by satellite photography, which coincide with surface ridge-lines, see A. H. Smith, *MWR*, 96 (1968), 315. On the formation of a square cloud pattern in Oklahoma and Kansas, interpreted by A. J. Abdullah, *JAS*, 23 (1966), 445, in terms of a pressure-jump/squall-line mechanism, see R. A. Brown, *JAS*, 24 (1967), 305.

20. V. J. Oliver and M. B. Oliver, *MWR*, 91 (1963), 621. J. Kuettner, *T*, 11 (1959), 267. G. A. Corby, *W*, 9 (1954), 312, and *QJRMS*, 80 (1954), 631. B. Haurwitz, *ANYAS*, 48 (1949), 722. L. F. Whitney, Jr., *MWR*, 89 (1961), 447. J. C. Freeman, Jr., *MWR*, 91 (1963), 630. On cloud streets, see *W*, 21 (1966), 284, for a photograph. For cloud lines over the Mediterranean that were not cloud streets, see *W*, 20 (1965), 163. For photographic measurements of cloud bands caused by shear and decreased stability, see J. H. Reuss, *W*, 22 (1967), 204; for a theory that indicates cloud bands should be orientated in the direction of wind shear, see G. E. Hill, *T*, 20 (1968), 132.

21. L. B. Leopold, *JM*, 6 (1949), 312, and 7 (1950), 165. For a linear cloud possibly initiated by one of the peaks of the Pennine hills, England, see *W*, 21 (1966), 320.

22. E.g., R. Noble, *MM*, 82 (1953), 28. For examples of the value of photographs taken during jet passenger flights, and later compared with satellite photographs, see E. L. Honodel, *MWR*, 92 (1964), 105, on the Honolulu–San Francisco route, also J. Vederman and R. Nagatani, *MWR*, 95 (1967), 657, on the Los Angeles–Honolulu route.

23. A. Court, M. Gutnick, and H. A. Salmela, *W*, 8 (1963), 34. G. L. Arnold, *TAGU*, 34 (1953), 189. W. C. Spreen and I. Solomon, *BAMS*, 39 (1958), 261. O. Essenwanger and G. Haggard, *JAM*, 1 (1962), 560. E. de Bary and F. Möller, *JAM*, 2 (1963), 806.

24. For general accounts of the processes involved in phase changes of water, see: P. M. Saunders, *W*, 11 (1956), 103; F. H. Ludlam, *W*, 11 (1956), 187; B. J. Mason, *W*, 14 (1959), 81; W. G. Durbin, *W*, 13 (1958), 143, and 16 (1961), 71 and 113; *PM*, pp. 203–31, 235–49, 260–70. On the structure of giant hailstones, see K. A. Browning, *QJRMS*, 92 (1966), 1. On precipitation processes, see: T. Bergeron, *Un. Geod. Geophys. Int.* (1935), 156. E. G. Bowen, *Austr. J. Sci. Res.*, 3, series A (1950), 193. F. H. Ludlam, *T*, 7 (1955), 277. For a mathematical framework joining the processes of condensation, water-droplet collection, advection, and freezing in clouds, see E. X. Berry, *JAS*, 26 (1969), 109. For observations of fluctuations of meteorological variables in clouds, see B. Ackerman, *JAM*, 6 (1967), 61. A distinction must be made between ice *crystals* and ice *nuclei*. For the relative proportions of ice crystals and ice nuclei in different types of cloud, see S. C. Mossop and A. Ono, *JAS*, 26 (1969), 130, and P. V. Hobbs, *JAS*, 26 (1969), 315. The concentration of ice crystals can be several orders of magnitude greater than that of ice nuclei: the ratio decreases with decreasing cloud-top temperature, approaching unity for a cloud-top temperature of $-25°C$. Glaciated altostratus contains approximately the same concentration of ice crystals and ice nuclei; cumulus and stratocumulus contain (at temperatures above $-10°C$) 10^3 times as many ice crystals as ice nuclei, is measured in New South Wales, Australia.

The term "*oreigenic*" has been suggested by T. Bergeron (1964) as a replacement for "orographic" as the adjective denoting clouds, precipitation, and other atmospheric processes owing their genesis to mountains under the airflow; see F. W. Decker, *MWR*, 95 (1967), 699. On the water balance of orographic clouds in southern California, see R. D. Elliott and E. L. Hovind, *JAM*, 3 (1964), 235.

25. J. S. Sawyer, *W*, 11 (1956), 375. R. D. Elliott and R. W. Shaffer, *JAM*, 1 (1962), 218. P. Williams, Jr., and E. L. Peck, *JAM*, 1 (1962), 343. For studies of orographic precipitation, see: R. T. Small, *WW*, 19 (1966), 204, on Washington state; J. Warner, *T*, 19 (1967), 456, on Hawaii; J. F. Nagel, *NOTOS*, 10 (1961), 59, on Table Mountain, South Africa; and B. W. Thompson, *W*, 21 (1966), 48, and D. E. Pedgley, *loc. cit.*, p. 187, on Mount Kenya. According to D. E. Pedgley, *MM*, 98 (1969), 97, the effect of the Ethiopian highlands on the diurnal variation of monsoon rainfall extends horizontally some hundreds of kilometers into the Sudan.

26. R. M. Poulter, *QJRMS*, 62 (1936), 49. P. B. Storebø, *JM*, 16 (1959), 600. For the use of non-radioactive tracers in following the motions within orographic cumulus over Mount Olympus, Greece, see B. D. Kyriazopoulos, G. C. Livadas, and P. N. Dimotakis, *GPA*, 64, no. 2 (1966), 248.

27. L. Alpert, *BAMS*, 36 (1955), 64. T. Ohtake, *JAM*, 2 (1963), 594. On warm rain, see: R. L. Lavoie, *T*, 19 (1967), 348, on Hawaii; R. G. Semonin *et al.*, *T*, 20 (1968), 227, on Hawaii; and E. X. Berry, *JAS*, 24 (1967), 688, on the mechanism.

28. Durbin, *op. cit.* (1961). V. J. Schaefer, *JAM*, 1 (1962), 481, and 2 (1963), 684. For a review of precipitation processes, see R. R. Braham, Jr., *BAMS*, 49 (1968), 343. For a device designed to provide a continuous record of cloud droplets impinging on it, see L. F. Clague, *JAM*, 4 (1965), 549. For details of an airborne cloud-particle collector, see P. A. Spyers-Duran and R. R. Braham, Jr., *JAM*, 6 (1967), 1108. Work in China, reported by Jaw Jeou-jang, *T*, 18 (1966), 722, suggests that the process of growth of cloud drops is a stochastic one; i.e., it must be studied by means of probability concepts (the *statistical theory of precipitation*). Circular or oval holes in a cloud sheet provide problems in explanation. For examples, see *WW*, 21 (1968), 194 and 238, on Vandenburg Air Force Base; and H. M. Johnson and R. L. Holle, *BAMS*, 50 (1969), 157, on two simultaneous holes at Miami. *Fall-streak holes*, on which see F. H. Ludlam, *W*, 11 (1956), 89, and 19 (1964), 90, are caused by ice crystals falling out of a cloud layer, thus leaving a hole in it; they are often associated with aircraft. For regions where it is theoretically impossible for clouds to form, see: G. F. Schilling, *JGR*, 69 (1964), 3663; K. E. McDonald, *loc. cit.*, p. 3669; and V. E. Moyer, *JGR*, 70 (1965), 247. Cloud is formed over the Victoria Falls of the Zambezi river by adiabatic expansion of air rising out of the 300-foot chasm into which the water drops; see *W*, 20 (1965), 286 and 297.

29. See P. H. Taft, *BAMS*, 45 (1964), 749, for a description of stratospheric clouds over the southwest United States. L. Jacobs, *MM*, 83 (1954), 115. G. A. Bull and D. G. James, *MM*, 85 (1956), 293. For an example of stratospheric cloud in the southwestern United States, see P. H. Taft, *BAMS*, 45 (1964), 748. On the synoptics of dust clouds, see: C. M. Stephenson, *W*, 24 (1969), 126; J. C. Gordon and R. Murray, *MM*, 93 (1964), 106, on Cyprus; and M. Delsi, *MM*, 96 (1967), 50, on the Sudan, in relation to cold fronts.

30. F. E. Volz and R. M. Goody, *JAS*, 19 (1962), 385.

31. J. S. Farquharson, *MM*, 81 (1952), 341.

32. F. H. Ludlam, *T*, 9 (1957), 341. G. Witt, *T*, 9 (1957), 365. E. Hesstvedt, *T*, 14 (1962), 290. Noctilucent cloud particles have been tentatively identified as: thermospheric dust, by S. Chapman and P. C. Kendall, *QJRMS*, 91 (1965), 115; interplanetary particles, by U. Shafrir and M. Humi, *JAS*, 24 (1967), 577; or probably of extraterrestrial origin, by C. L. Hemenway, R. K. Soberman, and G. Witt, *T*, 16 (1964), 84.

33. G. Witt, *T*, 14 (1962), 1. For a review of the problem of noctilucent clouds, see J. Paton, *MM*, 93 (1964), 161; on their morphology, see W. L. Webb, *JGR*, 70 (1965), 4463. For observations, see: J. Hallett, *W*, 22 (1967), 66, in Britain; A. D. Christie, *JAS*, 23 (1966), 446, in Canada; R. A. Hamilton, *MM*, 93 (1964), 201, in midwinter; and F. H. Ludlam, *W*, 20 (1965), 186. On the genesis and distribution of noctilucent clouds, see A. D. Christie, *JAS*, 26 (1969), 168; on the source of waves in them, see C. O. Hines, *JAS*, 25 (1968), 937.

34. B. Fogle, *W*, 20 (1965), 374. For records of the occurrence of noctilucent clouds, see: J. Paton, *MM*, 94 (1965), 180, for the world in 1964; B. Fogle, *op. cit.*, for the southern hemisphere; *MM*, 97 (1968), 193, for North America in 1964–1966; and W. Schröder, *JGR*, 72 (1967), 1971, for Germany from 1885–1965. For records for western Europe, see J. Paton, *MM*, 95 (1966), 174, for 1965; *MM*, 96 (1967), 187, for 1966; and *MM*, 97 (1968), 174, for 1967. See also A. M. Hanson, *W*, 18 (1963), 142.

35. R. S. Scorer, *W*, 19 (1964), 115. For details of mother-of-pearl clouds, see J. Hallett and R. E. J. Lewis, *W*, 22 (1967), 56.

36. F. H. Ludlam, *W*, 7 (1952), 300 and 382.

37. J. H. Conover, *JM*, 17 (1960), 532. J. S. Sawyer and B. Illett, *MM*, 80 (1951), 277. On jetstream cirrus, see F. R. Valovcin, *JAM*, 7 (1968), 817, on their radiative properties. For an example of dense cirrus in a subtropical jet, see *W*, 19 (1964), 293.

38. D. G. James, *MM*, 86 (1957), 1. For details of the sharp edges on cirrus canopies in tropical storms, see E. S. Merritt and R. Wexler, *MWR*, 95 (1965), 111. On cirrus over Malaya during the southwest monsoon, see R. F. Zobel and S. G. Cornford, *MM*, 95 (1966), 65. For observations of the vertical distribution of cirrus along the California–Honolulu air route, see M. E. Graves, *MWR*, 96 (1968), 809.

39. "*MO* discussion," *MM*, 88 (1959), 74. C. S. Durst, *MM*, 82 (1953), 79.

40. R. S. Scorer, *W*, 10 (1955), 281, and *BAMS*, 47 (1966), 197. F. R. Jones, *MM*, 84 (1955), 93. H. Appleman, *BAMS*, 34 (1953), 14. On contrails, see W. C. Livingston, *W*, 24 (1969), 56, see also R. A. Reinking, *W*, 23 (1968), 171 and 520, for their effect on insolation in the western United States. For details of *ship contrails*, i.e., anomalous cloud lines appearing on satellite photographs, which appear to represent buoyant exhaust plumes from ships, see J. H. Conover, *JAS*, 23 (1966), 778, and C. J. Bowley, *JAS*, 24 (1967), 596.

41. H. H. Dunning and N. E. La Seur, *BAMS*, 36 (1955), 73. P. E. Phillips, *MM*, 83 (1954), 279. R. S. Scorer, *MM*, 82 (1953), 279.

42. H. Appleman, *BAMS*, 34 (1953), 397.

43. E.g., see R. M. Poulter, *W*, 3 (1948), 232. For photographs of distrails, see *W*, 22 (1967), 454, and *W*, 23 (1968), 496.

44. E.g., see J. S. Marshall, *CD*, p. 126, on the effects of general lifting or subsidence on convective overturning, and C. F. Van Thullenar, *CD*, p. 103, on the pressure difference between a cumulus cloud and its environment. On the convection-vorticity correlation, see *WAF*, II, 155. For general accounts of cumulus, see *W*, 23 (1968), 255, 339, and 384. On cumulus characteristics in South Dakota, see J. H. Hirsch and C. L. Schock, *JAM*, 7 (1968), 882. On groups of cumuli, see D. C. Gaby, *MWR*, 95 (1965), 203, on the use of cumulus lines on satellite photographs as aids to the construction of streamlines in equatorial sea areas where conventional data is sparse, and V. G. Plank, *T*, 18 (1966), 1, on wind conditions associated with patternform and non-patternform cumuli in Florida, and *JAM*, 8 (1969), 46, on the size distribution of cumuli populations from air photos in Florida.

45. M. Glass and T. N. Carlson, *JAS*, 20 (1963), 397. For a technique for classifying cumulus clouds based on photogrammetry, see C. E. Anderson, *CD*, p. 50. For a study of the roots of orographic cumuli, see R. R. Braham, Jr., and M. Draginis, *CD*, p. 1. T. Fujita, K. A. Styber, and R. A. Brown, *JAM*, 1 (1962), 26. O. Johnson, *W*, 14 (1959), 212. J. Warner and J. W. Telford, *JAS*, 20 (1963), 313. On orographically induced cumulus, see H. D. Orville, *JAS*, 22 (1965), 700, on photogrammetric study of cumulus initiation over the Santa Catalina Mts., Arizona, *JAS*, 25 (1968), 385, for a mathematical model of mountain cumulus, and *loc. cit.*, p. 1164, for a mathematical model of upslope winds associated with mountain cumulus. For aircraft traverses of mountain cumulus over the San Francisco Peaks in Arizona, see C. J. Todd, *JAS*, 21 (1964), 529. For details of " artificial " cumulus, see E. N. Lawrence, *Marine Observer* (Jan. 1963), p. 33, on some initiated by ships at sea, and D. G. Morris, *BAMS*, 49 (1968), 1054, on some initiated by static firing of a Saturn V rocket.

46. R. J. Murgatroyd, *MM*, 83 (1954), 208. D. G. James, *QJRMS*, 79 (1953), 425. B. Woodward, *CD*, p. 28. F. H. Ludlam, *QJRMS*, 79 (1953), 430. On circulatory gust cells associated with cumulus in England, see A. Watts, *W*, 21 (1966), 188.

47. J. S. Malkus, *W*, 8 (1953), 291.

48. *Ibid.* For cumuli over the tropical oceans, see: J. S. Malkus and H. Riehl, *T*, 16 (1964), 275, on structure and distribution over the Pacific Ocean; R. L. Holle, *JAM*, 7 (1968), 173, on the climatology of cumuli populations; and M. Garstang and N. E. La Seur, *BAMS*, 49 (1968), 627, on a Barbados experiment on the effects of island and sea on cumulus convection.

49. H. Stommel, *JM*, 4 (1947), 91. For observations of cumulus entrainment, see P. W. Sloss, *JAM*, 6 (1967), 878, on Bemidji, Minn. For theories of cumulus development, see A. B. Fraser, *QJRMS*, 94 (1968), 71, for a mathematical model of the cumulus cycle, and K. Haman, *T*, 19 (1967), 33, on cumulus convection above an isolated heat source. For a theory of compensatory downward motions in the clear areas between individual cumuli, see R. Asai and A. Kasahara, *JAS*, 24 (1967), 487. On the relation between vertical motion and cumulus convection in a hurricane, see W. M. Gray, *MWR*, 95 (1967), 53.

50. E. L. Harrington, *JM*, 15 (1958), 127. For a study of cumulus convection in Florida, which combines both land and water tendencies, see V. G. Plank, *CD*, p. 109. B. Ackerman, *JM*, 16 (1959), 191. C. W. Newton and H. R. Newton, *JM*, 16 (1959), 483. See also C. W. Newton, *CD*, p. 135, for the dynamics of the system. For details of an unusual cumulus, in the form of a flat conical mound 60 miles in diameter, reaching to 7.5 km, which formed over turbid water in the Bay of Bengal, see J. D. Isaacs, *JGR*, 67 (1962), 2076. On the effect of vertical wind shear on cumulus growth, see N. S. Bhaskara Rao and M. V. Dekate, *QJRMS*, 93 (1967), 363, for India, and C. O. Erickson, *MWR*, 92 (1964), 283, on TIROS photographs of Florida.

51. B. Ackerman, *JM*, 13 (1956), 302. T. R. Morris, *JM*, 14 (1957), 281. S. Twomey and P. Squires, *T*, 11 (1959), 408. For details of cloud particles, see J. D. Marwitz and A. H. Auer, Jr., *JAM*, 7 (1968), 448, on nuclei spectra and updrafts beneath cumuli bases in the High Plains of Colorado and South Dakota; see also P. B. McCready, Jr., and D. M. Takeuchi, *JAM*, 7 (1968), 591, on droplet characteristics and precipitation initiation in Arizona.

52. P. Das, *JAS*, 21 (1964), 404. L. J. Battan, *JAS*, 22 (1965), 79.

53. P. M. Saunders, *MM*, 88 (1959), 233. For a photogrammetric study of cumulus growth in Wagga Wagga, Australia, see K. Higuchi, *JAS*, 22 (1965), 440. For details of glaciation of a cumulus at −4°C off the south coast of Australia, see S. C. Mossop, R. E. Ruskin, and K. J. Heffernan, *JAS*, 25 (1968), 889.

54. See R. Wexler, *CD*, p. 129, concerning cold clouds over the tropical oceans. F. H. Ludlam, *W*, 7 (1952), 199, and 11 (1956), 187. I. C. Browne, G. J. Day, and F. H. Ludlam, *MM*, 84 (1955), 72. P. M. Saunders, *op. cit.* (1959). R. D. Reynolds, *BAMS*, 38 (1957), 518. On shower clouds, see: K. A. Browning *et al.*, *QJRMS*, 94 (1968), 498, on air motion within a cloud over Pershore, England; K. A. Browning and D. Atlas, *JAS*, 22 (1965), 678, on initiation of precipitation over Oklahoma City; and L. J. Battan, *JAS*, 22 (1965), 79, on precipitation factors in Arizona. On showers, see W. D. Summersby, *MM*, 96 (1967), 41, on the climatology of ocean weather ships I and J; on shower activity in northwesterly airstreams over England and Wales, see C. A. S. Lowndes, *MM*, 95 (1966), 1, 80, 248, and 264.

55. For an example of a cumulonimbus over Sumatra reaching tropopause height (55,000 feet) in the region of the intertropical front, see R. Frost, *MM*, 81 (1952), 241. On history of cumulonimbus study and diagrams of models, see *AMM*, 5, no. 27 (1963), 3–17. R. S. Scorer and F. H. Ludlam, *QJRMS*, 79 (1953), 94. At Elmdon, Warwickshire, on May 18, 1952, according to A. W. Berry, *MM*, 81 (1952), 247, two small flat cumuli grew 9,000 or 10,000 feet upward within 5 minutes.

For general accounts of cumulonimbus, see *W*, 24 (1969), 31 and 65. For an example of the coexistence of cumulonimbus and haze at Yeovilton, England, see *W*, 20 (1965), 319. On cumulonimbus convection, see F. H. Ludlam, *T*, 18 (1966), 687. On circulations in large sheared cumulonimbus, see C. W. Newton, *T*, 18 (1966), 699. For an example of cumulus heads with anvil "puffs" in Cyprus, see G. J. Jefferson, *MM*, 94 (1965), 23.

56. R. S. Scorer, *W*, 8 (1953), 198. R. King, *MM*, 86 (1957), 57. For an example of a torrent in southern England, on July 1, 1953, see G. J. Day, *MM*, 82 (1953), 342; the cumulonimbus grew from a single cloud to a gigantic one 10 miles wide, and the maximum rainfall recorded from it was 1.64 inches at Cobham. The Liverpool–Birkenhead area appears to cause cumulonimbi to develop north of Chester above the widest part of the Mersey estuary; see M. H. O. Hoddinott, *W*, 11 (1956), 260. For details of updrafts beneath cumulonimbi in Colorado, see A. H. Auer and W. Sand, *JAM*, 5 (1966), 461. On the deduction of updrafts from hailstone structure, see A. E. Carte, *JAM*, 7 (1968), 1041. For details of the environments in which hailstones grow, see K. A. Browning, *MM*, 96 (1967), 202. For an example of hail from a heavy thunderstorm at a low-level station near the equator, in Kuching, Sarawak, see A. Stemmler and P. M. Stephenson, *MM*, 94 (1965), 236.

57. J. S. Malkus and C. Ronne, *T*, 6 (1954), 351. See also J. S. Malkus, *CD*, p. 65. J. S. Malkus and R. T. Williams, *AMM*, 5, no. 27 (1963), pp. 59–64.

For a convective index of cumulonimbus development in Alberta, see W. K. Sly, *JAM*, 5 (1966), 839. On available energy in growing cumulonimbus, see R. A. Brown, *JAS*, 24 (1967), 308.

58. P. Squires, *CD*, p. 44. Malkus and Ronne, (1954), *op. cit.* J. Levine, *JM*, 16 (1959), 653. Malkus and Williams, *op. cit.* (1963). On physical conditions in the summit areas of intense cumulonimbi in Oklahoma, see W. T. Roach, *QJRMS*, 93 (1967), 318

59. Malkus, *CD*, p. 65.

60. For many articles on stratocumulus, see *MAB* and *MGAB* subject indexes. On physical aspects of stratocumulus, see S. G. Cornford, *MM*, 95 (1966), 292; on ice crystals in stratocumulus, see J. B. Stewart, *MM*, 96 (1967), 23. For examples of stratocumulus, see *W*, 21 (1966), 58, on Jan Mayen islands and R. F. Zobel and S. G. Cornford, *MM*, 95 (1966), 65, on Malaya.

61. D. G. James, *MM*, 85 (1956), 202, and *QJRMS*, 85 (1959), 120; see also "*MO* discussion," *MM*, 86 (1957), 230.

62. James, *op. cit.* (1956 and 1959). For a mathematical model of mixed layers under a strong inversion topped with turbulence cloud (e.g., stratocumulus in the Californian marine layer off Oakland), see D. K. Lilly, *QJRMS*, 94 (1968), 293.

63. B. J. Mason and B. P. Howarth, *QJRMS*, 78 (1952), 226. T. Okita, *T*, 14 (1962), 310. C. P. Mook, *BAMS*, 36 (1955), 490. On precipitation from layer clouds in the English Midlands, see J. B. Stewart, *QJRMS*, 90 (1964), 287. According to P. B. Storebö, *T*, 20 (1968), 239, raindrop scavenging of low-level clouds, which have undergone little lifting, is an important precipitation initiator in western Norway.

64. K. E. Woodley, *MM*, 84 (1955), 25. For a theory of the formation of layer clouds, based on turbulence theory, see L. T. Matvejev, *T*, 16 (1964), 139. For a mathematical model in which stratus forms in layers, separated by clear air, in a stable atmosphere subjected to gravity waves, see A. J. Abdullah, *MWR*, 95 (1967), 189. For a mathematical model of the nocturnal development of air-mass stratus in south central Texas, see L. L. LeBlanc and K. C. Brundidge, *JAM*, 8 (1969), 177.

65. "*MO* discussion", *MM*, 87 (1958), 151. M. H. Freeman, *MM*, 91 (1962), 357. W. R. Sparks, *MM*, 91 (1962), 361. On low stratus (haar) at Leuchars, Scotland, see L. L. Alexander, *MM*, 93 (1964), 397, on tidal effects, and *MM*, 94 (1965), 292.

66. N. E. Davis, *MM*, 81 (1952), 231.

67. F. A. Barnes, *W*, 4 (1949), 187. T. M. Thomas, *W*, 11 (1956), 183. P. G. F. Caton, *MM*, 86 (1957), 162. C. S. Durst, *The Meteorology of Airfields*, *MOSP*, no. 507 (1949), pp. 16-23.

68. In particular, see the *MO Aviation Meteorology Reports*. See also *MAB* and *MGAB* abstracts under the heading "Synoptic Climatology" for various areas. For details of the effect of the Aleutian Islands

on mesoscale motions in oceanic stratus, as indicated by satellite photographs, see W. A. Lyons and T. Fujita, *MWR*, 96 (1968), 304. For a mathematical model of synoptic-scale variations in low stratus, see J. P. Gerrity, Jr., *MWR*, 95 (1967), 261.

69. R. S. Scorer, *SP*, 46 (1958), 75.

70. A standing wave-cloud over Pershore, Worcestershire, is described by R. W. Dillon, *W*, 15 (1960), 310. T. M. Thomas, *W*, 18 (1963), 166, describes the chinook arch cloud of the Great Plains of the United States. D. J. George, *W*, 14 (1959), 233. F. M. Bancroft, *MM*, 83 (1954), 57. F. H. Ludlam, *W*, 11 (1956), 187. For examples of wave clouds, see: J. H. Conover, *W*, 19 (1964), 79, from satellite and air photos of Cape Breton island; T. A. M. Bradbury, *W*, 20 (1965), 16, for Berkshire, England; *W*, 20 (1965), 57, for Sutherland, Scotland; and *W*, 24 (1969), 142, on wave-cloud shapes induced by conical mountains. For examples of wave clouds with lines or billows, see *W*, 21 (1966), 97 and 172. For examples of altocumulus varieties, see *W*, 21 (1966), 286.

71. R. M. Poulter, *MM*, 86 (1957), 245. On billow clouds, see: R. S. Scorer, *QJRMS*, 77 (1951), 235; K. E. Woodley, *W*, 7 (1952), 148; F. H. Ludlam, *QJRMS*, 93 (1967), 419. For an example of pileus, see A. H. Thompson, *W*, 19 (1964), 319. On the formation of billow clouds, see J. D. Woods, *QJRMS*, 94 (1968), 209, and see Ludlam, *op. cit.*, on their relation to CAT. For examples of billow clouds, see W. E. Pennell, *W*, 19 (1964), 226, and *W*, 21 (1966), 172. For billows in medium-level clouds, see *W*, 20 (1965), 118, and *W*, 21 (1966), 439. For examples of funnel clouds, see: R. J. Donaldson, Jr., and W. E. Lamkin, *MWR*, 92 (1964), 326, on a tornado in Oklahoma City; *W*, 21 (1966), 218, on Co. Waterford, Eire; and T. P. Leary, *BAMS*, 50 (1969), 251, on cone-shaped clouds succeeding a tornado funnel cloud at Omaha, Nebr.

CHAPTER 7

Visual Climate and Optical Climatology

Visibility is an atmospheric property that has a precise, although extremely complex, relationship to the Earth's surface where it is observed. It is thus a weather element of great geographical interest. In meteorological usage, the term visibility means both the condition of being visible, and the "visual range." The latter term was introduced by Bennett in 1930 to mean the distance at which something can be seen, or, in other words, the clearness with which objects stand out from their surroundings. As atmospheric obscuration of light increases, the visibility of an object is reduced, until finally it becomes zero.

The IMO in 1937 adopted Bergeron's definition of good visibility. In pure air, the atmosphere is extremely clear and transparent, so that distant objects and their details stand out in full relief from their background with great hardness and distinction, as in stereoscopic vision, without any softening veil, even at distances of 10 km or more. The IMO definition of visibility, as the maximum horizontal distance at which prominent objects (trees, houses, etc.) can be seen and recognized, is difficult to apply in practice, because whether or not the objects are recognized depends on whether they are familiar to the observer. Hence the American definition is usually preferred: visibility is the maximum distance at which an object can be seen against its background.

Visibility is one of the least understood of the climatic elements, despite its enormous practical importance and the existence of a considerable amount of optical theory. Information about changes in visibility is often of critical significance in aviation, and to geographers and others concerned with the "visible landscape," visual range is the most important of all the meteorological elements.

Although the curvature of the Earth's surface limits the horizontal distance at which topographical features can be seen from a given point, the normal distances visible at the Earth's surface are very much less than this limiting distance, because of the presence of the atmosphere. The various constituents of the atmosphere, water droplets, dust, smoke and other pollution particles, condensation nuclei, and the air molecules themselves, determine the optical visibility through it. Different airstreams, different air masses, and different topographies, geographical locations, and land uses have characteristic visibility ranges associated with them, depending on the season and time of day. The determination of the normal visibility range (not necessarily the same as the arithmetic average of observed visibilities at a station for a long period) for each of these categories is a problem with widespread practical ramifications, but we are as yet far from its solution.

Most of the visibility phenomena in the atmosphere are produced by particles much larger than air molecules, or by variations in the density of adjacent air layers. If visibility depended only on the scattering of light by air molecules, the horizontal visual range at sea level would always be several hundred kilometers.

Liquid water droplets are the most important scattering agents: they vary in radius from 10^{-6} to 10^{-1} cm, increasing in size as the relative humidity of the air increases, and more and more water vapor condenses on them. Particles with radii of 0.5 micron or less (air molecules, condensation nuclei, and smoke particles, for example) show selective scattering of visible light, scattering blue light more strongly than red. This scattering results in their appearing bluish by reflection, giving haze its typically blue colour. In contrast, fog is usually white or colorless, because it consists of water droplets of radii greater than 0.5 micron; droplets of this size scatter white light not selectively, but totally. Precipitation and dust particles have the same effect.

Meteorological visibility is defined as the greatest distance at which an object of specified characteristics can be seen and identified by an observer with normal sight under normal conditions of daylight illumination.[1] *Daylight visibility* is then defined as the distance at which dark objects, seen against the horizon sky as background, are just identifiable. It is estimated by noting the furthest object that can be seen and identified out of a series of preselected objects at known distances from the observer. These objects normally are situated from 10 to 20 yards away to the furthest distance at which a suitable object is available. They should subtend an angle of more then $\frac{1}{2}°$ in width and height (i.e., 4.6 feet at a distance of one mile), but not more than $5°$ in width, at the observer's eye.

Daylight visibility observations need not necessarily be made against a sky background, so long as the background is appreciably further away than the object. For a background 1.5 times as distant as the object, the difference in visual range from that for a background of the horizon sky is less than 4 per cent.

Night visibility is defined as the distance at which the standard objects used for daylight visibility estimations would be just distinguishable if it were daylight, and the degree of atmospheric transparency were unchanged. Its estimation is based on the visual perception of lights of known strengths, at known distances, expressed in terms of equivalent daytime visibility. Estimates from ships at sea are made in a similar manner.

Daylight and night visibilities are usually estimated in a horizontal direction. For visibility in an oblique direction, air to ground or vice versa, atmospheric properties vary with height, and the objects used are not viewed against a sky background, but against terrestrial backgrounds varying from almost perfect reflectors of light such as snowfields, to almost perfect absorbers such as dark forests or rock surfaces. Hence the visual ranges will be different.

At night, for example, with a uniform fog or haze layer extending to half the height of the observer, and with clear air above, an ordinary light on the ground can be seen by the observer when it is 1.5 to 1.8 times as far from him (actual, not horizontal, distance) as the furthest distance at which it is visible to another observer on the ground. If the air beneath the observer is divided into two equal layers, the lower with a horizontal visibility twice that of the upper, then, when the ground-level visual range is $1\frac{1}{2}$ miles or less, a light can be seen by the air observer for only 0.7 to 0.8 of the distance at which the same light can be seen on the ground. In each case, the discrepancy between oblique and horizontal visibilities increases as the candlepower of the light increases, and as the surface visibility range decreases.

Slant visibility is an oblique visual range specially designed for aeronautic purposes.[2] It is defined as the distance from the pilot's eye to the farthest point on the ground, farthest ground marker, or farthest approach or runway light that he can see. It describes the amount of ground or ground lighting visible to him, from the farthest visible point to the nearest visible point before the ground is cut off by the cockpit of the aircraft. Slant visibility can thus be measured either by its length, or by the angle it subtends at the observer's eye. It can be estimated by observing a light or lights carried by a tethered balloon, or by employing two intersecting searchlight beams, one vertical, the other at an angle to the vertical. Assuming the beams do not pass through fogs of different densities, the slant visibility may then be found approximately by varying the intensity of the vertical beam until the two beams visually match at the point of intersection.

There appears to be little correlation between slant and horizontal visibility. For example, with a horizontal visual range of 1,200 feet, the slant visual ranges observed from heights of 50, 100, and 150 feet amount to 900, 900, and 1,800 feet respectively. If there is a layer of horizontally stratified homogeneous fog on the ground, then the slant visual range varies with the height of the observer and the horizontal visibility range. If the density of the fog increases upward, the slant visibility decreases with height, and vice versa.

Vertical visibility is almost always less than horizontal visibility, particularly in radiation fog, because condensation in up-currents causes the water droplets to increase in size with height. Radiation fogs are often thickest in their upper layers, in which case the slant visibility decreases with height. However, fogs are usually far from uniform horizontally, an additional complication.

Horizontal visibility often varies considerably with direction, depending on surface winds. Visibilities are generally much lower for winds from the land than for sea breezes, due to the fact that condensation nuclei are more numerous, and more hygroscopic, in continental than in maritime air. With continental air, the visibility range depends on relative humidity at all humidity levels. With maritime air, atmospheric opacity decreases with decreasing relative humidity down to 70 per cent, below which the opacity is practically independent of the relative humidity. Explanation of this fact is not easy.

At Leeuwarden, for example, the condensation nuclei in maritime air appear to be in droplet form at high humidities, the radii varying with the excess of relative humidity over 70 per cent. Below 70 per cent humidity, the nuclei are solid, with constant radii. In continental air, the nuclei are pure solution droplets. Thus the nuclei transported to Leeuwarden by sea trajectories are mixed, consisting of insoluble matter with a coat of soluble hygroscopic substance; those brought via a land trajectory are completely soluble. However, the opposite is true at Frankfurt: maritime air contains fewer insoluble substances than continental air.[3]

These types of visual range involve partly subjective estimates, especially of night visibility. Accurate instrumental visibility measurements are thus very necessary. An understanding of the standard instrumental techniques demands, however, some knowledge of the optical theory of visibility on which they are based.

The Theory of Visibility

As yet, there is no perfect theory of meteorological optics. All the existing theories assume for simplicity that visual space is Euclidean, whereas our binocular vision presents us with a Riemannian manifold. The geometry of our visual space should be hyperbolic, rather than the geometry of Euclid.[4]

In other words, observation shows that visual space in the atmosphere is not a simple transformation of physical space. This is illustrated by measurement of the apparent shape of the sky. The daytime sky appears to be a flattened vault, which is flattest under cloudy conditions. A moonless night sky, by contrast, appears to be arched. This effect may result in mistakes when one is estimating cloud amount visually, the overhead cloud amount being underestimated and the horizon cloud amount overestimated. The distance of the terrestrial horizon and the general level of atmospheric brightness are additional factors.

As is shown in Figure 7.1, the apparent shape of the sky may be measured by determining the angle of elevation θ of the estimated position of point P, which bisects the imaginary arc ZH extending from zenith Z to horizon H. The flatter the sky appears, the smaller θ becomes. Clouds increase the apparent flatness, but the increase is negligible if the cloudiness increases beyond four-tenths (see Appendix 7.1).

Psychological factors enter into the problem, because a cloud layer at 2 km above the ground should produce a θ of 1°, but observation makes this angle out as between 20 and 30°; that is, the sky does not appear as flat as it ought to according to Euclidean geometry. Observation gives a θ of 27.8° for altocumulus and altostratus cloud layers for seven-tenths cloudiness, and 29.0° for 10-tenths of stratus. According to simple geometry, the opposite should be the case (see Figure 7.2).

The apparent shape of the sky is also affected by the distance to the terrestrial horizon: θ varies from 32.6°, if the horizon is 0.4 km distant, to 29.2°, if it is 12 km away. Variations in visual range have much less effect, but the apparent shape of the sky has a marked influence on visibility phenomena. For example, it causes overestimation of the height and steepness of mountains, causes an apparent enlargement of the sun and moon when they are near the horizon, and makes circular haloes appear elliptical. It also results in large errors in the estimation of the angular altitude of the sun: estimates of 45° are often given for an actual altitude of 25°.

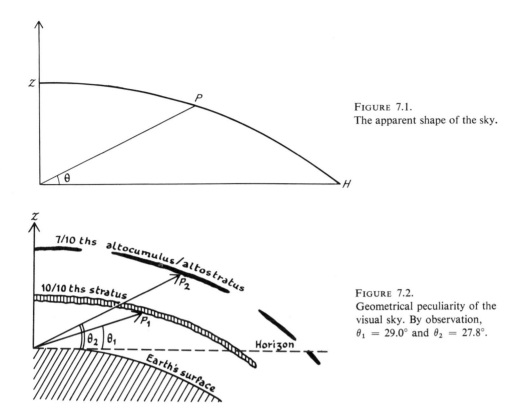

FIGURE 7.1.
The apparent shape of the sky.

FIGURE 7.2.
Geometrical peculiarity of the visual sky. By observation, $\theta_1 = 29.0°$ and $\theta_2 = 27.8°$.

Observation shows that the more distant an object is from the observer, the brighter it appears. This increase in the apparent brightness of an object with increasing distance from the observer is due to the sunlight and skylight scattered toward the observer by the atmospheric suspensoids in the line of vision. The angular distribution of the scattered light depends on the size of the suspended particles and the spatial distribution of the incident light.

For an object viewed against the horizon sky, a distance will finally be reached at which the brightness of the object is approximately the same as the brightness of the sky background. The object will then become invisible. This limiting distance is the *meteorological visibility* or *visual range*. The concept was first stated mathematically by Koschmieder in 1924, and the principle marks the beginning of the optical theory of atmospheric visibility.

A beam of light passing through the atmosphere is diminished in intensity by (a) scattering by air molecules, nuclei, and small water droplets, (b) diffuse reflection by the larger solid and liquid particles in suspension, and (c) absorption by the solid particles of dust, smoke and salt in the air. The total diminution (ΔE) in the density of luminous flux (E) of a parallel beam of light passing through a distance Δd of the atmosphere is given by

$$\Delta E = -\sigma E \, \Delta d$$

where σ, the *extinction coefficient* of the atmosphere, can be regarded as being made up of two parts: the first, representing the effect of scattering and diffuse reflection, is termed the *scatter coefficient*; the second, representing the effect of absorption, is termed the *absorption coefficient*. In air free from smoke particles, the latter is negligible, and the extinction coefficient approximately equals the scatter coefficient.

The maximum distance at which a given object can be seen depends on nine factors. The most important of these is the state of the atmosphere, represented by the extinction coefficient. The other factors include the refracting powers of the object and the background, the elevation of the sun, the general level of illumination, the angular size of the object, the angular separation of the sun and the object, and, obviously, the standard of eyesight of the observer. In the measurement of visibility, therefore, the effect of these factors must either be eliminated or standardized.

According to Koschmieder and Gotz, a measure known as the *standard visibility*, V, may be defined as equal to $\frac{3.912}{\sigma}$. The standard visibility so defined is the visibility computed from theory for a black object and a constant visual threshold. It is sometimes termed the *meteorological range*, which is defined as that distance for which the contrast transmittance of the atmosphere is 2 per cent. Standard visibilities may also be computed for nonblack (i.e., reflecting) objects viewed both against the horizon sky and against reflecting backgrounds. (See Appendix 7.2.)

The most important contributions to the extinction coefficient are made by salt nuclei and smoke particles. The number of these per unit volume of air varies with wind direction, and shows marked annual variation. As windspeed increases, the number of smoke particles in unit volume decreases, but the number of salt nuclei increases. As the relative humidity drops, the number of smoke particles also decreases, but the number of salt nuclei does not appreciably change.

For coastal locations, the extinction coefficient increases as windspeed and relative humidity increase. For inland locations, the extinction coefficient increases when the wind is in the direction of a smoke source, even 100 to 150 miles distant, and the increase is greater the lighter the winds. With high relative humidities, the extinction coefficient is, roughly, inversely proportional to the distance from the smoke source.

The extinction coefficient has a dimensionality of $length^{-1}$, and hence is not easy to visualize. An illustration of its effect may be given thus: if light is transmitted through a medium in which the extinction coefficient is uniform and equal to 1, its intensity would be reduced by nearly two-thirds in each unit distance (1 cm in the cgs system, and 1 m in the mks system). The atmosphere is very rarely as obscure as this, and hence the extinction coefficient is rather a large unit.

An alternative unit, the *nebule*, was introduced by Gold. This is easier to visualize than the extinction coefficient because, for example, ten nebules means ten times as much obscuring matter as one nebule, and 100 nebules reduces the light transmitted to one thousandth of its original intensity. For a daylight visibility of 10 yards, there will be 105 nebules between the object and the observer; and visibilities of 100, 1,000, and 10,000 yd correspond to 85, 67, and 53 nebules respectively. It is not possible to set up such simple equivalents between extinction coefficients and visibilities. In terms of nebules, the meteorological range, V, is defined by

$$V = 80\left(\frac{D - d}{m - n}\right),$$

in which m, n, are the number of nebules required to bring two similar objects, at distances d, D, from the observer $(D > d)$, to the limit of visibility. (See Appendix 7.3.)

The extinction coefficient depends on the wavelength of the transmitted light; its value for red light is different from that for yellow light, for example. In water fogs consisting of droplets of more than 2 microns in radius, there is little variation in σ with wavelength, but with small droplets or with smoke haze the extinction is appreciably less for red light. In polluted fogs, extinction is least for yellow light. The value of σ usually decreases with elevation, because of atmospheric stratification. Hence objects chosen as visibility standards should not be at greatly different elevations.

To determine the visibility ranges appropriate to different atmospheric conditions, therefore, the extinction coefficient must be determined, or the meteorological range V must be found. The extinction coefficient can be deduced from an application of Mie's theory, which in practice involves determining the number and size of obscuring particles—water droplets, salt and smoke particles, etc.—in a known volume of air; this is not a method to use on a routine basis. However, Mie's theory does enable some useful facts to be deduced. For example, it shows that fog droplets of radius 10 microns or more scatter twice the amount of light that would be expected intuitively. (See Appendix 7.4.)

From Mie's theory, the amount of light scattered by a water drop of radius r is $2\pi r^2 I$, where I is the intensity of the incident light. Because of diffraction, the amount of light scattered is greater than the incident light. Thus the illumination from a sky overcast by clouds consisting of large water droplets is actually greater than if the sky was clear.

As an illustration of the application of theory, consider conditions typical of a moderate fog: a droplet density of 10^{-6} gram of water per unit volume of air and an average droplet radius of 2×10^{-3} cm; in a distance of 200 m the atmospheric opacity would be 217 nebules and the transmitted light 3×10^{-7} of the incident light. The daylight visibility in such a fog would be 60 m, and at night a 10-candlepower lamp would be just visible in darkness at slightly over 100 m.

At night, in both theory and practice, conditions are more complicated, because of the variable adaptation of the observer's eye, and at any time the problem of *contrast* is difficult. The visibility of any object depends on the contrast between it and its background, and contrast is not easy to define or to measure. Contrast is a function of three variables: the brightness of the object, the brightness of the background, and the solid angle subtended by the object at the eye. The standard measures of contrast, however, do not take all these variables into account.

The two most well-known contrast measures, those of Middleton and Gold, are in effect models that allow specific conditions to be deduced. According to Middleton's model, for an object whose visual threshold is 0.02 for viewing direct against the ordinary horizon sky, if the brightness decreases to 10^{-3} candles per m², the threshold then becomes equal to unity and the object disappears. According to Gold's model, a very large object about 10 times as bright as the background sky can be seen up to 20,000 yards in daylight (without much variation between full daylight and twilight) when the meteorological range is 10,000 yards. After twilight, there is a rapid fall in the visibility of the object, to 10,000 yards in starlight and to 8,000 yards in overcast starlight. (See Appendix 7.5.)

One of the main difficulties in visibility specification, the observer's visual acuity, has been taken into account successfully in the study of slant visibility. The minimum illuminance that an observer can perceive varies for a given person from time to time, and also of course varies from observer to observer. Hence any value of this threshold has an associated probability of occurrence, and it is possible to determine its statistical distribution, remembering that this distribution will vary with background and with ambient light.[5]

The visual range may be found from Allard's law, which involves both the *visual illuminance threshold* of the observer (i.e., the minimum brightness of a light source that an individual observer can detect) and the *transmissivity* of the atmosphere, the latter being the fraction of the original beam of light remaining after its passage through the atmosphere from source to observer. The transmissivity depends on the extinction coefficient, and can be accurately measured by a transmissometer. Transmissometers operating from baselines of 500 to 750 feet were installed at almost 100 airfields in the United States in 1959. By siting them near the touchdown points, data can be provided for both runway visual range and approach-light contact height, both of which form important aeronautic visibility parameters. (See Appendix 7.6.)

From the cumulative distributions of each selected class of visual illuminance thresholds, a single threshold can be chosen, and the probability that an observer will have a threshold equal to or less than the selected value can be stated. Alternatively, given a particular probability value, the slant visual range can be computed using Allard's law, which may be restated to enable the atmospheric light transmission for a specific direction and distance to be determined. The effect of clouds on visibility range can then be taken into account, as well as the influence of snow, haze, fog, and smoke. (see Appendix 7.7.)

Nighttime visibility determinations pose many problems, both practical and theoretical. Because the general illumination is very low, the extinction coefficient increases almost indeterminably. Using the definition that visibility at night is the maximum distance at which a particular light is visible, it is possible to compute a standard visibility in terms of the intensity of this light. An *equivalent daylight visibility* may then be defined by expressing the visibility of a black object against the horizon sky in daylight in terms of the visibility of a light of known intensity at night in an atmosphere with the same scattering coefficient. (See Appendix 7.8.)

The standard visibility computed for nighttime conditions depends very much on the intensity of the observed light. The visibility in full moonlight is approximately 20 per cent of that in daylight in the same atmosphere, and a light of 100 candlepower just visible at 200 yards has an equivalent daytime visibility of 110 yards. In fact, for visual ranges of less than 1,000 yards, almost any light source will give a visibility considerably greater than the corresponding daytime visibility. This explains the apparent increase in visibility frequently reported between late afternoon and evening under conditions of poor visibility.

The visibility of point sources of light is much greater than that of diffuse extended sources (see Appendix 7.9), and the threshold density of luminous flux for large diffuse sources is about 10,000 times greater than that for a point source. These facts are clearly important in the selection of lights for night visibility observation. Low-intensity, noncollimated lights should be selected, street lights and obstruction lights, for example; beacons and lighthouses should be used only if they are very distant.

Visibility of lights at night depends on the presence or absence of other bright lights in the field of view, as well as on the factors already described as influencing daylight visibility observation. It is especially important to use direct vision for night observations, because the sensitivity of the eye for indirect vision (i.e., looking to one side of the light, instead of directly at it) is greater than for direct vision, for all except red light.

One must allow several minutes at least for the eye to become completely adapted to the dark before observing the lights. The visual threshold of the eye for night observations, i.e., the illumination produced at the eye by a light so faint that it can only just be seen, varies from person to person and from time to time in the same individual. On the average, it equals the illumination produced by a lamp of 0.15 candlepower at a distance of 1 km from the observer in perfectly clear air. Studies of night observations of runway lights by different observers have revealed how much subjectivity is involved. If two observers, both of whom have the standard visual threshold noted above, record the distance at which a given light can be seen, then the standard deviation of their visual acuity is approximately $14\,d$ per 160 yards, where d is the smaller visual range observed. If the observers have different visual acuities, the factor 14 has to be increased two or three times (see Appendix 7.10).

Climatological Representation of Visibility Observations

Standard observations of visibility in published form are almost entirely limited to subjective estimates of objects in daylight, or of lights at night. The official instructions for observers, recommended by the WMO Washington Conference in 1947, specified that the visibility categories should proceed in steps of 20 m from 20 to 200 m, and then in steps of 200 m to 16 km. Thus the international code is made up of several arithmetic progressions, although theory indicates that visibility categories ought to vary in a geometric progression.

Table 7.1, A, shows the current categories recognized for synoptic visibility reports. The selection of objects for use as daytime standards should be based on the fact that if perfectly black objects elevated above the horizon are employed, the visual range

TABLE 7.1, A
Visibility-reporting categories for synoptic purposes
(British Meteorological Office practice)

First Code Figure[a]	Second Code Figure									
	0	1	2	3	4	5	6	7	8	9
5	5[b]	—	—	—	—	—	6	7	8	9
6	10	11	12	13	14	15	16	17	18	19
7	20	21	22	23	24	25	26	27	28	29
8	30	35	40	45	50	55	60	65	70	over 70
9	<50	50	200	500	1,000	2,000	4	10	20	50 or over

[a] Code figures 01–50 give the actual visibility in tenths of kilometers. Code figures 51 through 55 are not used. Code figures 90 99 are used when visibility cannot be determined with sufficient accuracy to justify the use of the lower code figures. Figures in italic are in meters; all others are in kilometers.

[b] The figures in the table give the horizontal visual range. If the actual horizontal visibility is between two of the distances given in the table, the code figure for the lower distance is reported. (Taken from the M.O.'s *Introduction to the Daily Weather Report*, 1969.)

TABLE 7.1, B
Visibility-reporting categories for climatological purposes
(*British Meteorological Office practice*)

Object[a]	Standard distance	Permissible variations
Fine scale	(meters)	(yards)
A	20	20–24
B	40	40–48
C	100	100–120
D	200	200–240
E	400	400–480
F	1,000	1,000–1,200
Coarse scale	(kilometers)	(miles)
G	2	$1\frac{1}{8}$–$1\frac{3}{8}$
H	4	$2\frac{1}{4}$–$2\frac{3}{4}$
I	7	4–$4\frac{3}{4}$
J	10	$5\frac{5}{8}$–$6\frac{7}{8}$
K	20	$11\frac{1}{4}$–$13\frac{3}{4}$
L	30	$16\frac{3}{4}$–$20\frac{1}{2}$
M	40	22–28

[a] All climatological and synoptic stations are required to compile a list of objects at these standard distances, to ensure uniformity in visibility reporting for synoptic purposes. (Taken from *MOOH*.)

so estimated will be practically independent of the elevation of the sun and its angular separation from the selected objects. The objects should also be well-distributed in azimuth so that variations in visibility with direction can be reported. Wherever possible, enough objects should be selected to allow the following degrees of precision in visibility estimation: to the nearest 20 yards for visibilities up to 220 yards; to the nearest furlong for 220 yard to 1 mile visibility; to the nearest quarter-mile for 1 to 2 miles visibility; to the nearest half-mile for 2 to 5 miles visibility; and to the nearest mile for visibilities of 5 to 10 miles. Above 10 miles visibility, it should be possible to estimate correctly to the nearest code figure.

For synoptic purposes, it is necessary to select a large number of objects if the required degree of precision is to be obtained. For climatological stations, however, only 13 objects are required, located at distances as indicated in Table 7.1, B.

When visibility estimates are being statistically summarized, undue emphasis will be given to the larger visual ranges if reported occurrences of each code figure are merely counted. To counteract this effect, one must either multiply each frequency by the reciprocal of the range of distances it represents, or use the mean extinction coefficient to weight the frequencies.[6]

Because visibility is a local phenomenon, world maps of visibility climatology are of little value, although climatological maps for smaller areas can be of some use (see Figures 7.3 and 7.4). Isopleth diagrams (see Figures 7.5 and 7.6) showing average visibilities at different times of day in various months, for individual stations, are much used in geographical work on visibility variations.*

---

* E.g., see C. S. Durst, *Meteorology of Airfields, MOSP*, 507 (1949), 5 and 12–13. For climatological mean maps of visibility data for the United States, see R. G. Eldridge, *JAM*, 5 (1966), 277.

For microgeographical visibility studies, the normal visual estimates of synoptic work are not reliable enough; instead instrumental techniques must be used. No perfect instrument for such work is yet available; in fact, there is plenty of scope for further experiment in developing these instruments.

In view of the subjective sources of error inherent in many of the published climatic statistics on visibility, instrumental determinations of atmospheric visual range are much to be desired at benchmark stations. Instruments for visibility measurement may be divided into the following classes: (a) instruments that measure the extinction coefficient, including Gold's visibility meter, Bergmann's visibility meter, and the photoelectric visibility meter; (b) instruments that measure the scatter coefficient of the atmosphere, including the Loofah hazemeter and the Beuttell visibility meter; (c) other instruments, such as the transmissometer, the Waldrum range meter, and the Shallenberger–Little visibility meter.† Essentially, visibility meters for daytime use either compare brightness, and so function as *telephotometers*, or sample the scattering produced by a small volume of air, and so function as *nephelometers*; visibility meters for nighttime use employ uncollimated light sources and the principle of the *comparison photometer* (see Appendix 7.11). The basic procedure during daylight is to estimate the scattering coefficient, by determining a contrast threshold induced by the light scattered by the air; the procedure at night is to estimate the scattering coefficient by measuring the transmitted flux from a light source against a dark background. Thus, apparently identical visual ranges as measured by day and by night are very likely not at all identical. The main source of error is that, although brightness contrasts are the basic property measured at both times, for the nighttime measurements, both the target object and its background are illuminated by lights, but since objects the size of the typical daytime targets are too large to be illuminated artificially, the lights themselves must be used as targets.

Some Visual-Range Phenomena

Visual-range phenomena such as fog or haze are very complex to study. A thorough examination and description of the fog habits of a locality must concern itself with such topics as nocturnal cooling, turbulent diffusion, air-mass modification, heat and water balance, boundary-layer atmospheric dynamics, and synoptic meteorology, in addition to simple observation.

Haze is a visibility phenomenon that exists when the visual range is between 1,110 and 2,200 yards. Hazes may consist of water droplets, of dust or smoke particles, or of organic material, and they often are white or whitish, brown, or blue in colour. Smoke haze is not particularly common, except very locally. Smoke plumes from chimneys often drift several miles downwind without merging into a natural haze layer. Dust haze usually has the color of the soil or other unconsolidated surface from which its constituent particles have arisen, and is not normally very widespread. Blue haze, on the other hand, is very extensive. Haze in general is the most important visual attribute of the atmosphere in fine-weather situations. A haze layer often appears on only one-half the horizon; the haze appears white if the observer is facing toward the sun and viewing the haze layer from above, but purplish-brown if the observer is facing

---

† For details of visibility measurement using backscattered light, see H. Vogt, *JAS*, 25 (1968), 912.

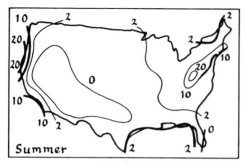
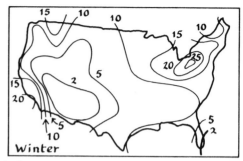

Percentage of time visibility is less than 2.5 km.

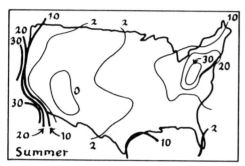
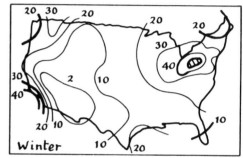

Percentage of time visibility is less than 5.0 km.

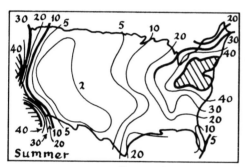
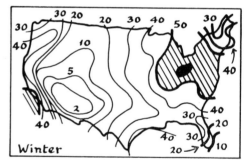

Percentage of time visibility is less than 10.km.

away from the sun. Several haze layers may often be present in anticyclones, the major, elevated layer coinciding with the inversion layer, and surface haze layers developing because of heating effects during the day.[7]

Atmospheric *blue haze* is of special climatological interest. Observations from aircraft soon convince one of the ubiquity of this phenomenon: at some height above the ground, usually between 5,000 and 10,000 feet, an increase of only a few dozen feet in the observer's altitude will result in a phenomenal increase in visual range by several orders of magnitude. Even when visibility at the ground is dozens of miles, such a level in the atmosphere may usually be found, often coinciding with the peplopause. The blue haze below this level is considered by Went to be composed of submicroscopic particles, smaller than the wavelength of light, of volatile organic substances emitted into the air by plants. Examples of these volatiles include terpenes (for example, pinene, which produces the typical aroma of a pine forest, and the substances

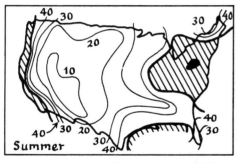
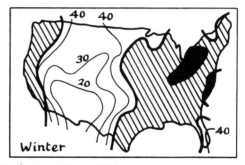

Percentage of time visibility is less than 20 km.

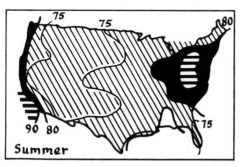
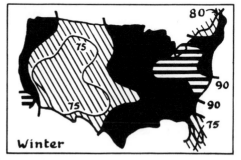

Percentage of time visibility is less than 40 km.

⬀ 50-80   ⬤ 80-90   ⊜ Over 90

FIGURE 7.3.
Cumulative visibility frequency maps for the United States, 1948-1958 (after R. G. Eldridge).

producing the aromatic smells of sagebrush and of various Mediterranean vegetation types), sesquiterpenes, acids, simple amines, esters, and such aromatic compounds as benzaldehyde, coumarin, and skatol. A fantastic amount of such volatile material is released by plants into the atmosphere, probably on the order of 1.75×10^8 tons per year for the Earth as a whole. The material disappears from the air layer nearest the ground within a very few hours, and, according to Went's hypothesis, is carried upward, if there is convection, to accumulate as *veil clouds* in the lower part of the inversion layer. These clouds are very dark, being grey if the observer is flying through them and almost black if he observes them in optical cross section, and are usually very thin, although in the tropics they may develop many layers. The dark coloring of the veil clouds, and also the dark linings of towering cumuli or cumulonimbi, appears not to be due to shadows but to matter other than water vapor that condenses in the middle and lower troposphere.[8]

Experimental evidence is available that light, passing through air containing an organic vapor, will cause the formation of a blue haze. For example, the passage of concentrated light from a high-pressure mercury or carbon arc through low concentrations of nitrogen oxide or a terpene produces a diffuse blue haze within a few minutes. The blue soon changes into white, and the white haze gradually comes to consist of small white droplets, which ultimately are visible to the naked eye. The energy for the formation of the haze from the organic vapor comes from the light. A similar

346

Average number of days per year on which
afternoon visibility is less than 4,400 yards.

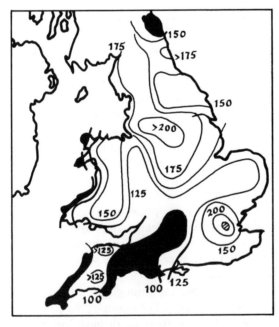

Average number of days per year on which
visibility in less than 6.25 miles.

FIGURE 7.4.
Afternoon visibility frequencies in England and Wales (after L. P. Smith, *MM*, 90, 1961, 355).

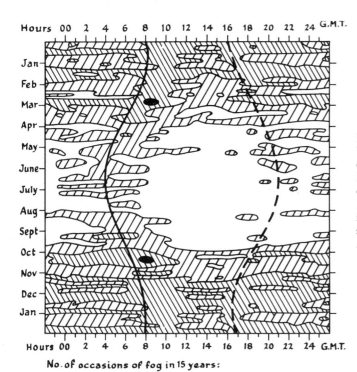

FIGURE 7.5.
Variation (annual and diurnal) of fog at Liverpool airport, England, 1945-1960. Fog is defined as a visibility of less than 1,100 yards. The solid line indicates sunrise, the broken line sunset. (After G. J. Bindon, *MM*, 91, 1962, 162.)

No. of occasions of fog in 15 years:
● 18 and over   ⬤ 6-18   ⬤ 1-6   ○ Less than 1

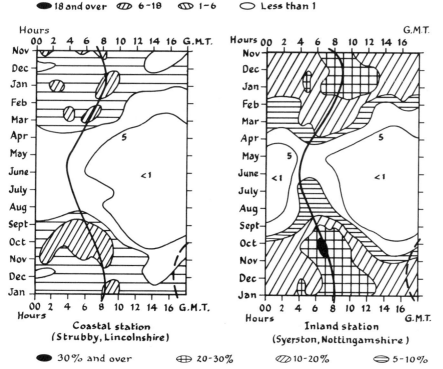

Coastal station
(Strubby, Lincolnshire)

Inland station
(Syerston, Nottingamshire)

● 30% and over   ⊞ 20-30%   ⬤ 10-20%   ⬤ 5-10%

FIGURE 7.6.
Comparison of fog frequency at two stations in eastern England, 1956–1964. The solid line indicates sunrise, the broken line sunset. (After F. J. Smith, *MM*, 96, 1967, 77.)

effect is observed in the atmosphere under summer anticyclonic conditions. Haze forming during the day becomes less dense at night, and nocturnal cooling produces a very shallow inversion layer, only a few meters deep. All the volatile matter given off by vegetation during the night accumulates in this shallow inversion layer; when the sun's rays reach this layer the following morning, blue haze begins to form, and the visual range near the ground begins to decrease, often quite rapidly. If much organic matter has accumulated during the night, the blue haze may soon change into a white haze, and later develop into a ground fog.[9]

Veil clouds and blue hazes apparently form only over vegetated areas. The white hazes characteristic of oceanic areas are composed of sodium chloride particles derived from the evaporation of sea water. Deserts are characterized by yellow dust hazes; semideserts and scrub areas with aromatic vegetation give rise to very diffuse blue haze. Polar or snow-covered areas give rise to white hazes composed of ice crystals.

The question of how the volatile materials disappear from the atmosphere is difficult. They may be washed down by rain and snow, becoming attached to the clay particles in the soil, which are carried by runoff into rivers and then into deltas; and since the organic matter is bituminous or asphaltic in nature, it may become the source of petroleum in oil-bearing shales. Alternatively, the materials may be burned within the atmosphere before they return to the surface of the Earth. For example, intense concentrations of the volatile matter in the dark parts of thunderclouds may somehow result in the separation of electrical charges: there is some evidence that thunderstorms are more frequent where the vegetation is more active. The funnel cloud of a tornado is very dark in color, and part of the energy maintaining the vortex may be chemical, originating in the burning of dense veil clouds in the upper part of the funnel.

That there is so much speculation in the above discussion on blue hazes at least proves that such an observationally simple phenomenon as continental haze may be very complex, both physically and geographically.[10]

Fog is a visibility phenomenon that exists when the visual range is less than 1,100 yards. It may consist of water droplets, ice crystals, water and solid matter, or ice and solid matter. Fogs may be classified in terms of the physical processes that give rise to them. The main types are: *radiation fog*, in which the air is cooled below dew-point temperature by radiational cooling of the underlying surface; *advection fog*, in which a relatively warm, moist mass of air is advected over and cooled by a colder surface; *steam fog*, which is formed by evaporation of water from a relatively warm surface into the overlying atmosphere; and *thaw fog*, which is formed by the cooling of warm air passing over a surface of thawing ice or snow. Fog is a cloud in contact with the ground, but to define it as stratus in contact with the Earth's surface is not really accurate, because stratus is featureless and uniform by definition, but fog is turbulent and may vary in density both optically and temporally. Measurements of fog density form useful climatological quantities.[11]

Historical data enable the average patterns of fog behavior at a single station to be described, but these patterns are of no great use in the vast areas of the world for which observations of visibility, if there are any, are unreliable. Gaps in the observing network may be filled in to some extent by determining the probability of fog for different surface wind directions, and then adjusting the conditional probabilities in the light of local geographical conditions, but the only sound basis for interpolation,

● 100 and above   ◍ 50-90

Figure 7.7.
Mean annual frequency of heavy fog. Number of days per year with visibility of 1/4 mile or less.
(After R. L. Peace, Jr.)

either geographical or temporal, is to use physics. However, fog, probably more than any other climatic element, must be studied in terms of both geography and physics (see Fig. 7.7). Because of this, objective prediction schemes for fog cannot be generalized without allowing for geographical effects. For example, fog is sometimes more liable to develop in the evening at Miami if the air is dry than if it is saturated; the relationship between fog and dew point is sometimes direct, at other times inverse. The critical predictor is derived from streamline analysis, and depends on the location of the hyperbolic point in advance of a cold front. Even apparently imperceptible geographical features may have an important influence on fog; for example, although the slope of the East Anglian ridge is not obvious to the casual observer at Stradishall (385 feet above sea level), it is quite enough to cause fog to form because of cooling during the slow ascent of saturated air in southerly or east-southeasterly airstreams. Fog may be general on the east and southeast coasts of England in these airstreams, yet the air will be clear on the northwestern (i.e., leeward) side of the ridge below 400 feet.[12]

The application of turbulence theory to the mixing of two saturated air masses at different temperatures provides a useful physical method of fog prediction. The fog density may be estimated from the equation of eddy exchange, and different boundary conditions of integration determine whether the type of fog will be radiative or advective. The theory indicates that the fog will be much denser if the radiative heat loss is from a surface that steadily rises from the ground to the top of the fog layer.[13]

Synoptic methods of fog prediction usually use an air-mass property, termed the *fog point*, which may be regarded as conservative within areas subject to a fairly uniform rate of heating. The concept of a fog point obviously applies only to water fogs. It appears to be a constant for a given air-mass, except near coasts with onshore winds. Local variations in the density of radiation fog are due to variations in the rate of nocturnal cooling at the ground, the latter variations arising from topographic and soil differences. Differences within an area in the time at which radiation fog sets in are due to geographically induced variations in the cooling rate of the lower atmosphere, rather than to variations in the fog point. For example, fog forms more frequently near the borders of the Fen country in East Anglia than in its center, and usually clears first on the borders, particularly on their downwind portion, because the Fen soil (a dark peat) cools more rapidly than the adjacent clay soil. The cooling differential gives rise to temperature discontinuities of about 1°F along the boundaries between the peat and the clay hillocks.[14]

The fog point of water fogs is associated with a specific screen-level temperature, but the latter should not be defined as the temperature at which fog forms, because one must distinguish smoke fogs, which characteristically have a visual range of 600 to 1,000 yards, in comparison with the less than 500 yards visibility typical of radiation fogs, and which are not closely associated with rises in relative humidity to 95 per cent or more, as water fogs are. The fog point may be obtained from tephigrams on which are plotted representative ascents through the air mass. Radiation fog may also be predicted from observations of the values of dew-point and dry-bulb temperature at sunset.[15]

The rate of visibility deterioration in fog is an important quantity. In water fog, visual range decreases abruptly a certain time after the formation of the fog; this time is related to the degree of wetness of the ground and the magnitude of the land track of the air mass. At Exeter Airport, visibilities almost always drop to 500 yards or less within one hour after fog forms. At Cardington, the visual range decreases very abruptly soon after fog forms to as little as 50 yards, and visibilities of less than 200 yards are the rule rather than the exception. At London Airport, intermediate visibilities (due to smoke pollution) are quite common, and the visual range of 330–440 yards, marking the division between water fogs and smoke fogs, is relatively infrequent. In general, once radiation fog forms inland, the visibility drops very quickly to between 800 and 1,000 yards; this visual range is very frequent at airfields in the English Midlands and East Anglia. Once this range is attained, further deterioration in visibility is delayed for about three hours, after which very rapid deterioration sets in. Steep slopes to windward delay the formation of radiation fog at high-level stations, whereas fog formation is accelerated if a light wind blows up a gentle slope. Aerodynamic properties influence radiation-fog formation over high ground: lee waves cause variations in the depth of the foggy inversion layer, and may result in undulations of the fog top, and gradual changes in the wavelength of the wave system may cause the alternating pattern of foggy and clear areas to move horizontally.[16]

Radiation fog forms more readily in stable air than in unstable air. This fact may be made use of in a fog-prediction method that provides the following rules: (1) if the lapse-rate throughout the lowest 5,000 feet of the atmosphere exceeds 4°F per 1,000 feet then the fog point is given by the potential dew point (the dew point that air at some specific altitude would have if it were lowered adiabatically to the ground) at 850 mb; (2) if a more stable layer is present below 850 mb, the potential dew point at that level equals the fog point; (3) if the air is stable from the surface upward, then the fog point is given by the surface dew point; and (4) in all intermediate cases, the potential dew point at the intermediate levels equals the fog point. There are exceptions to these rules, but in general the method gives the correct fog point within 1°F on two-thirds of all trials in Britain.[17]

The time at which radiation fog clears varies with the time of year and the depth of the fog, but in general the fog should clear when insolation has caused surface temperatures to rise enough to produce at least a saturated-adiabatic lapse-rate from the surface to the top of the fog. The time at which the critical surface temperature will be reached may be predicted from a tephigram on which is plotted the upper-air sounding for the previous night. Figure 7.8 shows the times at which the required surface energy will have been received in an average year. This method assumes there is no cloud during the night or early morning. The arrival of a stratocumulus sheet overhead during the night causes clearance of the fog before sunrise, unless the initial difference in temperature between soil surface and overlying air at screen level is very small. The development of a layer of low or medium cloud above the fog during the day also hastens its dissipation.[18]

Steam fog or *sea smoke* may form in winds varying from calm to gale force; the air temperature must be from 5 to 40°C lower than the surface temperature of the water, depending on the windspeed. Before steam fog can form, the air-sea temperature difference must exceed a threshold whose value depends on the humidity of the air and the temperature and salinity of the water. The threshold is lowest for moist air over cold water with a very low salinity. Steam fogs are characteristic of the western part of the North Atlantic in winter, because of the contrast, within a relatively short distance, of warm ocean surface waters (the Gulf Stream Drift) and the cold land surface. Their depth varies from 1 to 1,500 meters, depending on the degree of turbulence in the air within the fog. The equations of turbulent transfer show that two

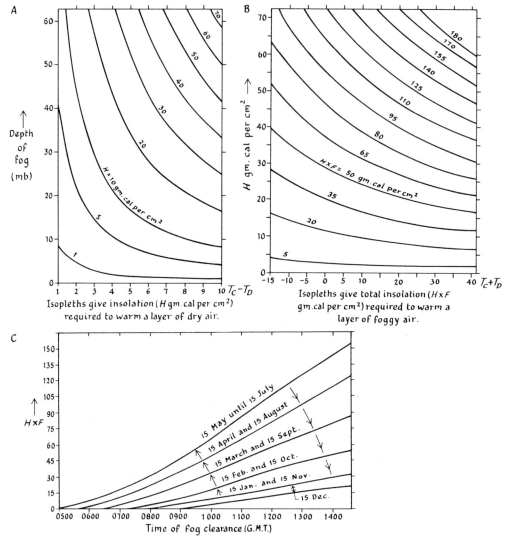

FIGURE 7.8.
Prediction graphs for clearance of radiation fog (after J. A. Barthram).
Procedure:
1, obtain H from A;
2, obtain $H \times F$ from B;
3, draw horizontal line on C from appropriate $H \times F$ value to relevant date curve, and read off clearance time on ordinate.

T_D and T_C are the surface temperatures in °C at dawn and at the time the fog clears, respectively.

masses of unsaturated air at different temperatures may, when mixed together by turbulent diffusion, produce a supersaturated (i.e., foggy) mass of air.[19]

Thaw fogs have been the subject of controversy. In Norway, fogs have been observed to dissipate above a snow surface, whereas in Britain they frequently form and persist above a snow cover. In Norway, for example, at Oslo, the air above the snow cover may be much drier than in Britain because of the föhn effect, but this is not the full explanation. For example, mist persists on frosty mornings in England, whatever the height and intensity of the sun, until all the hoar frost has disappeared from the ground surface, after which visibility rapidly improves. From the Norwegian observations, two rules have been derived: (a) fog will not form above a surface covered with melting snow; and (b) if a fog is carried by advection over a surface of thawing snow or ice, the fog will dissipate. The British observations indicate that fog will form over melting snow, provided the dew point of the air in the fog is above 0°C. If the dew point is below freezing, fog will only form if the snow surface is frozen; if the snow or ice is thawing, fog will not form over it, and any fog advected in from adjacent areas will dissipate as stated in rule b.[20]

Man-made fog is often a serious problem. The natural visual range of the atmosphere may be reduced by the spreading out of the dark smoke emitted from jet aircraft because of incomplete combustion of jet fuels. Visibilities in the vicinity of airfields with numerous movements of jet aircraft may often be only one-fifth of what they would be under purely natural conditions.[21]

The reduction of visibility by the formation of *smog* is the most obvious example of man-made fog. The emission of smoke into the atmosphere produces very low visual ranges with relatively low relative humidities; e.g., a visibility of 200 yards has been observed in London with a low relative humidity. Geographical and temporal variations in the smoke content of the atmosphere are well-established for certain areas. The fogs that form where this smoke content is high are usually slow to clear, for several reasons. For example, the solid smoke particles in the atmosphere do not evaporate, and therefore the fog persists long after the time for the clearance of a radiation fog has passed. If the fog consists of water droplets containing minute smoke particles, then the rate of evaporation of the fog droplets is less than that of clear water droplets, so that clearance is delayed. A drop in relative humidity that would result in the clearance of water fog has little effect on the smoke concentration, and may actually increase the density of the fog under certain conditions.[22]

Smog consists not only of smoke particles, but also of such substances as sulphur dioxide, carbon dioxide, carbon monoxide, and various halogens, such as chlorine and fluorine. The smoke comes mainly from incompletely combusted coal in domestic fires, the carbon dioxide from the combustion of coal, the carbon monoxide mainly from the exhaust of motor vehicles, and the sulphur dioxide mainly from industrial plants. The London smog of December 1952 covered 450 square miles to a depth of 500 feet; it contained approximately 380 tons of smoke, 370 tons of sulphur dioxide (not including smoke or sulphur particles either attached to or dissolved in the water droplets), 200,000 tons of liquid water, and 750,000 tons of water vapor. In addition, the smog prevented the evaporation of a water film that coated the ground and the objects on it and that weighed about 500,000 tons. The smog was practically a closed system from day to day, but to maintain it for several days, a considerable exchange of heat, moisture, and solid matter had to take place. The general meteorological effect of such a smog is to produce an artificial, low-level inversion, quite distinct from the natural subsidence inversion of the anticyclones in which the smog normally occurs.[23]

Smoke abolition by law has reduced the intensity and frequency of smogs in some areas (for example, Pittsburgh and St. Louis, and many cities in Britain), but preventing householders from burning coal and factories from emitting sulphur dioxide has not helped in other areas, especially Los Angeles. The Los Angeles smog is completely different chemically from the London variety: it is an ozonide produced by the action of petrol vapor. The smog initially has the form of a bluish haze, which later becomes dense blue; such a haze may be produced experimentally by mixing olefin vapors (i.e., the unsaturated aliphatic hydrocarbons that are found in petrol) with dilute ozone. In Los Angeles, a similar process is caused by nitrogen oxides, which normally are found in polluted air over cities, and which catalyze the oxidation of olefins in the presence of sunlight to produce ozonides and peroxides. Thus the smog associated with the automobile-dominated city of today consists of ozonized olefins, unlike the smoke and sulphur smog of the cities formed by the Industrial Revolution. The difference is important; the ozonized smog has a distinctive odor, is highly toxic, and can cause severe eye irritations and considerable damage to plants.[24]

Sky and Landscape Colors and Atmospheric Optical Effects

The shade of blue that is observed in a cloudless sky depends on the origin of the prevailing air mass. The sky appears blue because of *Rayleigh scattering* by the air molecules. The Rayleigh theory assumes that the air molecules are homogeneous spherical particles of equal size, whose radii are much smaller than the wavelength of the incident radiation from the sun. It then shows that the scattering coefficient of the air molecules varies inversely as the fourth power of the wavelength of the incident radiation. The shorter blue and violet waves of the solar spectrum are more readily scattered by the molecules than are the other colors; consequently, the sky appears to be blue to the observer (see Appendix 7.12). The Rayleigh theory includes only primary scattering, and the observed sky colors show deviations from the theoretical ones at certain specific points in the sky and sun elevations. The theory of multiple scattering devised by Sekera is a theoretical treatment of radiation that is scattered more than once before it reaches the observer's eye; it agrees with the observations better in the higher atmosphere, but not near the ground, where there are natural aerosols.[25]

When the cloudless sky is not a deep blue, but a somewhat diffuse blue, natural aerosols are present in various sizes. Their effect on sunlight is described as *Mie scattering*. The theory of scattering devised by Mie allows the intensities of the scattered light to be computed for different particle sizes, although a computer is necessary to compute intensities for all wavelengths and all types of aerosol (see Appendix 7.13). Because Rayleigh and Mie scattering take place together, a complete spectrum of shades of blue may be observed in the cloudless sky, the shades corresponding to different aerosol sizes and concentrations, which in turn depend on the type of air mass. One can construct a *cyanometric scale*, consisting of a series of colors varying from deep ultramarine to pure white, which, when compared with the bluest spot in the sky (usually 90° to one side of the observer if he faces away from the sun), will indicate the type of air mass prevailing. Linke's cyanometric scale is logarithmic and has 14 shades. Neuberger introduced a simpler scale with six shades.[26]

In the early nineteenth century, Arago discovered that light from the clear sky is polarized. The degree of polarization varies considerably, depending partly on the number and size of scattering particles in the atmosphere—volcanic dust is especially important—and partly on position in the sky. There are certain *neutral points* (see

Figure 7.9) at which the polarization disappears: for example, Arago's neutral point (20° above the antisolar point, i.e., the point diametrically opposite the sun on the celestial sphere), Babinet's neutral point (20° above the sun), and Brewster's point (20° below the sun). Arago's point of maximum polarization is 90° from the sun in its vertical plane.[27]

By measuring the degree of polarization with a *photo-polarimeter*, one can estimate the relative size and concentration of scattering particles in the atmosphere. By measuring the position of the neutral points with a *polariscope*, and then determining their angle of elevation with a theodolite, one can estimate the relative degree of turbidity of the atmosphere. By counting the number of neutral points, one can estimate the visual albedo of the ground, since with strong light reflection from the ground, new neutral points appear. Slight fluctuations in the position of the neutral points, and short-period fluctuations in the degree of polarization of a clear sky, are normally a prelude to cloud formation. The polarization of sky light can be used to determine whether a cloud is precipitating, when it cannot be seen at night.[28]

Combinations of air-molecule scattering with scattering by aerosols of various sizes and concentrations produce the countless types of light and color we observe in the sky and landscape. The observation and explanation of these colors is a fascinating study; it has been neglected by geographical climatologists, although it has much to offer them.[29]

Whiteout, which is typical of polar regions, is a sky-color phenomenon of special interest and importance. Snow-covered terrain and overcast sky combine to make the observer's distant field of view appear to be completely featureless, and it is impossible to tell where the horizon is. All shadows or contrasts disappear, and perspective effects have no meaning. White objects are completely invisible against their background, but small dark objects a mile or more away are clearly seen.[30]

Whiteout occurs in pure clean air, and is completely independent of the normal meteorological controls of visual range. The main factor in whiteout formation is the

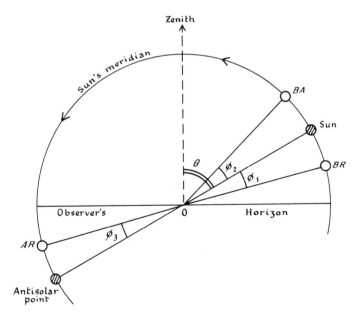

FIGURE 7.9.
O makes the position of the observer and θ is the zenith angle of the sun. Angle φ is about 20°. As the sun rises above or sinks below the horizon, angle φ_3 increases from 20° to about 23°. In the situation shown in the diagram, Arago's neutral point (*AR*) would be "invisible" to the observer, but Babinet's (*B A*) and Brewster's (*B R*) neutral points would be apparent. (After H. Neuberger.)

presence of a snow surface with a high reflecting power for visible light. The human eye is incapable of distinguishing discontinuities in surface brightness of less than 2 per cent; consequently, with a combination of overcast sky and snow surface of equal brightness, the horizon becomes invisible. Also, the optical uniformity of the overcast increases as snow accumulates on the dark surface of the ground and the latter's reflectivity gradually increases.

Atmospheric refraction of light gives rise to phenomena that are of considerable interest to the climatologist, not only as climatic elements in themselves, but also as indicators of the local physical state of the atmosphere.[31] The optical refractive index of air depends on its density: the denser the air, the smaller the angle of refraction of the light ray (see Figure 7.10, A). Consequently, the average lapse-rate of air layers not in immediate contact with the ground may be found from the "curvature" of the light ray. For example, for a light-ray path 30 km long, an apparent deviation of 1 second of arc in zenith distance corresponds to a change in lapse-rate of 0.04°C per 100 m. For steep lapse-rates, fluctuations of the light beam due to convection currents decrease the precision of the measurements.

Terrestrial refraction is a large-scale refraction phenomenon that causes an apparent expansion of the horizon and a decrease in its angular depression. Thus an observer at point P, in the horizon plane PH (see Figure 7.10, B), will see the horizon at D if the light rays are rectilinear, and there is no atmosphere. However, because of the

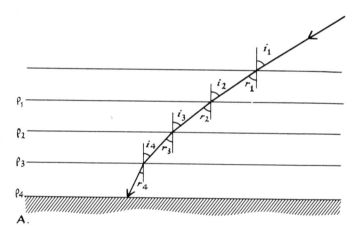

FIGURE 7.10.
Atmospheric refraction.
A. Assuming that the atmosphere consists of parallel layers of uniform density, the densities of the layers decreasing upwards, then $\rho_4 > \rho_3 > \rho_2 > \rho_1$ and $r_1 > r_2 > r_3 > r_4$, where r is the angle of refraction of a light ray of incident angle i.
B. Terrestrial refraction $= \alpha - \beta$, where α is defined as the geodetic depression.

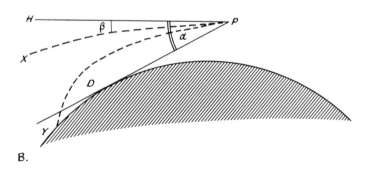

atmosphere and its attendant density stratification, the horizon appears to him to be in the direction *PX*, tangent to the curved ray *PY*; i.e., he sees the horizon at *Y*. The theory of refraction, applied to a model atmosphere consisting of a series of concentric, equidistant, isopycnic surfaces, enables the degree of terrestrial refraction to be determined in terms of barometric pressure, lapse-rate, and the temperature and refractive index of the air (see Appendix 7.14).

Scintillation is due to variations in density within the atmosphere, and has several components. Scintillation may cause a point-source object to suffer from the following effects: (a) unsteadiness in position; (b) color changes, for example, white light exhibits its chromatic components separately if it passes through a scintillating atmosphere, so that the object appears to be continually changing color; (c) fluctuations in intensity, i.e., the object seems to vanish and reappear. With an extended-source object, the shape and appearance of the object may appear to change, and shadow bands may suddenly develop across it. Typical observations show, for a collimated light source situated 1 m above the ground and distant 1 km from the observer, that strong scintillation produces from about 5 to 9 fluctuations per second in the light reaching the observer from the source. Scintillation may vary from one to 50 fluctuations per second. If the object viewed is nonluminescent and reflects light diffusely, then scintillation causes its image to appear to boil, producing *atmospheric shimmer*. Depending on the density of the air mass through which the object is seen, the apparent lateral displacement of vertical lines on such an object observed at a distance of several hundred meters varies between 1 and 5 seconds of arc.[32]

Various refraction phenomena produced by density gradients may be recognized. If the temperature lapse-rate is smaller than normal, i.e., if the decrease in atmospheric density with height is greater than normal, terrestrial refraction increases and *looming* occurs; i.e., objects that are usually beyond the horizon become visible (see Figure 7.11, A). The opposite effect, *sinking*, occurs if the lapse-rate is abnormally shallow, i.e., if the density decrease with height is abnormally small (see Figure 7.11, B). If the refractive index of the air decreases with height more rapidly than a linear function would, *towering* occurs, and the apparent height of the object is increased. If the density of the air increases with height, the apparent height of the object is decreased. If the refractive index decreases with height less rapidly than a linear function would, the light rays from the upper part of a distant object have a different "curvature" from those from the lower part of the object, and *stooping* occurs; i.e., the angle subtended by the object at the observer's eye becomes smaller, and the object appears to be elevated.[33]

Mirages develop when the atmospheric density stratification is such that light rays from the object reach the observer along several different paths. An image that appears above the object it represents is called a *superior mirage*; an image that appears below the object is termed an *inferior mirage*. If the image occurs to one side of the object— i.e., if vertical isopycnic surfaces are present, for example, near strongly heated vertical walls of rock or other suitable material—it is called a *lateral mirage*. The *Fata Morgana* is a mirage involving a complicated, distorted image of distant objects due to a complex density distribution.[34]

There is little doubt that mirages are real. An inversion produces a lens, roughly comparable to an optical lens, within the atmosphere. The mirage is projected on the lens, without enlargement. Observations in the Arizona desert provide examples of two types of mirage. The first type is the familiar image of a pool of water on a road

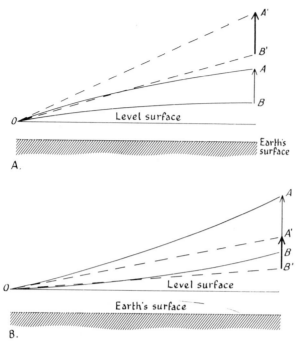

A.

B.

FIGURE 7.11.
Refraction by inversions and instability layers (after J. C. Johnson).

A. If a strong temperature inversion exists near the ground, i.e., if the density decrease with height is stronger than usual, the observer O will see an object AB at a higher angle of elevation than its true angle (*looming*), and the apparent height of the object, $A'B'$, will be increased (*towering*). If $\angle A'OB'$ is less than $\angle AOB$, then its apparent height, $A'B'$, will be shorter than its true height (*stooping*). Curvature of the Earth's surface is ignored.

B. If the lapse rate near the ground is highly unstable, i.e., if air density increases with height, then an observer O will see an object AB in position $A'B'$, i.e., at a lower angle of elevation than its true one (*sinking*). Note that stooping, as well as sinking, is present in this example.

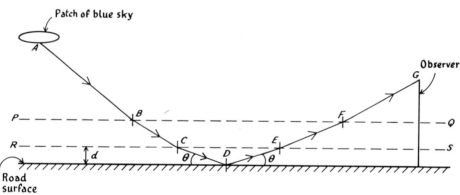

FIGURE 7.12.
Formation of a road mirage. RS represents the upper surface of the layer of very hot air immediately above the road surface. PQ represents the upper surface of the layer of warm air above the latter. A ray of light from a patch of sky, AB, will be refracted along BC, and then along CD. When angle θ is less than a certain critical angle (see S. E. Ashmore, W, 10, 1955, 336, for mathematical details), the ray CD will be subject to total reflection along DE. Thus, to the observer, the patch of blue sky will appear to be situated at and around the point D. For a road mirage at Wrexham, North Wales, distance d was found to be about 0.75 inch and angle θ to be 0.0792°.

surface (see Figure 7.12). This proves to be the image of a patch of blue sky refracted through a lens resting on the ground. The second type is much more extensive, involving a lens that forms well above the Earth's surface (often forming a "saddle" above a mountain ridge) and that can refract light several hundred miles, particularly if several such lenses form. In this second type, light enters the lens from the side, so that the image of a town, an oasis, or whatever, may be projected up one side of a mountain and down the other. The image remains sharp, whatever the distance over which it is transmitted, and is not magnified. Such lens phenomena demonstrate that there must be density anomalies in atmospheric inversions that cannot be detected by conventional techniques of vertical temperature-profile measurement.[35]

Atmospheric refraction effects allow some otherwise impossible measurements of certain meteorological phenomena to be made precisely. For example, the *schlieren* optical technique makes otherwise invisible phenomena easily seen, using differences in the refractive index of the air along the light ray (see Appendix 7.15). By this means "model" temperature inversions, produced above the surface of an ice rink by buoyant convection plumes generated by small thermocouple heaters, may be observed with considerable precision. The buoyant plumes cause variations in the refractive index of the air in the light path because of small-scale thermal turbulence. For the model to be realistic, the Grashof number (a measure of the intensity of turbulence associated with free convection) of both model and prototype situation, and their geometries, must be approximately the same (see Appendix 7.16). Measurements of scintillation and atmospheric shimmer above a level snow surface at night prove that one can estimate visual resolution conditions in stable thermal stratifications from standard climatic data. Visual resolution above a snow surface is good at all wind-speeds under an overcast or broken sky, i.e., in the absence of thermal stratification, and deteriorates as the average vertical temperature gradient increases. Visual resolution deteriorates as windspeeds increase to 5 miles per hour under a clear sky or scattered clouds, and then gradually improves as windspeeds increase still further.[36]

Useful information about other atmospheric suspensoids is revealed by optical phenomena in which the sun or the moon forms the source of light. *Rainbows*, i.e., arcs of color around the antisolar point, are produced on sheets of water droplets. *Halos* are due to refraction and reflection of sunlight or moonlight by ice crystals.[37]

The elementary theory of the rainbow, proposed by Descartes, depends on the refraction and internal reflection of light rays in a water drop (see Figures 7.13 to 7.15). In the drop, there must be a ray of minimum deviation; i.e., the originally parallel and equidistant light rays emerge from the drop concentrated to one side of the ray of minimum deviation, whose angle of elevation depends on the color of the light and on the refractive index of water, which varies with the wavelength of light (see Appendix 7.17). The theory devised by Airy regards the problem as one of the diffraction of waves of light through the water drops. This theory shows that the primary and supernumerary bows change in color as the diameters of the drops producing them vary from 0.05 to 2 mm (see Appendix 7.18). Consequently, changes in the appearance of the bows may be used as indicators of changes in the size of the water droplets in the atmosphere. The rainbow is usually seen in showery weather when the sun is not too high in the sky and is behind the observer; any changes in its colors may therefore provide information on the possibility of extensive precipitation from the clouds. The position in which a rainbow should occur can be determined graphically by constructing cones about the line joining the sun and the observer.[38]

Different types of halo may be observed (see Figures 7.16 to 7.19), depending on the size, shape, density, and orientation of the ice crystals in the atmosphere, and on whether or not they oscillate while falling (see Appendix 7.19). The *common halo*, 22° in radius, is typically associated with cirrostratus, but is also observed with fine-weather cirrus; it is produced by refraction of light in hexagonal ice crystals. Mock suns or *parhelia*, of 22° radius, appear as bright patches, sometimes brilliantly colored,

360

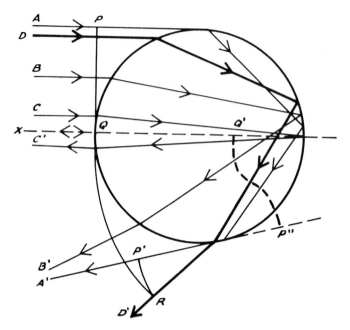

FIGURE 7.13.
Course of light rays through a raindrop (after W. J. Humphreys). Plane of figure passes through centers of raindrop and sun. X is the axial ray, i.e., the light ray from the sun through the center of the drop. D is the Descartes ray. Consider only the rays entering the drop above the axial ray: the deviations of the rays incident on the drop between the axial and Descartes rays are greater than those of D, hence their exits are between those of X and D. Deviations of rays incident on drop above the Descartes ray (e.g., ray A) must also be greater than the deviation of D, and although their exit from the drop is beyond that of ray D, they ultimately come between D and X (e.g., ray A'). Addition of further incident rays will show that the most rays leave the drop between rays A' and D'. The final effect is that the originally plane wave-front (PQ) of the incident rays becomes, after passing through the drop, a cusped wave front PRP'. The latter will appear to be the virtual wave front $Q'P''$. The latter gives rise to interference effects which result in the appearance of supernumary rainbows.

FIGURE 7.14.
Cone construction for primary rainbows.
Angle θ is about 42°, observer is at 0.

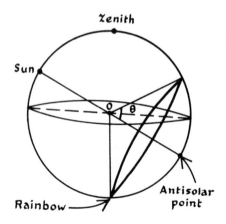

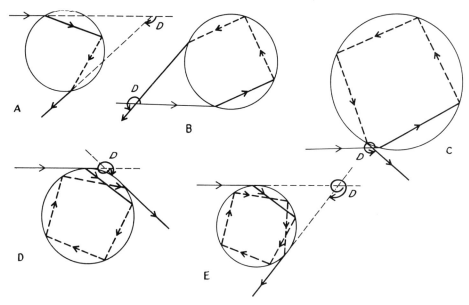

FIGURE 7.15.
Change of light direction in different types of rainbow. The lighter arrows are incident rays;
The heavier arrows are refracted rays; The broken line arrows are reflected rays.
D is the angle of deviation of the ray after passing through a spherical raindrop (see Table 7.1).
The sizes of the drops are arbitrary.

TABLE 7.1
Values of D in Figure 7.15.

		Violet light	Red light
A.	Primary rainbow	$\pi -$ 40° 36′	$\pi -$ 42° 22′
B.	Secondary rainbow	$2\pi -$ 126° 24′	$2\pi -$ 129° 36′
C.	Tertiary rainbow	$2\pi -$ 37° 52′	$2\pi -$ 42° 30′
D.	Quaternary rainbow	$3\pi -$ 131° 26′	$3\pi -$ 137° 10′
E.	Quinary rainbow	$3\pi -$ 45° 50′	$3\pi -$ 52° 5′

on either side of, and level with, the sun when the axes of the ice crystals are vertical; they appear on the common halo when the sun is low, but recede from it as the sun rises in the sky. There are numerous other types of halo, all of which may be explained physically in terms of the angles between the faces of the different forms of ice crystal. Their description requires angular estimations, for which a pilot-balloon theodolite is preferred, but if one is not available, approximate estimations may be made by holding a ruler graduated in centimeters at arm's length and assuming that each centimeter on it represents 1°. This equivalence is exact if the ruler is held at 22.6 inches (57.3 cm) from the eye.[39]

Coronae are one or more sets of colored rings produced by diffraction when the sun or the moon shines through relatively thin clouds of water or ice, the droplets or crystals being about the same size as the wavelength of light. The diameter of coronae is usually much less than that of haloes, being usually only a few degrees, but, in

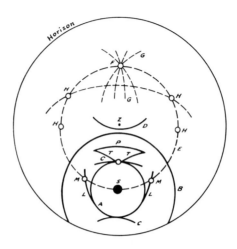

FIGURE 7.16.

Halo phenomena, location of main features. Refraction phenomena, often brilliantly colored, are shown by thick black lines; reflection phenomena, always white, are shown by thinner lines. This is a composite diagram: usually, only a few of the features described will be visible. S is the sun, and Z the zenith. Halo phenomena that may be observed include: A, the 22° halo, and B, the 46° halo, due to refraction in hexagonal ice crystals at angles of 60° and 90°, respectively; mock sun or parhelia, M, due to the presence of large numbers of ice crystals with vertical principal axes (the two lower parhelia are usually situated on the 22° halo when the sun is low in the sky, but recede from the halo as the sun rises, and may be connected with it via the arcs of Löwitz, L); arcs of contact, C, and tangent arc, T; the circumzenithal arc, D (a transient arc or circle in the form of a miniature rainbow, red toward the sun, formed by refraction through plate- or umbrella-shaped crystals with vertical axes); the Parry arc, P (very rare); the parhelic circle, E (a white ring due to reflection from the side faces of crystals with vertical principal axes) and the anthelion, F (a mock sun, rarely seen) with anthelic arcs, G, and paranthelia, H.

general, the smaller the droplets or the ice needles, the greater the diameter of the corona (see Appendix 7.20). Being smaller in diameter, coronae appear very near the luminary, and are therefore visible much more often around the moon than around the sun, unless the brightness of the latter is much reduced. The distinction between halos and coronae does not lie in their angular diameters, however, but rather in the sequence of colors and the presence of an *aureole*. The latter is the brownish-red inner portion of a corona plus the bluish-white region between the ring and the luminary. The aureole is usually wide, and contrasts with the narrow red ring that forms the inner portion of a halo. Very large diameter coronae, termed *Bishop's rings*, are formed by diffraction through volcanic dust in the atmosphere. The radius of the inner of the Bishop's rings is about 10°, and that of the outer ring about 20°, when the sun is high in the sky; both radii increase as the sun approaches the horizon.

The *anticorona*, *glory*, or *Brockenspecter* is a set of one or more colored rings that the observer sees around the enlarged shadow of his own head on a bank of fog or on the upper surface of a cloud situated below him on the side away from the sun (see Appendix 7.21). The physical explanation of the phenomenon is somewhat complex, but, in general, if it consists of multicolored rings, then the fog or cloud consists of water droplets, whereas if the colors are very faint or white, then the fog or cloud consists of ice crystals. As with the corona, the angular radius of a prominent ring in a glory is related to the size of the water or ice particles producing it.[40]

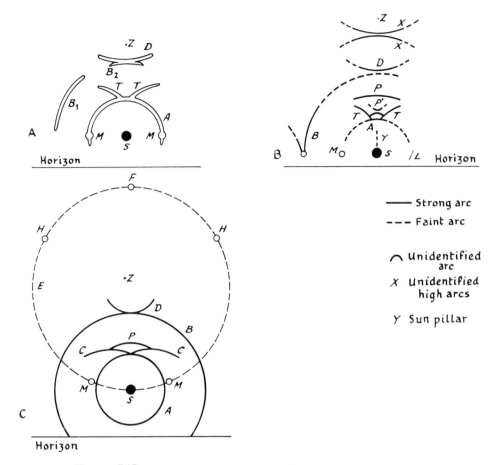

FIGURE 7.17.
Examples of halo displays.
    A. Bracknell, Berks. 7:00 P.M. G.M.T. on May 11, 1965; B_1, B_2, are parts of a 46° halo.
    B. A composite diagram of the same Berkshire halos, using data from several observing points.
    C. Edinburgh, Scotland, March 2, 1954
Letters are as in legend to Figure 7.16. (After P. A. Davies, E. C. W. Goldie, J. M. Heighes, and R. E. Lacy.)

The apparent color of the sun or moon is an important indicator of the type of matter in the atmosphere. If this color differs from normal, then some absorbing or refracting material must be present in the air between the sun and the observer. The physical explanation of this effect may sometimes be quite involved. For example, *blue sun* is caused by scattering and absorption of sunlight by a cloud or dust layer whose particles are (a) rather large compared with those in the clear atmosphere, (b) fairly uniform in size, and (c) at least 0.05 mm in diameter. But a complete explanation of the phenomenon must integrate the refraction effects over the area of the sun's disc, whose different parts have different properties as light sources. Blue sun was produced in large areas of Europe and eastern North America during late September 1950 by particles, at 10,000 to 13,000 m above the Earth's surface, that originated in extensive forest fires in Canada.[41]

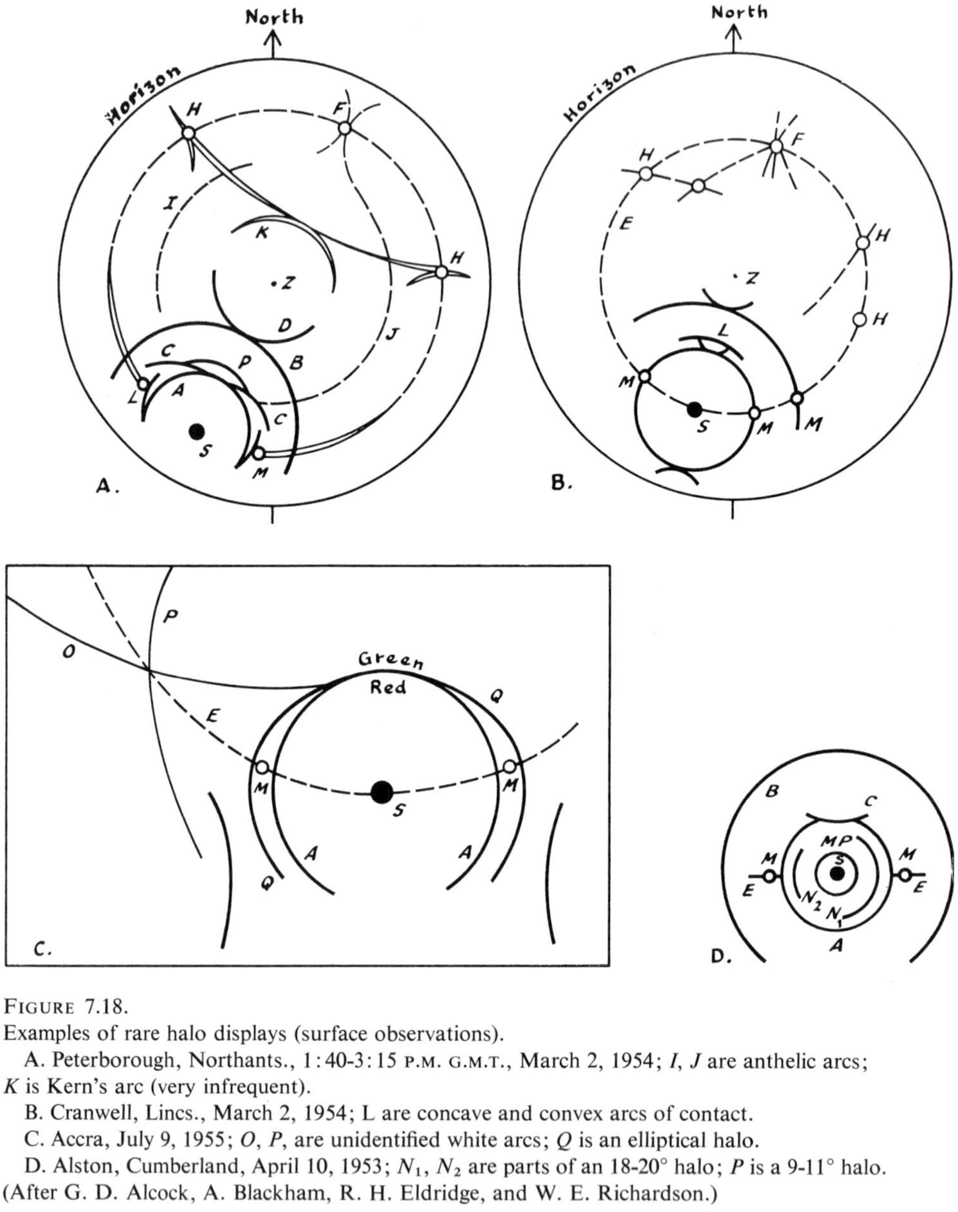

FIGURE 7.18.
Examples of rare halo displays (surface observations).
   A. Peterborough, Northants., 1:40-3:15 P.M. G.M.T., March 2, 1954; *I*, *J* are anthelic arcs; *K* is Kern's arc (very infrequent).
   B. Cranwell, Lincs., March 2, 1954; L are concave and convex arcs of contact.
   C. Accra, July 9, 1955; *O*, *P*, are unidentified white arcs; *Q* is an elliptical halo.
   D. Alston, Cumberland, April 10, 1953; N_1, N_2 are parts of an 18-20° halo; *P* is a 9-11° halo.
(After G. D. Alcock, A. Blackham, R. H. Eldridge, and W. E. Richardson.)

Some very interesting effects may occasionally be observed during a few seconds at sunset or sunrise. The *purple light*, which appears soon after sunset above the point where the sun has disappeared, is caused by the presence of fine volcanic ash in the upper atmosphere (see Appendix 7.22). The *green flash*, in which the upper portion of the sun appears a brilliant green during the final two or three seconds before its disappearance at sunset, takes place when there is an unusually clear atmosphere.[42]
   Many other sky colors and optical effects might be mentioned. Those of special interest to the climatologist are those that enable him to estimate visually the physical

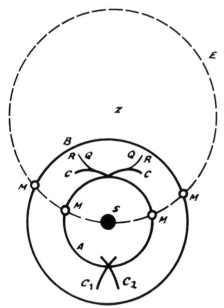

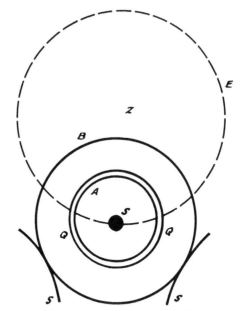

7:30 A.M., G.M.T., April 23,1959.   10:00 – 11:00 A.M., G.M.T., April 23,1959.

FIGURE 7.19.
Examples of rare halo displays (aircraft observations). *R*, Secondary parhelic circle caused by reflection of light by ice crystals from bright spot where *A* meets C. C_1, C_2, tangential arcs of contact to 22° halo, caused by refraction of light through crystals whose principal axes lie or oscillate in a plane both normal to the vertical plane through the sun and the observer, and also inclined to the horizon at an angle equal to the elevation of the sun (24°). *Q*, elliptical (circumscribed) halo, formed as for features C_1, C_2, when solar elevation is 45-50°. *S*, infralateral arcs, caused by refraction through mutually perpendicular faces of hexagonal ice crystals with horizontal principal axes. (After Singleton, Kerley, and Smith.)

state of the atmosphere and of the particles suspended in it. Such visual observations are often much more useful, because they are more immediate and local, than measurements made at the bench-mark stations. Furthermore, the "visual climatology" of the atmosphere is well worth study in its own right. In such a study we come into contact with local weather lore, which, given the visual acuity of many of the observers, provides many geographical problems of local climate to be solved by physical and mathematical techniques. Even if local weather sayings prove to have no statistical significance as weather prediction rules, an accurate observation of only *one* unusual sky color or optical effect is a fact that needs explanation, which may lead to the discovery of scientific principles of universal application. Not only must the occurrence of the color or effect on the day it was observed be explained, but its failure to appear on other days must be accounted for too.[43]

The climatology of atmospheric light is a very important, though neglected, study. Implicit throughout this book, and indeed most climatological or meteorological textbooks, is the idea that atmospheric phenomena belong either to the daytime or to the nighttime. The overriding control of the former is the net excess of incoming solar radiation over outgoing terrestrial radiation, whereas the nighttime phenomena are controlled primarily by the intensity of outgoing long-wave radiation. But the division into day and night takes no account of the twilight intervals around sunrise and sunset. Almost one-quarter of the population of the world is living and working in the "twilight zone" at any given time. With high-speed air travel, an aircraft may follow the twilight and remain within it for several hours, and it is commonplace to leave New York in the evening twilight and arrive in London in the early morning twilight of the following day without flying through the "deep night" for more than an hour or two. Clearly, *twilight climatology*, i.e., the study of the atmospheric phenomena typical and atypical of the sunrise and sunset periods, should be a very interesting and useful field of knowledge. A great deal of work in this field has been carried out in the U.S.S.R.[44] It is an intriguing field from several points of view; it combines optical climatology with physical climatology, and it deals with phenomena that must be strictly periodic astronomically, and that remain as entities for long times, even though they exist at any one spot on the surface of the Earth for only a few minutes at most.

Notes to Chapter 7

1. For a review of meteorological visibility, see *MOHSI*, pp. 352–73; *PM*, pp. 65–101, and W. E. K. Middleton, *Vision Through the Atmosphere* (Toronto, 1952). For details of standardized measurements of natural daylight illumination, see *Light and Lighting*, 54 (1961), 71. For natural daylight illumination data, see F. A. Godshall, *JAM*, 7 (1968), 499, on the northeastern United States, and S. M. Taylor and L. P. Smith, *MM*, 90 (1961), 289, on the British Isles. On the objective prediction of meteorological visibility, see V. D. Jack, *MM*, 95 (1966), 114.

2. "M.O. discussion," *MM*, 86 (1957), 84. On the estimation of probabilities of clear lines of sight through a cloudy atmosphere (i.e., line of-sight penetration between opaque clouds) on the basis of standard observations of clouds, see I. A. Lund, *JAM*, 4 (1965), 714, and *JAM*, 5 (1966), 625; see also R. J. Kauth and J. L. Penquite, *JAM*, 6 (1967), 1005.

3. T. J. Buma, *BAMS*, 41 (1960), 357. On the estimation of visual range from aerosol size distributions, see R. F. Pueschel and K. E. Noll, *JAM*, 6 (1967), 1045. On visual range in precipitation in Kinloss, Scotland, see S. M. Ross, *MM*, 96 (1967), 19.

4. For the fundamentals of meteorological optics, see: H. G. Houghton, *HM*, p. 242; W. E. K. Middleton, *op. cit.* (1952), and *CM*, p. 91; H. Neuberger, *CM*, p. 61; *PM*, pp. 65–99; E. Gold, *QJRMS*, 81 (1955), 128. A Riemannian space is one that is locally Euclidean, but the local Euclidean spaces at different points do not coincide. A simple description of the geometry of G. F. B. Riemann (1826–1866) may be found in E. T. Bell, *Men of Mathematics*. A rigorous treatment of the Riemann system is given in C. E. Weatherburn, *Riemannian Geometry and Tensor Calculus* (London, 1938).

5. T. O. Haig and W. C. Morton III, *AFSG*, no. 102 (1958).

6. R. M. Poulter, *QJRMS*, 63 (1937), 31. H. L. Wright, *QJRMS*, 65 (1939), 411.

7. See "*MO* discussions," *MM*, 81 (1952), 174, and 87 (1958), 51. See also *HM*, p. 912.

8. F. W. Went, *N*, 187 (1960), 641, and "Atmospheric hazes," *Penguin Science Survey B* (London, 1964), p. 212.

9. Went, *op. cit.* (1960).

10. For optical theory concerning thin layers of atmospheric haze and the brightness of the horizon. see F. E. Volz, *T*, 17 (1965), 166.

11. *WAF*, II, 107–18. For details of a fog-density indicator, see J. R. Bibby, *MM*, 86 (1957), 117. For a classification of fogs on the basis of the concentration of liquid water in the atmosphere, in terms of turbulent eddy-exchange theory, see C. B. Taylor, *T*, 18 (1966), 86. For water-droplet-size distributions in haze and fog, see R. G. Eldridge, *JAS*, 23 (1966), 605. On ice fogs in Fairbanks, Alaska, in relation to winter pressure patterns, see S. A. Bowling, T. Ohtake, and C. S. Benson, *JAM*, 7 (1968), 961. According to H. Dolezalek, *Reviews in Geophysics*, 1 (1963), 231, atmospheric electrical properties show significant variations in the hours immediately preceding the formation of fog. For an example of this "atmospheric electric fog" on Whiteface Mountain, New York, see A. W. Hogan and J. R. Hickman, *GPA*, 60, no. 1 (1965), 176. For details of automatic fog-warning equipment, employing submicrosecond light pulses, see F. Früngel, *BAMS*, 45 (1964), 597.

12. For the pattern of fog behavior at Heathrow Airport, London, see D. C. Evans, *MM*, 86 (1957), 333. For fog behavior at Manchester Airport, see E. R. Thomas, *MM*, 88 (1959), 104. For a determination of fog probability for various wind directions at Leeuwarden, the Netherlands, see T. J. Buma, *MM*, 89 (1960), 161. On streamline analysis, see R. B. Carson and R. C. Hardy, *JAM*, 2 (1963), 351. P. S. Griffiths, *MM*, 88 (1959), 331. For maps of the frequency of heavy fogs in the United States, see R. L. Pearce, Jr., *MWR*, 97 (1969), 116. On winter fogs at Miami, see R. E. Newell, *JAM*, 3 (1964), 342. For fog frequencies at inland stations in England, see F. E. Dinsdale, *MM*, 97 (1968), 314. For a comparison of fog frequencies at inland and coastal stations in eastern England, see F. J. Smith, *MM*, 96 (1967), 77. For unusual fog situations, see A. B. Fraser, *W*, 21 (1966), 92, on an inversion plus thermals in Wyoming, and *W*, 21 (1966), 16, on coastal fog coexisting with cumulus in Wester Ross, Scotland.

13. B. Rodhe, *T*, 14 (1962), 49.

14. "M.O. discussion," *MM*, 87 (1958), 51. W. E. Saunders, *MM*, 79 (1950), 213, and *MM*, 87 (1958), 65; G. A. Corby and W. E. Saunders, *MM*, 81 (1952), 225. M. H. Freeman, *MM*, 91 (1962), 350. For details of a temporary improvement in visibility in Shawbury, Shropshire, caused by glaciation of a water fog, see D. J. George and R. Hill, *MM*, 95 (1966), 121.

15. Corby and Saunders, *op. cit.* (1952). Saunders, *op. cit.* (1950). H. C. Belhouse, *BAMS*, 42 (1961), 349.

16. W. E. Saunders, *MM*, 86 (1957), 362. D. C. Evans, *MM*, 86 (1957), 333. E. Evans, C. J. M. Aanansen, and T. E. Williams, *MM*, 87 (1958), 33. W. R. Sparks, *MM*, 91 (1962), 232.

17. J. Briggs, *MM*, 79 (1950), 343.

18. J. A. Barthram, *MM*, 93 (1964), 51. C. J. Kennington, *MM*, 90 (1961), 70. W. E. Saunders, *MM*, 89 (1960), 261. On predicting the clearance of fogs in eastern and central England, see T. D. D. Jennings,

C. J. Macey, and B. L. Giles, *MM*, 94 (1965), 301. For a method of forecasting the time of clearance of radiation fog, see C. J. Kennington, *MM*, 90 (1961), 70, and D. J. Heffer, *MM*, 94 (1965), 259.

19. P. M. Saunders, *QJRMS*, 90 (1964), 156. See also the classic work of G. I. Taylor, in G. K. Batchelor, ed., *The Scientific Papers of Sir Geoffrey Ingram Taylor* (Cambridge, Eng., 1960), II, 1, 51. For micrometeorological measurements in ocean fogs on Great Duck Island, off the coast of Maine, see F. M. Weiner, J. H. Ball, and C. M. Gogos, *JGR*, 66 (1961), 3974.

20. C. S. Durst *et al.*, *MM*, 88 (1959), 334; the Norwegian rules are Petterssen's.

21. J. E. McDonald, *JAM*, 1 (1962), 391.

22. "M.O. discussion," *MM*, 87 (1958), 51.

23. A. R. Meetham, *W*, 10 (1955), 103. W. C. Turner, *W*, 10 (1955), 110.

24. For details, see Went, *op. cit.* (1964). For details of damage to tobacco leaves at Washington, D.C., by ozone, see R. C. Wanta and W. B. Moreland, *MWR*, 89 (1961), 289. For examples of the effect of the Clean Air Act of Parliament on visibility characteristics in England, see: J. E. Atkins, *MM*, 97 (1968), 173, for Manchester airport; G. A. Corfield and W. G. Newton, *MM*, 97 (1968), 204, for Finningley; and M. H. Freeman, *MM*, 97 (1968), 214, for London airport, Heathrow. For year-to-year variations in the frequency of fog at London airport, see P. J. Wiggett, *MM*, (1964), 305. For the connection between a sharp smog bank and fog development in Riverside, California, see D. F. Leipper, *BAMS*, 49 (1968), 354.

25. J. W. S. Rayleigh, *Phil. Mag.*, 41 (1871), 107, 274, and 447, and 47 (1899), 375. Z. Sekera, *Handbuch der Geophysik* (Berlin, 1957), II, 288. On radiation scattered by natural aerosols, see K. Bullrich, *AGP*, 10 (1964), 99.

26. G. Mie, *Ann. Physik*, 25 (1908), 377. *HM*, p. 912.

27. H. Neuberger, *BAMS*, 31 (1950), 119. The *antisolar point* is the point 180° from the sun on the line from the sun through the center of the Earth.

28. For measurement techniques for polarization of sky light, see Z. Sekera, *CM*, p. 79. For the theory of visual albedo, see *PM*, pp. 88–90. For using polarization to detect precipitation, see D. L. Jorgensen, *JM*, 10 (1953), 160. The polarization of sunlight reflected from the sea may be used to determine extreme values of the slope of the sea surface, see J. Pandolfo, *JGR*, 67 (1962), 4303.

29. For a general introduction, see J. Paton, *W*, 6 (1951), 195. The standard work is M. J. Minnaert, *Light and Color in the Open Air* (Bell, 1940).

30. S. Fritz, *W*, 12 (1957), 345.

31. For reviews of refraction, see H. Neuberger, *CM*, p. 61, and *PM*, pp. 1–29.

32. On atmospheric scintillation, see W. J. Humphreys, *Physics of the Air* (New York, 1940), p. 462. On shimmer, see *PM*, p. 18.

33. *PM*, pp. 16–18.

34. *PM*, pp. 19–21. See also Humphreys, *op, cit.* (1940), pp. 470–75. For examples of mirages in La Encantada, Gulf of California, see R. L. Ives, *W*, 23 (1968), 55, and *BAMS*, 40 (1959), 192.

35. C. M. Botley, *W*, 20 (1965), 22. J. Gordon, *Ann. Report Smithsonian Inst., 1959*. For other accounts of mirages (in the Salt Lake desert), see R. L. Ives, *J. Franklin Inst.*, 245 (1948), 457. For details of a road-mirage, see S. E. Ashmore, *W*, 10 (1955), 336. Although mirages are refraction phenomena, there are records of *reflection mirages*; see C. M. Botley, *op. cit.* On the effects of reflection from different types of natural surface on the angular and spectral distribution of skylight, see K. L. Coulson, *JAS*, 25 (1968), 759. On the optical reflection properties of natural surfaces, see K. L. Coulson, G. M. Bouricus, and E. L. Gray, *JGR*, 70 (1965), 4601. For details of a meter for measuring visual albedo, see S. J. Bolsenga, *JAM*, 7 (1968), 939. For details of banded reflections from the sea surface (i.e., a dark blue border to the sea horizon), which are frequent on the Atlantic coast of Florida, see F. O. Rossmann, *W*, 15 (1960), 409.

36. T. V. Crawford and A. S. Leonard, *JAM*, 1 (1962), 251. E. Ryznar, *JAM*, 2 (1963), 526.

37. For general reviews, see: *PM*, pp. 175–99; Humphreys, *op. cit.* (1940), pp. 476–556; *MOOH*, pp. 131–32. For details of experimentally produced optical phenomena, see J. O. Mattson, *W*, 21 (1966), 14.

38. On Descartes' theory, see C. B. Boyer, *The Rainbow: From Myth to Mathematics* (New York, 1959), pp. 200–232. On Airy's theory, see Humphreys, *op. cit.* (1940), pp. 493–95. On the construction, see C. D. Walshaw, *W*, 17 (1962), Schools Suppl., p. 49. On the theory of rainbows and cloudbows, see W. V. R Malkus, *W*, 10 (1955), 331. According to J. E. McDonald and B. M. Herman, *BAMS*, 45 (1964), 279, the terms "interference theory" or "wave-optical theory" are correct, rather than "diffraction theory," for the physical theory of rainbows. For descriptions of unusual rainbows, see: R. A. Brown, *W*, 14 (1959), 36, for a nocturnal rainbow; P. E. Phillips, *MM*, 84 (1955), 381, for a complete lunar rainbow; D. Roberts, *W*, 23 (1968), 292, for a vertical rainbow or *reflection primary*; and R. K. Pilsbury, *MM*, 82 (1953), 55, and L. E. Watson, *MM*, 82 (1953), 218, for rainbows in which arcs were observed curving in the opposite direction to the usual bows. For examples of supernumary fogbows at subfreezing temperatures, see R. A. Brown, *MWR*, 94 (1966), 47. For examples of cloudbows, see J. E. McDonald, *W*, 17 (1962), 243, for a gigantic horizontal cloudbow observed in the North Pacific, which formed a luminous arc spanning many tens of miles, and F. A. Gifford, *MWR*, 92 (1964), 84, for a horizontal cloudbow over western Europe.

39. Humphreys, *op. cit.* (1940), pp. 501–36. *PM*,

pp. 185–89. *MOOH*, p. 126. For points concerning the photography of halos, see G. D. Roth, *W*, 8 (1953), 249. On the frequency of halo phenomena at Oxford, 1881–1951, see J. G. Balk, *MM*, 81 (1952), 263, and in Germany, see G. D. Roth, *W*, 7 (1952), 248. According to the latter, the relation of halo phenomena to coming weather changes depends upon geographical conditions; for example, halos are reliable symptoms of future precipitation in Holland, but unreliable indicators at Munich. For descriptions of halo occurrences, see R. Scutt, *W*, 15 (1960), 309, with colored sketches, and P. A. Davies, *W*, 21 (1966), 138. For a tropical example at Accra, see R. H. Eldridge, *W*, 11 (1956), 20. For ancient observations of halos made in China, see Ho Ping-Yu and J. Needham, *W*, 14 (1949), 124. For descriptions of important halo displays in Britain, see: W. E. Richardson, *MM*, 82 (1953), 277, for four simultaneous concentric halos in Cumberland; R. E. Lacy, *W*, 9 (1954), 206, for Edinburgh; G. A. Jones and K. J. Wiggins, *W*, 19 (1964), 289, for Odiham, Hants.; R. K. Pilsbury, *MM*, 94 (1965), 315, for Bracknell, Berks.; and E. C. W. Goldie and J. M. Heighes, *W*, 23 (1968), 61 and 294, for Berkshire. For an example from Fairbanks, Alaska, see J. R. Blake, *MWR*, 94 (1966), 599. For complex halo displays in Britain, see *MM*, 83 (1954), 186, and C. M. Botley, *W*, 22 (1967), 386, on the classic Hereford display of April 8, 1233. For a complex halo display in the tropics, Singapore, see J. M. Bayliss, *MM*, 83 (1954), 57. For examples of halo phenomena around the moon, see J. Hallett, *W*, 20 (1965), 51. For observations of rare halos, see: R. S. Scorer, *W*, 18 (1963), 319, for an 8° halo; E. C. W. Goldie, *W*, 19 (1964), 328; and H. Neuberger, *BAMS*, 49 (1968), 1060, for a circumscribed halo. For examples of rare mock suns, see: R. Scutt, *MM*, 82 (1953), 218, for a mock sun to 32° halo, due to refraction through pyramidal ice crystals; H. E. Landsberg, *MWR*, 92 (1964), 352, for a double mock sun to one side of the sun; M. Minnaert, *W*, 21 (1966), 250, and *W*, 22 (1967), 37, for a one-sided mock sun at 3° 30′ in the Timor Sea; C. M. Botley, *W*, 22 (1967), 260, for mock suns close to the sun, which could be lateral mirages. The "sub-sun," due to refraction through ice crystals with their flat surfaces horizontal, is a bright spot (seen from above) on the upper surface of an ice cloud; see G. J. Jefferson, *MM*, 92 (1963), 254, J. T. Houghton, *W*, 23 (1968), 292, and A. B. Fraser, *W*, 24 (1969), 160. For examples of arcs of contact, see: S. W. Visser and C. T. J. Alkemade, *QJRMS*, 82 (1956), 92, for arcs of the rare 24½° halo; R. S. Scorer, *QJRMS*, 89 (1963), 151, on a white horizontal arc through the antisolar point, due to hexagonal ice crystals with vertical axes; and J. H. Chaplan, *W*, 21 (1966), 147, and C. M. Botley, *W*, 11 (1966), 265, for an arc of the rare 46° halo, or arc of Galle. For the arcs of Löwitz, see A. J. Tomkins, *W*, 14 (1959), 224, and R. Scutt, *W*, 23

(1968), 525. For the Parry arc, see C. M. Botley, *W*, 10 (1955), 343. For an example of a sun pillar, see A. W. E. Barber, *MM*, 85 (1956), 155. For unusual refraction phenomena, see C. S. Durst and G. A. Bull, *MM*, 85 (1956), 237, on a bowler-hat-shaped "cloud" with a mushroom-like head, due to refraction of light rays from a distant cumulonimbus cloud through an inversion layer, and *W*, 22 (1967), 18, for details of anticrepuscular rays, i.e., beams of sunlight, passing through gaps in cloud cover, that converge toward the antisolar point.

40. B. Ray, *Proc. Indian Assoc. Cult. Sci.*, vol. 8 (1923), part 1. For an example of a solar corona, see *W*, 19 (1964), 339. For a Bishop's ring phenomenon observed in the southern hemisphere after a volcanic eruption on Bali, see F. Burdecki, *W*, 19 (1964), 113. According to G. C. Bridge and S. G. Cornford, *W*, 22 (1967), 300, glories are usually circular, but may be elliptical or even hyperbolic when the sun is low in the sky.

41. A. Ångström, *T*, 3 (1951), 135. E. M. Elsley, *W*, 6 (1950), 1. H. Wexler, *WW*, 10 (1950), 12. W. Jenne, *T*, 3 (1951), 129. On light and color in the Earth's shadow during eclipses, see J. E. Hansen and S. Matsushima, *JGR*, 71 (1966), 1073, on the theory. For details of a pale green glow observed in the southern sky over Aden, followed by a russet-colored band of light low down on the horizon, both probably due to moonlight reflected from dust layers in the atmosphere, see N. F. Hirst, *MM*, 87 (1958), 278.

42. Humphreys, *op. cit.* (1940), p. 466.

43. On other colors and effects, see *ibid.* and *MOOH*. For examples of weather lore, see J. Claridge, *The Country Calendar, or the Shepherd of Banbury's Weather Rules*, newly annotated by G. H. T. Kimble (London, 1946), also Richard Inwards, *Weatherlore*, ed. E. L. Hawke (London, rev. ed., 1950). For an account of optical effects observed in Antarctica, see D. J. George, *MM*, 87 (1958), 289.

44. For a comprehensive account, see G. V. Rozenberg, *Twilight: A Study in Atmospheric Optics*, trans. R. B. Rodman (New York, 1966). For examples of work in the United States, see J. H. Joseph, *JGR*, 72 (1967), 4020, on the use of twilight to detect noctilucent clouds from satellites. See also P. Gouin, *W*, 23 (1968), 70, and J. R. C. Young, *W*, 23 (1968), 340, for dark blue radiating bands observed over Ethiopia during twilight, which progressively cover the entire visible hemisphere; according to L. R. Pittwell, *W*, 24 (1969), 121, these are peculiar anticrepuscular rays, which may occasionally be seen elsewhere.

PART III

Geographical Climatology

CHAPTER 8

Geographical Climatology

In the previous chapters of *Techniques of Climatology*, we have been concerned with climatology as a science, i.e., with the question of how to order the mass of facts that climatological observations bring to light. In conformity with normal scientific procedure, we have regarded climatology as the study of a causal system or set of systems, and have interpreted the facts of climate in terms of mathematical structures or physical models. The mathematical structures are necessary because science aims at full understanding of the systems it studies, and prediction of the systems with mathematical exactitude is an essential step in the acquisition of full understanding. Obviously, full understanding of a system implies perfect control of it; unless we can predict accurately the behavior of a system, we cannot hope to control it. Mathematical structures and physical models may not, however, explain completely all the observed facts of climate, and it is still legitimate to ask the climatologist to supply the full facts about climate, even when he cannot explain them. It is the function of geographical climatology to describe and interpret the *actual* (not the predicted or expected) facts of climate.

It is a valid criticism of the previous chapters of this book to point out that they are only concerned with those climatic facts that fit into the view of climate as an abstract, theoretical system. In this chapter we shall be concerned with all the facts, as they exist (rather, as we think they exist, since there are limits to the accuracy of our observations and of our cartographic and statistical means for expressing the observations) in the actual world. The usual method of presenting these facts is in the form of maps or verbal statements, or in the form of the statistical tables from which the maps or verbal statements were derived. Such a method of presentation may seem simple and obvious, and the geography student may be tempted to say that it provides him with all that he needs to know about weather and climate, but such an attitude would be extremely naïve. The climatologist who interests himself in the mathematics and physics of his subject soon discovers that mapping the climatic facts, or making simple verbal inferences from a set of climatic figures, may take him on some very profound (to him) excursions into the fields of mathematical physics and statistics.

The first requirement for a climatologist is that he should know how his "working substance," the atmosphere, ought to behave, and be able to explain its normal behavior. The preceding chapters in this book, and those in *Foundations*, are really concerned with this requirement, and because it is a rigorous one, involving a difficult literature, it has been treated at considerable length. The second requirement for a climatologist is that he should know where to find climatic facts when he needs them, and it is convenient if these facts are available in an easily digested form. For many years, geographers have been concerned with the presentation of climatic facts in such a form, and have deliberately avoided the use of mathematical terminology. Consequently, the literature on this second requirement is very easy to follow, and only brief references to it are necessary.

A convenient starting point for a review of actual climates is provided by the maps in a generally available climatic atlas, supplemented by the maps in a recent general textbook on climatography.[1] It is the purpose of this chapter to direct the reader to sources that provide a more detailed description of actual climates than are available in the maps of such publications.

World Studies of Actual Climates

Map scales obviously limit the amount of information that can be incorporated in small-scale global maps. Many studies describe the distribution of precipitation in much greater detail than could be given in a world map of manageable proportions, and Figure 8.1 illustrates some of the comparisons these make possible.[2] Other studies deal with temperature distribution; those describing the temperature distributions within urban areas are of special interest.[3] Similar comparisons, between detailed maps based on "all" the observations in a limited area and more generalized maps that cover a much larger area and so must be based on a few "representative" stations in the limited area of the detailed study, may be made for sunshine, cloud, insolation, and winds.[4] All such comparisons emphasize that theories of how climate influences nonmeteorological phenomena should not be based on a map at only one scale. In fact, such theories should not be based on a map at all, but rather on a mathematical or physical model that demonstrates a logical relationship between the relevant climatic elements and the phenomenon under test.[5]

Maps of climatic elements not usually included in atlases may often be found in the periodical literature, for example, maps of soil temperatures, dew, and various measures of precipitation effectiveness.[6] Potential evapotranspiration has been a favorite derived element for detailed regional mapping, as has water balance decided

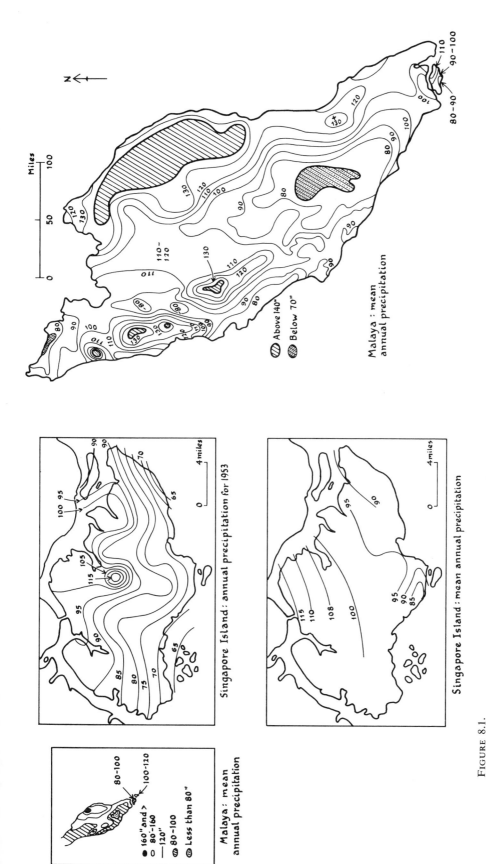

Miles

Malaya : mean
annual precipitation

Above 140"
Below 70"

Singapore Island : annual precipitation for 1953

Singapore Island : mean annual precipitation

Malaya : mean
annual precipitation

160" and >
80–160
120"
80–100
Less than 80"

80–100
100–110

FIGURE 8.1.

Annual precipitation distribution over Malaya. The purpose of this figure is to illustrate the importance of scale considerations (both space and time) in interpreting geographical distributions. (After I. E. M. Watts and W. L. Dale.)

on the basis of evapotranspiration.[7] One world map of moisture index, refreshingly, goes back to first principles, i.e., to Dalton's law of evaporation (see Appendix 8.1).

World maps of combinations of the climatic elements must involve careful considerations of physics. By ignoring such considerations one can produce such maps quite simply, but one then has difficulty in demonstrating exactly what it is that these maps depict. Application of the Köppen system of climatic classification in detail to small areas has been popular, for obvious reasons, but the difficulty here is that logically one should continue the process until one is dealing with microclimatic observations in a very minute area, in which case climatic types must be distinguished both vertically and horizontally. The same applies to studies in which the Thornthwaite systems are employed to determine very small-scale climatic regions.[8]

The importance of the general circulation of the atmosphere as a control of climatic characteristics in different parts of the world has been taken up in many studies. For example, major "static" climatic regions have been derived from streamline and air-mass maps of the world, and the trades and the trade-wind regions of the world have been described in terms of simple statistics derived from ship observations. The interrelationship between trades and monsoons has been described on the basis of aerological data, and the extent of the doldrums has been defined statistically and their weather described. A useful function has been performed by synthesizing on a regional basis the results of investigations by meteorologists and theoretical climatologists; in particular, a synthesis for the westerlies may be noted, and others for the tropics in general, for the southern South Pacific, and for the tropical Pacific Ocean. The significance of the stratospheric circulation in the interpretation of climatic facts at ground level has also been demonstrated synthetically.[9]

Fluctuations of the general circulation on a worldwide scale have received considerable attention, for example, in terms of the climatic consequences in different regions of the observed variation in monthly or seasonal pressure patterns, or of the general effects of climatic fluctuation on sea ice, glaciers, tree rings, and so on.[10] Some studies have attempted to find a correlation between circulation fluctuations and variations in some specific climatic element in a given area: for example, an attempt was made to determine whether or not the general circulation has varied in a manner that would harmonize with the observed temperature increase throughout polar regions.[11] Solar influence has been preferred as an instigator of climatic fluctuation, and the manner in which given stages of the sunspot cycle produce effects of differing magnitude in different regions has been mapped.[12] General circulation patterns in past geological epochs have been estimated from palaeontological studies. For example, the presence of warm-type fauna and flora in the Arctic indicates that the present latitudinal zonation of climates did not exist in all geological periods, and that there must have been more or less worldwide periods of warmth. Such investigations are complex, because palaeomagnetic effects and continental drift must be allowed for, but, nevertheless, reconstructions of the climatic conditions during the Permian, during the Pleistocene, and especially during the last glacial maximum of the latter have been made.[13]

Studies of regional climatic fluctuation, i.e., investigations of the time-variation in a given climatic element in a specific area, are useful, since they prove that the stable-looking distributions displayed in the world maps of atlases are very unstable if viewed in terms of geological ages. In fact, these maps are "climate-weather maps" in relation to the average pattern (or "astronomical climate") of the meteorological state of the Earth throughout geological time. Elements for which such studies have been made include temperature, precipitation, winds, and snowfall, and there have been numerous general regional studies of climatic fluctuations and their consequences.[14] All these studies have implications for students of climatic classification, for what the climatic taxonomists are trying to classify is not a stable phenomenon, and static climatic regions based on mean values correspond to no real feature of nature.

The different latitudinal belts of the world have different climatic problems, which are quite distinct from their meteorological problems. In polar regions, for example, a type of climate may be recognized (the polar marine climate) whose existence is overlooked in the classic schemes of climatic classification. An important control of polar climates is the albedo of the Earth's surface, and the distribution of albedo variations over Arctic regions has important consequences. An important feature of polar climates is the prevalence of gales, and interesting comparisons can be made between the Arctic and the Antarctic in terms of gale frequency. In tropical regions, an important distinction is between the arid tropics—i.e., the hot deserts, such as the Sahara, the Kalahari, the Atacama, the Sonoran, the Thar, and the Australian—and the humid tropics, the latter posing a problem in the definition of their limits. In middle latitudes, the problem as to what exactly constitutes a temperate climate is not easy to solve, although it is possible to demonstrate mathematically, by means of radiation theory, that middle latitudes should have superior climates in terms of solar heating.[15]

Regional Studies of Actual Climates

The climate of the British Isles is a fascinating one for the climatologist to analyze. British weather is so variable, and the British seasons so uncertain, that one sometimes wonders whether or not Britain does have a climate, but there is some synoptic evidence that natural seasons may be distinguished in terms of weather types, so that there is some justification for speaking of a British climate. One thing is certain; the vertical rate of change of climatic conditions in Britain—as indeed in other temperate Atlantic-type climates—is very significant, and very slight changes in temperature result in widespread effects. For example, a 2–3°F increase or decrease in summer mean temperatures in Britain could account for all the principal vicissitudes in climate and related phenomena during the retreat of the Pleistocene ice from its maximum extent.[16]

The topographical diversity of the small area that comprises the British Isles has a number of climatological implications. The first one concerns the mapping of the climatic elements. It is a commonplace observation, for example, that the map of annual precipitation for the British Isles closely resembles the map of elevation of the land surface. This is not surprising, since for many years it has been customary to draw isohyets as though they depended on elevation alone. A knowledge of the meteorological processes controlling rainfall soon convinces one of the inadvisability of such a procedure. Although it is perfectly correct to state, for example, that Wales is, in general, characterized both by wetness and by high ground, the relationship between the two is not a simple one. Under certain synoptic or atmospheric situations, high ground may be much drier than an adjacent plain. Second, because of the rapid altitudinal rate of change of the British climate, the upland areas are especially interesting. In climatic terms, the tundra is not far away from the higher British mountains, despite the hundreds of miles of horizontal distance to the true tundra of the northern hemisphere. The complicated microtopography characteristic of the British uplands forms an excellent natural laboratory for the study of local site effects; for example, daily range of temperature rather than minimum temperature appears to be the criterion for how a valley situation will effect local climate. Considerable differences in thermograph records from hill and valley stations may be noted for differences in altitude as little as 300 feet, for example, in the Pennines. The climatic uniqueness of certain areas, which may be anything but obvious from a study of their synoptic climatology, is brought out by a study of mean data and, in particular, by a visual study of their vegetation; a good example is Galloway, which, although in a northerly part of the British Isles, has mean monthly temperatures similar to Oporto or Santiago and mean monthly precipitation totals similar to Funchal, and consequently is able to support palms and tree ferns.[17]

The climate of British towns and cities has received a considerable amount of attention. Sometimes the climate of the area in which the town or city is situated has been described, and the differences between rural and urban areas within it have been noted; sometimes special stations have been established to provide statistics for a local climatography of the city area, or temperature and humidity traverses have been made across the built-up area to assess the extent of its "heat island." The effects that London's urban area might have on the distribution of thunderstorm rainfall have also been investigated.[18]

The synoptic climatology of British seasons is of special interest, because it involves an intermingling of maritime, continental, and altitudinal effects on the general circulation. As a consequence, the occurrence of seasons of extreme severity can rarely be explained with mathematical exactitude: good examples are provided by the cold winter of January–March 1947, and the summer of 1954 in Scotland.[19]

Interesting synoptic studies of tornadoes in Sweden and of pressure waves in Scandinavia deal with phenomena that are not generally considered typical of the European climatic scene. The tornado study is exceptionally valuable, comprising estimates of the kinetic energy and energy sources of the tornadoes as well as descriptions of the geographical circumstances. The pressure waves were generated as a result of the explosion at Oslo on December 19, 1943.[20]*

Different parts of Europe are characterized by geographical circumstances that give rise to very distinct climatological problems. For example, the deep incision of the Alpine valleys has a profound effect on the insolation received on their floors. Severe ice conditions in the Baltic may be so serious for shipping that synoptic reporting of the extent of the ice is necessary. Since the mean extent of ice in the Åland archipelago increases and decreases each winter and spring, respectively, by passing through almost identical stages, one can derive a mathematical expression that will indicate at what time the sea will probably become coated by ice. In fact, for the Baltic generally, one can develop a mathematical theory, based on Newton's law of cooling, that describes the critical threshold of heat conduction from the sea surface, when the latter

---

* The explosion was heard in Sweden, at 100 to 300 km, but not at some closer distances to Oslo. This is an example of an acoustic property of the atmosphere that gives rise to zones of audibility separated by zones of inaudibility. See Sir N. Shaw, *Manual of Meteorology* (Cambridge, Eng., 1942), III, 35–52, for an account of the physics of this property.

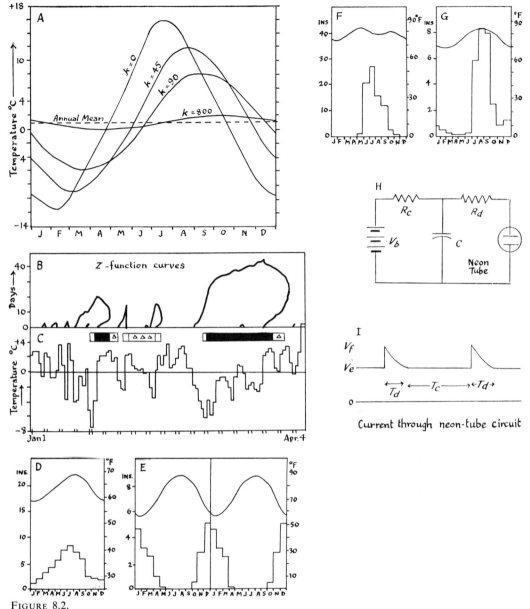

Figure 8.2.

Some investigations in physical climatology A-C, Ice formation in the Baltic in relation to air temperatures. A, Normal air temperatures and tau (τ) curves for different values of Rodhe's constant. B, Daily values of Rodhe's z-function at Eggegrund, January 1 to April 4, 1949. C, Daily values of air temperature at Eggegrund, January 1 to April 4, 1949, and ice formation in Gäule Road:

□ Young ice ■ Fast ice △ Drift ice

D-G, Electrical analogy to some temperature and rainfall regimes. D, Sao Paulo (2,690 feet); upland savanna type. E, Tiberias (653 feet); Mediterranean type. F, Bombay (37 feet), Monsoon (pulsed) type. G, Mazatlan (23 feet); mixed rainfall regime. H-I, A simple pulse generator. C, capacity of condenser (farads). R_c, Charging circuit resistance (ohms). R_d, Neon-tube circuit resistance. V_b, Voltage of charging battery. V_e, Extinction voltage of neon tube. V_f, Firing voltage of neon tube.

$$T_c = 2.202 \, R_c C \log_{10} \frac{V_b - V_e}{V_b - V_f}.$$
$$T_d = 2.30 R_d C \log_{10} V_f / V_e.$$

(After B. Rodhe and R. L. Ives.)

approaches freezing temperatures, such that sea ice forms (see Figure 8.2 and Appendix 8.1). In Finland, the accumulation and the decreasing of snow in pine-dominated forests, the measurement of humidity and temperature above crops, and the observation of local air motions have been the subjects of useful studies.[21]

In Iceland, the high land crossing the country from northwest to southeast forms a very effective " weather divide," separating warm, moist air from the Atlantic to the south and west from cool, dry air from the Arctic Ocean to the north and east. The weather patterns on either side of this divide differ considerably even when a single type of air mass covers the country. On this effect is superimposed the effects of the movements of the Arctic front, which causes great fluctuations in precipitation and temperature from year to year and from place to place. As a consequence of these two effects, the climate of Iceland varies greatly both regionally and temporally, with important implications for agriculture. The variation in the extent of ice covering the Icelandic seas has significant implications for the climate of much of the northern hemisphere.[22]

Observations of the peculiar meteorological conditions associated with surfaces of inland ice have been made on the Hoffellssander, a glacial outwash plain on the margin of Vatnajökull, and on Langjökull, both in Iceland. The former showed very marked katabatic winds during nighttime, during which ablation of the ice sheet ceased, with warm conditions during the day despite cold northerly winds, because adiabatic warming of the cold air descending from the summit of Vatnajökull to the coastal plateau produced a well-developed föhn effect. Sudden increases in air temperature at screen level above the ice surface were characteristic of Langjökull. The increases were about 1 or 2°C per hour, and were partly due to release of latent heat. A correlation with ice-cap size is evident, for the small magnitude of the Langjökull (a small ice cap with an area of 1,000 km²) effect contrasts with the sudden increases in temperature of about 3°C in Sweden (the Mikka glacier), 4°C in Baffin Island (the Barnes ice cap, area 5,900 km²), and 15–20°C over the central Greenland ice cap. Upslope surface winds, i.e., opposite in direction to what one expects for a flow of air cooled from below, are a feature of the Mikka glacier, with the true downslope wind some 30 to 40 m above the ice surface; a similar effect has been noted on the Fröya glacier in northeast Greenland, where the upslope wind was only 2–4 m deep, and on the Kårsa glacier, where it was up to 10–50 m thick. Summertime observations in Norway, on the Skagastöl glacier of the Jotunheimen, indicate that the glacier cools the air flowing over it by 3–4°C, and that this air has an extremely low viscosity. Interpretation of the observations in the light of turbulence theory shows that the atmospheric boundary layer above the glacier (which is up to 100 m deep) has an abnormally stable stratification in summer, and the streaming air layer is stressed from both the ice surface and from the air above; thus, despite the low viscosity, the wind-speed within the surface layer remains low (see Appendix 8.3).[23]

The provision of climatological information for undeveloped countries has much engaged the attention of WMO, UNESCO, and other international organizations in recent years. Africa provides an excellent example. Despite the information available in two excellent atlases, there are numerous gaps in our knowledge of the distribution of even the basic elements in the continent, particularly those, such as evaporation and evapotranspiration, that pose instrumental difficulties, or elements, such as soil temperature, that are very hard to generalize in terms of geographical distribution. The climates of certain African countries have been described in terms of daily weather developments, which adds life to the abstract mean patterns in the climatic atlases, and elements such as precipitation have been interpreted in weather terms. Other studies have dealt with weather phenomena, such as dust storms, that tend to be overlooked in continent-wide climatic descriptions, despite their obvious importance to local inhabitants.[24]

Synoptic studies of weather and climate in the southern half of Africa are interesting as tests of the methods of analysis originally developed for the northern-hemisphere westerlies, and as illustrations of the inadequacy of the "mean-value school" of climatology. The old theory of the climate of South Africa stated that the entire country (except the Cape Peninsula, which experienced rain from northwesterly winds) received rain from southwesterly winds from April to October and from southeasterly winds from November to May. However, this theory did not explain the fact that most of Durban's rain came with southwest winds and most of East London's rain with northeast winds. The seasonal wind change was regarded as a consequence of the change from a permanent low-pressure area over the continent in summer to a high-pressure area in winter; i.e., a monsoonal mechanism was predicted. Before the Second World War, South African weather was explained as due to the interaction of three air masses—Egyptian air from the north, trade-wind and monsoon currents from the South Atlantic, and the southeast trade from the Indian Oceans—that converged near the center of Africa between latitudes 5° and 15°S, and gave rise to one of the rainiest regions of Africa. The absence of rain in winter was attributed to the presence of a stable anticyclone over the eastern half at least of the African plateau.[25]

By 1935, aerological observations made clear that there was virtually no difference, winter or summer, between the atmospheric circulations at the Earth's surface and 3,000 m above it; i.e., a weak anticyclone exists both summer and winter, centered over northern Natal and the southeast Transvaal, and the monsoon mechanism is not applicable. It was also recognized that the three-air-mass theory could not explain the marked diurnal variation in surface wind on the African plateau, over which westerly winds prevail during the greater part of the day. Bilevel weather maps produced during the Second World War indicated that distinct circulations were characteristic of the coastal areas and the plateau, the two circulations being separated by a discontinuity that broke down gradually to the north, so that the southeast trades could move far inland in Southern Rhodesia and the northern Transvaal. The sea-level chart was used for the depiction of conditions in coastal areas, and the chart of the 1,200 or 1,250 geopotential-decameter levels for conditions over the plateau. It was found that the surface wind and pressure distribution reflected the upper winds; i.e., the latter came down to the surface, and the variation of wind with height was largely due to diurnal variation in temperature and turbulence. By 1955, the concept of the general circulation over South Africa was one in which a weak anticyclone forms the only persistent pressure feature. The anticyclone is weaker in summer than in winter; during the latter it moves north and brings westerlies to South and Middle Cape Province. The center of the anticyclone, which lies over northern Natal and southeastern Transvaal at a height of 2 km, is displaced westward with height, so that at 4–6 km it lies over central-northern Southwest Africa. The northward displacement of the anticyclone during the winter occurs at 2–4 km. The special weather features of South Africa—for example, hot spells, summer rains, and winter cold snaps—are interpreted in terms of departures from the mean circulation.[26]

Southern Africa possesses local weather systems that are very much a consequence of the topography of the country. These include the Berg wind, which blows directly from plateau to coast as the only west wind over the lands below the Drakensberg escarpment. The Berg wind may be interrupted during the day if the diurnal wind variation near the plateau edge is great, because the normal backing of the wind from east to northwest or west associated with the latter breaks down the pressure gradient necessary to maintain the Berg wind. Other topographically induced wind systems include a particularly thoroughly analyzed group of local valley and mountain wind systems and nocturnal winds in Natal, the former developing into large-scale regional air movements, and sea-breeze systems that are strongest on the west coast of South Africa. Further north in the continent one finds other examples of topographic effects. For example, the central Rift Valley of Kenya is relatively dry, and much of the precipitation that falls is usually described in elementary textbooks as scattered local showers due to short-lived convection currents set up by minor topographic details, i.e., random effects in the meteorological sense. In fact, over 90 per cent of the showers are in an organized form, so that they must be linked to large-scale circulation features.[27]

Synoptic weather analysis in Australia began in 1893, and led to the description of the synoptic climate of the continent south of latitude 20°S in terms of a series of rapidly moving anticyclones following each other across the continent from west to east with great regularity. Such a description could be based only on surface charts until 1937. In 1939, pilot-balloon ascents became available, and air-mass and frontal concepts began to be applied; in 1942, radiosonde ascents began. The main results of air-mass analysis were the recognition of coastal fronts and subsidence fronts. The latter form between a surface layer of air and air descending from above. They limit convection and result in stratiform cloud. They are not stable synoptic features and fluctuate rapidly in position. The coastal front begins as a simple sea-breeze front but may move considerable distances inland, especially in summer. A further type of front, the meridional front, was recognized as developing between two adjacent stationary anticyclones.[28]

In 1944, emphasis on the methods of tropical—particularly streamline—analysis in the interpretation of Australian weather began to appear. Frontal analysis sometimes produced inconsistent results, but a more useful pattern was found to emerge when pressure and streamline maps were used in conjunction. It was noted, for example, that under steady conditions a cross-isobaric airflow occurred from high to low pressure, the angle between wind direction and isobars increasing with decreasing latitude, becoming 90° at the equator. Convergence was associated with cyclonically curved isobars in such a steady state, and weather analysis became a search for areas of convergence and, therefore, bad weather. The effects of mountain barriers on convergence were found to extend to considerable horizontal distances; for example, cloud banks were observed 50–100 miles to windward of the topographic feature producing them.

Australia is of a convenient size to present a peculiar climatological problem concerning differential heating. Differences in the rate of heating of the continent and its surrounding seas result in the production of one or more domes in the isobaric surfaces above the 5,000-foot level, so that a semipermanent anticyclone is produced between 5,000 and 15,000 feet over northern Australia. Another peculiar problem is posed by tropical cyclones in the northern portions of the continent.[29]

The weather and climate of South Island, New Zealand, is largely controlled by the southern-hemisphere westerlies. Because of this control, cloud and precipitation is heavy on the west coast but light to the east of the main mountain backbone of the island. In North Island, conditions are more complicated. The configuration of the land surface becomes important, in that surface winds are much channeled and diverted by the shape of the mountain massifs and by the gaps through them. Because of this, for example, gales are a prominent feature of the Cook and Foveaux Straits.[30]

Despite the importance of its westerlies, New Zealand weather is only under the control of westerly winds for 40 per cent of the year. The 20 per cent of the year under easterlies and southeasterlies is important for the distribution of hailstorms, and southerly winds are associated with snowfall, particularly in South Island. The key to the synoptic climatology of New Zealand is that the subtropical high-pressure belt of the southern hemisphere, which is centered between latitudes 35° and 40°S, and covers northern New Zealand and the northern part of the Tasman Sea, is not continuous, but consists of a series of semipermanent anticyclones that move slowly east-northeast across the country, at the rate of one or so per week. The anticyclones bring fine, calm weather to New Zealand, with frosts in winter. When an anticyclone moves off to the east of New Zealand, the prevailing northwest winds freshen, and orographic cloud forms to the west of the Southern Alps, gradually extending northward to cover western parts of the country. The low-pressure troughs between the anticyclones often contain cold fronts, which bring low cloud and rain, and the wind changes from its normal northwesterly direction to west, southwest, south, or southeast, depending on the exact position of the trough line.[31]

The area between Australasia and Southeast Asia, the Southwest Pacific, presents an especially difficult climatological problem, that of tropical cyclones. There is a continuous spectrum of tropical disturbances in the Southwest Pacific, with intensities ranging from barely detectable (5–10 knots) to hurricanes (100 knots or more). The convention is to regard disturbances with maximum windspeeds of force 9 (45 knots) or more as tropical cyclones, and those with maximum winds of less than force 8 (35 knots) as tropical depressions. The cyclones occur both in winter and in summer; some form (especially from December to March) in the I.T.C.Z.; others, especially in winter (June to September) form in wave-like perturbations of the prevailing southeast trades. These perturbations are quite distinct from those that give rise to easterly waves. They do not require deep easterly currents in which to develop, and usually appear in trades that are much less than 10,000 feet in vertical depth. The average annual frequency of these tropical cyclones is 3.6, which is less than that of tropical cyclones in the North Atlantic, and much less than that of tropical cyclones in the North Pacific.[32]

Two climatological phenomena are of special interest in Southeast Asia: the monsoon and the main tropospheric jetstream. We noted in *Foundations* that the so-called monsoon effect cannot alone explain the observed sequence of weather in the monsoon countries of Asia. In fact, the seasonal climatic features of these countries result from large-scale changes of the atmospheric circulation associated with shifts in the position of the jetstream, for which the existence of the Tibetan Plateau is critical. The highlands of Central Asia cause a splitting of the zonal westerlies in winter. The northern branch flows north of the Tibetan Plateau and is relatively weak and unstable. The southern branch is located south of the Tibetan Plateau and is much more stable, usually lying over the southern slopes of the Himalayas and being responsible for winter precipitation in northern India-Pakistan and southern China via the regenerating and steering effect it exerts on eastward-moving perturbations. The strong subsidence that is a feature of the equatorward side of this southern jet gives rise to dry seasons in winter and spring. The convergence zone where the two branches of the jet come together in the lee of the Tibetan Plateau is an important cyclogenetic area: cyclones originating here are important producers of precipitation in southern China.[33]

The droughts characteristic of the cooler months of the year in peninsular India and peninsular Southeast Asia are controlled by the presence of strongly subsident air to the south of the southern branch of the jet. When the latter disappears in early June to take up its "normal" position north of the Tibetan Plateau, there is a rapid northward surge of equatorial air over Southeast Asia, bringing the summer rainy season. This is the "Burst of the Monsoon" in India and the beginning of the *Bai-u* rains in Japan and southern China. The premonsoon months, from mid-March until the Burst of the Monsoon, are the most critical ones of the Indian climatic year. During this period, showers from storms and from "nor'westers" are very important for agriculture. Nor'westers are typical of eastern India-Pakistan, particularly of Bengal, where their frequency is about 10–30 per year. They are connected with the prevailing situation in which dry air from the northwest lies above moist air from the southwest or southeast, and may be associated with the development of a concavity in the zone of separation between the two air masses.[34]

The climatic seasons of Japan may be divided into the winter monsoon, the early summer (Bai-u) rains, and the autumn typhoons. The southern branch of the Southeast Asian double-jet structure lies over southern Japan during the winter monsoon season. The northern branch of the jet causes a blocking anticyclone to develop over the Sea of Okhotsk and a cut-off low over Manchuria. The blocking anticyclone generates a surface anticyclone that causes cold northeasterly (i.e., winter monsoon) air to flow over Hokkaido and northern Honshu. The sudden displacement of the southern branch of the jet to its summer position north of the Tibetan Plateau in late May or early June marks the commencement of the Bai-u. The latter then shifts northward over eastern Asia in stages. Although the rainy seasons in China and Japan appear nearly simultaneously, they have different climatological and meteorological structures. For example, the relation between the northward shift of the Bai-u frontal zone and the time of appearance of the rainfall maximum is different in China, southwestern Japan, and eastern Japan. The location of the North Pacific subtropical high-pressure cell is highly significant for local rainfall in Japan during the Bai-u: if it is strong or located well to the north, the frontal zone is displaced northward and drought occurs in southwestern Japan. A sudden reappearance or intensification of the jet over the south coast of Japan occasionally results in heavy precipitation at the end of the Bai-u system.[35]

The high population density, topographical diversity, and agricultural problems of Southeast Asia have long made its weather and climate subjects of intense interest for geographers. This interest has sometimes been manifested in attempts to describe the climates of the countries of Southeast Asia in synoptic terms, at other times in studies of synoptic entities that are of considerable importance as local weather controls, for example, line-squalls in Malaya. Squalls along the west coast of Malaya during April to November, termed *sumatras*, take the form of more or less continuous lines of towering cumulus or cumulonimbus 200 miles or more in length, except where the cloud continuity is broken by terrain effects. They usually accompany moving airstream boundaries. Most of them form in the Malacca Straits, probably initiated by land breezes along the Malayan coast that undercut conditionally unstable air over the Straits. Their direction of movement bears no relation to the low-level winds, and is independent of changes in pressure distribution. Most of them move with the wind at the 6,000-foot level, either with the clouds at lower levels constantly dissipating in the west and reforming to the east, or by means of a process in which cumulonimbus downdrafts move eastward with the wind at 6,000 feet and undercut the latter at ground level to form another cumulonimbi series downwind. During the period May–September, *sumatra line-squalls* strike the coast between Singapore and Port Swettenham with 50-miles-per-hour wind gusts, and cause temperature drops of 5°F in 5 minutes with or without rainfall. These line-squalls occur at night or early morning in homogeneous airstreams. They take the form of lines of almost continuous towering cumulus or cumulonimbus (often thunderstorms), and their intensity depends on the curvature of the shoreline. Both sumatra line-squalls and sumatras are most intense with concave shorelines: the land breezes converge toward the center of curvature, thus undercutting and uplifting the warm air offshore. The diurnal variation of summer rainfall in Malaya is especially interesting in that the precipitation pattern cannot be explained by a combination of afternoon convectional heating and the effect of orography on the prevailing low-level westerlies. In fact, five distinct diurnal precipitation patterns may be recognized, due to interactions between the synoptic flow (i.e. the southwesterlies) and local land and sea breezes and anabatic and katabatic winds.[36]

The climate of eastern Asia, especially that of China, is of obvious interest for western climatologists. Since the winter climate of eastern Asia is not significantly correlated with variations in the northern-hemisphere zonal index, one must define a local index if one is to interpret synoptically the observed weather sequences. Despite occasional collections of research papers in English by Chinese climatologists, very little is available on China's climate in western geographical sources, except for elementary discussions of such topics as air masses, precipitation, climatic regions, and the length of the growing season.[37]

Studies of the climate of the Soviet Union have been dominated by applications of the concepts of heat and water balance as developed by Budyko. Although the mathematical validity of the Budyko technique has been questioned, it has been taken up on a very large scale by Soviet geographers and climatologists, especially for its applications in nonclimatic problems and the transformation of nature by man. The classification of Russian climates has also been based on concepts derived from the Budyko approach, for example, evaporability. Other classifications have been attempted, for example, on the basis of the march of precipitation.[38]

Russian climates obviously bring to mind winter cold and snow. A useful climatic concept, that of *warm-core winters*, has been developed in the U.S.S.R. In a warm-core winter, one winter month is warmer than the adjacent months, whereas during a coreless winter, temperatures decline to the coldest month and then increase. That the cold pole of the northern hemisphere is in northeast Siberia is an oft-quoted statement in elementary climatology, but in fact it is a "cold zone" rather than a cold pole. The snow cover of the Moscow region has been recognized as one of the dynamic forces in the formation of its landscape.[39]

Much of the U.S.S.R. is hot and arid as well as cold and snowy. *Sukhovey*, or hot, dry spells, are characteristic of some areas, and dust storms are typical of the steppes of western Siberia and Kazakhistan. In Southwest Asia, the sirocco, a phenomenon associated with migrating cyclones from the Mediterranean, brings the hot, dry characterisics of the Saharan and Arabian Deserts, plus high winds and very poor visibility. Synoptic study shows that the sirocco of Lebanon and Syria is quite distinct from the sirocco of Egypt.[40]

The last decade has seen the development of considerable interest in climatology in Japan, especially in the synoptic climatology of Japan, for example, the synoptics of rainfall, of snowfall, of winds, and of the formation of orographic rain-bands.[41] There has also been much activity in local climatic studies, many of which have concentrated on the production of synoptic maps of the distribution of temperature or humidity, radiation, or winds in city or town areas.[42] These have mainly been descriptive studies, but occasionally physical interpretations have been attempted. For example, the distribution of temperature over Kumagaya City has been shown to be closely related to a parameter, derived from Brunt's night-cooling equation, whose value depends on the physical properties of the construction material (see Appendix 8.4). The effect of urban expansion on city temperatures has received some attention, as has the nature of the vertical temperature profile above an urban area.[43]

Studies of rural areas in Japan give some indication of the effect of surface cover on temperature and humidity distributions. For example, in similar topographic situations, bare fields, bushland, and cultivated fields are 0.6–1.0°C warmer in mean air temperatures than forests, grasslands, and marshes. Bushlands, paddy fields, and forests have 5–10 per cent higher relative humidities during the daytime in late summer than bare fields and grasslands. Interesting examples of mesoscale climatic situations have been made that illustrate, for example, the importance of sea breezes as well as surface temperature discontinuities as mesoclimatic divides. The heat budget of the lower atmosphere has been shown to be a significant mesoclimatic influence in the Kanto Plain. The influence of local topography on wind characteristics has received considerable attention. For example, measurements up to a height of 5 m above the ground in flat open land and on the sides and floor of a small valley in Nagano Prefecture showed that the ratio of valley windspeed to open-land windspeed changed with the variation in windspeed, the relation being negative during the day but positive at night for weak general (synoptic) winds. Windspeeds were always weaker on the shadow sides of the valley than over the flat open surface, especially for the steeper valley sides. Differences in the local topographic situation were found to have an effect on windspeed similar to those produced by variations in the roughness parameter or zero-plane displacement in the logarithmic law of wind variation with height.[44]

Some interesting regional studies of climate have been made in North America. In California for example, the well-known summer fogs of the San Francisco Bay area have been analyzed, and the role of the open pine forests of the Sierra Nevada in ensuring that this snow-covered area experiences relatively high daytime temperatures has been elucidated.[45] Despite a deep and persistent snow cover, the central Sierra of California exhibits abnormally high air temperatures at its crest. This warmth is not advected in from other areas, but is created in the crest region by the peculiar heat-balance characteristics of the local landscape, especially by the heat exchange between the pine forests and the snow cover. A high intensity of solar radiation is received in the Sierras, mainly because of the high frequency of anticyclonic airflow at 700mb, and the warmth resulting from this insolation is stored in the forest canopy. The heat given off from the latter is then trapped between the concave ground below and a subsidence inversion above, and so remains in the local air.

Death Valley has special climatological interest, and the winter temperatures of desert mountain ranges in the western United States provide important illustrations of unusual "island" effects. The climate of the Sonoran Desert has been described in considerable detail, including the visibility phenomena and optical effects its atmosphere produces, and it has been demonstrated how the phase relationships of its temperature and rainfall may be represented in analogue form by means of a neon-tube oscillator circuit (see Figure 8.2, D) to almost any desired degree of accuracy (see Appendix 8.5). The climate of the central North American grassland, with its very erratic precipitation and temperature, and consequently severe drought risk, has been explained in terms of the center of mean atmospheric divergence, which is located at the eastern base of the Rocky Mountains. The existence of this center implies that a continental airstream must prevail over the western part of the Great Plains in every month of the year. During years of major drought, the continental airstream extends further east than usual. The economic consequences of the subtropical climate of Florida have been described in simple terms, and, at the other extreme, the cold summer of 1816 in the northeastern United States, and its consequences, has been interpreted in terms of cold waves originating from the accumulation of cold air over the upper Mackenzie basin when the North Pacific was warmer than usual. Other studies have been concerned with the detailed subdivision of a state in terms of simple inspection of the mean climatic data, and with case histories of important weather events, for example, the New England hurricane of September 1944, and their human and economic consequences.[46]

The occurrence and consequences of chinook winds in the Colorado high plains have been recorded. Chinooks vary greatly within a small area, so that winds of 50 miles per hour may be reported at one station, and little or no wind at a station a few miles away. The highest windspeeds are usually found in leeward eddies, which are partly due to the existence of the semipermanent low-pressure trough in the lee of the Rockies, and chinooks are known to "jump" across intermontane basins from a few to several hundred miles in width, leaving a windless zone in the bottom of the basin. The effects of chinooks on temperature may be spectacular once or twice per season, but they have little effect on the climatic means. For example, the annual mean maximum temperature at Denver (1930–1948) was raised only 0.23°F as a result of chinooks. Although chinooks may cause a considerable amount of snow to disappear (8 inches were removed by a single chinook in May 1935), the average amount of snow removed by chinooks per year is not large. One important effect is that intense static electricity is generated during chinook spells, leading to the accumulation of an electric charge on wire fences large enough to kill livestock, the stalling of automobile engines, erratic radio communication, and extraordinary transmission of ordinary sonic communication; train whistles to the west of the Colorado Front Range may be heard 80 miles downwind, on the eastern side of the range.[47]

Climatological studies dealing with Central and South America and published in English are not very common in geographical literature. Two important exceptions are the work of McGill University on certain aspects of Caribbean climatology, and analyses of precipitation and water balance. The variability of Mexican precipitation,

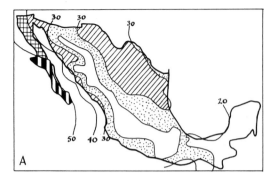

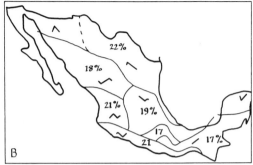

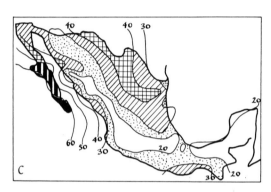

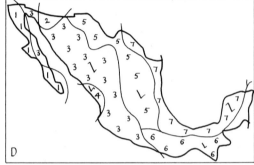

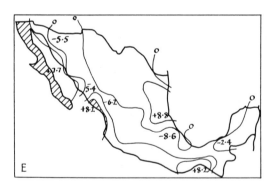

FIGURE 8.3.
Precipitation variation in Mexico (after C. C. Wallén). A, Coefficient of variation of annual precipitation (per cent). B, Generalized secular (annual) precipitation changes 1920-50. C, Relative interannual variability of annual precipitation (per cent). D, Fluctuations of monthly relative interannual variability over the year (January, April, July, October).

    1: Maximum, October; minimum, January-April.
    2: Maximum, October; minimum, July (October > April).
    3: Maximum, April; minimum, July (October < April).
    4: Maximum, January; minimum, July (October > April).
    5: Maximum, January; minimum, July (October < April).
    6: Maximum, January; minimum, July (October ≤ April.)
       (smaller variations over the year).

for example (see Figure 8.3), has been described in detail; the main features of rainfall in Baja California and Central America have been evaluated; and the water balance of the Lake Maracaibo Basin has been studied.[48]

Canadian weather and climate presents several distinct problems, corresponding to the broad geographical zones into which the country may be divided. Important studies have been made in each of these zones. For example, the nature and origin of the Canadian dry belt, which covers 50,000 square miles and is the largest of the dry regions of the Great Plains of North America, has been explained in terms of the atmospheric general circulation. The occurrence of drought in the Canadian Northwest has been analysed, and water-balance requirements in this region have been evaluated. Important studies of sub-arctic climates have been made at Schefferville in Labrador, and the most obvious features of a forest microclimate have been measured approximately (see Figure 8.4) in British Columbia. The Canadian Arctic is clearly of great climatological interest (see Figure 8.5); topics such as the heat and moisture exchange above the surface of Penny Icecap in Baffin Island, and its importance for ablation, and the effect of the Barnes Icecap, also in Baffin Island, on the lower atmosphere, have been investigated in great detail. One interesting feature is that the layer of air immediately above the surface of ice islands in the

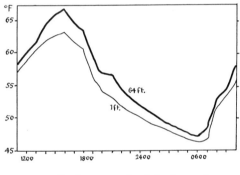

Hourly mean dry-bulb temperatures

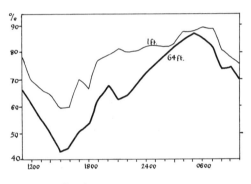

Hourly mean relative humidities

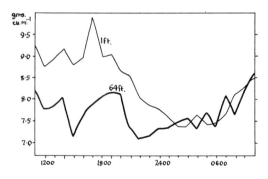

Hourly mean absolute humidities

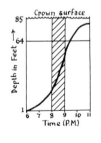

Depth of katabatic wind

Shaded area shows
dew formation

FIGURE 8.4.
Microclimatic observations in British Columbia, made by standard thermometers, August 26 to September 19, 1953, in forest on northern shore of Stuart Lake, B.C. Thermometers on north-facing side of spruce trunks, at heights of 1 foot and 64 feet above the ground.
(After F. G. Hannell.)

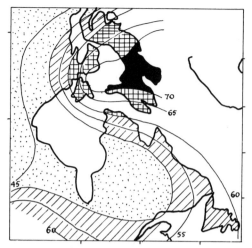

A   Percentage frequency of
cyclonic airflow in January

■ More than 70% ▦ 60 to 70%

▨ 50 to 60% ▤ Less than 50%

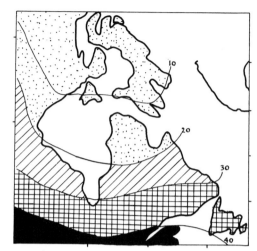

B   Percentage of occasions in January
with a front within 150 mi.

■ More than 40% ▦ 30 to 40%

▨ 20 to 30% ▤ Less than 20%

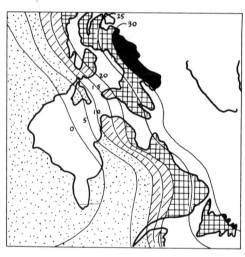

C   Percentage frequency of Polar
maritime air in July

■ More than 30% ▦ 20 to 30%

▨ 10 to 20% ▤ Less than 10%

FIGURE 8.5.
Occurrence of synoptic features in northeasternCanada. (After F. K. Hare.)

Canadian Arctic often has an adiabatic or even superadiabatic lapse-rate. This un-expectedly very unstable layer is usually relatively thin and has a stable layer above it, but it can persist for long periods of time. It appears to be produced by the warming effect of the surrounding sea combined with large values of wind shear, and in extreme cases the resulting lapse-rates have been observed to be seven times the adiabatic lapse-rate. Microclimatic profiles of temperature and humidity in the North American Arctic are very sensitive and require careful measurement, as observations at Point Barrow in Alaska indicate.[49]

The Antarctic continent is another major area of the globe that poses intriguing climatological problems. For example, unknown orographic features in the con-tinent have been predicted on the basis of observed weather patterns that could be induced by the presence of mountain barriers. The isolation of Antarctica has always suggested that it might have a distinct climatology of its own, especially because of its ice cover. However, although distinct local effects do occur, both the synoptic climate of the Antarctic region, and the observed variations of its climatic elements, such as temperature, indicate that the geography of the continent is not as important as the general atmospheric circulation in the genesis of its weather and climate.[50]

Regional comparisons of climate may be of considerable value as indicators of the existence of "problem" climates. For example, useful comparisons have been made of dry western littorals—where there are too few stations to allow explanations of the local climates in synoptic terms—and of tropical western margins of the Northern Oceans. Air-temperature profiles normal to coastlines in different regions have also been compared. Such investigations are especially valuable if the situations to be compared are each interpreted in terms of a general theoretical structure common to all of them. Comparisons of the climates of widely separated countries, for example Argentina and Australia, may be of value in practical fields such as agriculture, as well as in a purely scientific context, and comparisons in terms of analogues, for ex-ample, water-balance analogues, may be illuminating.[51]

Methodological Studies in Geographical Climatology

For many years, expecially until the early years of the second World War, geographical studies of climate were based merely on superficial inspection of climatic data. Largely owing to the wartime service of many geographers as meteorologists with the armed forces, the last twenty years have witnessed an increasing interest in climatological methodology, which first took the form of a reaction against the Kendrew-Miller type of regional climatic description, in favor of description in terms of air-mass climatology. This particular reaction stemmed from its protagonists' ex-perience as weather forecasters in a certain area, and some interesting regional de-scriptions resulted, but the next step, of basing geographical climatology on air-mass analysis, although logical, was unwise, because in the meantime the meteorologists had developed newer and more powerful concepts than the synoptic one, as was realized by the very few geographers who had obtained training and experience in "real" (i.e., theoretical) meteorology, as distinct from the very mixed empirical-theoretical-rule-of-thumb world of the weather forecaster. The important distinction betweeen synoptic and dynamic climatology in the climatologists' sense became apparent by the early 1950's, although it has been confused by the tendency of some geographers to term synoptic climatology "dynamic climatology," in contrast to the Kendrew-Miller type of climatology, which is regarded as "static climatology." Since the expression "dynamic meteorology" refers to the investigation of meteorological phenomena by means of the principles of dynamics, it seems advisable to use the expression "dynamic climatology" only when speaking of the investigation of climatological phenomena in terms of dynamic principles, and to use the term "synoptic climatology" for the study of climate in terms of air masses and fronts. The Russian concept of "complex climatology" is basically only a form of synoptic climatology, although it does go beyond what is regarded as synoptic climatology in the meteorological sense, in that it takes into account microclimatic effects and geographical influences. By contrast, the concept of "topoclimatology" has some claim to embrace a distinct type of climatological investigation, i.e., one in which the emphasis is on the reactions between microclimatic, synoptic, and geographic influences. In such work, sound physical reasoning is essential, and the concept of energy balance is fundamental.[52]

Despite the increasing interest in climatological methodology among geographers, there is no evidence of a distinctly geographical approach to pure climatology. Considerable intellectual energy is often expended in explaining in oversimplified verbal terms meteorological concepts that are essentially mathematical and cannot be expressed other than mathematically. The concept of climate is a difficult one, and simplification is full of dangers.[53]

In the prewar years, climatic classification constituted the chief climatological activity of geographers: good climatic geographers were supposed to be those with good memories, and as late as 1950 a system of climatic classification was devised that was based on mnemonic principles. Although many geographers still practice climatic classification, they at least consider its purpose, and the limitations of the arithmetic mean as a basis of classification are understood. The limitations of the climatic record are such that climate can never be regarded with mathematical exactitude as a stable phenomenon that may be divided into quasipermanent discrete types, but it is perfectly feasible to apply mathematical techniques of analysis to the climatic record for the purpose of determining the validity of categories of weather or climate that have been defined on *a priori* grounds. The analysis of variance provides a sound basis for such a procedure (see Appendix 8.6). Making local additions to the climatic record may be a useful and instructive exercise; students or school pupils have been used effectively in such exercises.[54]

The effective portrayal of climatic data has been of perennial interest among geographers. For map portrayal, the isopleth technique is still the most common. Despite elementary discussions of isopleth accuracy, however, the fact that an isopleth map is a mathematical structure, and therefore has certain implications, has received little or no attention in the literature. The geographical disadvantages of climatic maps reduced to mean sea level may be avoided by reducing the climatic data to levels that have some meaning in a given area. Instead of one map of, say, mean temperature for an area, several maps will be required (see Figure 8.6), but the amount of geographical information that can be obtained from the same climatic data is greatly increased. With graphic portrayal, considerable care is necessary. For example, rainfall dispersion diagrams have been popular for many years, but there is some doubt that they are valid in terms of mathematical statistics.[55]

Statistical analysis of climatic data is an obvious necessity in many geographical studies, but the literature must be read very critically. Precipitation has been a

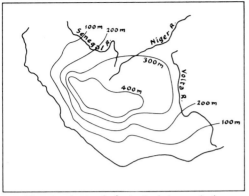

Elevation (meters) of principal streams beds.

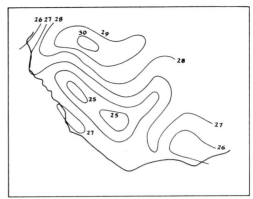

Annual mean temperature (°C)
reduced to stream-bed surface.

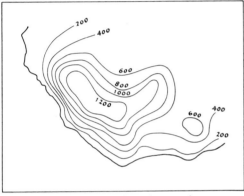

Elevation (meters) of interfluve surfaces.

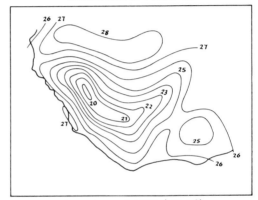

Annual mean temperature (°C)
reduced to interfluvial surface.

FIGURE 8.6.
Reduction of annual mean temperatures to significant levels for the Guinea Highlands, West Africa (after W. E. Howell).

popular element for statistical attention. Studies range from subjective interpretations of the obvious features exhibited by the raw data, through investigations involving elementary statistical concepts, such as skewness and variation, regression, correlation, and residual mass curves, to more elaborate techniques, such as harmonic analysis.[56] Harmonic analyses of the annual march of precipitation over the United States (see Figures 8.7 and 8.8) and over Canada, and a classification of types of rainfall seasonality regime in the tropical southwestern Pacific, are of special interest.[57] Extreme-value theory has been applied to precipitation data for Japan and for Australia, and also to temperature data for the United States.[58] Other statistical studies of temperature have dealt with the standard deviation of monthly temperatures in Anglo-America (see Figure 8.9) and accumulated temperatures in the British Isles.[59] A study of summer mean temperatures in Sweden by periodographic techniques indicates that the periodic components only account for 10 to 20 per cent of the observed variation in temperature. Consequently, the possibility of accurate prediction of the mean temperature of a given summer by means of past history is very slight, but this is partly a defect of the periodogram technique. Spectral analysis is a much surer

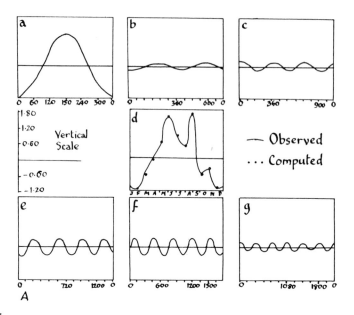

FIGURE 8.7.
Harmonic analysis of precipitation in the United States: I (after L. H. Horn and R. A. Bryson).

A, Harmonics for Madison, Wisconsin: a, first harmonic; b, second; c, third; d, comparison between observed and computed precipitation; e, fourth harmonic; f, fifth; g, sixth. Abscissae give time in months; ordinates give deviations of monthly precipitation means from the arithmetic mean for the 12 months (in inches).

B, Percentage of total precipation variability obtained from the first three harmonics.

C, Phase angle of the first harmonic:

0–90°,   maximum, April 15 to January 15,
minimum, October 15 to July 15;

90–180°, maximum, January 15 to October 15,
minimum, July 15 to April 15;

180–270°, maximum, October 15 to July 15,
minimum, April 15 to January 15;

270–360°, maximum, July 15 to April 15,
minimum, January 15 to October 19.

D, Phase angle of the fourth harmonic. Dates of maxima:

0°, 8 February, May, August, November;
30°, 1 February, May, August, November;
60°, 22 January, April, July, October;
90°, 1 January, April, July, October;
120°, 8 January, April, July, October;
150°, 1 January, April, July, October;
180°, 22 December, March, June, September;
210°, 15 December, March June, September;
240°, 8 December, March, June, September;
270°, 1 December, March, June, September;
300°, 22 November, February, May, August;
330°, 15 November, February, May, August;

In map D, areas within the same category experience maxima of the precipitation cycle, which repeats itself 4 times every year, on the same dates. I.e., areas marked solid black have maxima on: January 15 and 22; February 1 and 8, April 15 and 22, May 1 and 8, July 15 and 22, August 1 and 8, October 15 and 22, and November 1 and 8. To determine the magnitudes of these maxima, refer to Figure 8.8, D.

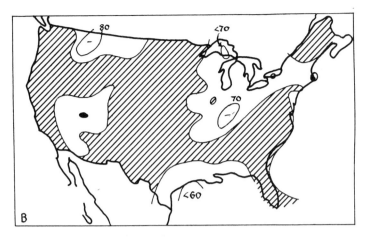

B

▨ Over 90%   ■ Less than 50%

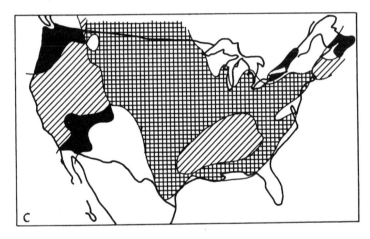

C

▨ 0-90°   ■ 90-180°   �𝕀𝕀 180-270°   ▦ 270-360°

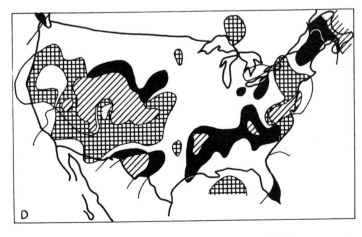

D

▨ 0-90°   ■ 90-180°   □ 180-270°   ▦ 270-360°

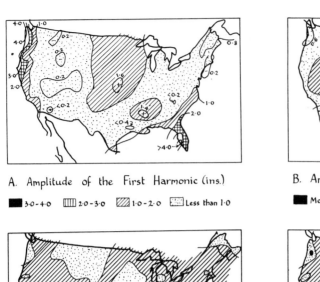

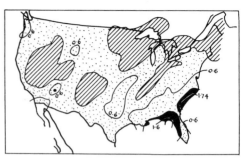

A. Amplitude of the First Harmonic (ins.)

■ 3·0-4·0   ⫼ 2·0-3·0   ▨ 1·0-2·0   ▦ Less than 1·0

B. Amplitude of the Second Harmonic (ins.)

■ More than 1·0   ▦ 0·2-1·0   ▨ Less than 0·2

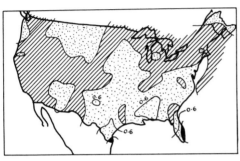

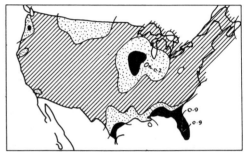

C. Amplitude of the Third Harmonic (ins.)

■ More than 1·0   ▦ 0·2-1·0   ▨ Less than 0·2

D. Amplitude of the Fourth Harmonic (ins.)

■ More than 0·4   ▦ 0·2-0·4   ▨ Less than 0·2

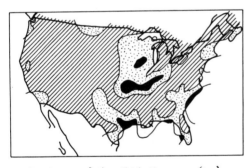

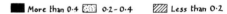

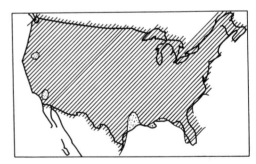

E. Amplitude of the Fifth Harmonic (ins.)

■ More than 0·4   ▦ 0·2-0·4   ▨ Less than 0·2

F. Amplitude of the Sixth Harmonic (ins.)

▦ More than 0·2   ▨ Less than 0·2

FIGURE 8.8.
Harmonic analysis of precipitation in the United States: II (after L. H. Horn and R. A. Bryson).

guide, and an exhaustive study of the power spectra of all available climatic data for New Zealand is of special interest in this context, as is a statistical model that explains the cross-spectra of New Zealand climatic data in terms of energy flow and available potential energy. When all search for hidden periodicities in the climatic data for an area proves futile, recourse must be made to probabilities. In any case, the probabilities inherent in temperature and other climatic records deserve far more attention than they receive from geographers.[60]

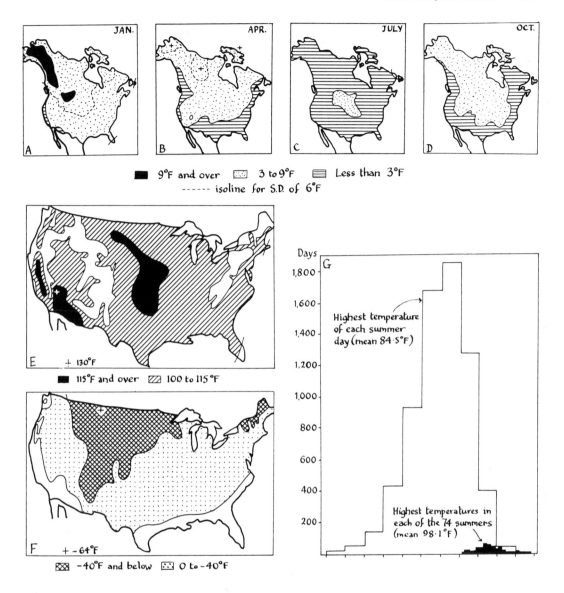

FIGURE 8.9.
A–D, Standard deviation of temperature in North America. E–F, Temperature extremes to be expected in the United States: E, highest temperature in 100 years; F, lowest temperature in 100 years. G, Temperature frequencies, Washington, D.C., 1872–1945 (June–August). (After A. Court and A. R. Sumner.)

Trends in statistical analysis naturally find their way, after an appropriate lag, into the geographical literature. Sometimes the resulting exercises are depressing for climatology as a science—for example, the attempt to prove that climatic change is a random series—and sometimes they bring new life to climatic classification, as did a multivariate approach to climatic regionalization of the United States. Advances in mathematics take much longer to have an impact on climatic classification; for example, what is apparently the first application of set theory to the Köppen classification, in English, did not appear until 1966.[61]

It is always stimulating to read climatological studies that are both geographical and meteorological in content, i.e., which deal with actual climates in a manner that is a contribution to science in its own right. Such studies are appearing in increasing numbers in geographical literature. Examples include: an interpretation of the data on typhoons associated with easterly waves in the Pacific and Atlantic areas in terms of resonance theory (see Appendix 8.7); a study of microfluctuations of temperature in and above a potato field and over fallow ground in Sweden in terms of turbulence theory; and an analysis of regional heat-budget values for northern Canada, which employs the Elsasser chart and heat-balance theory to compute regional estimates of the Bowen ratio that may be used for predicting air-mass modifications. A mathematical investigation of the amount of solar radiation in a zonal area betweeen two parallels of latitude proves that cool-highland agriculture is only possible in a region of intense summer heating because of the extremely high rate of concentration of solar radiation into a very short period. The altitude of the zenithal sun has also been used as the basis for a survey of sun-baking and its effects, and a refreshingly original approach has been made to the problem of climatological time-series on the basis of the possible effect of planetary tides on the atmosphere, although unfortunately without any confirmatory physical evidence.[62]

Other geographical work with a theoretical basis has dealt with the relations between the temperature lapse-rate in the lower half of the troposphere and some meteorological elements at ground level, and the problem of the estimation of potential evapotranspiration. The diurnal variation of winds over Hong Kong has been interpreted in terms of the equations of motion and harmonic analysis, and it appears that the diurnal variation of summer precipitation in that area is closely related to the wind variation. The work of the Thornthwaite Laboratory of Climatology has always been characterized by a sound theoretical approach, not only in the water-balance studies and vertical wind measurements already mentioned, but also in such investigations as an analysis, based on turbulent diffusion principles, of the time-height variations in the micrometeorological factors attending radiation fog near Centerton, New Jersey.[63]

Local winds as influenced by topography have been the subject of investigations in Japan that have some theoretical interest.[64] For instance, measurements of vertical-windspeed profiles, made at heights of up to 5 m above the ground on flat open land and on the side slopes and bottom of small valleys dissecting the flat horizontal surface in Nagano Prefecture, show that there is not necessarily a single profile, even in meteorological situations that should give rise to a logarithmic profile. In both the middle and the lower portion of the valley side slope, wind maxima occurred 2 to 3 m above ground level. Eddies developed when the general windspeed decreased, but no eddies occurred in valleys with very gentle sides. The effect of the eddies was to cause an air current directed up the side slope, so that windspeeds sharply decreased just below the upper edge of the side slope, then remained constant for a certain distance down the side slope. When this distance equaled the horizontal upper width of the valley, the windspeed increased again, regardless of the depth of the valley. Wind conditions in the deep valleys were found to resemble those in the rear of a thick windbreak; those in shallow valleys resembled the situation in the rear of a thin windbreak. The critical factor is the presence of eddies in the deep valleys and the absence of eddies in the shallow valleys.

Applied Climatology

The geographical literature of climatology contains many descriptions of the effect of climate on various human activities, and on various nonclimatic phenomena in specific areas. This is to be expected, for the main object of the geographer's investigations is the interrelations between different phenomena or groups of phenomena, and not one category of phenomena alone. Many of these descriptions are concerned with the effects of climate on what may be loosely termed economic life; others deal with the effects of climate on organic life or on a specific type of operation; still others deal with applications of climatology to various fields of pure science.

Climate may be regarded as a natural resource, and an important application of climatology must then be to determine whether or not this resource is being properly utilized. The climatic resource is of obviously critical significance in relation to other natural resources, in particular, water resources, and considerable attention has been paid to this question in various countries, for example, in Britain and the United States, and especially in the arid and semiarid parts of the world. Other regions of the globe have too much water in one season and too little in another, the humid tropics for example, and the phasing of the annual curves of temperature, precipitation, and evaporation requires very careful study in such regions. Reservoirs may have a measurable effect on the local climate of the area in which they are established, as do shelterbelts and windbreaks, and it is claimed that, by careful siting of these and other elements of the cultural landscape, a transformation of the ecology, and hence of human economy, has been accomplished in certain parts of the Soviet Union, mainly by controlling the water component of the energy-balance equation.[65]

The effect of climate on economic life may be relatively unobtrusive, or it may be catastrophic, as with floods. Studies of floods, their meteorological causes, and their economic consequences, are available for many areas, such as Britain and especially New Zealand. The climatic contribution to famine has been described for various countries; India is the obvious example, but countries such as Scotland have their problems too. The Great Plains of North America represent large areas whose economic prosperity is critically linked with climate, particularly as it affects crop yields. The unobtrusive, and constant, effect of climate, however, is in many ways the most telling.[66]

Agriculture is clearly the economic activity most directly linked with climate, and climatological attention in this direction has followed several lines. First, investigations of the relationship between agriculture and specific climatological processes have been made. Studies of moisture conservation, on the basis of energy balance, have been very popular, for example in the U.S.S.R., in Africa, in New Zealand, and Japan.[67] Moisture conservation studies are also frequently based on the concept of potential evapotranspiration as defined by Thornthwaite, because it may be applied very readily.[68] Microclimatological processes have formed the basis for other agro-climatological studies, for example, in Scandinavia.[69] A second line of approach has involved the elucidation of the relation between the yield or growth of a particular crop, in a given area, and some climatic element or elements. Examples include evapotranspiration and sugar cane in Barbados, rainfall and rice in Japan, winter temperatures and citrus fruit in Florida, and weather and spring wheat in the northern Great Plains of the United States.[70] A third line of approach is the analogue method, in which areas of similar climate are delimited in countries X and Y, and the assumption is made that if crop Z grows well in a climatic area in country X, then there is no climatic reason why crop Z should not prosper in a similar climatic area in country Y. The analogue method has been applied, for example, to compare the crop ecologies of the Ukraine, Palestine, and Transjordan with those of North America.[71]

The importance of climate as an influence in pastoral agriculture has also received some attention from geographers, for example, dairying in Anglesey, and cattle production in northern Australia.[72]

Climate as a control or influence in the activities of organic life is the field of the bioclimatologist. This field has always been a well-tilled one, since the days of the ancient Greeks, through the period between the two World Wars, when Huntington and Markham produced their well-known but scientifically unsophisticated works, to the present day, when it is recognised that training in biology and especially in physiology is prerequisite to mastery of the field.[73] Physiological climatology as a distinct study is now firmly established, mainly on the basis of laboratory experiments designed to measure the reaction of human subjects to climatic stress, and has a sound theoretical foundation.[74] Geographical applications of physiological climatology have included the determination of the most suitable type of clothing for wear in different parts of the world, studies of indoor comfort and wind chill in southern Australia and northern Canada, respectively, and investigations of the climatic factors that are (or should be) taken account of by architects in, for example, the humid tropics and the Adelaide area.[75] Bioclimatic studies have an obvious significance in problems of acclimatization and the establishment of health resorts, and even very general descriptions of the most obvious aspects of the relation between climate and man in specific areas—for example, China, the U.S.S.R., and the Mediterranean and Middle East—have their uses when precise physiological studies of the climates of these areas are lacking.[76] Bioclimatic investigations of nonhuman organisms may also be found in the geographical literature: for example, the effects of climatic change on animals and plants, the relationships between climate and vegetation in the Villavicencio region of eastern Colombia, or the demarcation of broad vegetation zones, such as those of the Boreal forests of eastern Canada, which appear to be related to the zonation of climate.[77]

Climatology may be applied to many forms of man-controlled operations. An example on a small scale is the planning of fuel requirements for the heating of a house. In moderate-sized towns in areas of relatively subdued relief (Reading, for example), local variations in heating requirements may be expected to be on the order of 20 to 30 per cent as a result of local climate characteristics. On a larger scale, a correlation between the magnitude of the present climatic fluctuation in Britain and its fuel requirements may be demonstrated. The range of variation in the total demand for fuel for heating in Britain in the 1950's (a consequence of the climatic fluctuation) was approximately equal to the amount of fuel imported. Climatology has been used in the determination of the most suitable sites for the location of nuclear reactors, and in the assessment of soil tractionability, i.e., the properties of a soil type (as determined by its moisture content) that determine its reaction to the movement of heavy vehicles, such as tanks or tractors, across its surface. Climatic information is of obvious value in aviation operations, ranging from the assessment of the frequency of visual conditions under which light aircraft may be operated locally, to the evaluation of the use of pressure patterns from trans-Atlantic flights. Climatic intelligence is clearly vital for operations in wartime. Current proposals for weather modification, both on local and on continental scales, raise very profound geographical questions that require careful thought.[78]

Viewed in the broadest terms, applied climatology also includes the application of climatology to various "pure" forms of knowledge. The application of climatology to geography, in particular, to regional description, is an obvious example. Although the climatic sections of most regional textbooks of geography are likely to disillusion the genuine climatologist who intends working in the field of geography, one can point to some texts that show awareness of the existence of climatology as a science.[79] There will always be a place for books on the climate of a specific country that in effect present the facts without comment, because they provide an excellent starting point for work by scientists.[80] Of the pure sciences that have benefitted from climatological assistance, palaeoclimatology and glaciology are obvious examples with geographical connotations.[81] Geographical literature also contains examples of the use of climatology by geomorphologists; for example: studies of the effect of tropical cyclones on coastal morphology in Mauritius; the relationship between winds, waves, and beach changes along the Durham coast; the origin of asymmetric valleys in the Chilterns; the orientation of sand dunes in Tasmania; the filling and drying of Lake Eyre; the regionalization of freeze-thaw activity in North America in terms of critical temperature thresholds; and the controlling influence of climatic factors in the distribution of certain frozen-ground phenomena.[82] Climate has long been recognized as a main determinant of the nature of the geomorphic cycle, and in recent years the concept of climatic geomorphology has given rise to important investigations.[83] For many years, climatology has been applied to the solution of problems in the fields of biology, ecology, medicine, architecture, agricultural science, and so forth. Many of these problems are also geographical problems; see the standard textbooks for an appreciation of these.[84]

There remains one final topic for discussion: is applied climatology a distinct field of study? In answering this question, the distinctions between climatology, meteorology, and geography must be kept firmly in mind. Meteorology is a purely general science, concerned with the laws of atmospheric behavior, and is as applicable to Mars, Jupiter, Venus, or any other planets with atmospheres as it is to the Earth. When we apply meteorology to the Earth's atmosphere, we are dealing with climatology. There is also a climatology of Venus, a climatology of Mars, and so on. But the unqualified term "climatology" refers to the Earth alone. To explain fully the observed facts of climate on the Earth, we find that meteorology alone is insufficient, doubly so if we wish to use the observed facts in some practical manner instead of merely explaining them. Consequently, some element of geographical empiricism—in particular, some recognition of the uniqueness of any given site on the Earth's surface—is always necessary when dealing with problems of applied climatology. However, the scientists will always attempt to reduce the degree of empiricism to a minimum, provided the labor involved in translating the problem into a theoretical structure is not out of proportion to the resulting gain over an empirical approach to the final result.

The current literature in applied meteorology is very rigorous and exhaustive, and the geographically inclined climatologist may wonder if his special interest is really necessary.[85] The answer is very much in the affirmative, as may be illustrated by two examples.

The field of agroclimatology has been well-studied for many years, so that two approaches, the classical and the recent, may be discerned. The classical approach is still valid. It involves establishing empirically—by statistical analysis of the variation in time and area of the relevant climatic data—the existing interrelations between crop development and atmospheric factors. The recent approach involves using the general laws of meteorology to predict what the atmospheric environment ought to be in a particular site, and then to predict how a physical model of the crop in question ought to react in the site.[86] The classical approach may result in an agroclimatic classification of land areas, but it will not be of practical value unless the climatic parameters that are mapped are defined in terms of the relevant physical processes of plant physiology and soils, and are also valid meteorologically.[87] The recent approach may result in a very detailed description of the climatic environment— possibly in much more detail than can be verified by actual measurement, owing to instrumental imprecision—but again this will be useless for practical purposes unless the extent to which the actual site differs from the idealized one (which must be assumed for the purpose of the theoretical analysis) is established by observation. Consequently, both a theoretical analysis and an observational program, the errors of the latter being analyzed in statistical terms, are essential for a valid study.

The assessment of frost-risk to agriculture and how to minimize it is a good example of a problem in applied climatology that requires both geographical and meteorological expertise. Two types of frost-risk may be envisaged: that associated with air frost and that associated with ground frost. Air frosts are usually widespread and a function of the synoptic situation; ground frosts are influenced very greatly by the incidence of cold-air drainage and katabatic flow. The principles of micrometeorology (e.g., those used to study night cooling under clear skies in Chapter 4) may be used to predict the frequency of both types of frost in a given area, and also to estimate the depth of katabatic flows. Both predictions will involve the assumption of idealized site conditions, and the way in which the predictions must be modified to fit the actual site must be determined by observation, making use of spot observations, traverses, or statistical evaluations based on bench-mark stations with constants in the equations to represent geographical factors.[88] A completely separate, and complementary, approach is to map the extent of frost-risk in an area by means of visual inspection of frost damage to fruit trees, etc.; after some experience, the extent of damage likely to result from frosts of a given intensity can be predicted with fair accuracy from a large-scale topographic map.[89]

Once the frost incidence is known, the efficacy of known frost-prevention methods in the given situation may be evaluated. For example, the efficiency of heaters and wind machines or of frost control by irrigation sprinkling may be estimated theoretically, and the protection obtainable under different meteorological situations may be predicted accurately.[90] Such predictions, however, refer to idealized sites, and empirical investigations at the actual site are necessary for a valid estimate to be made. For the climatologist's report to the farmer to be complete, questions of pure geography are also involved. If wind machines are proposed, where is the power to come from? If sprinklers are to be used, where is the water to come from? The cost of the proposed installations will be required, and an idea of how the complete system will fit into the economy of the farm. This should indicate to the budding applied climatologist that knowledge of many fields other than meteorology is necessary if he wants to provide a report that will be useful to the farmer.

In all agro-climatological investigations, indeed, in all work in applied climatology, the phenomenon, process, or operation to which climatology is to be applied needs to be analyzed as a geographical system before meteorological complications are introduced. Quite frequently, a problem that appears to be purely climatological will have other aspects that need to be clarified before the climatologist may isolate his special interest. For example, C. W. Thornthwaite was originally engaged as a consulting climatologist by Seabrook Farms in New Jersey to advise on irrigation, on the basis of his studies of water balance.[91] However, he soon discovered that he first had to solve a completely different problem: namely, how to ensure an even flow of vegetable produce to the Farm's canneries, so that periods of hectic activity requiring large quantities of water, separated by slack periods requiring little or no water, were avoided. This was accomplished by means of an investigation on operational research lines, in which the growth rates of the different types of crop plant were analyzed in terms of their climatic requirements.[92] The times of planting or transplanting the crops could then be scheduled, knowing the local climate, in such a manner as to enable them to be ready for the cannery on the most convenient day.

A second example of the importance of both meteorological and geographical knowledge in applied climatology is provided by air pollution. It is obvious that factories and other pollution sources should be sited so as to constitute minimum danger to public health. The concentration of pollution emitted from a given type of source in an idealized situation may be predicted very accurately by means of turbulence theory, and the effect of certain topographic or microtopographic situations on the pollution pattern may also be determined theoretically.[93] However, it is not feasible to produce a theoretical prediction system for each and every topographic situation, so that empirical observation at the site in question is always necessary as a confirmation or extension of the predictions. It is possible in principle, and at great expense, to set up an observational program that would enable the pollution potential of every part of the United Kingdom to be rigorously evaluated, both theoretically and practically. The engineering side of the question is sufficiently understood for clean air to be provided anywhere it is required, at a price. Unfortunately, the funds that can be made available for cleaning the atmosphere are severely limited, as, of course, they are for anything else. The availability of funds and other economic resources is in itself a geographical matter, and because of its necessary existence, it is more prudent to approach the question of air pollution in a given area by means of a combination of a little theory, a little observation, and a little geography, than by means of a scientifically rigorous and exhaustive investigation on the basis of turbulence theory.

Two qualifications are necessary at this point. First, our concern here is with applied climatology, not with pure climatology. There is every justification for a scientifically rigorous and exhaustive investigation of air pollution as a contribution to pure science, i.e., as a contribution to the advancement of turbulence theory. But such an investigation should be recognized for what it is, and not as a solution of the air-pollution problem in a given area. Second, there is no justification for empirical geographical studies that purport to solve the pollution problem, or any other problem, in a certain region on the basis of maps constructed without reference to the findings of theoretical meteorologists. Empiricism without theoretical foundation must always be inconclusive, and theory without the backing of confirmatory observation is useless as a contribution to applied climatology.

Notes to Chapter 8

1. E.g., the *Oxford Atlas*, C. Lewis and J. D. Campbell, eds. (Oxford, 1951), and H. J. Critchfield, *General Climatology* (Englewood Cliffs, N.J., 2d ed., 1966).

2. See, e.g.: K. M. Jensen, *Geogr. Tidsskrift*, 59 (1960), 103, for Denmark; I. E. M. Watts, *MTJG*, 7 (1955), 66, for Singapore Island; R. S. Dick, *JTG*, 12 (1958), 32, for Queensland; W. L. Dale, *JTG*, 14 (1960), 11, for Malaya; J. C. Pugh, *IGU* (Washington, D. C., 1952), p. 280, for Nigeria; and N. V. Harris, *NZG*, 19 (1963), 25, for Fiji.

3. See, e.g.: K. M. Jensen, *op. cit.* (1960), for Denmark; W. L. Dale, *JTG*, 17 (1963), 57, for Malaya; T. J. Chandler, *G*, 46 (1961), 295, for London; T. Sekiguti, *TJC*, 1, no. 1 (1964), 6 and 17, for Yonezawa, Yamagata, and Ogaki in Japan; I. Kayane, *loc. cit.*, p. 48, for Tokyo; and T. Kawamura, *TJC*, 1, no. 2 (1964), 74, for Kumagaya.

4. See, e.g.: F. Lindholm, *GA*, 37 (1955), 1, for sunshine and cloud in Sweden; W. L. Dale, *JTG*, 19 (1964), 20, for sunshine in Malaya; M. Schloss, *GR*, 52 (1962), 389, for cloud cover in the Soviet Union; J. A. Davies, *Nigerian Geog. J.*, 9 (1966), 85, for solar radiation in Nigeria; C. B. McIntosh, *NZG*, 14 (1958), 75, for winds in New Zealand; and M. Yoshino, *TJC*, 1, no. 1 (1964), 60, for winds in Tokyo.

5. The well-known works of Ellsworth Huntington, *Civilization and Climate* (New Haven, Conn., 1924) and *Mainsprings of Civilization* (New York, 1945), and of S. F. Markham, *Climate and the Energies of Nations* (Oxford, 1942), were excellent examples of poor scientific method in this context.

6. Jen-hu Chang, *AAAG*, 47 (1957), 241. D. Ashbel, *GR*, 39 (1949), 291. J. Setzer, *GR*, 36 (1946), 247, and P. R. Crowe, *GS*, 1 (1954), 44.

7. E.g., S. Erinc, *GR*, 40 (1950), 224, applied the concept of potential evapotranspiration to climatic mapping in Turkey. See also the classic paper by C. W. Thornthwaite, *GR*, 38 (1948), 55, and the analysis of the instability factor in Thornthwaite's potential evapotranspiration term by L. Curry, *CG*, 9 (1965), 13. See also P. R. Crowe, *GS*, 4 (1957), 56, for an alternative, simplified approach to evapotranspiration. The concept of water balance has been discussed in a climatic context by J. R. Mather, *PC*, vol. 14,

no. 3 (1961), as well as by Thornthwaite and Mather, *PC*, vol. 8, no. 1 (1955). Using the Thornthwaite system, the water balance of the Earth has been analysed by T. E. A. van Hylckama, *PC*, vol. 9, no. 2 (1956), that of the Mediterranean and Black Seas by D. B. Carter, *PC*, vol. 9, no. 3 (1956), and that of the Delaware basin by Carter and van Hylckama, *PC*, vol. 11, no. 3 (1958). Other areas studied in terms of the Thornthwaite procedure for water-balance analysis include, *inter alia*: India by V. P. Subrahmanyam, *AAAG*, 46 (1956), 300; Malaya, by S. Nieuwolt, *JTG*, 20 (1965), 34; New Zealand, by B. J. Garnier, *NZG*, 7 (1951), 43; and Queensland, by A. D. Tweedie, *AG*, 6 (1956), 34. A study of water balance in New South Wales by N. S. McDonald, *Monograph Series*, no. 1 (Univ. of New Zealand, Dept. of Geography, Armidale, New South Wales), employs the schemes of Penman, Blaney and Criddle, and Veihmeyer, as well as the Thornthwaite system. For maps of rainfall intensity in Australia, see J. N. Jennings, *AG*, 10 (1967), 256.

8. A. A. Miller, *TIBG*, no. 17 (1951), p. 15. E.g., see J. R. Villmow, *AAAG*, 42 (1952), 94, for the position of the Da/Db boundary in the eastern United States, where it is shown that the year-to-year movements of this boundary during 1920–49 reached an extreme value of nearly 700 miles along a single meridian. See also S. Gregory, *Erdkunde*, 8 (1954), 246. For regional application of the first Thornthwaite system, see, *inter alia*, B. J. Garnier, *AAAG*, 36 (1946), 151, for New Zealand. For regional application of the 1948 Thornthwaite system, see, e.g.: Thornthwaite, *GR*, 38 (1948), 55, for the United States; Jen-hu Chang, *AAAG*, 45 (1955), 393, for China; S. Erinc, *AAAG*, 39 (1949), 26, for Turkey; A. M. McFarland and R. R. Osburn, *AAAG*, 41 (1951), 237, for the Argentine; B. R. Schulze, *SAGJ*, 40 (1958), 31, for South Africa; and D. B. Carter, *PC*, vol. 7, no. 4 (1954), for Africa and India. For an evaluation of the 1948 system, see, e.g., Jen-hu Chang, *AAAG*, 49 (1959), 156.

On the Köppen classification of climate, see: A. A. Wilcock, *AAAG*, 58 (1968), 12, for a review; E. Fukui, *JPC*, 2 (1965), 34 and 43, on secular movements of Köppen's climatic boundaries in eastern

Asia and the North Pacific, and *loc. cit.*, p. 26, on the effect of data paucity on the boundaries of Köppen's regions in eastern Asia; and T. Sekiguti, *loc. cit.*, p. 8, on core regions in Japan, in which Köppen's boundaries are relatively stable. On the expression of secular fluctuations of climatic boundaries in terms of "climatic years," see, for the western Pacific, T. Aoyama, *JPC*, 3 (1968), 91. On the Thornthwaite classifications of climate, see L. Curry, *CG*, 9 (1965), 13, and A. J. Freile and T. A. Leavy, *The Professional Geographer*, 19 (1967), 244, for a nomogram.

9. J. R. Borchert, *AAAG*, 43 (1953), 14. P. R. Crowe, *TIBG*, no. 15 (1949), p. 38, and no. 16 (1950), p. 23. H. Flohn, *PIGU* (Washington, D.C., 1952). P. R. Crowe, *TIBG*, no. 17 (1951), p. 23. F. K. Hare, *GR*, 50 (1960), 345. D. Kerr, *CG*, no. 14 (1959), p. 17. L. Curry, *NZG*, 16 (1960), 71. L. Curry and R. W. Armstrong, *GA*, 41 (1959), 245. F. K. Hare, *GR*, 52 (1962), 525.

10. H. H. Lamb and A. I. Johnson, *GA*, 41 (1959), 94, and 43 (1961), 363. See also H. H. Lamb, *G*, 46 (1961), 208. See, *inter alia*, G. Manley, *GR*, 41 (1951), 656, for a general review; also H. W. Ahlmann, *GV*, 112 (1949), 165. D. J. Schove, *SGM*, 66 (1950), 37, deals in an elementary manner with tree rings and summer temperatures, A.D. 1501–1930.

11. S. Petterssen, *GC*, pp. 212–21.

12. E.g., see S. W. Visser, *PIGU Lisbon*, p. 650, on solar activity and heat waves in the Netherlands, and B. L. Dzerdzeyerski, *SG*, 5 (March 1964), 37, on the atmospheric general circulation as an essential link in the Sun-climatic change system. H. C. Willett, *GC*, p. 295.

13. E.g., see: D. J. Schove, A. E. M. Nairn, and N. D. Opdyke, *GA*, 40 (1958), 216, on the Permian; K. W. Butzer, *GA*, 39 (1957), 48, on the Pleistocene general circulation in relation to pluvial periods in the Mediterranean, *loc. cit.*, p. 105, in relation to low-latitude conditions, and *CG*, 8 (1964), 125, for a reconstruction of Pleistocene palaeoclimates in the Kurkur Oasis; and H. C. Willett, *GA*, 32 (1950), 179, on the last glacial maximum.

14. E.g., D. J. Schove, *GA*, 36 (1954), 40, described summer temperatures in northern Scandinavia as determined from tree-ring analysis; G. H. Liljequist, *GC*, pp. 159–77, applied periodographic techniques to summer temperatures in Sweden; J. Keränen, *Fennia*, 75 (1952), 5, discussed temperature changes in Finland during the last hundred years; J. Eythorsson, *GA*, 31 (1949), 36, analyzed temperature variations in Iceland; and T. Segota, *Erdkunde*, 20 (1966), 110, described temperature changes in Central Europe during the Quaternary. See, e.g., H. Landsberg, *GC*, p. 125, on temperature at New Haven, Conn.; and L. G. Polozora and Y. S. Rubinshteyn, *SG*, 5 (March, 1964), 3, on northern-hemisphere temperatures. S. Gregory, *GJ*, 122 (1956),

346, dealt with regional variations in the trend of annual rainfall over the British Isles, and E. C. Barrett, *TIBG*, no. 35 (1964), p. 55, examined local variations in rainfall trends in the Manchester region. See, e.g., H. Arakawa, *JTG*, 17 (1963), 34, on precipitation in Japan; J. H. Vorster, *SAGJ*, 39 (1957), 61, on precipitation in South Africa; and D. J. Schove, *GA*, 44 (1962), 303, on winds in Europe. N. K. Kononova, *SG*, 4 (March 1964), 42, showed that the duration and areal extent of cyclonic activity in eastern Siberia increased between 1906–1915 and 1944–1953, particularly in the north, while the stability of the anticyclonic regime in southern Siberia and over the Arctic basin declined. J. G. Potter, *CG*, 5 (1961), 37, discussed the changes in seasonal snowfall exhibited by Canadian cities. G. Manley, *GJ*, 117 (1951), 43, described the range of variation of the British climate. F. G. Hannell, *AS*, 12 (1956), 373, and F. Oldfield, *TIBG*, no. 28 (1960), p. 99, investigated climatic fluctuations in the Bristol area (based on climatic data) and in lowland Lonsdale (based on pollen analysis), respectively. I. Hustich *et al.*, *Fennia*, 75 (1952), 117, described climatic fluctuations and their consequences in Finland.

15. J. A. Shear, *AAAG*, 54 (1964), 310, on polar marine climate, P. Larsson, *GR*, 53 (1963), 572, on albedo, and A. Morrison, *CG*, 8 (1964), 72, on the relation between albedo and the tone of air photographs in the Canadian sub-Arctic. H. H. Lamb, *GJ*, 123 (1957), 287, covers polar gales. F. R. Fosberg, B. J. Garnier, and A. W. Küchler, *GR*, 51 (1961), 339, discuss the tropics. H. P. Bailey, *GA*, 42 (1960), 1, and *GR*, 54 (1964), 516, discusses temperate climates. On the radiation theory, see E. Fukui, *TJC*, 1 (1964), 76.

16. For a discussion of the British seasons in synoptic terms, see H. H. Lamb, *QJRMS*, 76 (1950), 393, and *W*, 8 (1953), 131 and 176. G. Manley, *Climate and the British Scene* (London, 1952), gives a very readable introduction. See also his papers in *GR*, 35 (1945), 408, and *GJ*, 117 (1951), 43.

17. The original principles for the construction of isohyetal maps of Britain were given by H. R. Mill in *British Rainfall* (London, 1908), p. 140, and *QJRMS*, 34 (1908), 65, and extended by J. Glasspoole, *British Rainfall* (London, 2d ed., 1928), p. 271. For a method for the construction of precipitation maps for hilly regions see E. L. Peck and M. J. Brown, *JGR*, 67 (1962), 681. J. Oliver, *G*, 43 (1958), 151, discussed the rainfall of Wales as a factor in its geography, and in *GA*, 44 (1962), 293, described the thermal regime of upland peat bogs in a maritime temperate climate. J. L. Davies, *G*, 37 (1952), 19, described some effects of aspect on valley temperatures in southern Cardiganshire. A. Garnett, *loc. cit.*, p. 24, discusses the Pennines, and R. M. Adam, *SGM*, 62 (1946), 24, discusses Galloway.

18. E.g., P. Hamilton, *SGM*, 79 (1963), 74, described the climate of Aberdeen. See, in particular, the climatic section of the British Association regional surveys; for example, F. A. Barnes on Nottingham (1966), R. G. Barry on Southampton (1964), A. Garnett on Sheffield (1956), S. Gregory on Liverpool (1953), C. A. Halstead on Glasgow (1958), and F. G. Hannell on Bristol (1955). See W. G. V. Balchin and N. Pye, *QJRMS*, 73 (1947), 297, on the Bath area; and T. J. Chandler, *GJ*, 47 (1962), 279, and B. W. Atkinson, *TIBG*, no. 44 (1968), p. 97, on London.

19. H. H. Lamb, *SGM*, 71 (1955), 14. R. T. Cornish, *G*, 32 (1947), 67. See H. H. Lamb, *GJ*, 133 (1967), 445, on Britain's fluctuating climate in relation to the general circulation and to human activities. See also: E. N. Lawrence, *SGM*, 85 (1969), 64, on mountain and valley winds in the Pennine hills, at Helmshore, Lancs.; S. E. Durno and J. C. C. Romans, *loc. cit.*, p. 31, on the evidence for variations in altitudinal zonation of climate in northern Britain, since the Boreal Period, provided by peat deposits; and G. Manley, *Glaciers and Climate* (Stockholm, 1949), p. 179, on variations in height of the snowline in Britain.

20. M. Båth, *GA*, 27 (1945), 266. F. Lindholm, *loc. cit.*, p. 241. On Scandinavia, see: S. Lindqvist, *GA*, 50, series A (1968), 79, on the heat-island effect in the city of Lund, Sweden; and Y. T. Gjessing, *Norsk Geografisk Tidsskrift*, 22 (1968), 200, on the relation between meteorological conditions and the freezing of Lusterfjord, Norway.

21. The painstaking study of insolation and relief in the Alps by A. Garnett, *TIBG*, no. 5 (1937), suffers from the disadvantage that the effects of atmospheric absorption on incoming radiation are ignored. On the Baltic, see E. Palosuo, *Fennia*, vol. 77, no. 1 (1953), and B. Rodhe, *GA*, 37 (1955), 141. On Finland, see: M. Seppänen, *Fennia*, 86 (1961), 1; J. O. Mattson, *LSGA*, no. (1961), 16; and I. Y. Ashwell, *GA*, 45 (1963), 152.

22. V. H. Malmström, *AAAG*, 50 (1960), 117, described the influence of the Arctic Front on the climate and crops of Iceland. See I. I. Schell, *GA*, 43 (1961), 354, on Icelandic sea-ice.

23. A. Sundborg, *GA*, 37 (1955), 176. I. Y. Ashwell and F. G. Hannell, *GA*, 41 (1959), 67. E. Sjördin, *GA*, 39 (1957), 54. C. C. Wallén, *GA*, 31 (1949), 275.

24. The two atlases are P. D. Thompson, *The Climate of Africa* (Oxford, 1965), and S. P. Jackson, *Climatological Atlas of Africa* (Lagos, Nairobi, 1961), the latter being the first of the regional climatic atlases that will constitute the WMO World climatic Atlas. The instrumental difficulties involved are evident from, e.g., J. A. Davis, *GA*, 48A (1966), 139, and *Nigerian Geog. J.*, 8 (1965), 17, on Nigeria. Problems of soil temperatures in the arid tropics,

with special reference to Khartoum, are described by J. Oliver, *JTG*, 23 (1966), 47. Detailed information is particularly desirable in African countries with considerable altitudinal variation: for an example (the climates of Ethiopia), see H. Suzuki, *Ethiopian Geogr. J.*, 5 (1967), 19. J. H. Hubbard, *GS*, 3 (1956), 56, described daily weather at Achimota, near Accra. R. Miller, *G*, 37 (1952), 198, examined the climate of Nigeria in terms of synoptic weather. N. B. Bowden, *JTG*, 19 (1964), 1, briefly described the dry seasons of intertropical Africa and Madagascar. See J. H. Hubbard, *GS*, 1 (1954), 69, for a note on the rainfall of Accra, F. W. Oliver, *GJ*, 108 (1947), 221, on Egyptian dust storms, and K. Tato, *Ethiopian Geogr. J.*, 2 (1964), 28, on rainfall in Ethiopia in relation to the general circulation.

25. P. A. MacGregor, *SAGJ*, 37 (1955), 41.

26. J. A. King and H. van Loon, *SAGJ*, 41 (1959), 62, analyzed the weather of the winters of 1957 and 1958 in synoptic terms.

27. P. D. Tyson, *SAGJ*, 48 (1966), 13. S. P. Jackson, *SAGJ*, 36 (1954), 13. M. B. Oliver, *GA*, 38 (1956), 102. On Africa, see K. Nakarawa, *GRTMU*, no. 2 (1967), p. 49, and no. 3 (1968), p. 43, on the climate of East Africa in relation to the equatorial westerlies, and *JPC*, 4 (1967), 61, on the urban heat island of Nairobi; see also P. D. Tyson, *W*, 19 (1964), 7, and 20 (1965), 115, on the Berg wind of South Africa.

28. W. J. Gibbs, *AG*, 5 (1945–50), 47.

29. For details, see: W. W. Moriarty, *CTP*, no. 7 (1955); F. A. Berson, D. G. Reid, and A. J. Troup, *CTP*, no. 8 (1957) and no. 9 (1958). See E. A. Fitz-Patrick, *JTG*, 14 (1960), 29, on the implications of these cyclones for precipitation in eastern Australia.

30. I. E. M. Watts, *NZG*, 3 (1947), 115.

31. I. E. M. Watts, *NZG*, 1 (1945), 119.

32. J. W. Hutchings, *NZG*, 9 (1953), 37, gives a detailed account.

33. See G. T. Trewartha, *Erdkunde*, 12 (1958), 205, for full details.

34. S. B. Chatterjee, *IGU*, Japan (1957). See M. M. Yoshino, *JPC*, 2 (1965), 70, on the relation between precipitation and frontal frequencies, and E. Fukui, *loc. cit.*, p. 38, on the existence of a short dry period that interrupts the wet summers of the region. On the Indian summer monsoon, see Jen-hu Chang, *GR*, 57 (1967), 373; also see C. Ramaswamy, *T*, 8 (1956), 26, on the role of the subtopical jetstream in the development of large-scale convection over northern India and Pakistan in the months preceding the onset of the southwest monsoon.

35. See M. Yoshino, *TJC*, 3, no. 1 (1966), 24, for a general synoptic account; also R. Saito, *IGU*, Japan (1957), p. 173. The stages of northward shifting of the Bai-u are described in detail in synoptic terms by M. Yoshino, *TJC*, 3, no. 1 (1966), 72, 88, 115.

36. E. H. G. Dobby, *GR*, 35 (1945), 204, gave an

elementary account of winds and fronts in Southeast Asia, and I. P. Gupta and R. N. Tikkha, *Nat. Geogr. J. India*, 8 (1962), 215, described the characteristics and frequency of weather types in Uttar Pradesh. See I. E. M. Watts, *JTG*, 3 (1954), 1, on Malayan line squalls, and C. S. Ramage, *JTG*, 19 (1964), 62, on the precipitation patterns. On Malaysia, see S. Nieuwolt, *AAAG*, 58 (1968), 313, on diurnal rainfall variations, and *JTG*, 20 (1965), 34, on evaporation and water balances; see also Chia-lin Sien, *JTG*, 27 (1968), 1, on local rainfall patterns in Selangor. On Singapore, see S. Nieuwolt, *JTG*, 22 (1966), 30, on its urban microclimate, and *JTG*, 27 (1968), 23, on its daily weather.

37. Jen-hu Chang, *AAAG*, 49 (1959), 156, defines a local index. The periodicals *Acta Meteorological Sinica* and *Acta Geographica Sinica* (Peking) are valuable sources of research papers. J. R. Borchert, *AAAG*, 37 (1947), 169, presented a map of China's climates based on air masses and the Köppen system, and Jen-hu Chang, *G*, 42 (1957), 142, compiled air-mass maps of China (including Manchuria). See, e.g., A. Lu, *GR*, 37 (1947), 88, on precipitation, and C. S. Chen, *MJTG*, 5 (1955), 26, on the climatic regions of Taiwan, and *AG*, 7 (1957), 59 on the growing season.

38. See *inter alia*, M. I. Budyko, *SG*, 2 (April 1961), 3, and 3 (May 1962), 3, on the heat balance of the Earth's surface, and M. I. L. L'vovich, *SG*, 2 (April 1961), 14, and 3 (December 1962), 37, for the water balance. For a criticism, see A. S. Monin, *SG*, 5 (April 1964), 3, and the reply by Budyko, *loc. cit.*, p. 18. A. Z. Grigor'yev, *SG*, 2 (September 1961), 3, discussed the heat and moisture balance and geographic zonality. O. A. Drozdov, *SG*, 2 (February 1961), 12, described the importance of the moisture circulation for natural processes, particularly in agriculture. V. S. Mezentsev, *SG*, 5 (May 1964), 24, examined the relation of the natural moisture balance of the West Siberian Plain to the Lower Ob' problem. For a very general treatment of the transformation of nature, see M. I. Budyko and I. P. Gerasimov, *SG*, 2 (February 1961), 3. On climate classification, see A. A. Grigor'yev and M. I. Budyko, *SG*, 1 (May 1960), 3, and V. K. Chukreyev, *SG*, 2 (November 1961), 20.

39. Y. S. Rubinshteyn, *SG*, 3 (November 1962), 14. Y. P. Parmuzin, *SG*, 1 (Jan.–Feb. 1960), 40. Y. A. Nefed'yeva and G. D. Rikhter, *SG*, 3 (June 1962), 47.

40. P. E. Lydolf, *AAAG*, 54 (1964), 291. K. F. Zhirkov, *SG*, 5 (May 1964), 33. T. Sivall (1957) gave a complete analysis of the siroccos. For details of a climatological traverse across the Dead Sea lowlands, from Jerusalem to Jericho, see P. Beaumont, *G*, 53 (1968), 70.

41. See: M. Yoshino, *TJC*, 1, no. 2 (1964), 34, on precipitation on the Kanto Plain; M. Mizukoshi, *loc. cit.*, p. 46, on precipitation associated with Bai-u

fronts; and T. Kawamura, *loc. cit.*, pp. 1 and 14, on the winter monsoon and on the synoptics of winter precipitation in Hokkaido. T. Sekiguti *et al.*, *TJC*, 2, no. 2 (1965), 12, present a climatology of heavy rains in the Tokyo area, and M. Mizukoshi, *loc. cit.*, p. 21, described the extreme values of daily precipitation in Japan. T. Kawamura, *TJC*, 1 no. 2 (1964), 39, discussed the distribution of heavy rainfall associated with the Kanogawa typhoon on the Izn Peninsula. T. Sekiguti and M. Yoshimwa, *TJC*, 2, no. 2 (1965), 29, analyzed the geographical pattern of daily snowfalls in Hokuriku. T. Kawamura, *TJC*, 1, no. 2 (1964), 10, described surface winds over Hokkaido in terms of synoptic types. T. Sekiguti, T. Yoshitama, and M. Yagi, *loc. cit.*, p. 61, studied the formation of oro-graphic rain-bands in the Kanto Plain, and T. Sekiguti, *TJC*, 2, no. 2 (1965), 1, presented a theoreti-cal evaluation of the orographic component of ty-phoon rainfall in Japan.

42. For temperature, see: T. Sekiguti, *TJC*, 1, no. 1 (1964), 6, 13, and 17 for Yonezawa, Yamagata, and Ogaki; M. Takahashi, *loc. cit.*, p. 45, for Ogaki and Kumagaya; I. Kayane, *loc. cit.*, p. 48, for Tokyo; T. Nishizawa, *TJC*, 1, no. 2 (1964), 71, T. Kawamura, *loc. cit.*, p. 74, and M. Mizukoshi, *TJC*, 2, no. 2 (1965), 41, for Kumagaya; H. Shitara, *TJC*, 1, no. 2 (1964), 79, for Hiroshima; and T. Sekiguti, *loc. cit.*, p. 90, for Ina. See also M. Nakahara, *PIGU*, Japan (1957), p. 166, and T. Sekiguti, *loc. cit.*, p. 188, for a comparison between built-up areas and parks, etc. On humidity maps for urban areas, see T. Sekiguti, *TJC*, 1, no. 2 (1964), 90, for Ina, and K. Sasakura, *TJC*, 2, no. 2 (1965), 45, on Tokyo. T. Sekiguti *et al.*, *TJC*, 1, no. 1 (1964), 35, examined the geographical distribution of solar radiation in the Tokyo area, and also, *TJC*, 1, no. 2 (1964), 90, in and around the small city of Ina. On winds, see M. Yoshino, *TJC*, 1, no. 1 (1964), 60 and 68, for Tokyo, Niigata, and Noshiro.

43. See: E. Fukui, *TJC*, 1, no. 2 (1964), 53, and I. Kayane, *loc. cit.*, p. 67, for Tokyo; T. Sekiguti *et al.*, *TJC*, 1, no. 1 (1964), 29, for Ogaki and Yonezawa; and H. Shitara, *TJC*, 1, no. 2 (1964), 82, for Hiro-shima. See S. Nieuwolt, *JTG*, 22 (1966), 30, on Singapore.

44. For surface cover effects on temperature and humidity, see T. Sekiguti, *TJC*, 2, no. 1 (1965), 36 and 46, and *PIGU* Japan (1957), p. 188. On meso-climatic divides, see H. Shitara, *TJC*, 2, no. 2 (1965), 57, and I. Kayane, *TJC*, 1, no. 2 (1964), 27. See also I. Kayane, *loc. cit.*, p. 21, and *Tokyo Kyoiku Daigaku Sci. Rep.*, no. 9, (1966), p. 125, for an application of a mathematical description based on the changing rate of temperature variation at a given place. On the heat budget in the Kanto Plain, see I. Kayane, *TJC*, 2, no. 21 (965), 33. On the effects of local topography on winds, see, e.g., M. Yoshino, *TJC*, 2, no. 1 (1965), 58, 60, 77, and no. 2 (1965), 47.

For synoptic studies of Japanese weather and climate, see: I. Maejimi, *GRTMU*, no. 2 (1967), p. 77, on natural seasons and weather singularities; T. Yazawa, *loc. cit.*, p. 71, on regionalities of pressure variation; and K. Nakamura, *GRTMU*, no. 1 (1966), p. 149, on anomalous southerly winds present on the Japan Sea coast during the northwest—i.e., winter—monsoon. On the synoptics of the Shurin season, see J. C. Kimura, *GRTMU*, no. 1 (1966), 113, no. 2 (1967), 105, and no. 3 (1968), 63. On synoptic climatology, see: T. Yazawa, *loc. cit.*, p. 97, on the Izu Islands; H. Shitari, *JPC*, 3 (1966), 47, on local weather areas in the Tohoku district; K. Nakamura, *JPC*, 4 (1967), 47, on the southerly weather type in winter in the Hokwiku district; and T. Sekiguti and S. Inoue, *loc. cit.*, p. 71, on heavy rains in the Kanto area.

For local studies of Japanese weather see: H. Fukui, *JPC*, 1 (1964), 14, on the Aizu and Inawashiro basins; I. Maejima, *GRTMU*, no. 3 (1968), p. 107, on the water balance of the Tone basin; H. Shitara, *JPC*, 1 (1964), 48, on a statistical "front" affecting temperature and winds in the Ofsanbongi coastal plain; T. Kawamura, *JPC*, 3 (1966), 28, on winter surface-wind systems in central Japan; H. Shitara, *JPC*, 4 (1967), 78, on a lake breeze on Lake Inawashiro, *loc. cit.*, p. 36, on the effect of the Watari Hills on nocturnal cooling, and *loc. cit.*, p. 60, on the mesoclimatic divide in the Sanbongi Plain; T. Nishizawa and S. Yamashita, *JPC*, 4 (1967), 66, on solar radiation attenuation at Tokyo and Osaka; and J. Sasakura and M. Mochizuki, *JPC*, 1 (1964), 35, on the visibility of Mt. Fuji, as seen from Shizuoka City, in relation to weather.

45. C. P. Patton, *UCPG*, vol. 10, no. 3 (1956). D. H. Miller, *UCPG* vol. 11 (1955), and *GR*, 46 (1956), 209.

46. A. Court, *GR*, 39 (1949), 214. R. F. Logan, *GR*, 51 (1961), 236. J. R. Borchert, *AAAG*, 40 (1950), 1. R. B. Carson, *Econ. Geog.*, 27 (1951), 321. J. B. Hoyt, *AAAG*, 48 (1958), 118. R. J. Russell, *AAAG*, 35 (1945), 37, discusses the climates of Texas. C. F. Brooks and C. Chapman, *GR*, 35 (1945), 132.

47. R. L. Ives, *AAAG*, 40 (1950), 293. On North America, see R. G. Barry, *Geogr. Bulletin*, 9 (1967), 79, on the seasonal location of the Arctic front, and I. Matsuda, *GRTMU*, no. 3 (1968), p. 81, on jetstream "seasons." See also L. Williams, *AAAG*, 57 (1967), 579, for evidence that very high temperatures at a desert station at Yuma, Arizona, are due to air in the layer between 850 and 650 mb being brought down to ground level by afternoon convection, or by a föhn-like effect produced by the surface pressure pattern.

48. E.g., see W. D. Rouse and D. Watts on aspects of the climate of Barbados, in *Climatol. Research Series*, no. 1 (Montreal, 1966). For Mexico, see C. C. Wallén, *GA*, 37 (1955), 55. For Baja California, see J. R. Hastings and R. M. Turner, *GA*, 47A (1965),

204. For Central America, see W. H. Portig, *GR*, 55 (1965), 68. On Lake Maracaibo, see D. B. Carter, *PC*, vol. 8, no. 3 (1955). See S. L. Hastenrath, *JTG*, 25 (1967), 24, on climatic fluctuations and associated changes in lake levels in Central America. On the West Indies, see D. G. Tout, *McGill Climatol. Bull.*, no. 2 (1967), p. 29, and B. K. Basnayake, *McGill Climatol. Bull.*, no. 3 (1968), 21, on solar radiation at Barbados; see also R. E. Randall, *loc. cit.*, p. 23, on aerial salt at Barbados.

49. On the Canadian dry belt, see J. R. Villmow, *AAAG*, 46 (1956), 211. On drought, see M. Sanderson, *GR*, 38 (1948), 289, and on potential evapotranspiration, Sanderson, *GR*, 40 (1950), 636. See the papers published by the McGill University Sub-Arctic Research Laboratory. On forest microclimate, see F. G. Hannell, *TIBG*, no. 22 (1956), p. 73. On Baffin Island, see S. Orvig, *GA*, 36 (1954), 197, and 33 (1951), 166. On Arctic ice islands, see R. W. Longley, *CG*, 6 (1962), 143. On Arctic microclimate, see J. R. Mather and C. W. Thornthwaite, *PC*, vol. 9, no. 1 (1956), and vol. 11, no. 2 (1958). See W. M. Wendland, *Geogr. Bulletin*, 9 (1967), 1, on net radiation in Canada. For urban climatic studies, see: T. R. Oke, *McGill Climatol. Bull.*, no. 3 (1968), p. 36, on Montreal; O. J. Diduch and C. E. Klaponski, *McGill Climatol. Bull.*, no. 5 (1969), p. 26, on gusts in Montreal; T. R. Oke, *loc. cit.*, pp. 1 and 35, on the urban heat island; and R. F. Fuggle, *McGill Climatol. Bull.*, no. 4 (1968), p. 31, on longwave radiation flux and the urban heat island. On the Canadian Arctic, see R. B. Sagar, *Geogr. Bulletin*, no. 22 (1964), p. 13, and *Geogr. Bulletin*, 8 (1966), 3, for temperature and radiation observations on the Gilman glacier (Ellesmere Island) and the Barnes ice cap.

50. H. H. Lamb, *GJ*, 111 (1948), 48. For a general discussion of weather variations in the Antarctic see H. H. Lamb and G. P. Britton, *GJ*, 121 (1955), 334. M. E. Sabbagh, *GA*, 45 (1963), 52.

51. For a comprehensive account of one view concerning what constitutes a "problem" climate, see G. T. Trewartha, *The Earth's Problem Climates* (Wisconsin, 1962). For detailed comparisons of world climates based on vegetation regions, see G. R. Rumney, *Climatology and the World's Climates* (New York, 1968), pp. 112–642. P. E. Lydolf, *AAAG*, 47 (1957), 213. Jen-hu Chang, *AAAG*, 52 (1962), 221. J. B. Leighly, *AAAG*, 37 (1947), 75. J. A. Prescott *et al.*, *GR*, 42 (1952), 118. H. J. Critchfield, *NZG*, 22 (1966), 111.

52. The reaction was expressed, for example, by F. H. W. Green, *TIBG*, no. 13 (1947), p. 47, and by E. M. Frisby and F. H. W. Green, *TIBG*, no. 15 (1949), p. 143. See S. R. Eyre, *G*, 35 (1950), 155, for a forecaster's impression of weather between the Tasman Sea and the Blue Mountains of Australia. Basing geographical climatology on air-mass analysis was advocated, e.g., by A. A. Miller, *G*, 38 (1953), 55. On

the distinction between synoptic and dynamic climatology, see: F. K. Hare, *GA*, 39 (1957), 87, and *AAAG*, 45 (1955), 152; A. Court, *AAAG*, 47 (1957), 125; and P. E. Lydolf, *AAAG*, 49 (1959), 120. For a comprehensive review of the physical concept of energy balance in a geographical context, see D. H. Miller, *AGP*, 11 (1965), 175; also, for an illustration of the value of heat balance concepts in topoclimatological studies, see *Geographia Polonica*, 2 (1964), 69. For topoclimatological studies at Mont St. Hilaire, Quebec, see B. J. Garnier, *CG*, 12 (1968), 241, who estimates the variation in direct solar radiation due to topography. See also: B. J. Garnier and A. Ohmura, *JAM*, 7 (1968), 796, for the theory involved; A. Ohmura and R. F. Fuggle, *McGill Climatol. Bull.*, no. 4 (1968), p. 1951, on energy balance over short grass in an orchard; and D. S. M. Munro, *loc. cit.*, p. 31, for a comparison of evaporation measurements at Lac Hertel.

53. For relevant examples, see: P. R. Crowe, *TIBG*, no. 36 (1965), p. 1; F. K. Hare, *G*, 51 (1966), 99, and 50 (1965), 229; and G. B. Tucker, *G*, 46 (1961), 198.

54. See A. Davies, *SGM*, 66 (1950), 4, for the mnemonic classification. On the purpose of classification, see, in particular, C. W. Thornthwaite, *GR*, 38 (1943), 233; see also F. K. Hare, in L. D. Stamp and S. W. Wooldridge, eds., *London Essays in Regional Geography* (London, 1951), pp. 111–34, and A. Marshall, *AG*, 6 (1954), 3. See the note by J. A. Collins, *loc. cit.*, p. 7, on the limitations of the arithmetic mean, and C. F. Brooks, *AAAG*, 38 (1948), 153, on those of the climatic record. On student participation in exercises, see, e.g.: W. G. V. Balchin, *G*, 33 (1948), 128; P. A. Jones, *G*, 39 (1954), 182; R. H. C. Carr-Gregg, *G*, 46 (1961), 307. On climatic classification, see: Y. Ambe, *GRMTU*, no. 1 (1966), 139, for a climatic classification for the world based on variability of seasonal character of climate; T. Sekiguti, *JPC*, 2 (1965), 1, on synthetic climatic regions in Japan; and E. Fukui, *JPC*, 3 (1966), 999, for an index describing the degree of development of the Mediterranean-type climate. For areas such as Eastern Europe, for which climatic descriptions in the English language are not easily available, even elementary climatic classifications have their uses; for an example, see W. Okolowicz, *Geogr. Polonica*, 14, (1968), 119, on Poland. On southeast Europe, see M. Hess, *Geographia Polonica*, 14 (1968), 133 on vertical climatic zones in the eastern Alps and western Carpathians, and *Geogr. Polonica*, 13 (1968), 57, on the determination of climatic conditions in the western Carpathians.

55. On isopleth accuracy, see, e.g., D. I. Blumenstock, *AAAG*, 43 (1953), 288. See W. E. Howell, *AAAG*, 39 (1949), 12, on reduction to meaningful levels. For a description of rainfall dispersion diagrams in general, see W. H. Hogg, *G*, 33 (1948), 31,

and for an illustration of their use in describing the rainfall in part of the United States, see P. R. Crowe in Stamp and Wooldridge, *op. cit.*, p. 71. For a criticism from a mathematical viewpoint, see the discussion following P. R. Crowe in *QJRMS*, 66 (1940), 285. On isopleths, see F. Stearns, *AAAG*, 58 (1968), 590, for a method for estimating the quantitative reliability of isoline maps.

56. On subjective interpretation of raw data, see S. S. Visher, *GR*, 37 (1947), 106; see also his *Climatic Atlas of the United States* (Cambridge, Mass., 1954) for an exhaustive extension of the same procedure to all available elements. By the same technique, W. J. Maunder, *NZG*, 13 (1957), 151, analyzed the diurnal variation of New Zealand rainfall. On skewness and variation in Sweden, see K. E. Bergsten, *LSGA*, no. 1 (1950). On regression for New Zealand, see W. J. Maunder, *NZG*, 18 (1962), 184. On correlation for Malaya, see S. Nieuwolt, *Erdkunde*, 20 (1966), 169. On residual mass curves for Japan, see T. Sekiguti, *TJC*, 1, no. 2 (1964), 56. See L. C. Nkemdirim, *CG*, 12 (1968), 248, on the estimation of areal rainfall in upland watersheds.

57. For the United States, see L. H. Horn and R. A. Bryson, *AAAG*, 50 (1960), 157. For Canada see M. A. Sabbagh and R. A. Bryson, *AAAG*, 52 (1962), 426. On the southwestern Pacific, see E. A. Fitzpatrick, D. Hart, and H. C. Brookfield, *Erdkunde*, 20 (1966), 181.

58. M. Mizukoshi, *TJC*, 1, no. 2 (1964), 43. G. H. Dury, *AGS*, 2 (1964), 21. A. Court, *GR*, 43 (1953), 39.

59. A. R. Sumner, *GR*, 43 (1953), 50. S. Gregory, *TIBG*, no. 20 (1954), p. 59.

60. G. H. Liljequist, *GC*, p. 149. J. N. Rayner, *An Approach to the Dynamic Climatology of New Zealand* (Christchurch, N.Z., 1965), and *CG*, 11 (1967), 67. E. Lackey, *IGU*, Washington, (1952), p. 274.

61. L. Curry, *AAAG*, 52 (1962), 21. D. Steiner, *Tijdscrift van het Koninklijk Nederlandsch Aardjkskundig Genootschap*, 82 (1965), 329. J. A. Shear, *AAAG*, 56 (1966), 508.

62. J. O. Mattson, *LSGA*, no. 31 (1965). R. A. Bryson and P. M. Kuhn, *Canadian Geographical Bulletin*, no. 17 (1962), 57. E. Fukui, *TJC*, 1, no. 1 (1964), 76. R. P. Beckinsale, *GR*, 35 (1945), 596. C. J. Bollinger, *PIGU*, Lisbon (1949), p. 529.

63. G. Arnason, *GA*, 27 (1945), 254. J. L. H. Sibbons, *GA*, 44 (1962), 279. For an appreciation of the empirical approach to the measurement of potential evapotranspiration, see R. C. Ward, *G*, 48 (1963), 49. P. C. Chin, *JTG*, 17 (1963), 46. C. W. Thornthwaite, W. J. Superior, J. R. Mather, and F. K. Hare, *PC*, vol. 14, no. 1 (1961); Thornthwaite, Superior, and Mather, *PC*, vol. 14, no. 2 (1961); J. R. Mather, *PC*, vol. 14, no. 3 (1961); Thornthwaite and Mather, *PC*, vol 8, no. 1 (1955). F. K. Davis, *PC*, vol. 10, no. 1 (1957).

64. M. Yoshino, *TJC*, 2, no. 1 (1965), 60.

65. A. A. Miller, *AS*, 13 (1956), 56. For examples from the U.S.S.R., see F. F. Davitaya, O. A. Drozdov, and Y. S. Rubinshteyn, *SG*, 1 (June 1960), 11. For a general account, see W. G. V. Balchin, *GJ*, 124 (1958), 476. For the United States, see P. Meigs, *GR*, 42 (1952), 346, and, for Florida, R. B. Marcus and D. Mookherjee, *G*, 47 (1962), 268. On arid and semiarid lands, see, in particular, the UNESCO Symposia: Arid Zone Program, Reviews of Research, and Proceedings of Symposia, especially nos. I, II, VII, X, XI, and XVIII. On the humid tropics, see, for example, the work of the Humid Tropics Commission of the IGU. On the effects of reservoirs on climate, see: S. L. Vendror and L. K. Malik, *SG*, 6 (December 1965), 25; K. N. Dyakomov and A. Y. Reteyum, *loc. cit.*, p. 40; M. I. L'vovitch, *SG*, 3 (December 1962), 12; and S. I. Zhakov, *SG*, 5 (March 1964), 52. On water resources and their conservation, see: P. Crabb, *G*, 53 (1968), 282, on South Australia; and M. I. L'vovitch, *SG*, 10, no. 3 (1968), 95; A. M. Grin and N. I. Koronkevich, *loc. cit.*, p. 118, and N. N. Dreyer, *loc. cit.*, p. 137, on the Soviet Union. On the design of catchment areas for water balance studies in England, see R. C. Ward, *GJ*, 133 (1967), 495. On the agricultural potential of the humid tropics, in relation to climate, see Jen-hu Chang, *GR*, 58 (1968), 334.

66. For general descriptions, see, e.g.: J. A. Steers, *GJ*, 119 (1953), 280, on the east-coast floods of 1953; F. H. W. Green, *SGM*, 74 (1958), 48, on the Moray floods of July and August 1956; and F. A. Barnes and H. R. Potter, *East Midland Geogr.*, no. 10 (1958), p. 3, on a flash flood in Derbyshire. See J. D. Coulter, *NZG*, 22 (1966), 22, and, in particular, J. N. Rayner and J. M. Soons, *NZG*, 21 (1965), 12, where the financial consequences, as well as the meteorological causes, of a storm are assessed. See, e.g., J. D. Wood, *SGM*, 81 (1965), 5, on the complicity of climate in the 1816 Depression in Dumfriesshire; A. R. Wannop, *SGM*, 71 (1955), 23, on the consequences of the weather of 1954 for the Scottish farmer; and K. Walton, *SGM*, 68 (1952), 13, for a historical account of climate and famines in northeast Scotland. See, e.g., W. Calef, *AAAG*, 40 (1950), 267, on the winter of 1948–49 in the Great Plains, and L. Hewes, *AAAG*, 48 (1958), 375, on climatic causes of wheat failure in Nebraska, 1931–1954. L. Curry, *GR*, 42 (1952), 367, for example, describes the time-scheduling of various economic activities in relation to climate. See D. C. Weaver, *JTG*, 27 (1968), 66, on the hurricane as an economic catalyst in Grenada, West Indies. See J. F. Rooney, Jr., *GR*, 57 (1967), 538, on the urban snow hazard in the United States.

67. S. A. Sapozhnikova and D. I. Shashko, *SG*, 1 (September 1960), 20, described in general terms the agroclimatic conditions deciding the distribution and specialization of Soviet agriculture; V. P. Popov *et al.*, *SG*, 2 (September 1961), 16, indicated how to raise the productivity of agriculture in the Ukraine on the basis of heat and moisture-balance concepts; U. L. Rauner, *SG*, 3 (June 1962), 40, explained the role of the heat balance of the forests in the Moscow region; and A. V. Pavlov and G. P. Ustenko, *SG*, 7 (February 1966), 44, studied the heat balance and radiation regime of corn crops. A. H. Bunting, *G*, 46 (1961), 283, discussed general problems of agriculture in tropical Africa; J. S. Whitmore, *SAGJ*, 39 (1957), 5, described the influence of climatic factors on agricultural development in Sotuh Africa; and J. S. Oguntoyinbo, *Nigerian Geog. J.*, 8 (1965), 83, examined the relation between climate and the cane-sugar industry of Nigeria. On climate and agriculture, see M. J. Maunder, *CG*, 12 (1968), 73, for a review of agroclimatological relationships, and Jen-hu Chang, *The Professional Geographer*, 20 (1968), 317, for a review of progress in agroclimatology. L. Curry, *GR*, 52 (1962), 174, described the climatic basis of intensive grassland farming in the Waikato, and W. J. Maunder, *NZG*, 22 (1966), 55, related the climatic variation experienced in New Zealand to variations in national income by means of a multiple regression model. On Japan, see, e.g., M. Ichikawa, *TJC*, 3, no. 1 (1966), 19.

68. C. W. Thornthwaite, *AAAG*, 37 (1947), 87. See also Jen-hu Chang, *Erdkunde*, 19 (1965), 141.

69. See in particular the exhaustive studies of J. O. Mattson, *LSGA*, no. 16 (1961), no. 35 (1966), and no. 36 (1966), on the microclimatology of potato crops.

70. J. S. Oguntoyinbo, *JTG*, 22 (1966,) 38. M. Yoshino, *TJC*, 3, no. 1 (1966), 24. See also H. Shitara, *TJC*, 2, no. 2 (1965), 71. W. C. Found, *CG*, 2 (1965), 63. E. M. Frisby, *TIBG*, no. 17 (1951), examined the problem of forecasting spring-wheat yield in the northern Great Plains on the basis of weather data.

71. M. Y. Nuttonson, *GR*, 37 (1947), 216 and 436.

72. F. A. Barnes, *TIBG*, no. 21 (1955), p. 137. D. S. Simonett, *AG*, 6 (1953), 15.

On the relation between crop production and climate, see J. S. Oguntoyinbo, *Nigerian Geog. J.*, 10 (1967), 43, for cotton in Nigeria.

73. See, in particular, E. Huntington, *Civilization and Climate* (New Haven, Conn., 1924) and *Mainsprings of Civilization* (New York, 1945). See also G. Manley, *GR*, 48 (1958), 98, on the revival of climatic determinism. S. F. Markham, *Climate and the Energy of Nations* (Oxford, 1942).

74. See D. H. K. Lee, *AAAG*, 43 (1953), 127, and W. H. Terjung, *JTG*, 22 (1966), 49. See also W. H. Terjung, *AAAG*, 56 (1966), 141, for a classification of the bioclimates of the United States in terms of physiological climatology, and A. M. Shul'gin, *SG*, 1 (September 1960), 67, for a Russian view of bioclimatology. On the physical approach to human bioclimatology, see K. J. K. Buettner, *CM*, p. 1112. On bioclimatology, see W. R. D. Sewell, R. W. Kates, and L. E. Phillips, *GR*, 58 (1968), 262, for a

review of geographical work on the human response to weather and climate. For maps of physioclimatic stresses, see W. H. Terjung, *GR*, 57 (1967), 225, on the United States, *JTG*, 22 (1966), 49, on the Sudan, and *GA*, 50, series A (1968), 173, on Africa. For a note on the relationship between crime in Los Angeles and the Santa Ana wind, see W. H. Miller, *The Professional Geographer*, 20 (1968), 23.

75. D. H. K. Lee and H. Lemons, *GR*, 39 (1949), 181. A. Marshall, *AGS*, 1 (1965), 115. M. K. Thomas and D. W. Boyd, *CG* (1957), p. 29. D. H. K. Lee, *GR*, 41 (1951), 124. A. Marshall, *AG*, 9 (1963), 13.

76. R. G. Bowman, *GR*, 39 (1949), 311, described acclimatization in New Guinea. See *Geographia Polonica*, 2 (1964), 79, for the use of local climatic classification in the establishment of health resorts. On China, see Chi-yun Chang, *AAAG*, 36 (1946), 44. On the U.S.S.R., see P. A. Letunov *et al.*, *SG*, 1 (August, 1960), 32. On the Mediterranean and Middle East, see H. H. Lamb, *AS*, 25 (1968), 103, on the climatic background to the birth of civilization, and I. Blake, *AS*, 25 (1969), 409, on climate and society in Palestine before 3,000 B.C.

77. E.g., D. J. Crisp, *GJ*, 125 (1959), 1, described these effects in Europe. See also C. G. Johnson and L. P. Smith, *The Biological Significance of Climatic Changes in Britain* (New York, 1965). M. Bates, *GR*, 38 (1948), 555. F. K. Hare, *GR*, 40 (1950), 615.

On climate and vegetation, see: J. R. Mather and G. A. Yoshioka, *AAAG*, 58 (1968), 29, on vegetation distribution; H. Onodera, *GRTMU*, no. 2 (1967), p. 123, on the effects of the winter monsoon in the Oou Mts. of Japan; and R. A. Bryson, *Geogr. Bull.*, 8 (1966), 228, on air masses and streamlines in relation to the boreal forest in Canada.

78. See M. Parry, *AS*, 13 (1957), 326, on Reading. G. Manley, *loc. cit.*, p. 324. C. W. Thornthwaite, J. R. Mather, and F. K. Davis, Jr., *PC*, vol. 13, no. 1 (1960). C. W. Thornthwaite *et al.*, *PC*, vol. 7, no. 3 (1954). G. Cunningham, *AAAG*, 42 (1952), 247. W. Warntz, *GR*, 51 (1961), 187. C. R. Burgess, *GJ*, 114 (1948), 235. For a discussion of some of these problems, see W. R. D. Sewell, ed., "Human Dimensions of Weather Modification," *Univ. Chicago Dept. Geog. Res. Publ.*, no. 105 (1966).

79. Among the first examples were: G. T. Trewartha, *Japan* (Wisconsin, 1945); J. H. Wellington, *South Africa*, vol. I (Cambridge, 1955); and C. A. Fisher, *Southeast Asia* (London, 1964).

80. E.g., see: B. J. Garnier, *The Climate of New Zealand* (London, 1958); A. A. Borisov, *Climates of the U.S.S.R.*, ed. C. A. Halstead, trans. R. A. Ledward (Edinburgh and London, 1965); and T. J. Chandler, *The Climate of London* (London, 1965).

81. E.g., R. G. Barry, *GA*, 42 (1960), 36, applied synoptic climatology to the palaeoclimatology of Labrador-Ungava. G. Hattersley-Smith, *GA*, 45 (1963), 139, made climatic inferences from firn studies

in Ellesmere Island, and G. Østrem, *GA*, 48, series A (1966), 126, developed methods for determining the climatic snowline from the limits of glaciation in British Columbia and Alberta. See also the *Journal of Glaciology*, various papers.

82. W. G. McIntire and H. J. Walker, *AAAG*, 54 (1964), 582. C. A. M. King, *TIBG*, no. 19, (1953) p. 13. C. D. Ollier and A. J. Thomasson, *GJ*, 123 (1957), 71. J. O. Jennings, *loc. cit.*, p. 474. C. W. Bonython and B. Mason, *GJ*, 119 (1953), 321. L. Williams, *AAAG*, 54 (1964), 597. P. J. Williams, *GA*, 43 (1961), 339.

On the relationship between climate and freeze-thaw processes, see: K. Hewitt, *CG*, 12 (1968), 85, for the Karakoram Himalaya; K. C. Arnold and D. K. Mackay, *Geogr. Bull.*, no. 21 (1964), p. 123, for data problems; and F. A. Cook and V. G. Raiche, *Geogr. Bull.*, no. 18 (1962), for freeze-thaw cycles at the soil surface at Resolute, Northwest Territories, Canada.

83. A. A. Miller, *G*, 46 (1961), 185. See L. Holzner and G. D. Weaver, *AAAG*, 55 (1965), 592, for a review of this field. On climate as a control of erosion, see J. M. Soons and J. N. Rayner, *GA*, 50, series A (1968), 1, on microclimate in the Southern Alps of New Zealand; see also I. Douglas, *TIBG*, no. 46 (1969), p. 1, on the efficiency of denudation systems in the humid subtropics. On climatic gemorphology, see: G. H. Dury, *AG*, 10 (1967), 231; I. P. D. Jungerius, *Geogr. Bull.*, 9 (1967), 218, on the influence of Pleistocene climates on the development of pediments at Cypress Hills, Alberta; and A. T. Grove, *GJ*, 135 (1969), 191, on landforms and climatic change in the Kalahari and Ngamiland.

84. For introductions, see: J. E. Aronin, *Climate and Architecture* (New York, 1953); A. and V. Olgyay, *Design with Climate* (Princeton, N.J., 1963); S. Licht, ed., *Medical Climatology* (New York, 1964); S. W Tromp, ed., *Biometeorology* (New York, 1962); S. W. Tromp, ed., *Medical Biometeorology* (New York, 1963); Jen-yu Wang, *Agricultural Meteorology* (Milwaukee, 1963); G. Z. Ventskevich, *Agrometeorology* (Jerusalem, 1961).

85. See, in particular, *JAM* and *AMM*, also the *Journal of Agricultural Meteorology* (Tokyo), the *International Journal of Biometeorology* (Amsterdam), *Agricultural Meteorology* (Amsterdam), and various *Technical Notes* of the WMO.

86. For examples, see R. M. Cionco, *JAM*, 4 (1965), 517, on a mathematical model for airflow in a vegetative canopy, and B. Slavik, *AMM*, 6, no. 28 (1965), 149, on the supply of water to plants.

87. E.g., see C. C. Wallén's essay review, *Agric. Met.*, 3 (1966), 263, of Thran and Broekhuizen's *Agroclimatic Atlas of Europe*, vol. I (Amsterdam, 1965).

88. For examples of spot observations, see E. N. Lawrence, *MM*, 81 (1952), 65 and 137. For an

illustration of the importance of local geographical factors as determinants of low minimum temperatures in Norfolk, see J. Oliver, *MM*, 95 (1966), 13. For a chart for calculating longwave radiation, based on Brunt's night-cooling equation, see W. Bach, *McGill Climatol. Bull.*, no. 2 (1967), 45. For an example of the effect of a valley site in Herefordshire, Eng., on minimum temperatures, see E. N. Lawrence, *MM*, 85 (1956), 79. See D. E. Angus, *UCSCM*, p. 265, and T. V. Crawford, *AMM*, 6, no. 28 (1965), 81 for details on statistical evaluations. For an interesting example of the prediction of net radiation balance at the surface, on the topoclimatic scale in southern California, based on both physical theory and geographical parameters, see W. H. Terjung *et al.*, *AMGB*, 17, series B (1969), 21.

89. For details on procedure, see R. Bush, *Frost and the Fruit Grower* (New York, 3d ed., 1947), pp. 28–51.

90. J. A. Businger, *AMM*, 6, no. 28 (1965), 74, describes the principles involved.

91. For details, see C. W. Thornthwaite: *WW*, vol. 4, no. 2 (April 1951); *PIGU*, Washington (1952), p. 290.

92. For details, see C. W. Thornthwaite, pp. 368–80, in J. F. McCloskey and F. N. Trefethen, eds., *Operations Research for Management* (Johns Hopkins, 1954).

93. For details, see *MAE*; also P. J. Meade, *WMO Tech. Note*, no. 33, and *AGP*, 6 (1959), 331. See E. W. Bierly and E. W. Hewson, *JAM*, 1 (1962), 383, for examples. For an example of an empirical approach to the topographic factor in air pollution in Sheffield, Eng., see A. Garnett, *TIBG*, no. 42 (1967), p. 21.

Postscript

Techniques of Climatology has been concerned with climate as a physical problem. Its main task has been to demonstrate that climatology is not merely a collection of facts about weather and climate, classified in a manner sometimes scientifically valid, sometimes arbitrary. Instead, it is a physical science with sound foundations. These foundations have formed the subject-matter of the book. The average reader will, no doubt, have received some grounding in the purely descriptive side of climatology, i.e., world climatic regions and climatic types, at school. I hope that such a reader will have concluded, upon reaching this Postscript, that a real understanding of climate requires a knowledge of physics, which reveals that many supposed problems of world climatic regions and climatic types are trivial, if not illusory.

Once the reader has grasped the essentially physical nature of climatology, he will encounter another problem, that of mathematics, for here the climatologist differs from the meteorologist. The latter approaches the study of weather and climate as a physicist, well-versed in the use of his primary tool, mathematics, and consequently tends to investigate only those problems that are directly amenable to analysis by mathematical techniques. The climatologist, on the other hand, approaches the study of weather and climate in terms of the phenomena. His inspiration comes from personal experience of the variety of weathers and climates the Earth has to offer, and from a conviction of the importance of meteorological influences on human life and activities. To the climatologist, only those branches of meteorological physics that are relevant to the study of actual weather situations are of interest, and only those parts of mathematics that are necessary to the understanding of these branches of physics are of use to him—hence the highly selective treatment of meteorological physics and meteorological mathematics in these pages.

The geographer who wishes to carry out climatological work that will be a lasting contribution to human knowledge must become proficient in the mathematical analysis of the phenomena that interest him, for without mathematics, prediction is impossible, and without prediction, science cannot advance. The geographical literature contains numerous examples of climatological investigations that came to nought because of a lack of mathematical insight. Likewise, the physicist and the mathematician should not neglect the writings of geographers, naturalists, and other empiricists, for these may reveal climatological problems that the analytical approach may completely overlook. The geographer is very much concerned with problems of balance between different, often conflicting, interests, and may be in a position to outline the most profitable area to which the high-powered techniques of mathematics and physics may be applied. Meteorological literature is full of contributions to pure climatology that, although physically refined and mathematically elegant, are all but useless to mankind, because they deal with situations that rarely, if ever, occur on the real Earth. The actual world of weather and climate is vastly more complicated than the abstract world defined by the computer-generated maps of the numerical weather-prediction expert, even though his contribution does provide a sound foundation for the study of actual weather and climate, and hence for climatology.

1.1. According to *MC*, p. 242, a climatically coherent region is one within which the variability of differences between measurements of the same type of phenomena (i.e., rainfall, temperature, etc.), taken at two stations within the region, increases *continuously* as the distance between the two stations increases, and within which the variability of the quotients of measurements increases continuously as the difference in elevation between the two stations increases.

1.2. Following *MOHSI*, p. 178, the required equations are (see Figure A1.2):

$$w \cos \phi = v \cos(\theta + \alpha) - q \cos \theta,$$

$$w \sin \phi = v \sin(\theta + \alpha) - q \sin \theta,$$

where θ is the compass heading of the aircraft, q is its true airspeed, v is its groundspeed, α is its angle of drift, and w is the true windspeed.

1.3. For example, one must compensate for the fact that the aircraft gradually ascends along its run as fuel is consumed and its total weight decreases. Time must be read to within 10 seconds to obtain an accuracy of 2 km in the position of the aircraft, assuming a groundspeed of 450 knots.

1.4. The vernier principle enables linear or angular magnitudes to be read very much more accurately than is possible with a scale subdivided in the ordinary manner; it depends essentially on the fact that the eye can estimate the coincidence of two

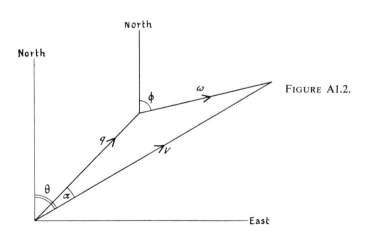

FIGURE A1.2.

graduations more readily than it can judge distances. See *MOHSI*, p. 26, for an example.

1.5. According to *Hygrometric Tables* (M.O. pamphlet no. 265, 4th ed., 1940), the original Regnault formula is

$$x = f - Ap(t - t'),$$

where x is the vapor pressure of the air at the temperature (t') of the wet bulb; f is the saturation vapor pressure at t'; A is a constant that depends on (a) the speed of the air moving past the wet bulb, (b) the value adopted for the latent heat of evaporation, and (c) the thermometric scale adopted; p is the atmospheric pressure; and t is the dry-bulb temperature; other psychometric formulae for ventilated psychrometers have been derived since Regnault, e.g., by A. F. Spilhaus, *Trans. Roy. Soc. South Africa*, 24 (1936), 185, by R. B. Montgomery, *JM*, 5 (1948), 113, and by D. J. Bouman, *T*, 6 (1954), 399. The original Regnault equation was derived in M. V. Regnault, *Comptes-rendu Acad. Sci. Paris*, 20 (1845), 1127 and 1220. The various forms of the Regnault formula adopted by different meteorological services are as follows.

British formulae: for observations made in thermometer screens,

$$x = f - 0.444(t - t') \text{ for wet bulb} > 32°F,$$

$$x = f - 0.400(t - t') \text{ for wet bulb} < 32°F;$$

for observations made with ventilated psychrometers,

$$x = f - 0.37(t - t') \text{ for wet bulb} > 32°F,$$

$$x = f - 0.33(t - t') \text{ for wet bulb} < 32°F.$$

American (Marvin) formula for ventilated psychrometers:

$$x = f - 0.367(t - t')\left(1 + \frac{t - 32}{1571}\right).$$

Austrian (Pernter) formula:

$$x = f - 0.364(t - t')\left(1 + \frac{t' - 32}{1098}\right).$$

Indian (Blanford) formula:

$$x = f - 0.437(t - t')\left(1 + \frac{t' - 32}{1098}\right).$$

Swedish (Rolf) formula:

$$x = 0.9737f - 0.331(t - t').$$

France, Germany, and Norway have used the Regnault formula, with the values 0.44, 0.368, and 0.441, respectively, for the constant A, in the Angot, Assmann, and Bjirkeland formulae, respectively. The three values of A are for wet bulb $> 32°F$.

1.6. According to A. F. Spilhaus, *JM*, 5 (1948), 161, the drop-size distribution is given by

$$\ln \frac{\bar{D} R_D}{R} = -k_0^2 u^2,$$

where $\bar{D}$ is the median diameter of the drops, and k_0^2 is a constant for all rains;

$$R_D = \frac{1}{6} \pi N D^3 v$$

in which N is the number of drops per unit volume, D is a drop diameter, and v is its terminal velocity (attained when the gravitational force on it is balanced by the drag of the air against it), given by $v^2 = K^2 D$, K being defined by

$$K^2 = \frac{2}{3} \rho_w g / \rho C_d,$$

in which ρ, ρ_w, represent the densities of air and water, respectively, and C_d is the drag coefficient for a spherical raindrop; finally,

$$u = \sqrt{\frac{D}{\bar{D}} - 1}.$$

The distribution devised by A. C. Best, *QJRMS*, 76 (1950), 16, combined all raindrop-size measurements published up to that time into a single function, described by

$$1 - F_x = \exp\left[- \left(\frac{x}{a}\right) \right]^n, \, a = A I^p, \, W = C I^q,$$

where F_x is the fraction of liquid water in the air that consists of drops with diameters of less than x, x is a drop diameter, I is the rate of rainfall, W is the liquid-water content of the rainfall per unit volume of air, and A, C, n, p, q, are constants for each set of data, decided by the method of least squares. R. M. Griffith, *JAS*, 20 (1963), 198 and 624, found that n varies with the most likely nucleus concentration at the observing site.

The distribution devised by J. S. Marshall and W. M. Palmer, *JM*, 5 (1948), 165, states that

$$N_D = N_0 e^{-\Lambda D},$$

where there are $N_D \delta D$ drops with diameters between D and $D + \delta D$ in a unit volume of space; N_0 is the value of N_D for $D = 0$ ($N_0 = 0.08$ cm^{-4} for any intensity of rainfall); and Λ is defined either by $\Lambda = 41 R^{-0.21}$ per cm, where R is the rate of rainfall in mm per hour, or by $\Lambda = 3.75/D_0$, where D_0 is the median volume diameter.

1.7. Following J. R. Probert-Jones, *QJRMS*, 88 (1962), 485, the radar equation may be expressed in the form

$$P_r = \frac{\pi^3}{16 \log_e 2} \cdot \frac{P_0 h}{\lambda^2} G^2 \theta_1 \phi_1 \frac{1}{R_2} \left| \frac{\varepsilon - 1}{\varepsilon + 2} \right|^2 \sum r^6,$$

where θ_1, ϕ, are the radar-beam widths to -3 decibels for one-way transmission; P_0, P_r, represent the transmitted and received powers, respectively; R is the range of the radar; h is the pulse-length in space; r is the radius of the drop reflecting the beam; ε is the dielectric constant; and G is the gain of the aerial system, i.e., the ratio of power radiated by an isotropic antenna necessary to produce a given field strength at a given distance to power radiated by the directional antenna producing the same field strength at the same distance in the direction of maximum transmission. The derivation of the equation assumes that the actual beam may be represented by a sharp-edged conical beam of uniform intensity from the axis outward to the points at which the beam intensity falls to half its power along the axis. Then, $G = \pi^2/\theta_1\phi_1$.

Assuming that the radar beam is entirely filled with precipitation, and that none of the energy of the beam is absorbed by atmospheric gases or water, than the average power, $\bar{P}_r$, received by the radar from the precipitation is given by

$$\bar{P}_r = \frac{C|K|^2 Z}{r^2},$$

where C is a constant for the radar apparatus; r is the range of the instrument; $|K|$ is the dielectric factor (0.93 for water, 0.20 for ice); and $Z = \Sigma\ D^6$, i.e., the sum of the sixth powers of the diameters of the precipitation particles per unit volume of air. The equation for $\bar{P}_r$ therefore takes no account of attenuation, which reduces the power received by the radar. Empirical correction factors to allow for attenuation (for 3.2-cm radar) are as follows:

rain of 1 mm per hour, 0.025 db per mile;
rain of 10 mm per hour, 0.50 db per mile;
rain of 100 mm per hour, 10.0 db per mile;
clouds, 0.28 db per mile per gm per m^3;
atmospheric gases at 1,000 mb, 0.05 db per mile;
atmospheric gases at 500 mb, 0.01 db per mile.

See K. L. S. Gunn and T. W. R. East, *QJRMS*, 80 (1954), 798, for details. P. M. Austin, *JM*, 4 (1947), 121, showed how, by assuming all the drops were the same size, a measurement of the approximate size of the raindrops could be found from the attenuation.

Rainfall rate, R, is given by $Z = 200R^{1.6}$ for most types of rain, where Z is in mm^6 per m^3 of air, R in mm per hour; $Z = \sum D^6$, as above, because of Ryde's fundamental formula, for which see p. 169 in Ryde's work cited in note 77 on p. 74 of this book. *Rain density*, M, is given by $M = 72R^{0.88}$ or $Z = 24,000M^{1.82}$ (M in gm per m^3, M being the liquid-water content). Both of these parameters are defined for small samples of rain taken at the ground. For falling raindrops,

$$\log Z = 0.072(\log R)^2 + 1.317(\log R) + 2.569,$$

for Illinois precipitation. Geographical variations are important. For Hawaii, $Z = 290R^{1.41}$ (nonorographic rain), $Z = 16.6R^{1.55}$ (orographic rain within the cloud), or $Z = 31R^{1.71}$ (orographic rain at the cloud base). These formulae work for rains consisting of droplets of approximately equal sizes. If the drops are of different sizes, then regional variations appear (see G. B. Foote, *JAM*, 5, 1966, 229): in winter, in Australia, $Z = 127R^{2.287}$; in England, $Z = 630R^{1.45}$, and in Japan, $Z = 700R^{1.6}$; in summer in Arizona, $Z = 520R^{1.81}$.

The *radar rainfall* rate, R_r, is defined by

$$R_r = \left(\frac{Z_e}{200}\right)^{\frac{1}{1.6}},$$

where Z_e is the *equivalent radar reflectivity*. It gives the correct measure of the actual rainfall rate, provided that the precipitation particles (a) consist of liquid water, (b) obey the law $Z = 200R^{1.6}$, (c) fall in air which has no vertical component of motion. Z_e is the quantity displayed on the radar screen, and is defined by

$$Z_e = \frac{\bar{P}_r r^2}{0.93c} = \frac{|k^2|}{0.93} z.$$

For rain, Z_e is a direction measure of Z. Allowances may be made that enable Z_e to be transformed to the actual rainfall rate when updrafts occur or when $z \neq 200R^{1.6}$.

For aggregate snowflakes, $Z = 2000R^{2.0}$, $M = 250R^{0.90}$, and $Z = 0.0096M^{2.2}$, where R, M, are in terms of water equivalent. For single crystals of snow, the rain relation $Z = 200R^{1.6}$ works quite well, but for melting snow and hail the relation is complex.

For the mathematics, and for measurements of *precipitation volume*, see G. B. Walker, L. S. Lamberth, and J. J. Stephens, *JAM*, 3 (1964), 164.

1.8. Following *MOHSI*, p. 190, if the vane is exposed to a varying wind of direction $\dfrac{\theta_0 + \theta \sin 2\pi t}{\tau}$, where θ_0 is the mean wind direction, θ is the amplitude of the wind direction, and τ is its period of variation, then the vane will execute oscillations described by

$$\phi = \phi_0 e^{\frac{-t}{\lambda}} \sin\left(\frac{2\pi t}{T} + \alpha\right) + \frac{\theta}{A} \sin\left(\frac{2\pi t}{\tau} - \beta\right),$$

where α is obtained from

$$\phi = \phi_e e^{\frac{-t}{\lambda}} \sin\left(\frac{2\pi t}{T} + \alpha\right),$$

which describes the oscillation of a wind vane in a wind of unvarying direction, ϕ_e; ϕ is the deviation of the angle of the wind vane from the true wind direction ϕ_0; T is the period of oscillation; β is an empirical constant; and $A = f(\lambda, T, \tau)$, with minimum value when τ is nearly equal to T. After the free oscillations have been damped out, the ratio of the amplitude of the forced oscillations to that of the varying wind direction is $\dfrac{1}{A}$, which may be computed for a particular type of instrument; β then gives the amount by which the forced oscillations lag behind the true wind oscillations.

1.9. The restoring force equals $\frac{1}{2}cA\rho v^2$, where c is a constant depending on the size and shape of the plate and A is its area, ρ is air density, and v is windspeed.

1.10. Bernouilli's equation may be expressed as

$$p_T = p_s + \frac{1}{2}\rho v^2 = k,$$

where p_T in the total head of pressure and k is constant; i.e., p_T consists of two components, a static pressure, p_S, and a velocity head, $\frac{1}{2}\rho v^2$, where ρ is air density and v is windspeed; v may be deduced from measurements of the difference between p_T and p_S made by means of a pitot-static tube.

1.11. The anemometer factor equals the windspeed, v, divided by the speed of the cup centers, V; that is

$$v = a + bV + cV^2 + dV^3 + \ldots,$$

where a, b, c, d, are constants. The best instrument is one in which a, c, d, and the coefficients of higher powers of V are close to zero.

1.12. The critical factor, K, is defined by

$$K = \frac{0.55\rho R^2 r^2 Tv}{I},$$

where R is the radius of the circle described by the centers of the cups, r is the radius of the cups, T is the period of variation of the windspeed v, and I is the moment of inertia of the rotating components. As K decreases, the degree of overestimation of the windspeed increases.

1.13. For the mathematics, see R. M. Schotland, *JM*, 12 (1955), 386. The basic principle is that the speed of propagation of a sound wave in the atmosphere, c, is related to the absolute temperature and water-vapor content of the air through which it travels by the expression

$$c = 20.067\sqrt{T\left(1 + 0.319\,\frac{e}{p}\right)}$$

where c is in m per sec, T in $^\circ$K, and e and p are in the same units.

1.14. According to M.O. pamphlet no. 396, *The Measurement of Upper Winds by Means of Pilot Balloons* (London, 3rd ed., 1944), the computation is as follows. In Figure A1.14,A, which depicts the position of a balloon in the northeast quadrant at the ends of the first and second minute after release, $\angle C_1AB_1$ represents the angle of elevation e_1^0 of the balloon at the end of the first minute and $\angle C_2 AB_2$ represents that, e_2^0, at the end of the second minute. The actual distances traveled by the balloon are $\overline{AB_1}$ after the first minute, and $\overline{B_1B_2}$ between the first and second minutes, corresponding to horizontal distances of $\overline{AC_1}$ and $\overline{C_1C_2}$, respectively. From the horizontal-plane diagram (see Figure A1.14,B), the windspeed and wind direction at the end of the first minute may be found from $\overline{AC_1}$ and azimuth $\angle a_1^0$. At the end of the second minute, the balloon will have traveled $\overline{Y_2 C_2} - \overline{Y_1C_1}$ toward the east and $\overline{AY_2} - \overline{AY_1}$ toward the north, that is, $\overline{YC_2}$ and $\overline{C_1Y}$, respectively. The displacement $\overline{C_1C_2}$ of the balloon is then obtained from $\overline{YC_2}$ and $\overline{C_1Y}$, by

$$\overline{YC_2} = h_2 \cot e_2 \sin e_2 - h_1 \cot e_1 \sin a_1,$$

$$\overline{YC_1} = h_2 \cot e_2 \cos a_2 - h_1 \cot e_1 \cos a_1,$$

where h_1, h_2, are the heights of the balloon after the first and second minutes, respec-

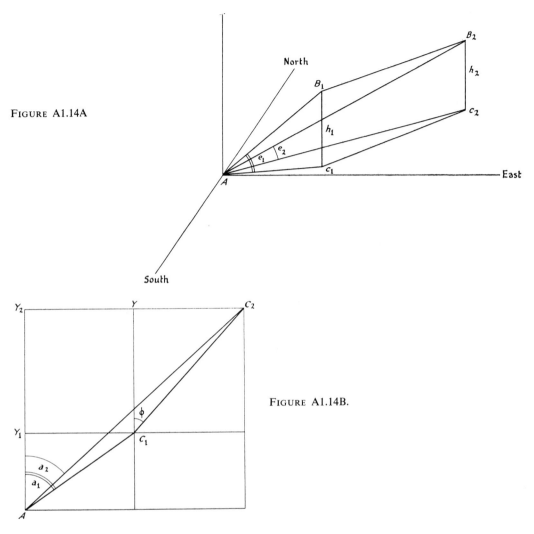

FIGURE A1.14A

FIGURE A1.14B.

tively. Since $\tan \phi = \dfrac{\overline{YC_2}}{\overline{YC_1}}$, and $\overline{C_1C_2} = \dfrac{\overline{YC_2}}{\sin \phi}$, where ϕ is the angle between north and the direction of travel of the balloon during the second minute, and h, e, and a, may be found at successive minutes, then ϕ and $\overline{C_1C_2}$ may be deduced. Thus the direction and velocity of the wind during the second minute may be found. A similar procedure

TABLE A1.14
Signs for quadrants

Azimuth	$\overline{XC}$	$\overline{AX}$
0–90°	+	+
90–180°	+	−
180–270°	−	−
270–360°	−	+

is followed for the third and successive minutes. Elevation must be read accurately, for an error of 0.1° in its measurement corresponds to an error of 250 feet in $\overline{AC_n}$ for $e = 10°$ and $\overline{AC_n} = 5$ miles.

1.15. According to *MOHUI*, p. 15, the errors are as follows. The error in computed wind is

$$\delta \mathbf{V} = \sqrt{(\delta v)^2 Q^2 + \frac{2v^2 T^2}{t^2}[(\delta e)^2(Q^2 + 1)^2 + (\delta a)^2 Q^2]},$$

The error in height determination is

$$\delta h = \frac{(\delta v)t}{\sqrt{2}}.$$

Here δv is the random variation in the mean rate of ascent of the balloon between consecutive soundings; $Q = \dfrac{\overline{\mathbf{V}}}{\bar{v}}$ is the ratio of mean vector wind to mean rate of ascent up to height h; T is the time taken to reach h; t is the time interval between consecutive soundings; and δe, δa, are the random errors in elevation e and azimuth a. See H. A. Ayer, *BAMS*, 39 (1958), 279, for an example of single-theodolite wind-finding errors.

1.16. Following *MOHSI*, p. 200, the principle involved is expressed in the form

$$RI^2 = (K + c\sqrt{V})(T - T_0),$$

where I is the electric current that must be passed through a cylindrical wire, held normal to an airstream of speed V and temperature T_0, in order to ensure that the heat loss from it remains constant; T and R represent the equilibrium temperature and the equilibrium resistance of the wire, respectively; K is a constant that depends on the heat-loss properties of the wire; and c is a constant that depends on the diameter of the wire and the physical properties of the air in contact with the wire.

1.17. Following A. Ångström, *T*, 13 (1961), 214, the concept of *turbidity* integrates the total effect of all scattering particles within the atmosphere (e.g., dust particles from the ground, water and ice particles from condensation, meteoric dust, hygroscopic salt nuclei) on radiation. The transmission of solar radiation through the atmosphere may be described by the expression

$$I_\lambda = S_\lambda[\tau_\lambda']^{m_1} \cdot [\tau_\lambda'']^{m_2} \cdot [\tau_\lambda''']^{m_3},$$

where I_λ is the solar radiation of wavelength λ reaching the Earth's surface, S_λ is the solar constant for wavelength λ, and τ_λ and m represent transmission factors and *relative air masses*, respectively. The relative air masses are theoretical air masses that correspond to the radiation-attenuating factors such as Rayleigh scattering (m_1), selective absorption by different atmospheric gases (m_2), and scattering and absorption by solid and liquid particles (m_3), the appropriate transmission factors for these attenuators being τ_λ', τ_λ'', τ_λ''', respectively. For solar elevations above 15°, $m_1 = m_2 = m_3 = m$, and

$$I_\lambda = S_\lambda[\tau_\lambda' \cdot \tau_\lambda'' \cdot \tau_\lambda''']^m,$$

which may be expressed in the form

$$\tau_\lambda = e^{-\frac{\beta}{\lambda^\alpha}},$$

in which α is the wavelength exponent (which may vary from 0.5 in a very polluted atmosphere to 1.6 in a very clear atmosphere, but usually is about 1.3), and β is thus defined as the *turbidity coefficient*; β has different values in different parts of the world, and has an annual variation at a given location.

1.18. Following *PM*, p. 5, the index of refraction for dry air, m, may be defined by

$$(m - 1) \times 10^6 = k\rho,$$

where ρ is the air density and k is constant. For moist air,

$$(m - 1) \times 10^6 = A \frac{p}{T} \left(1 + \frac{B}{T} \frac{e}{p}\right),$$

where p is atmospheric pressure, e is vapor pressure, T is absolute temperature, and A and B must be determined by experiment or theory.

1.19. W. A. Arvola, *BAMS*, 38 (1957), 212, defined B as $(n - 1) \times 10^6 + 0.012z$, where n is the refractive index for radio frequencies and z represents height (in feet) above the ground. B may be expressed in the form

$$B = \frac{77.8}{T} \left(p + \frac{4780}{T} e\right) + 0.012z,$$

in which T, p, and e are in $°K$, and mb.

1.20. For example, C. W. Spencer, *MM*, 81 (1952), 206, found that high field strengths for propagation of 89 mc radio waves over a 257-km path from Slough to Moorside Edge are twice as probable with anticyclonic curvature as with cyclonic curvature.

1.21. Following P. A. Sheppard, *AGP*, 9 (1962), 77, if the surface wind at a height Z_a (of about 10 m) above the surface of the Earth is V_s, then the surface stress τ_0 is given by

$$\tau_0 = C_d \rho V_s^2,$$

where C_d is the drag coefficient, defined by

$$C_d = \frac{k}{\log z_a/z_0}$$

in which k is von Kármán's constant and z_0 represents the aerodynamic roughness height of the surface. J. E. Vehrencamp, *EAFM*, p. 99, gives the equation

$$\bar{\tau}_0 = \frac{1}{2} \rho C_d \bar{V}_s^2$$

for the mean surface stress, when $\bar{V}_s$ is the mean surface windspeed.

According to H. H. Lettau, *AGP*, 6 (1959), 241, the surface windspeed V_s may be replaced by the surface geostrophic windspeed V_{gs} so that the equation becomes

$$C^{\frac{1}{2}}_{dn} = \frac{0.104}{[\log_{10}(C^{\frac{1}{2}}_{dn}R_0) - 2.24]},$$

where C_{dn} is the value of the drag coefficient for a neutral atmosphere, and $R_0 \left(\equiv \dfrac{V_{gs}}{fz_0} \right.$

where f is the Coriolis parameter $\Big)$ is the surface Rossby number.

1.22. Following D. J. Portman, *EAFM*, p. 26, the thermal diffusivity equation may be expressed as

$$k = \frac{\lambda}{\rho c},$$

where k is the thermal diffusivity (in cm^2 per sec), λ is the heat conductivity through the soil (in cal per sec per degree), ρ is the bulk density of the soil (in gm per cm^3), and c is the soil's specific heat (in cal per gm per degree). The product ρc (in cal per cm^3 degree) constitutes the *heat capacity* of the soil.

1.23. Following B. Solot, *MWR*, 67 (1939), 100, the amount of precipitable water (W_p inches) in an air column of unit cross section between pressure levels p_0 and p_z is given by

$$W_p = 0.0004 \int_{p_z}^{p_o} q \, dp,$$

where q is the specific humidity. It may be computed rapidly by means of a template or a nomogram; see A. K. Showalter, *MWR*, 35 (1954), 129, and K. R. Peterson, *BAMS*, 42 (1961), 119. C. H. Reitan, *MWR*, 88 (1960), 25, computed the mean values of precipitable water by means of the equation

$$W_p = \frac{10^2 \varepsilon}{g} \cdot \frac{e_s R_H}{p} \cdot \Delta p,$$

where W_p is in cm, ε is the ratio of the molecular weight of water vapor to that of dry air, e_s is the saturation vapor pressure (in mb) determined by using mean temperatures, R_H is the relative humidity, p is the pressure (in mb) at the point of observation, and Δp is the depth of the layer (in mb) that is taken to surround the point of observation for computational purposes. Usually, $\Delta p = 50$ mb.

1.24. According to J. Gentilli, *BAMS*, 36 (1955), 128,

$$T_{\min} - T_d = 117.45 - 63.45 \log R_H,$$

where $T_{\min}$ and T_d represent mean monthly minimum and dew-point temperatures respectively, and R_H is mean monthly relative humidity.

1.25. The *Bowen ratio*, β, derived by I. S. Bowen, *Phys. Rev.*, 27 (1926), 779, is defined by $\beta = \dfrac{H}{L_v E}$, in which H represents the turbulent heat flux and $L_v E$ is the latent heat flux, L_v being the latent heat of vaporization for water and E the upward flux of water vapor (i.e., the rate of evaporation). Following E. K. Webb, *AMM*, 6 (1965), 27, β may be determined either from

$$\beta = \frac{C_p \, \Delta \theta}{L \, \Delta q},$$

where Δ denotes a difference between measurements made at two heights, and the assumption is made that $K_E = K_H$, or from

$$\beta = \left[\frac{s+\gamma}{\gamma} \cdot \frac{\Delta\theta_w}{\Delta\theta} - 1 \right]^{-1},$$

where

$$\gamma = \frac{C_p}{L},$$

θ_w denotes the wet-bulb potential temperature, and S is the slope of the saturation specific humidity curve at a temperature midway between the two wet-bulb temperatures.

Following C. W. Thornthwaite and F. K. Hare, *AMM*, 6 (1965), 163, the energy-balance approach is described by the equation

$$E = \frac{R_n - S - P}{L_v(1 + \beta)},$$

where R_n is the net radiation (in ergs per cm^2 per sec), S is the heat flux into the soil in ergs per cm^2 per sec, and P is the rate of heat consumption by photosynthesis per unit area (in ergs per cm^2 per sec). E is in g per cm^2 per sec, and L_v in $^{\circ}T$ cal per g.

1.26. Following E. K. Webb, *AMM*, 6 (1965), 27, the Dalton formula may be written

$$E = \rho f V_s(q_{ws} - q_a),$$

where V_s is the rate of movement of the air above a water surface, q_{ws} is the saturation value of the specific humidity of the air above the water surface at the temperature of the latter, q_a is the actual specific humidity of the air, and f is an empirical constant.

The Thornthwaite-Holzman equation, on which see C. W. Thornthwaite and B. Holzman, *U.S.D.A. Tech. Bull*, no. 817 (1942), may be written

$$E = \frac{-\rho k^2(U_b - U_a)(q_b - q_a)}{[\ln(b/a)]^2},$$

where U_a, U_b, are horizontal wind velocities at heights a, b, above the Earth's surface; q_a, q_b, are specific humidities at these heights; k is von Kármán's coefficient; and the units are c.g.s. F. Pasquill, *QJRMS*, 75 (1949), 249, showed that the equation may be rewritten as

$$E = -k U_b(q_b - q_a),$$

where the constant k for a given surface must be established initially by measurement of the ratio $\dfrac{U_b}{U_a}$ in a given site.

1.27. A modified form of the equation devised by W. C. Jacobs, *Scripps Inst. Oceanog. Reports*, 6 (1951), 27, is given by D. W. Privett, *QJRMS*, 85 (1959), 424, as

$$E = 0.00587(e_s - e_a)w,$$

where w is the windspeed (in knots) at a standard height above the sea surface, E is in gm per cm^2 per day, e_a and e_s are in mb, and e_s is the saturated vapor pressure (corresponding to sea-surface temperature) multiplied by 0.98 to correct for salinity effects.

The formula devised by H. L. Penman, *TAGU*, 37 (1956), 43, and *Netherlands J. Agric. Sci.*, 4 (1956), 9, is

$$E = 0.35\left(0.5 + \frac{U_2}{100}\right)(e_s - e_a),$$

where U_2 is the run of the wind (in miles per day) at 2 m above the surface, and e_a is the vapor pressure at screen level (1.5 m); E is in mm per day, e in mm of mercury.

The formula devised by J. J. Marciano and G. E. Harbeck, *U.S. Geol. Survey Prof. Paper*, no. 269 (1954), p. 46, and no. 298 (1958), is

$$E = CU_a(e_s - e_a),$$

where $C = 1.39 \times 10^{-3}$ at sea level if $a = 4$ m and $C = 1.62 \times 10^{-3}$ if $a = 2$ m (C also varies with water-surface area, altitude, and latitude). Units: E is in cm every 3 hours, U_a in m sec, and e in mb.

1.28. The basic Penman equation, according to E. K. Webb, *loc. cit.*, is

$$L_v E = \frac{S}{S + \gamma} R_n + \frac{\gamma}{S + \gamma} LE_a,$$

in which $\gamma = \dfrac{C_p}{L_v}$, R_n is the net radiation, S is the slope of the saturation specific-humidity curve $\left(\text{i.e., } \dfrac{dq}{dT}\right)$ at a temperature midway between T_s (the temperature of the Earth's surface) and T_a (the air temperature), and E_a is obtained from

$$E_a = 0.35\left(0.5 + \frac{U_2}{100}\right)(e_s - e_a),$$

where $e_s - e_a$ represents the saturation deficit. An alternative statement of the Penman formula follows Thornthwaite and Hare, *AMM*, 6, no. 28 (1965), 170:

$$E_0 = \frac{\dfrac{\frac{de}{dT}}{\gamma_p} \cdot \dfrac{R_n}{L_v} + f(u)(e_a - e_d)}{\dfrac{\frac{de}{dT}}{\gamma_p} + 1}$$

where E_0 is the evaporation from a hypothetical open-water surface, $e_a - e_d$ is the saturation deficit, where the subscript a denotes screen level, γ_p is the psychometric constant $\left(\gamma_p = \dfrac{C_p p}{0.622}\right)$, where p is atmospheric pressure$\Big)$, and $f(u)$ is derived from Dalton's equation for an open-water surface expressed in the form

$$E_0 = f(u)(e_0 - e_d),$$

in which e_0 is the saturation vapor pressure at the temperature of the water surface, and e_d is the vapor pressure corresponding to dew point at the screen level;

$$f(u) = 0.35\left(0.5 + \frac{U_2}{100}\right) \text{ over open water,}$$

and

$$f(u) = 0.35\left(1 + \frac{U_2}{100}\right) \text{ over short turf.}$$

The resulting value of E_0 will be in mm per day. The ratio $\dfrac{de}{dT}$ is the reciprocal of the known slope of the curve of vapor pressure plotted against temperature.

1.29. G. Stanhill, *QJRMS*, 88 (1962), 80. For Israel, the relation is

$$E_a = 0.1469E + 0.1118,$$

where E is the water loss measured by a Piche evaporimeter and E_a is the aerodynamic term in the Penman formula.

1.30. Following Thornthwaite and Hare, *op. cit.* (1965), p. 172, the potential evapotranspiration (PE) in cm per month is given by

$$PE = 1.6l\left(\frac{10T_a}{I}\right)^m$$

where

$$I = \sum_1^{12}\left(\frac{\overline{T}_a}{5}\right)^{1.514},$$

T_a is the "actual" mean air temperature (in °C) for a month, $\overline{T}_a$ is the climatological normal temperature for each of the 12 months, I is the heat index, and m is a cubic function of the heat index.

1.31. Following Thornthwaite and Hare, *op. cit.* (1965), p. 174 *evaporability* (E_v) in the Budyko sense may be defined as

$$E_v = \rho D(q_s - q_z),$$

where

$$D = \frac{1}{\displaystyle\int_0^z \frac{1}{k_W}\,dz},$$

D being in cm per sec. Before E_v can be determined, the saturation specific humidity (and therefore the temperature T_s) of the moist evaporating surface must be known. This difficulty is eliminated by equating the net radiation corresponding to the air temperature, R_{na}, to R_n:

$$R_n = R_{na} - 4s\sigma T_a^3(T_s - T_a),$$

where s is the ratio of actual to black-body emissivity and T_a is the air temperature. Therefore,

$$R_a - S \equiv L_v \rho D(q_s - q_a) + (\rho C_p D + 4s\sigma T_a^3)(T_s - T_a),$$

where S is the heat flux into the soil. Since q_a is a known function of T_s, q_s can be eliminated from the original equation for E_v. Annual evaporability is

$$\sum_1^{365} E_v = 0.18 \sum_1^{365} (T - 10)$$

in the Budyko system, where T is the daily mean temperature, and the units are mm per year.

2.1. Following *HSM*, p. 261, the general expression for smoothing *m* times over *n* overlapping terms is

$$\frac{R'}{R} = \left(\frac{\sin \dfrac{n\pi}{L}}{n \sin \dfrac{\pi}{L}} \right)^m \qquad [A2.1]$$

where R is the unsmoothed range, R' is the range of unsmoothed values, and L is the length of the oscillation. If the time-series is known to contain a real oscillation that is to be retained after smoothing, then smoothing by overlapping means is not suitable unless $\dfrac{R'}{R}$ is very little less than unity.

2.2. In the formula for a running mean,

$$M = \frac{(a + b + c)}{3},$$

the value b receives too little weight and may cause a depression of the curve representing the series of running means. Putting $n = 2$ and $m = 2$ in Equation A2.1 gives

$$M' = \frac{(a + 2b + c)}{4},$$

which is the expression for a 3-unit *weighted running mean*. If M' does not remove the irregularities of the time-series sufficiently, then M' is repeated twice, giving:

$$M'' = \frac{(a + 4b + 6c + 4d + e)}{16},$$

which yields a 5-unit weighted running mean.

In the *method of differences*, successive differences are tabulated as follows.
Original series: $a, b, c, d, e, f, g, \ldots$.
First differences: $a', b', c', d', e', f', g', \ldots$, where $a' = b - a$, $b' = c - b$, $c' = d - c$, etc.

Second differences: a'', b'', c'', d'', e'', f'', g'', ..., where $a'' = b' - a'$, $b'' = c' - b'$, $c'' = d' - c'$, etc.

The differencing process is continued until the differences become small and fairly constant, with a definite trend. The mean value of the last difference is then found, and smoothed values of the original series are obtained by working back successively through the differences. The method assumes that the original series can be expressed by a polynomial; since most climatological time-series can be represented adequately by a third-order polynomial, it is rarely necessary to go beyond the third difference.

2.3. Besson's coefficient is evaluated as

$$R_B = \left(\frac{1 - p}{1 - p_1}\right) - 1,$$

where p is the general probability the event will happen, and p_1 is the probability that it will happen twice in a row.

In a long time-series of N days, the expected number of runs of an event that lasts for at least 1, 2, 3, 4, ... days, if there is no persistence, is given by Nqp, Nqp^2, Nqp^3, Nqp^4, ..., where p is the probability that the event occurs on a given day, and q is the probability that it does not occur on that day. The total number of expected runs if there is no persistence is equal to Nqp. In a short time-series of N days, the expected number of runs of n days in length, if there is no persistence, is

$$2p^n q + (N - n - 1)p^n q^2.$$

When N is large, this reduces to Nqp.

2.4.

$$R_A = \left(\frac{\dfrac{\sigma}{\sqrt{2}}}{\sigma_d}\right) - 1,$$

where σ_d is the standard deviation of the differences between each observation and the next succeeding observation in the time-series, and σ is the standard deviation of the observed series. The quantity $1 + R_B$ is also termed the *persistence ratio*. The 95% confidence limits of the persistence ratio are given by

$$\frac{1}{\left(1 \pm 1.96\sqrt{\dfrac{p}{Nq}}\right)},$$

symbols as defined in Appendix 2.3.

2.5. According to R. P. Waldo Lewis and D. H. McIntosh, *MM*, 81 (1952), 242, if the standard deviation of the original series is σ, then the standard deviation of a set of means of n consecutive terms is $\dfrac{\sigma}{\sqrt{n}}$ if there is no persistence, but is equal to

$$\left[\frac{\sigma^2}{n}\left(1 + \frac{2}{n}\{(n - 1)r_1 + (n - 2)r_2 + \cdots + r_{n-1}\}\right)\right]^{1/2}$$

if there is persistence, where r_1, r_2, r_3, ... represent the autocorrelation coefficients between the terms in the original series that are two terms, three terms, four terms, and so on, apart. If the original series is u_1, u_2, u_3, ..., u_N, where N is very large, then

$$r_1 = \frac{u_1 u_2 + u_2 u_3 + \cdots + u_{N-1} u_N}{(N-1)\sigma^2}$$

and

$$r_2 = \frac{u_1 u_3 + u_2 u_4 + \cdots + u_{N-2} u_N}{(N-2)\sigma^2}.$$

2.6. The *Fourier Series* for any variable x is

$$f(x) = \frac{1}{2} a_0 + (a_1 \cos x + b_1 \sin x) + (a_2 \cos 2x + b_2 \sin 2x)$$

$$+ (a_3 \cos 3x + b_3 \sin 3x) + \cdots + (a_n \cos nx + b_n \sin nx)$$

where a_0, a_1, b_1, a_2, b_2, a_3, b_3, ..., a_n, b_n, are constants and may take any value between $-\infty$ and $+\infty$. The function $f(x)$ has period of 2π; i.e., $f(x)$ extends from $-\pi$ to $+\pi$. The *Fourier coefficients*, a_n^*, b_n^*, of the function $f(x)$ are given by

$$a_n^* = \frac{1}{\pi} \int_0^{2\pi} f(x) \cos nx \, dx$$

and

$$b_n^* = \frac{1}{\pi} \int_0^{2\pi} f(x) \sin nx \, dx,$$

for all values of n. If *phase angles* are introduced, then the series may be expressed as a single series of sines or cosines, for example,

$$y_t = \frac{1}{2} a_0 + A_1 \sin(x + \phi_1) + A_2 \sin(2x + \phi_2) + \cdots + A_k \sin(kx + \phi_k), \quad [A2.2]$$

in which ϕ_1, ϕ_2, ϕ_k, are the phase angles and A_1, A_2, A_k, are the *amplitudes* of the harmonics; y_t is the value of the variable at any time t, and $\frac{1}{2}a_0$ is the arithmetic mean of all the values the variable takes. The phase angles determine the values of x at which the extremes of y occur. In climatology, x usually refers to time, so that $x = \left(\frac{2\pi}{L}\right)t$, where $t = 1, 2, 3, \ldots$, and L is the length of the period of the observational record.

2.7. Harmonic analysis may be carried out in two ways:

1. If the series to be analyzed consists of either 12 or 24 equidistant values, a printed schedule may be used, which simplifies the computation very considerably. For such schedules, see *HSM*, pp. 342–45, and *MC*, pp. 123–25. Schedules for other sets of values may also be drawn up to suit particular investigations.

2. If no schedule is available, the series must be analyzed mathematically, as follows. The curve representing the observed data, it is assumed, can be represented by the algebraic sum of a finite number of sine curves of the form

$$y_t = A \sin[c(t - t_0) + \phi],$$

where c converts time to radians; $c = \dfrac{360}{L}$. Since maximum values of y_t occur when

$c(t - t_0) + \phi = 90°$, and minimum values when $c(t - t_0) + \phi = 270°$, the phase angle ϕ is decided by the times of the maxima and minima on the curve representing the periodicity. The complete curve, synthesizing all the different periodicities, is given by rewriting Equation A2.2 as

$$y_t = a_0 + a_1 \cos x + a_2 \cos 2x + a_3 \cos 3x + \cdots$$
$$+ b_1 \sin x + b_2 \sin 2x + b_3 \sin 3x + \cdots,$$

in which a, b, are the *components* of the various harmonics, and may be determined from

$$a_k = \frac{2}{L} \sum_{t=0}^{L-1} (y_t \cos kx)$$

and

$$b_k = \frac{2}{L} \sum_{t=0}^{L-1} (y_t \sin kx).$$

When the components of each harmonic have been determined, their amplitudes and phase angles may be found from the relations

$$\phi = \tan^{-1}\left(\frac{a}{b}\right)$$

and

$$A = \sqrt{a^2 + b^2}.$$

The quadrant in which the phase angle lies is decided by the signs of a and b. If $\frac{a}{b} = \alpha$ and the angle whose tangent is α is found from the table to be β, then β will obviously lie between 0° and 90°. For other possibilities, see Table A2.7.

The analysis is continued until all the appreciable harmonics have been computed. To test significance, *Brooks and Carruthers' criterion* may be applied (see *HSM*, p. 336). When all the significant harmonics have been determined, they should be plotted as separate curves, superimposed on one another, then added together algebraically to test whether the synthetic harmonic curve approximates the observed curve closely enough for the purpose in hand.

TABLE A2.7
Quadrant of Phase Angle as a Function of the Signs of the Harmonic Components

a	b	ϕ	Quadrant
+	+	β	First
+	−	$180 - \beta$	Second
−	−	$180 + \beta$	Third
−	+	$360 - \beta$	Fourth

Amplitudes and phase angles of different periodic phenomena that have the same frequency may be compared conveniently by means of the *harmonic dial*, which is really a method for representing a sine curve vertically. The phenomena to be compared must have periodicities of equal length; for example, the diurnal waves of pressure and temperature at a station could be compared, or the semiannual precipitation variations at different stations could be contrasted. See *HSM*, p. 207, for details.

2.8. Fourier integral analysis may be employed to search for periodicities in data that are not apparently periodic. According to S. Goldman, *Frequency Analysis, Modulation, and Noise* (New York, 1948), the *Fourier integral* of the function $f(x)$ is given by

$$\int_0^\infty \{a(\omega)\cos \omega t + b(\omega)\sin \omega t\}\, d\omega,$$

where $a(\omega)$—the function a of ω—$b(\omega)$ are the *Fourier transforms* of $f(x)$, analogous to the Fourier coefficients in the normal Fourier series. The Fourier transforms are defined as

$$a(\omega) = \frac{1}{\pi}\int_{-\infty}^{+\infty} f(t)\cos \omega t\, dt,$$

$$b(\omega) = \frac{1}{\pi}\int_{-\infty}^{+\infty} f(t)\sin \omega t\, dt,$$

in which $f(t)$ represents the Fourier series. In all these expressions, ω is the angular frequency (i.e., the frequency with which a value occurs) in radians. The expression for the Fourier integral can be rewritten as

$$\frac{1}{\pi}\int_0^\infty S(\omega)\cos[\omega t + \phi(\omega)]\, d\omega,$$

where $S(\omega)$ and $\phi(\omega)$ can be found from the original observations, as

$$S(\omega) = \sqrt{\left(\int_{-\infty}^{+\infty} f(t)\cos \omega t\, dt\right)^2 + \left(\int_{-\infty}^{+\infty} f(t)\sin \omega t\, dt\right)^2},$$

$$\tan \phi(\omega) = \frac{-\displaystyle\int_{-\infty}^{+\infty} f(t)\sin \omega t\, dt}{\displaystyle\int_{-\infty}^{+\infty} f(t)\cos \omega t\, dt}.$$

In the computation, $f(t)$ represents the observed curve of values, and figures are taken at equidistant points along it to evaluate the integrals. A graph of $S(\omega)$ against ω, on rectangular coordinates, will then show the frequency of the observed curve.

2.9. If the mean values of the sections of the series, each of length U, are a, b, c, d, e, ..., then the differences $(a - b)$, $(c - b)$, $(c - d)$, $(e - d)$, ... are formed and means of the successive differences are found, i.e., $\frac{1}{2}\{(a - b) + (c - b)\}$, $\frac{1}{2}\{(c - b) + (c - d)\}$, $\frac{1}{2}\{(c - d) + (e - d)\}$, ... These latter means are plotted on the Y-axis against time on the X-axis to form the periodogram. If, by inspection, it appears that a periodicity of length C and amplitude B exists in the graph, then

$$L_1 = \frac{2CU}{2U + C},$$

and

$$L_2 = \frac{2CU}{C - 2U},$$

where C is the length of the periodicity that apparently exists in the data. The phases and amplitudes of the periodicities of lengths L_1 and L_2 are then calculated. The amplitude for either L_1 or L_2 is

$$\frac{B}{2} \cdot \frac{\pi U}{L} \cdot \frac{1}{\sin^3\left(\dfrac{\pi U}{L}\right)}$$

where B is the amplitude of the apparent periodicity. The phase angle for L_1 is $\dfrac{2\pi - 2\pi t_m}{C}$ and the phase angle for L_2 is $\dfrac{2\pi t_m}{C}$ where t_m is the time of the first maximum of the inferred periodicity. To discriminate between L_1 and L_2, a new value of U is taken, and the whole process is repeated. The two curves are then compared.

2.10. If S_{i-2v}, S_{i-v}, S_i, S_{i+v}, S_{i+2v}, are successive nonoverlapping sums of V, then the combined sum equals $-S_{i-2v} + 2S_i - S_{i+2v}$.

2.11. If the series is $x_1, x_2, x_3, \ldots, x_n$, the correlation coefficient is given by

$$\frac{\sum x_i \cdot x_{i+l}}{(n - l)\sigma^2},$$

where l is the lag period and σ is the standard deviation of the series.

2.12. A mean difference is given by the expression

$$\sum_{i=1}^{n-1} \frac{|x_i - x_{i-l}|}{n - l}.$$

2.13. The autocorrelation coefficient of order s is defined as the correlation coefficient between the series $y_1, y_2, y_3, \ldots, y_i$, and the series $y_{1+s}, y_{2+s}, y_{3+s}, \ldots, y_{i+s}$; y_2' is the autocorrelation coefficient between the series $y_1, y_2, y_3, \ldots, y_{n-2}$ and the series $y_3, y_4, y_5, \ldots, y_n$.

2.14. Fuhrich's criterion for reality is $\dfrac{1}{2\sigma^2}$ where σ is the standard deviation of the first two transformed values. The smaller this criterion is, the more likely it is that the periodicity is genuine. If the value of the criterion approaches $n/2$, there being n values in the original time-series, then the reality of the periodicity is very doubtful.

2.15. Following J. L. Holloway, *AGP*, 4 (1958), 351, the frequency responses of the different smoothing functions are as follows.

Normal curve smoothing:

$$e^{-2\pi^2\sigma^2 f^2},$$

where f is frequency and σ is defined by the analytic form of the smoothing function defined by

$$\omega(f) = \frac{e^{-\frac{t'}{2\sigma^2}}}{\sqrt{2\pi\sigma^2}}.$$

Equally weighted running means:

$$\frac{\sin \pi fT}{\pi fT},$$

where T is the filtering interval.

Exponential smoothing:

$$\frac{1 + 4\pi^2 f^2 \lambda^2 - 2\pi i f\lambda}{1 + 4\pi^2 f^2 \lambda^2},$$

where λ is the lag coefficient. The phase error of the smoothed series is

$$\tan^{-1} - 2nf\lambda.$$

Binomial smoothing:

$$\cos^n \pi f \Delta t,$$

where Δt is the data interval. In binomial smoothing, the weights are proportional to the coefficients of the expression $(p + q)^n$.

Running means of N terms each used successively M times:

$$\left(\frac{\sin N\pi f \Delta t}{N \sin \pi f \Delta t}\right)^M.$$

2.16. For full details of power-spectral analysis, see textbooks dealing with information theory or communications engineering, e.g., R. M. Reza, *An Introduction to Information Theory* (New York, 1961). For a general account of the mathematical basis, see J. Van Isacker, *AGP*, 7 (1961), 189. The term "power" is used in a mathematical sense; its formal definition is given by J. H. Lassing and R. H. Battin, *Random Processes in Automatic Control* (New York, 1956), as follows. For any real-valued function of time, $x(t)$—e.g., a curve representing the variation of temperature x with time t at a particular location—the quantity x^2 is defined as the *instantaneous power* associated with x. The total energy associated with this function is defined to be

$$\int_{-\infty}^{\infty} x(t)^2 \, dt$$

when this integral converges. The *power-spectral density* of the function $x(t)$ is then defined to be the mathematical expectation (see Appendix 2.20) of the spectral densities of the individual functions comprising $x(t)$, the spectral density itself being calculated from the Fourier cosine transform of $x(t)$. The *power-spectral density* associated with $x(t)$ is given by

$$\lim_{T \to \infty} \frac{1}{T} \left| \frac{1}{\sqrt{2\pi}} \int_{-T}^{T} x(t) e^{-i\omega t} \, dt \right|^2, \qquad [A2.3]$$

in which T and $-T$ represent the time limits of the function.

If the function $x(t)$ possesses a constant power-spectral density, it is defined to be a *white noise*. In Equation A2.3, the term

$$\frac{1}{\sqrt{2\pi}} \int x(t)e^{-i\omega t} \, dt,$$

if integrated between $+\infty$ and $-\infty$ rather than between T and $-T$, is the Fourier transform $a(\omega)$ of the function $x(t)$.

Mathematically, the power spectrum for a particular series (e.g., the curve representing the variation of temperature with time at a station) can be derived from the autocorrelation function for the series (i.e., the curve representing the values of the autocorrelation coefficients between different elements of the original series, for different lags), because the power spectrum and the autocorrelation function are both Fourier transforms. If $S(f)$ represents the power spectrum—i.e., the curve of the spectrum (f = frequency)—and $R(\tau)$ represents the autocorrelation function (τ = lag), then the appropriate transforms are

$$S(f) = 4 \int_0^\infty R(\tau) \cos 2\pi f\tau \, d\tau, \qquad\qquad [\text{A2.4}]$$

$$R(\tau) = \int_0^\infty S(f) \cos 2\pi f\tau \, df. \qquad\qquad [\text{A2.5}]$$

The autocorrelation curve can, of course, be easily constructed, and the power spectrum can be found by making a transform of it. This is facilitated by the fact that Equation A2.4 can be proved equivalent to

$$S(f) = \lim_{T\to\infty} \frac{1}{T} \left| \int_{-\frac{1}{2}T}^{\frac{1}{2}T} u(t)e^{-2\pi ift} \, dt \right|^2 \qquad\qquad [\text{A2.6}]$$

where T is the length of the observed series, and $u(t)$ represents the observed curve. Using lags of up to $\tau = T$ will thus enable the power spectrum to be found from Equation A2.6, which will give the estimates of the continuous spectrum at frequencies of $\frac{1}{T}, \frac{2}{T}, \frac{3}{T}, \frac{4}{T}, \frac{5}{T}$, etc. The observed series is assumed to be stationary (i.e., exhibiting no long-term trend), and its time average is assumed to be zero.

2.17. The classic work dealing with computational aspects of spectral analysis is R. B. Blackman and J. W. Tukey, "The measurement of power spectra," *Bell System Technical Journal*, vol. 37 (1958). For examples of meteorological computations, see H. A. Panofsky and R. A. McCormick, *QJRMS*, 80 (1954), 546. The Woodstock data is from H. A. Landsberg, J. M. Mitchell, and H. L. Crutcher, *MWR*, 87 (1959), 283.

Briefly, the analysis involves forming mean lag products $x_i \ldots x_{i+L}$ of the variable x at different time intervals L, subjecting these lag products to a cosine transform, and then smoothing the resulting values. A simplified analogy to a power spectrum may be completed by finding the *covariance* exhibited by the values in the series of observations for different lags. To produce such a spectrum, we draw up two columns of figures from the original data: column A gives the figures as actually recorded; column B gives the values m time units later, i.e., with a lag of m. If there are n values in each column, and x denotes the difference between each individual value in column A and the arithmetic mean for all the values in that column, and y has a similar meaning for column B, then the covariance is

$$\frac{\sum (xy)}{n}.$$

If, therefore, we wish to produce a simplified power spectrum of ten years' records of daily mean temperature at a station, we determine the covariance between temperatures on days separated by lags of 1, 2, 3, 4, 5, ..., up to, say, 500 days. A graph on rectangular coordinates, of covariance squared as ordinate against lag in days as abscissa, then provides the required spectrum. The chances are that the curve will be extremely irregular, and so some form of smoothing is desirable. Maxima in the curve represent periodicities in the data.

For accurate work, it is necessary to compute the power spectrum more rigorously. The first step involves transforming the original time-series into a series that has a zero average and no trends. This is done by *prewhitening* the original series according to the expression

$$\tilde{X}_t = X_t - 0.6X_{t-1},$$

in which X_t represents the original time-series ($t = 0, 1, 2, 3, ...$ spaced at unit intervals) and $\tilde{X}_t$ represents the series after prewhitening. In the transformed series, the index t commences at 1 instead of at 0. The second step involves calculating mean lagged products (with an adjustment for the mean) from the prewhitening series. A mean lagged product C'_r is defined by

$$C'_r = \frac{1}{n-r} \sum_1^{n-r} \tilde{X}_t \cdot X_{t+r} - \left(\frac{1}{n}\sum_1^n \tilde{X}_t\right)^2,$$

in which n is one less than the number of discrete observations, and $r = 0, 1, 2, 3, ..., m$, where m is the longest lag. At this stage it may be necessary to make further adjustment for the effect of a linear trend; see Blackman and Tukey, *op. cit.*, for details. C_r is the autocovariance at lag $r\,\Delta\tau$ where $\Delta\tau$ denotes the time interval between lags. The third step involves computing the finite cosine series transform V_r, which is given by

$$V_r = C'_0 + 2 \sum_{q=1}^{m-1} C'_q \cdot \cos\frac{qr\pi}{m} + C'_m \cdot \cos r\pi,$$

in which C'_0, C'_q, C'_m, are mean lagged products for $r = 0$, for $r = m$, and for $r = q$, respectively; V_r then forms the raw estimate of the power density of the variable at a frequency of f hertz, given (for equispaced, discrete data points) by

$$f = \frac{r}{(2m \cdot \Delta\tau)} = \left(\frac{r}{m}\right) f_N,$$

where

$$f_N = \frac{1}{(2 \cdot \Delta\tau)}$$

is termed the *Nyquist frequency* or the *folding* frequency. The fourth step involves smoothing the raw estimates. The "hanning" smoothing function is given by

$$U_0 = 0.5\ V_0 + 0.5\ V_r,$$
$$U_q = 0.5\ V_q + 0.25\ (V_{q-1} - V_{q+1}),$$
$$U_m = 0.5\ V_{m-1} + 0.5\ V_m,$$

in which U_0 is the smoothed power density for $r = 0$, U_m is the smoothed power density for $r = m$, and U_q is the smoothed power density for $r = q$, where q is any intermediate value of the variable. The final step involves correcting the smoothed estimates for prewhitening and correction for the mean. The corrected final power densities are

$$U_0' = \frac{U_0\, n}{(n-m)\left(\dfrac{1.36 - 1.20 \cos 2\pi}{6m}\right)},$$

$$U_q' = \frac{U_q}{\dfrac{1.36 - 1.20 \cos 2\pi r}{2m}},$$

$$U_m' = \frac{U_m}{1.36 - 1.20 \cos\left(1 - \dfrac{1}{6m}\right)2\pi},$$

The U_0' estimate applies to the range just above zero frequency; the U_m' estimate applies to the range just below a frequency of 0.5 cycles per observation; and the U_q' estimates apply to the range in the vicinity of the frequency of $\dfrac{r}{2m}$ cycle per observation.

An important qualification is that the autocovariance function C_r cannot be determined for arbitrarily long lags, and estimates of C_r cannot be made for lags longer than the length of the data record. In practice, it is not desirable to employ lags longer than 5 to 10 per cent of the length of the record.

As an example of a computational scheme for estimating power spectra, that of Landsberg, Mitchell, and Crutcher, *op. cit.* (1959), may be cited. The procedure involves the following steps.

1. Computing *serial products*, SP, given by

$$SP = \sum_{i=1}^{n-p} (X_i - \bar{X})(X_{i+L} - \bar{X}),$$

where i indicates any particular value in the observational series, which consists of n discrete data points, L is the lag, and $\bar{X}$ is the arithmetic mean of all the X_i values in the series.

2. Computing *covariances*, R, given by

$$R = \frac{SP}{n - L}.$$

3. Computing *covariance ratios*, R', defined by

$$R' = \frac{R}{R_0}$$

and

$$R'_L = \frac{R_L}{R_0}.$$

4. Computing *line powers*, *LP*, defined by

$$(LP)_0 = \left(\frac{1}{2m}\right)(R'_0 + R'_m) + \frac{1}{m}\sum_{L=1}^{m-1} R'_L,$$

$$(LP)_h = \left(\frac{1}{m}\right)R'_0 + \left(\frac{2}{m}\right)\sum_{L=1}^{m-1} R'_L \cos Lh\frac{\pi}{m} + \left(\frac{1}{m}\right)R'_m \cosh \pi,$$

$$(LP)_m = \left(\frac{1}{2m}\right)[R'_0 + (-1)^m R'_m] + \frac{1}{m}\sum_{L=1}^{m-1}(-1)^L R'_L,$$

where $(LP)_0$ is the first line power (i.e., the power density for minimum lag), $(LP)_m$ is the line power for maximum lag, and $(LP)_h$ is the line power for any intermediate lag h.

5. *Smoothing* the line powers, by

$$(LP)'_0 = 0.54\,(LP)_0 + 0.46\,(LP)_1,$$

$$(LP)_h = 0.54\,(LP)_h + 0.23\,\{(LP)_{h-1} + (LP)_{h+1}\},$$

$$(LP)'_m = 0.54\,(LP)_m + 0.46\,(LP)_{m-1}.$$

A graph of $(LP)'$ (where $r = 0, 1, 2, \ldots, h, \ldots, m$) against lag then forms the completed power spectrum.

2.18. It is necessary, at this point, to emphasize the importance of knowing the mathematical basis of even the most elementary types of statistical analysis before attempting to carry out an investigation. There are many books that provide the basic knowledge, e.g., *TYS*, and P. G. Hoel, *Introduction to Mathematical Statistics* (New York, 3d ed., 1962). In particular, the distinction between *discrete* and *continuous* quantities is important right from the start, because the various measures and parameters are defined in different ways, depending upon whether or not the quantity to be investigated is continuous. In general, if the quantity we are investigating is variable, it is termed a *variate* if it possesses a frequency distribution, and a *random variable*, if, for any given number k, the probability that the variable has a value equal to or less than k is at least theoretically calculable. A *discontinuous (discrete)* variate can take only some of the values in its range of variation, whereas a *continuous* variate may assume any of the values within its range of variation. For purposes of convenience, and also because no measurement is ever 100 per cent precise, it may be advantageous to group the observations of what we know to be a continuous variate into classes that correspond to equal subranges of the variate. If we then label each of the classes with the value corresponding to the midpoint of its class interval, in effect we have converted a continuous variate into a discrete one. The reverse process, the conversion of a discrete variate into a continuous one, is exemplified by the procedure of sampling, in which we smooth a histogram to obtain an estimate of the population from which we assume the observed sample has been drawn.

2.19. The arithmetic mean of a set of numbers $a, b, c, d, \ldots$ is $\dfrac{a + b + c + d + \cdots}{n}$ where there are n numbers. The simplicity of this definition should not be allowed to blind us to the implications of the real meaning of the term *arithmetic mean*. In the first place, according to *TYS*, p. 27, the correct definition of the arithmetic mean for a discrete variate is given by

$$N\bar{x} = \sum_{i=1}^{k} f_i x_i,$$

where $\bar{x}$ is the arithmetic mean of the variate x, and the variate x takes the values x_i with frequencies f_i respectively $(i = 1, 2, 3, \ldots, k)$, and the total frequency

$$N = \sum_{i=1}^{k} f_i.$$

The correct definition of the arithmetic mean for a continuous variate is given by

$$\bar{x} = \int_{-\infty}^{+\infty} x\phi(x)\, dx,$$

where $\phi(x)$ is the *probability density* of the variate, and is measured by the ordinate at x of the probability curve $y = \phi(x)$. In practice, we may find $\bar{x}$ for a continuous variate by measuring the area under the curve $y = \phi(x)$ and then constructing a rectangle of the same area. The physical meaning of the arithmetic mean is that it is the point on which the frequency distribution describing the variate would balance if this distribution was plotted on cardboard and then cut out. In other words, $\bar{x}$ is the ordinate of the point representing the centroid of the area of the histogram, for a discrete variate.

For a discrete variate, the *geometric mean*, $\bar{x}_g$, is defined by:

$$\bar{x}_g = \sqrt[n]{a \cdot b \cdot c \cdot d \cdot e \ldots},$$

where the variate consists of n values in the series $a, b, c, d, \ldots$.

The *harmonic mean*, $\bar{x}_h$ is defined by:

$$\bar{x}_h = n\left(\frac{1}{a} + \frac{1}{b} + \frac{1}{c} + \frac{1}{d} + \frac{1}{e} + \cdots\right)^{-1},$$

where the variate is discrete and consists of n values in the series $a, b, c, d, e, \ldots$.

2.20. Following *TYS*, p. 42, the *expectation* $\varepsilon(x)$ of the variate x may be defined as follows.

For a discrete variate,

$$\varepsilon(x) \equiv \sum_{i=1}^{n} p(x_i) \cdot x_i,$$

where $p(x_i)$ is the probability of the variate taking the value x $(i = 1, 2, 3, \ldots, n)$ and the values x_i are all mutually exclusive. This expression for $\varepsilon(x)$ reduces to that for $\bar{x}$. Consequently, the expectation of a discrete variate is its arithmetic mean.

For a continuous variate,

$$\varepsilon(x) \equiv \int_{-\infty}^{+\infty} x\phi(x)\,dx,$$

where $\phi(x)$ is the probability density defining the distribution of x.

2.21. See *HSM*, p. 35, for the mathematical definition of the various moments. The concept derives from a physical analogue to the statistical case. In general, following *TYS*, pp. 31 and 39, the *r*th *moment* of a frequency distribution about its mean, m_r, is defined, for discrete variates, as

$$Nm_r \equiv \sum_{i=1}^{k} f_i(x_i - \bar{x})^r,$$

and for continuous variates as

$$m_r = \int_{-\infty}^{+\infty} (x - m_1')^r \phi(x)\,dx,$$

the notation being the same as in Appendix 2.19;

$$m_r' = \int_{-\infty}^{+\infty} x\phi(x)\,dx \qquad\qquad [A2.7]$$

is the *r*th moment about $x = 0$, and $\bar{x}$ is m_1 about $x = 0$. The second moment is defined as the variance of the distribution if the first moment is zero, and consequently, for discrete variates, the *variance*, σ^2, is defined by

$$N\sigma^2 = \sum_{i=1}^{k} f_i(x_i - \bar{x})^2,$$

and for continuous variates, σ^2 is defined by

$$\sigma^2 \equiv m_2' - (m_1')^2$$

where m_1' and m_2' are defined by Equation A2.7.

The third moment, measuring the *skewness* of a continuous variate, is

$$m_3 = \int_{-\infty}^{+\infty} (x - m_1')^3 \phi(x)\,dx.$$

Skewness, β_1, is defined by

$$\beta_1 = \frac{m_3'}{m_2^3}.$$

For a discrete variate, *Pearson's measure of skewness* is

$$\frac{\bar{x} - \text{mode}}{\sigma},$$

where σ is the standard deviation of the distribution. The kurtosis, β_2, of a distribution may be defined by

$$\beta_2 = \frac{m_4}{m_2^2}.$$

2.22. The *mean deviation*, μ, is the arithmetic mean of the absolute values of the differences from average, d_i. Following *MC*, p. 38,

$$\mu = \frac{1}{n} \sum_{i=1}^{n} |d_i|.$$

The *coefficient of variability* (or of *variation*), *CV*, is defined by

$$CV = \frac{100\sigma}{\bar{x}},$$

where σ is the standard deviation of the distribution and $\bar{x}$ is its arithmetic mean. The *relative variability*, *RV*, is defined by

$$RV = \frac{100AV}{\bar{x}},$$

where *AV* is the average variability of the distribution, defined as

$$AV = \frac{\sum |x_i - \bar{x}|}{n},$$

where the distribution consists of n observations x_i. Unlike *CV*, *RV*, and *AV*, which are independent of the order or sequence of items in the observational series, the intersequential variability, *SV* (see *MC*, p. 52), depends on the sequence of the items as well as on their magnitude.

$$SV = \frac{|x_1 - x_2| + |x_2 - x_3| + \cdots + |x_{n-1} - x_n|}{n - 1},$$

where the series is $x_1, x_2, x_3, \ldots, x_n$.

The standard deviation enables departures from the mean to be classified as follows. Values beyond -3σ or $+3\sigma$ from the mean are extremely subnormal. Values between -2σ to -3σ or $+2\sigma$ to $+3\sigma$ are subnormal. Values between $-\sigma$ to -2σ or $+\sigma$ to $+2\sigma$ are fairly subnormal. Values between $-\sigma$ to $+\sigma$ are normal.

2.23. Following *HSM*, p. 40, σ corrected for the most likely difference between the mean of the sample and the true mean is the standard error, *SE*, of the individual observations;

$$SE = \sqrt{\frac{\sum (x_i - \bar{x})^2}{(n - 1)}}.$$

2.24. Chi-square is defined by the expression

$$\chi^2 = \sum_{i=1}^{k} \frac{(o_i - e_i)^2}{e_i},$$

where there are k observed frequencies, o_1, o_2, o_3, $\ldots$, o_k, corresponding to which there are k expected frequencies, e_1, e_2, e_3, $\ldots$, e_k. If $\chi^2 = 0$, there is perfect agreement between observed and expected frequencies.

If an experiment involving the calculation of a value of χ^2 is repeated many times, always on the assumption that the expected values remain unchanged, and a value of χ^2 is computed for each experiment, then a histogram of χ^2 values will be obtained. There is a frequency curve (and hence a probability curve) that approximates this histogram, described by

$$y = C(\chi^2)^{\frac{v-2}{2}} e^{\frac{-\chi^2}{2}}, \qquad [A2.8]$$

where $v = k - 1$ (i.e., v is the number of degrees of freedom) and C is a constant that depends on v and is chosen so that the total area under the curve defined by this equation is equal to unity. Equation A2.8 describes the distribution of χ^2 on the assumption that χ^2 is a discrete variate, and tables of χ^2 can be drawn up from it; see H. L. Alder and E. B. Roessler, *Introduction to Probability and Statistics* (San Francisco, 4th ed., 1968), pp. 205ff, for details. For the equation describing the χ^2 distribution (and involving the *Gamma function*) on the assumption that χ^2 is a continuous variate, see *TYS*, p. 204. Tables of χ^2 may be found in most textbooks of elementary statistics.

If $\bar{x}$ is the mean of a sample of k observations of a variate i ($i = 1, 2, 3, 4, \ldots, k$), then for a given $\bar{x}$ there can only be $k - 1$ independent values of x, because once we have accounted for $k - 1$ items of the sample, the kth value must of necessity be determined. Thus we say that the sample has $k - 1$ *degrees of freedom*, i.e., one less than the number of observations. In other words, following Alder and Roessler, *op. cit.* (1968), the number of degrees of freedom for a given set of conditions is the maximum number of individual measurements of a variable that can freely be assigned (i.e., calculated or assumed) before the remainder of the measurements are completely determined.

2.25. See *ibid.*, pp. 292–97, for tables of the probability integral.

2.26. For the *SE* of the original observations, see Appendix 2.23. According to *SM*, p. 40, if only a small number of observations (<20) are available, the expression

$$\sigma = \sqrt{\frac{\sum (x_i - \bar{x})^2}{(n - 1)}}$$

should be used instead of

$$\sigma = \sqrt{\frac{\sum (x_i - \bar{x})^2}{n}}.$$

The *SE* of a sample mean is

$$\sqrt{\sum \left\{ \frac{(x_i - \bar{x})^2}{n(n - 1)} \right\}}.$$

The *SE* of a variance is

$$\sigma^2 \sqrt{\frac{(2 + \beta_2)}{\sqrt{n}}}.$$

For a normal distribution, this reduces to

$$\sigma^2 \sqrt{\frac{2}{n}}.$$

For other parameters, see *HSM*, p. 58. The probable error of an observation is the value of the deviation from the mean that (in a normal distribution) will be exceeded on half the possible occasions. $PE = 0.6745 \, \sigma$ for an individual observation, and $PE = 0.6745 \, \sigma_{SE}$ for a mean value, where σ_{SE} is the standard error of the sample mean.

2.27. "Student's t" is essentially a significance test for sampling investigations in which the samples are small. It is based on the t-distribution. Following Alder and Roessler, *op. cit.* (1968), p. 138, if the probability of occurrence of a quantity t is given by

$$y = c\left(\frac{1 + t^2}{v}\right)^{\frac{-v+1}{2}}, \qquad [A2.9]$$

then the distribution of the quantity is termed a *student's t-distribution*, where v is the number of degrees of freedom, and c is a constant chosen to make the total area under the probability curve equal to unity. See *TYS*, p. 153, for an explicit statement of this probability function for t which involves the *Beta function*. As v (and therefore, the number of observations) becomes larger and larger, the curves given by Equation A2.9 approach the normal probability curve; t is defined by the expression

$$t = \frac{(\bar{X} - \bar{x})}{S_B},$$

where $\bar{x}$ is the mean of a variate that satisfies the normal distribution, $\bar{X}$ denotes the means of all possible samples of n observations taken of the variate, and S_B is the best estimate of the standard deviation of the mean of all possible samples of n observations, defined by

$$S_B = \frac{S}{\sqrt{n}},$$

where S is the best estimate of the standard deviation of the variate (i.e., of the "population") when the n observations in a sample are known. S is defined by

$$S = \sqrt{\frac{\sum x^2}{(n - 1)}},$$

where x is the deviation of an individual measurement x_i of the variate (in a sample of n observations) from the sample mean. The introduction of S and S_B is necessary, because for small samples (30 or fewer measurements of the variate), it is not permissible to assume that the standard deviation of a sample may be taken as an approximation of the standard deviation of the population. "Student's t" tables are based on the probability function of t for v degrees of freedom. They may be found in most elementary textbooks of statistics.

2.28. The log-normal transformation is given by $u = a + b \log(x + c)$, and may be fitted provided the mean, standard deviation and coefficient of skewness of the distribution are known. The reversed log-normal transformation is fitted in the same manner as the log-normal transformation, but signs are reversed. See *HSM*, pp. 102, 105, and 130.

2.29. *HSM*, p. 97. Alder and Roessler, *op. cit.* (1968), pp. 292–97, provide convenient tables of the probability integral. These tables are based on the *standard normal probability curve*

$$y = \frac{1}{\sqrt{2\pi}} e^{\frac{-z^2}{2}},$$

which is obtained by substituting $z = \frac{x}{\sigma}$ in the equation for the normal frequency curve; z is a measure of deviation from the mean in standard units. The tables give the value of y corresponding to given values of z.

2.30. Following R. L. Schlaifer, *op. cit.* (1959), pp. 95–113, when smoothing extremely sparse data, we should adjust the cumulative frequencies rather than the actual frequencies, because it is obviously contrary to common sense to use the actual frequencies to estimate the long-run ones if the actual data is sparse. It may be shown that if a sample of n observations is taken from some distribution and is arranged in order of size, then the kth observation is a reasonable estimate of the $\frac{k}{(n + 1)}$ fractile of the distribution. In essence, the procedure with extremely sparse data therefore involves fitting a smooth cumulative probability curve to the data by eye, and then, because the curve is actually stepped, not smooth, obtaining a *grouped* probability distribution from the curve. I.e., instead of reading off probabilities directly from the curve, we read off cumulative probabilities at the left-hand edges of brackets representing grouped values of the variate.

2.31. Bayes' theorem, on which see T. Bayes, *Phil. Trans. Roy. Soc. London*, 53 (1763), 376, has important philosophical implications in statistics. See I. J. Good, *The Estimation of Probabilities* (Cambridge, Mass., 1965) for an account of modern Bayesian methods; see also E. P. Epstein, *JAM*, 1 (1962), 169, and E. J. Gumbel, *Statistics of Extremes* (New York, 1958). Bayes' theorem is useful, in particular, in the revision of probabilities in the light of new information. For example, suppose our initial information is that on x days out of n days, at a station, fog is recorded; i.e., $P(F) = x/n$, where $P(F)$ denotes the probability of fog. We know also that out of these days on which fog is recorded, the daily mean temperature is $\geq 40°F$ on a days and $< 40°F$ on b days. Let W stand for days with temperature $\geq 40°F$. Then $P(W \mid F)$ is the probability that a day is warm given that it is foggy, and $P(C \mid F)$ is the probability that a day is cold given that it is foggy.

$$P(W \mid F) = \frac{a}{x}$$

and

$$P(C \mid F) = \frac{b}{x}.$$

Suppose that we now receive additional information, i.e., that on y of the original n days, the daily mean pressure at the station in question exceeds 1020 mb, and that of these y days, c have temperatures $\geqslant 40°F$ and d have temperatures $< 40°F$. If $P(H)$ stands for the probability of high pressure, $P(H) = \dfrac{y}{n}$, and the probability that a day is warm given that it has high pressure is $P(W \mid H)$. Similarly, the probability that a day is cold given that it has high pressure is $P(C \mid H)$. $P(W \mid H) = \dfrac{c}{y}$ and $P(C \mid H) = \dfrac{d}{y}$. By means of Bayes' theorem, we may now determine $P(F \mid W)$, i.e., the probability that a day is foggy given that it is warm, and also $P(F, W)$, i.e., the probability that a day is both foggy and warm. Bayes' theorem for this case may be stated as

$$P(F \mid W) = \frac{P(F, W)}{P(W)} = \frac{P(F) \cdot P(W \mid F)}{P(F) \cdot P(W \mid F) + P(H) \cdot P(W \mid H)}.$$

$P(F \mid W)$ constitutes the *a posteriori probability* of fog, and $P(F)$ constitutes the *a priori probability* of fog. If we believe that we may make a reasonable intuitive guess at the value of $P(F)$, i.e., by assuming that it closely corresponds to the relative frequency $\dfrac{x}{n}$, then we may apply Bayes' theorem with confidence. If, on the other hand, we believe that $P(F)$ is itself a conditional probability, the problem becomes much more complicated, and one becomes involved in the difference of opinion between the subjectivist (Bayesians) and objectivist (non-Bayesian) schools of statistics. See Schlaifer, *op. cit.* (1959), Good, *op. cit.* (1965), and Epstein, *op. cit.* (1962), for discussions of these viewpoints.

2.32. Following Court, *op. cit.* (1952), the statistics may be defined as follows:

1. The probability that an event will not occur in any of N trials is $\left(1 - \dfrac{1}{\overline{T}}\right)^N$, where the probability of occurrence of the event in a single trial is p, and $\overline{T} = \dfrac{1}{p}$ is defined as the *return period*.

2. The probability of at least one occurrence in N trials is

$$1 - \left(1 - \frac{1}{\overline{T}}\right)^N.$$

3. The probability of occurrence for the first time on the Nth trial is

$$\frac{(\overline{T} - 1)^{N-1}}{\overline{T}^N}.$$

4. The probability of exactly H occurrences in N trials is

$$\left(\frac{N!}{[H!(N - H)!]}\right) \frac{(\overline{T} - 1)^{N-H}}{\overline{T}^N}.$$

2.33. On this use of the χ^2 test, see *HSM*, pp. 154–60. The important point to note about the correlation coefficient is that it only has meaning when it is computed to show the relationship between two attributes of the same population; i.e., a correlation coefficient must be computed between the variates in a bivariate distribution. This ensures that the variables being correlated are related in a physical sense. See *TYS*, Chapter 6, for a discussion of the properties of discrete bivariate distributions, and pp. 228–38 for the properties of continuous bivariate distributions.

2.34. Following *TYS*, p. 99, the *covariance of x and y* is defined by

$$CV_{XY} = m'_{11} - \bar{x}\bar{y},$$

where $\bar{x}$, $\bar{y}$, are the means of x and y, and m'_{11} is the first moment of x and the first moment of y about $x = 0$, $y = 0$, respectively. Practically, following *HSM*, p. 211, the covariance of x and y is given by $CV_{XY} = \dfrac{\sum (XY)}{N}$, where X, Y, are the deviations of x and y from their respective means, and there are N pairs of observations.

2.35. See *HSM*, p. 218, for details of the procedure. Essentially, we have two regression lines, one giving the regression of X on Y and the other the regression of Y on X, which both pass through the sample mean. The equations of these lines are $x = ry$, $y = \dfrac{x}{r}$, and the angle θ between them is given by $\tan \theta = \dfrac{(1 - r^2)}{2r}$; see *TYS*, p. 104. In these equations, r represents the correlation coefficient. It should be noted that, unlike the correlation coefficient, which is a measure of *interdependence* of X and Y, the regression between X and Y measures their *dependence*.

2.36. See *HSM*, pp. 227, 235–40. The *rank correlation coefficient*, r_{rc} is given by

$$r_{rc} = \frac{1 - 6 \sum d^2}{N(N^2 - 1)},$$

where d is the difference between the rank numbers of two corresponding observations and there are N pairs of observations.

2.37. See *HSM*, p. 245. The difference between the use of multiple correlation coefficients (i.e., the technique of *multivariate regression*) and the use of partial correlation coefficients should be noted. Multivariate regression studies the influence on one variate in a distribution of the other variates it contains. Partial correlation assesses the interdependence of two of the variates in a multivariate distribution, after the influence of all the other variates in the distribution have been eliminated. For a trivariate distribution, the coefficient of multiple correlation of x_1 with x_2 and x_3, denoted by $r_{1.23}$, is given by

$$r_{1.23} = \sqrt{\frac{r_{12}^2 + r_{13}^2 - 2r_{12} \cdot r_{23} \cdot r_{31}}{1 - r_{23}^2}},$$

where r_{12} is the correlation coefficient between x_1 and x_2, etc.; see *TYS*, p. 123.

2.38. See *HSM*, Chap. 11, for a comprehensive account. The essential difference between a stereogram and a frequency (or correlation) surface is that the former is the three-dimensional representation of a discrete bivariate distribution, the latter of a continuous bivariate distribution. The stereogram is based on actual measurements, and constitutes a discrete sample. When we smooth the stereogram to obtain the population from which the sample was drawn, we transform the distribution from a discrete to a continuous one.

2.39. For estimating from frequency tables, see *HSM*, p. 183. The "module" of a wind is its speed. The "velocity" of a wind is a vector quantity, implying a direction as well as a speed. The module of the wind is thus the magnitude of its velocity. Following *HSM*, p. 177, the vector mean wind $\mathbf{V}_M$ is given as follows. Module of $\mathbf{V}_M$ is

$$\mathbf{V}_M = \frac{1}{N} \sqrt{(\textstyle\sum V_N)^2 + (\textstyle\sum V_E)^2},$$

and direction of $\mathbf{V}_M$ is

$$\tan \alpha = \frac{\sum V_E}{\sum V_N},$$

where $\sum V_N$, $\sum V_E$, are sums of the north and east components of the actual wind; i.e., if V is the speed of the wind on any one occasion, and θ is its direction measured clockwise from north, then $V_N = V \cos \theta$ and $V_E = V \sin \theta$. The vector sum $\sqrt{(\sum V_N)^2 + (\sum V_E)^2}$ is termed the *resultant wind*, and its direction is given by α.

2.40. Following *HSM*, p. 190, the equation for a frequency surface is, for a normal distribution,

$$z = \frac{N}{2\pi\sigma_x \sigma_y} \exp\left[-\frac{1}{2} \left\{ \left(\frac{V_x}{\sigma_x}\right)^2 + \left(\frac{V_y}{\sigma_y}\right)^2 \right\} \right],$$

where z is the height of the ordinate of the frequency surface. The surface refers to a vector $\mathbf{V}$, which may be separated into components V_x and V_y; σ_x, σ_y, are the standard deviations of V_x, V_y, respectively, and v_x, v_y, are the deviations of V_x, V_y, respectively from the mean. N is the number of vectors.

$$V = \sqrt{V_x^2 + V_y^2},$$

where $v_x = V_x - \bar{V}_x$ and $v_y = V_y - \bar{V}_y$.

2.41. The S.V.D., written as σ, is given by

$$\sigma^2 = \frac{\sum V^2}{N} - V_R^2,$$

where V is the module of the individual vector wind observations, V_R is the module of the resultant wind, and N is the number of observations. The *constancy q* of a set of vectors is given by

$$q = 100 \frac{V_R}{V_S},$$

where V_S is the *scalar mean wind*; i.e., V_S is the arithmetic mean of the modules of the vectors, and has no direction.

2.42. Following R. W. Lenhard, Jr., A. Court, and H. A. Salmela, *JAM*, 2 (1963), 99, R_S and R_T are defined as

$$R_S = \frac{(S_{ux} + S_{vy})}{\sqrt{(S_u^2 + S_v^2)(S_x^2 + S_y^2)}},$$

$$R_T = \frac{(S_{vx} - S_{uy})}{\sqrt{(S_y^2 + S_v^2)(S_x^2 + S_y^2)}},$$

where

$$S_u^2 = \frac{\sum (u_i - \bar{u})^2}{n},$$

$$S_{uv}^2 = \frac{\sum (u_i - \bar{u})(v_i - \bar{v})}{n},$$

$$\bar{u} = \frac{\sum u_i}{n}.$$

In these equations, u, v, and x, y, represent the orthogonal components of two vectors W, Z.

2.43. In effect, the *method of least squares* involves fitting a line to a set of n points in such a way that $\sum (y - y_e)^2$ has the smallest possible value, y being the ordinate of the actual value of the variate, and y_e being the ordinate of the estimated value of y read from the line of best fit. The equation of the line of best fit is (see Figure A2.43)

$$y_e = mx + c.$$

Consequently, the problem is to determine the constants m and c in the above equation, for a given sample of n pairs of values of x and y_e, in such a way that $\sum (y - y_e)^2$ is minimized. By means of differential calculus, it may be proved (see *TYS*, p. 103) that m and c must be the solution to the linear equations

$$\sum y_e = m \sum x + cn,$$

$$\sum xy_e = m \sum x^2 + c \sum x.$$

Following Alder and Roessler, *op. cit.* (1968), p. 180, the solution of these equations

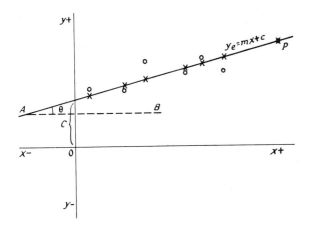

FIGURE A2.43.
Circles mark the observed value of variate y, crosses the estimated value of variate y (i.e., y_e) according to the least squares hypothesis; m is the tangent of angle θ, where AB is parallel to the x axis. Note that at P, observed and estimated values of y coincide.

is given by

$$m = \frac{\sum xy - n\bar{x}\bar{y}}{\sum x^2 - n\bar{x}^2}$$

and

$$c = \bar{y} - m\bar{x},$$

where x, y, represent corresponding pairs of measurements of the variables X and Y, and $\bar{x}$, $\bar{y}$, are the arithmetic means of all the X, Y, measurements, respectively. Substitution of these values of m and c above gives the required *least-squares estimate* of the relation between X and Y.

2.44. After taking logarithms, the decay curve $y = ax^{-b}$ becomes

$$\log y = \log a - b \log x.$$

Similarly, the growth or exponential curve $y = ae^{bx}$ becomes

$$\log y = \log a + bx \log e.$$

2.45. The *contingency ratio*, R_{ij}, is defined by

$$R_{ij} = \frac{O_{ij}}{E_{ij}},$$

where i, j, represent any specific combination of predictor and predictand classes, respectively, O represents the observed frequency, and E the expected frequency, of occurrence of the i, j, combination of classes.

The *normalized contingency ratio* R_{ij} is defined as

$$R'_{ij} = 1 + D'_{ij},$$

where

$$D'_{ij} = D_{ij}f_{ij},$$

$$D_{ij} = R_{ij} - 1,$$

$$f_{ij} = \frac{E_{ij}}{E_0},$$

D_{ij} representing the deviation from the "chance" value 1.00 for any given predictor-predictand class combination, and f_{ij} is the reduction factor for each class combination.

$$E_o = \frac{N}{a^2},$$

where a is the number of predictand classes. Normalizing is accomplished by computing the factor F_k for each contingency table, where

$$F_k = \sqrt{\frac{N_k}{N_o}},$$

there being N_k observations included in each contingency table k, and N_o is an arbitrary value of N, usually taken to be the largest value of N in all the tables. D_{ij} for each table is then multiplied by F_k to give $(D_{ij})_k$, and the final (twice-normalized) contingency ratios $(R_{ij})_k$ are given by

$$(R_{ij})_k = (D_{ij})_k + 1.$$

2.46. The elaborate procedure follows E. T. Whittaker and G. N. Robinson, *The Calculus of Observations* (London, 1924). See *HSM*, pp. 162–165, for details of the simple method. Suppose the observed point-values at A, B, and C are x, y, and z respectively (see Figure A2.46). Then P, Q, and R, must have the values

$$\frac{y+z}{2}, \frac{x+y}{2}$$

and

$$\frac{x+z}{2}$$

respectively. By a simple geometrical property of the triangle ABC, the lines AP, BR, and CQ must intersect at point S, and $AS = 2SP$, etc. Hence the value of the variable at S may be determined. The process may be repeated, including the new points S_1, S_2, S_3, etc., in the triangulation.

2.47. The *Thiessen polygon method*, on which see A. H. Thiessen, *MWR*, 39 (1912), 1082, helps measure precipitation averages for large areas: it is more accurate than a straightforward arithmetic mean, but makes no allowance for orographic effects. The method involves constructing perpendicular bisectors of straight lines that connect the rain gauge locations on a map. The constructed lines divide the area into polygons, each polygon enclosing a rain gauge, such that any point within a given polygon is nearer to its gauge than to any other gauge. The reading of each gauge is then weighted in proportion to the area of the associated polygon to obtain an areal estimate of precipitation amount.

2.48. See *MC*, pp. 258–70, for information on conventional schemes. Note that a knowledge of differential equations is necessary for valid plotting of isolines on a map. A favorite occupation of geographers is to make accurate maps and graphs of climatic

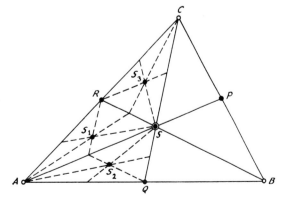

FIGURE A2.46.
P,Q,R, are the midpoints of *BC*, *AB*, and *AC*, respectively.
○ Original points of observation.
◉ Point value inferred by first order triangulation.
● Point value inferred by second order triangulation.

facts, and then to compare them subjectively. See S. Gregory, *QJRMS*, 80 (1954), 610, on the annual rainfall areas of southern England as an example. There is no need for this subjective element in the analysis, and to introduce it greatly reduces the scientific value of the results of the investigation. A climatic map is a type of mathematical structure, and numerous mathematical techniques enable us to specify the properties of such a structure, and hence to predict further climatic facts, without recourse to subjective judgements. In particular, isopleth maps involve gradients, which are vector quantities, and hence are amenable to study by vector methods.

2.49. The nomogram principle depends, usually, on the properties of similar triangles. Graduated scales are drawn on a sheet of paper so that a ruler laid across the scales will cut them in such a manner as to satisfy a given equation. For mathematical details, see *HM*, p. 137; for examples, see *MC*, p. 72, and W. W. Lamoreux, *MWR*, 90 (1962), 26, who reproduces a nomogram for the determination of free-water evaporation as a function of meteorological factors.

2.50. Averaging wind direction is difficult because our numbering system corresponds to points on a line, but direction corresponds to points on a circle. The correct average of directions 50° and 350° is 20°, whereas the simple average, i.e., $\dfrac{(50 + 350)}{2}$, comes to 200°. To obtain the correct average of a direction, one must transform the points on the circle to points on a line; i.e. the circle must be broken at 360°, otherwise, averaging over this breakpoint will give erroneous results.

3.1. J. C. Rodda, *W*, 17 (1962), 54, used annual precipitation data (*p* inches) from standard rain gauges, establishing correlations with altitude (*h* feet) above mean sea level of the site of each gauge and exposure (*e*), the latter being defined as the angle (in radians) extending to the circumference of a circle of 1 mile in radius centered on the gauge, in which no topographic feature was higher than or as high as the gauge. The general statistic model used was

$$p = A + Bh - Ce,$$

where *A*, *B*, *C*, had the values 46.88, 0.02, and 3.29, respectively, for 1958, and 34.91, 0.02, and 3.16, respectively, for 1959. For an example of an empirical statistical model of a severe rainstorm producing 10 inches of rain in southern Illinois, August 16–17, 1959, see F. A. Huff and S. A. Changnon, Jr., *JAM*, 3 (1964), 587.

3.2. Full details of the Π theorem may be found in H. Rouse, *Fluid Mechanics for Hydraulic Engineers* (New York and London, 1938). Originally devised by Buckingham in 1915, the derivation first assumes that any variable A_1 depends only on the independent variables $A_2, A_3, A_4, \ldots, A_n$, so that

$$A_1 = f(A_2, A_3, A_4, \ldots, A_n).$$

The variables may also be grouped in another functional relationship equal to zero:

$$f'(A_1, A_2, A_3, A_4, \ldots, A_n) = 0.$$

The Π theorem then states that if the *n* variables may be described in terms of *m* fundamental dimensional units, they may be grouped into $n - m$ dimensionless Π terms thus:

$$\phi(\Pi_1, \Pi_2, \Pi_3, \ldots, \Pi_{n-m}) = 0. \qquad \text{[A3.1]}$$

In each Π term there are $m + 1$ variables, only one of which needs be changed from term to term. Assuming that fluid motion depends only on linear dimensions and on the kinematic and dynamic characteristics of the flow, then

$$f'(a, b, c, d, V, \Delta p, \rho, \gamma, \mu, \sigma, e) = 0 \qquad \text{[A3.2]}$$

where *a*, *b*, *c*, *d*, are the linear dimensions of the object(s) determining the pattern of flow; V, ρ, γ, μ, e, represent the velocity, density, specific weight, dynamic viscosity, and elastic modulus of the fluid $\left(\dfrac{\gamma}{p}=g,\right.$ the acceleration due to gravity, and

$\dfrac{\mu}{\rho}=v$, the kinematic viscosity$\Big)$, Δp is the pressure gradient, and σ represents surface tension. Since three fundamental dimensions are used in fluid mechanics, Equation A3.1 must comprise eight terms with three variables (represented by length a, velocity V, and density ρ) common to each. That is,

$$\phi(\Pi_1, \Pi_2, \Pi_3, \Pi_4, \Pi_5, \Pi_6, \Pi_7, \Pi_8) = 0,$$

in which

$$\Pi_1 = \frac{a^{x_1} V^{y_1} \rho^{z_1}}{b},$$

$$\Pi_2 = \frac{a^{x_2} V^{y_2} \rho^{z_2}}{c}, \ldots,$$

$$\Pi_8 = \frac{a^{x_8} V^{y_8} \rho^{z_8}}{e},$$

where x, y, z, are to be determined. By replacing each variable by the corresponding combination of the dimensions L, T, and M or F, simultaneous linear equations result which show that

$$\Pi_1 = \frac{a}{b},$$

$$\Pi_2 = \frac{a}{c},$$

$$\Pi_3 = \frac{a}{d},$$

$$\Pi_4 = \frac{V^2}{\left(\dfrac{\Delta p}{\rho}\right)},$$

$$\Pi_5 = \frac{V^2}{a}\bigg/{\left(\dfrac{\gamma}{\rho}\right)}.$$

$$\Pi_6 = \frac{V_a}{\left(\dfrac{\mu}{\rho}\right)}.$$

$$\Pi_7 = \frac{V^2 a}{\left(\dfrac{\sigma}{\rho}\right)}$$

and

$$\Pi_8 = \frac{V^2}{\left(\dfrac{e}{\rho}\right)}.$$

These groups of dimensionless variables are substituted in order in Equation A3.2, giving

$$\phi\left(\frac{a}{b}, \frac{a}{c}, \frac{a}{d}, \frac{V^2}{\dfrac{\Delta p}{\rho}}, F, Re, W, C\right) = 0,$$

where F (the Froude number), Re (the Reynolds number), W (the Weber number), and C (the Cauchy number) correspond to Π_5, Π_6, Π_7, and Π_8, respectively.

$$F = \frac{V^2\rho}{a\gamma}; \qquad\qquad W = \frac{V^2 a\rho}{\sigma};$$

$$Re = \frac{V a\rho}{\mu}; \qquad\qquad C = \frac{V^2\rho}{e}.$$

If λ is the numerical value of the selected scale length in the model, then model and prototype will be similar if the model to prototype scales are combined as follows. For F:

$$\text{time} = \sqrt{\frac{\lambda}{\left(\dfrac{\gamma_s}{\rho_s}\right)}},$$

$$\text{velocity} = \sqrt{\frac{\lambda}{\left(\dfrac{\gamma_s}{\rho_s}\right)}},$$

$$\text{pressure} = \lambda\gamma_s$$

For Re:

$$\text{time} = \frac{\lambda^2}{\left(\dfrac{\mu_s}{\rho_s}\right)},$$

$$\text{velocity} = \frac{\left(\dfrac{\mu_s}{\rho_s}\right)}{\lambda},$$

$$\text{pressure} = \frac{\mu_s^2}{(\lambda^2\rho_s)},$$

where μ_s, γ_s, and ρ_s are the numerical values of the scales for various properties corresponding to a numerical value of λ.

3.3. The Euler number E is defined by

$$E = V\left(\frac{2\Delta p}{\rho}\right)^{-1/2},$$

where Δp is the differential pressure, V the velocity, and ρ the density of the air.

3.4. *Re* decreases with height for a given rate of balloon ascent. Many balloons have diameters and rates of ascent such that, at some part of their ascent, *Re* lies within the critical range for changeover from turbulent (below) to laminar (above) flow. Beyond this critical range, the rate of ascent considerably decreases. This makes it impossible to derive a simple formula connecting the rate of ascent of a balloon with its weight and free lift over all ranges of altitude. For high-altitude 2-kilogram balloons, the critical *Re* value occurs at a height of 10 to 15 km.

Equating the weight of a spherical hailstone to its drag (proportional to $C_d V^2$, where C_d is the drag coefficient and *V* its velocity), the change in the drag coefficient of the sphere in the critical *Re* range creates an upper limit to its rate of fall, giving its terminal velocity. This upper limit is approximately 200 feet per sec, which occurs with a mass of 1.5 lb, and which therefore determines the largest possible size for a hailstone. The change to turbulent flow occurs when the hailstone falls to a level for which *Re* has the critical value, and produces a sharp drop in the value of the drag coefficient, the turbulence reducing the fall of pressure immediately behind the hailstone.

3.5. The *Richardson number* is defined by

$$Ri = \frac{g\left(\dfrac{dT}{dz} + \Gamma\right)}{\left(\dfrac{d\bar{u}}{dz}\right)^2 T},$$

where $\bar{u}$ is the mean horizontal wind speed at height *z*, Γ is the appropriate adiabatic lapse-rate, and *T* is absolute temperature.

3.6. The *Mach number* is defined by $M = \dfrac{V}{a}$, where *V* is the velocity of the air and *a* is the speed of sound.

$$M = V\left(\frac{\gamma_q R}{\theta}\right)^{-1/2},$$

where θ is the static temperature, *R* the gas constant, and $\gamma = 1.4$ (the ratio of the specific heats of a gas). The *Rossby number* is defined by

$$Ro = \frac{U}{a\Omega},$$

where *U* is windspeed, *a* is a characteristic length, and Ω is a characteristic frequency (e.g., the Earth's angular velocity).

3.7. A classic reference for these instruments is L. Prandtl and O. G. Tientjens, *Applied Hydro- and Aeromechanics* (New York and London, 1934). The Pitot tube, introduced by Pitot in 1732, is a tube with its open end facing the air or fluid (see Figure A3.7,A). The velocity of flow along streamline *x-y* is decreased to zero and transformed into a pressure, according to Bernouilli's theorem. The point of contact *y* then forms a stagnation point, at which pressure $p(= p_0 V)$ is zero. From Bernouilli,

$$\frac{p_1}{w} + \frac{V^2}{2g} = \frac{p_0}{w},$$

where p_1 is the upstream pressure, *V* is the approach velocity, and *w* is the vertical component of velocity. The Pitot tube measures the stagnation head of pressure $\left(\dfrac{p_0}{w}\right)$, while the static head $\left(\dfrac{p_1}{w}\right)$ may be measured by means of a static hole in the wall of the tube, and *V* may thus be found. In the *Pitot-static tube*, a static pressure tube jackets the Pitot tube (see Figure A3.7,B). Static holes are located so that the slight pressure decrease caused in the flow by the tube is exactly compensated for by the slight pressure increase due to stagnation on the stem. These tubes may be made very minute, so they can be located within 0.02 inch of the fluid boundary; they are insensitive to yaw, and may enclose an angle as large as 19° with the stream flow, resulting in an error of only 1 per cent. See P. S. Barna, *Fluid Mechanics for Engineers* (London, 1957).

3.8. Expressed by

$$S = g\beta \frac{d^2}{U^2},$$

where β is the static stability, U is the windspeed, and d is a representative length.

3.9. See Prandtl and Tientjens, *op. cit.* (1934). The Prandtl manometer gives the true total head of pressure, provided $\frac{ur}{v} > 30$, where u is the air velocity, r is the radius of the mouth of the tube, and v is the kinematic viscosity of the air. Only short lengths of hypodermic tubing are used, to minimize the effects of drag and to avoid excessive vibration when the instrument is exposed in an airstream.

3.10. Let u_* be defined by

$$\frac{u}{u_*} = \left(\frac{T}{T_*}\right)^{1/2},$$

where u is a velocity greater than u_* measured in the unobstructed tunnel, $\frac{u}{u_*}$ is the velocity ratio, T is the observed wind shear and T_* is the threshold value of the wind

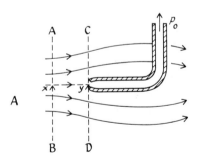

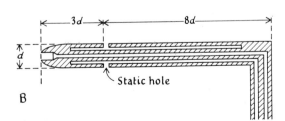

FIGURE A3.7.
Types of Pitot tube (after Barna).
A shows a Pitot tube. At *AB*, velocity $= V_1$, pressure $= p_1$; At *CD*, velocity $= 0$, pressure $= p$. Arrows show streamlines. B. shows a Prandtl tube, a combined Pitot static tube.

shear. Then T becomes T_* at the sand boundary, enabling u_* to be found; the effective velocity reduction is then equal to $100\left(1 - \dfrac{u}{u_*}\right)$ at this point.

3.11. Using the barrier height as the characteristic length, Re for the model came to between 5.3×10^4 and 8.4×10^4, and Re for the prototype between 48×10^4 and 88.2×10^4. For the comparisons the ratios $\dfrac{z}{H}$ and $\dfrac{V_f}{V_o}$ were plotted, z being the elevation of the point at which the velocities were measured above the tunnel, floor, or ground level, H the height of the snow fence, V_f the velocity aft of the fence, and V_o the velocity in an open tunnel or the velocity of the unobstructed wind in the field.

3.12. The velocity ratio $\dfrac{V}{V_0}$ is given by $\dfrac{V}{V_0} = \left(\dfrac{p}{p_0}\right)^{1/2}$, where V is the velocity at a given point with the model in position and V_o the velocity at the same point without the model in the tunnel, and p, p_0, are the corresponding values of the differential pressures.

3.13. The complete condition for similarity is that the following equation must be satisfied in both model and atmosphere:

$$\nabla^2 y_0' + \frac{1}{2}[(\nabla y_0')^2 - 1]\frac{d}{dy_0}\ln(u_0'^2\rho') - \frac{gL}{\bar{u}_2\bar{u}_0\rho'}\frac{d\rho'}{dy_0'}(y_0' - y') = 0,$$

in which

$$y_0' = \frac{y_0}{L}, \; x' = \frac{x}{L}, \; y' = \frac{y}{L}, \; u_0' = \frac{u_0}{\bar{u}}, \; \rho' = \frac{\rho}{\bar{\rho}}.$$

Here, u, v, are the velocities in the horizontal (x) and vertical (y) directions; $x_0 = u_0 t_0$, where u_0 is the upstream basic velocity and t_0 the time a given particle passes a certain point; $u_0(y_0)$ is the basic velocity distribution with height, y_0, being the height of a particle at time t; ρ represents density in the model, and potential density (i.e., density reduced to a standard pressure) in the atmosphere $\left(\text{in both cases } \rho = \dfrac{p}{y_0}\text{, so that}\right.$ $\dfrac{d\rho}{dt}$ is zero, and both the velocity and density distributions upstream depend only on $y\Big)$; L is a representative length, related to the horizontal dimensions of the orographic barrier; $\bar{u}$ and $\bar{\rho}$ are representative velocity and density, respectively, i.e., the average basic velocity or average density through the entire depth of the fluid or troposphere.

Long found by experiment that the middle term on the left-hand side of the similarity equation could be neglected. The condition for similarity then reduced to ensuring that

$$\frac{d\rho'}{dy_0'}\left(\frac{1}{F^2 u_0'^2 \rho'}\right)$$

has the same value in both model and prototype, where $F^2 = \dfrac{\bar{u}^2}{gL}$. Furthermore, the effect of the curvature of the vertical density profile was found to be negligible, provided the difference in density between the top and bottom of the fluid layer remained constant. Assuming a linear density gradient then reduces the similarity requirement to ensuring that the *modified Froude number Fi* is identical in both model and atmosphere; *Fi* is defined by

$$Fi = \frac{\bar{u}}{\left(g \frac{\Delta \rho}{\bar{\rho}} h\right)^{1/2}},$$

where h is the depth of the troposphere or model fluid.

3.14. The principle is that the image seen by the camera is derotated (hence there is no necessity for the camera itself to rotate) by optical means before it reaches the camera or observer, by means of a Thoma rotoscope, which uses the fact that the image seen along the longitudinal axis in a Dove (reversing) prism rotates at twice the rate of rotation of the prism about its axis.

3.15. Following *HSM*, pp. 70–84, the following parameters may be derived for the normal or nearly symmetric binomial distribution:

first moment $= np$,   standard error $= \sqrt{\dfrac{npq}{N}}$

second moment $= npq$,   standard deviation $= \sqrt{npq}$

third moment $= npq(q - p)$,   skewness $= \dfrac{(q - p)}{\sqrt{npq}}$

fourth moment $= npq[1 + 3pq(n - 2)]$,   kurtosis $= \dfrac{(1 - 6pq)}{npq}$

where p is the probability that an event will occur, $q = 1 - p$, there are n possible occasions on which the event may occur, and N trials are made.

The possible outcome of a series of binomial tests may be expressed very conveniently by means of *generating functions* for both probabilities and moments, on which see *TYS*, pp. 51–57. The probability-generating function of the Poisson distribution is given by $e^{m(t-1)}$, where $m = np$, and t is derived from the expression $(q + pt)^n$, which represents the probability-generating function of the binomial distribution. Thus the full statement of the Poisson distribution is

$$e^{m(t-1)} = e^{-m} \left(1 + \frac{mt}{1!} + \frac{m^2 t^2}{2!} + \cdots + \frac{m^r t^r}{r!} + \cdots\right), \qquad [\text{A3.3}]$$

which compares with the following full statement of the binomial distribution:

$$(q + pt)^n = p_n + n! \, (q^{n-1}) pt + \left[\frac{n!}{2!(n - 2)!}\right] q^{n-2} p^2 t^2 + \cdots$$

$$+ \left[\frac{n!}{x!(n - x)!}\right] q^{n-x} p^x t^x + \cdots + p^n t^n \, \ldots \, . \qquad [\text{A3.4}]$$

Each of the coefficients of t^x on the right-hand sides of Equations A3.3 and A3.4 represents the probability of 0, 1, 2, 3, ... occurrences of the event in question. The Poisson equation should be used for very rare events, the binomial in other cases.

3.16. The Poisson exponential series is given by

$$Ne^{-m}\left(1 + m + \frac{m^2}{2!} + \frac{m^3}{3!} + \cdots\right),$$

where $m = np$ and there are N trials out of a possible n. The terms give the frequencies 0, 1, 2, 3, ... occurrences in N sets of n. The parameters of the Poisson exponential of distribution are:

$$\text{third moment} = m, \qquad \text{fourth moment} = 3m^2 + m,$$

$$\text{standard deviation} = \sqrt{m}, \qquad \text{skewness} = \frac{1}{\sqrt{m}}, \qquad \text{kurtosis} = \frac{1}{m}.$$

3.17. The form of the integrated vorticity equation devised by J. G. Charney, *JM*, 6 (1949), 371 states that, in a barotropic atmosphere,

$$\frac{\partial \bar{\zeta}}{\partial t} = -\bar{V} \cdot \mathbf{V}(k\bar{\zeta} + \lambda), \qquad\qquad [A3.5]$$

where the bar denotes a vertical pressure average, i.e.,

$$\bar{x} = \frac{1}{p_0} \int_0^{p_0} x \, dp,$$

$\bar{\zeta}$ representing the mean relative vertical vorticity component, and $k = \dfrac{\bar{A}^2}{(\bar{A})^2}$. The basic assumption is that horizontal winds in the atmosphere, in large-scale synoptic systems, follow the law

$$\mathbf{V}(s_1, s_2, p) = A(p) \cdot \mathbf{V}(s_1, s_2, p_0),$$

where s_1, s_2, are orthogonal curvilinear distance coordinates on a spherical Earth, p is the vertical pressure coordinate, and p_0 is the mean surface pressure. In effect, this law states that winds at all heights in the atmosphere are parallel in direction. The main difficulty was how to integrate Equation A3.5, and this was overcome by J. G. Charney, R. Fjörtoft, and J. von Neumann, *T*, 2 (1950), 237. For a general discussion of the problem, see A. Eliassen, *T*, 4 (1952), 145, who provides a simplified picture. See also R. Fjörtoft, *T*, 4 (1952), 179, on the integration of the equation by computer, and P. D. Thompson, *T*, 6 (1954), 319, on integration by manual labor using a simple graphic process. J. S. Sawyer, *T*, 15 (1963), 336, described a completely different approach to integrating the vorticity advection equation, by means of a Lagrangian technique.

Following Charney, Fjörtoft, and von Neumann, the quasigeostrophic form of the vorticity equation may be written as

$$\frac{\partial}{\partial t}(\mathbf{V}_s^2 z) = \frac{\partial \eta}{\partial s_1}\frac{\partial z}{\partial s_2} - \frac{\partial \eta}{\partial s_2}\frac{\partial z}{\partial s_1}. \qquad\qquad [A3.6]$$

The right-hand side of the equation is, by definition, the surface spherical *Jacobian* of

η and z with respect to s_1 and s_2, and is usually written as $J_S(\eta, z)$; $\mathbf{V}_s^2$ is the surface spherical Laplacian operator, and η is the absolute vorticity, defined by $\eta = \left(\dfrac{g}{\lambda}\right)\mathbf{V}_s^2 z + \lambda$, where z is the height of the pressure level where divergence is nil, which is somewhere between 400 and 500 mb and in this case was taken to be 500 mb. The solution of Equation A3.6 may be obtained by solving for $\dfrac{\partial z}{\partial t}$ and extrapolating the motion forward in time, provided the boundary conditions are known.

Charney, Fjörtoft, and von Neumann also give details of the finite-difference procedure, and the manner in which it transforms Equation A3.6. The procedure involves first of all mapping the spherical surface of the Earth conformally onto a plane. If m is the magnification factor of the mapping, the Laplacian and Jacobian operators become transformed thus:

$$\mathbf{V}_s^2 = m^2\mathbf{V}^2 \quad \text{and} \quad J_s = m^2 J,$$

and Equation A3.6 then becomes

$$\frac{\partial}{\partial t}(\mathbf{V}^2 z) = J(\eta, z), \qquad [\text{A3.7}]$$

where

$$\eta = h\mathbf{V}^2 z + \lambda$$

in which

$$h = \frac{gm^2}{\lambda}.$$

Here $\mathbf{V}^2$ and J are the Laplacian and Jacobian operators measured on a plane surface in the usual fashion. The mapping is performed by means of the stereographic projection, with the mapping plane tangent to the Earth at the North Pole. On such a map, latitude ϕ and distance r from the Pole are related by

$$r = \frac{\cos\phi}{(1 + \sin\phi)},$$

where the unit of distance is the radius of the equator on the map,

$$m = 1 + r^2,$$

$$\lambda = \frac{2\Omega(1 - r^2)}{(1 + r^2)},$$

and

$$h = \frac{g(1 + r^2)}{2\Omega(1 - r^2)}.$$

If $\xi = \nabla z$, Equation A3.7 then becomes

$$\frac{\partial \xi}{\partial t} = J(\eta, z),$$

$$\eta = h\xi + \lambda,$$

$$\nabla\left(\frac{\partial z}{\partial t}\right) = \frac{\partial \xi}{\partial t}. \qquad [\text{A3.8}]$$

Provided the boundary is far removed from the prediction area, it is immaterial what values are assigned to z and ∇z on the boundary. The boundary condition chosen is $\frac{\partial z}{\partial t} = 0$ and $\frac{\partial \xi}{\partial t} = 0$ for $z_t \geq 0$, where z_t is the tangential derivative of z, taken in the direction that has the interior of the prediction area on its left. The finite-difference approximation is applied to Equation A3.8, resulting in the determination of $\frac{\partial \xi}{\partial t}$ and $\frac{\partial z}{\partial t}$ at a specific time for each of the grid points employed. (For the finite-difference approximation to the Laplacian, see Appendix 5.27 in *Foundations*). The finite-difference approximation to the Jacobian at point 0 (see Figure A3.17), where the axes are OX and OY as shown, is given by

$$J(x_1 y)_0 = \frac{(x_3 - x_1)(y_4 - y_2) - (x_4 - x_2)(y_3 - y_1)}{4d^2},$$

where x_3, y_3, are the coordinates in the OX, OY, directions, respectively (using an arbitrary origin), of the quantity S_3 at point 3, etc.

To obtain a prediction for time $n \, \Delta t$ ahead, where Δt is a small interval of time, the following extrapolations are performed n times for each grid point:

$$\xi(t + \Delta t) = \xi(t - \Delta t) + 2\Delta t\left(\frac{\partial \xi}{\partial t}\right),$$

$$z(t + \Delta t) = z(t - \Delta t) + 2\Delta t\left(\frac{\partial z}{\partial t}\right).$$

In these two equations, both terms on the right are known, and therefore the condition of ξ and z at time $t + \Delta t$ may be found. For the computational procedure in solving the barotropic equation, see *EDM*, pp. 193–95.

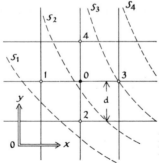

FIGURE A3.17.
Broken lines are isopleths of quantity S.

3.18. Following F. H. Bushby and M. K. Hinds, *QJRMS*, 80 (1954), 16, the *Sawyer-Bushby model* is described by the equations

$$\mathbf{V}^2\frac{\partial h'}{\partial t} + J(h_m, m^2 g\lambda^{-1}\mathbf{V}^2 h^{-1}) + J(h'_1 m^2 g\lambda^{-1}\mathbf{V}^2 h_m + \lambda) = \frac{4\pi_m\left(\dfrac{\mathbf{V}^2 h_m + \lambda^2}{m^2 g}\right)}{p_0 - p_1},$$

$$\mathbf{V}^2\frac{\partial h_m}{\partial t} + J(h_m, m^2 g\lambda^{-1}\mathbf{V}^2 h_m + \lambda) + \frac{1}{3}J(h', m^2 g\lambda^{-1}\mathbf{V}^2 h') = \frac{8\pi_m \mathbf{V}^2 h'}{3(p_0 - p_1)} = 0,$$

$$\frac{\partial h'}{\partial t} = \frac{m^2 g\lambda^{-1}J(h', h_m) - RA\Gamma_p \pi_m}{g},$$

where:

$$A \equiv \frac{1}{2}\frac{p_0 + 3p_1}{p_0 - p_1} - \frac{4p_0 p_1}{(p_0 - p_1)^2}\log_e\frac{2p_0}{p_0 + p_1};$$

$\mathbf{V}^2$, J represent the two-dimensional Laplacian and Jacobian operators referred to rectangular coordinates on a plane projection of the Earth's surface; m, the magnification factor, is $\sec^2\left(\dfrac{\pi}{4} - \dfrac{\phi}{2}\right)$, where ϕ is latitude; π is a measure of the vertical motion $\left(\pi = \dfrac{dp}{dt}\right)$; π_m is the value of π at pressure level p_m; p_0 is 1,000 mb; p_1 is 200 mb; p_m is the mean level of the atmospheric layer considered (in this case, $p_m = 600$ mb); h' is the thickness of the 1,000-mb to 500-mb layer; and $\Gamma_p = \dfrac{\partial T}{\partial p} - \Gamma/gp$, where Γ is the appropriate adiabatic lapse-rate.

3.19. According to J. S. Sawyer, *QJRMS*, 85 (1959), 31, the small-scale turbulent motion may be estimated from

$$\tilde{w}_0 = \frac{g}{\lambda}\left[\left\{\frac{\partial}{\partial y}(\tau_0 \cos\alpha) - \frac{\partial}{\partial x}(\tau_0 \sin\alpha)\right\} - \frac{\beta P}{g}V_{GN}\right],$$

where $\tilde{w}_0\left(\equiv \dfrac{dp}{dt}\right)$ is the vertical velocity at the top of the friction layer; τ_0 is the surface stress; β is the variation of Coriolis parameter with latitude; P is the depth of the friction layer; V_{GN} is the northward component of geostrophic wind velocity; and the surface stress is assumed to be along the direction of the surface wind at an angle α to the isobars. Following Sawyer, *op. cit.* (1959), the *form-drag* may be defined as the drag on the atmosphere exerted by the macroorography of the Earth's surface. If this drag is denoted by C_f, then

$$C_f = \pi\rho U_0^2 H^2 b^2\int_0^l k(l^2 - k^2)^{1/2}e^{-\frac{2b}{k}}\,dk.$$

where U_0 is the air velocity at the Earth's surface; l is Scorer's parameter (see Appendix 2.21 in *Foundations*); $\dfrac{2\pi}{k}$ denotes wavelength; and the profile of the topography is described by $\dfrac{Hb^2}{x^2 + b^2}$, as in Appendix 2.23 in *Foundations*.

3.20. Following G. P. Cressman, *MWR*, 88 (1960), 327, the final barotropic model is given by

$$\frac{\partial \zeta}{\partial t} + \mathbf{V} \cdot \nabla \eta - \frac{\mu \eta}{\psi} \frac{\partial \psi}{\partial t} + \left(\frac{\partial \zeta}{\partial t}\right)_m + \left(\frac{\partial \zeta}{\partial t}\right)_f = 0,$$

where ψ is the stream function derived from 500-mb heights on the basis of the relations $u = -\dfrac{\partial \psi}{\partial y}$ and $v = \dfrac{\partial \psi}{\partial x}$, $\bar{\psi}$ is the mean value of ψ at 500 mb; $\left(\dfrac{\partial \zeta}{\partial t}\right)_m$ represents the mountain effect; and $\left(\dfrac{\partial \zeta}{\partial t}\right)_f$ represents the surface-friction effect. Here $\mu = 1/K(1 - \varepsilon)$ where ε is the ratio of the density of the upper fluid to that of the lower fluid, and $K = \dfrac{h}{z}$, where h is the height of the interface between the two fluids and z is the height of a selected isobaric surface in the lower fluid.

3.21. The *mountain term* is given by

$$\left(\frac{\partial \zeta}{\partial t}\right)_m = \left(\frac{\eta}{p_g - p_t}\right) \mathbf{V}_g \cdot \nabla_{pg},$$

where p_g is the standard pressure at ground level; p_t is the pressure at mean tropopause level (i.e., 200 mb); $\mathbf{V}_g$ is the wind at ground level, given by $\mathbf{V}_g = \mathbf{V}\left[1 - 0.8\left(\dfrac{p_g - 500}{500}\right)\right]$ if p_g is in mb.

The *surface-friction term* is given by

$$\left(\frac{\partial \zeta}{\partial t}\right)_f = \frac{\eta w_H}{p_g - p_t},$$

where w_H is the vertical velocity (i.e., dp/dt) at the top of the friction layer, given by

$$w_H = \frac{g}{f}\left(\frac{\partial \tau_x}{\partial y} - \frac{\partial \tau_y}{\partial x}\right),$$

where τ_x, τ_y, are the components of surface stress that are related to the components of ageostrophic mass transport (M_x, M_y) by the relations $\tau_x = f M_y$ and $\tau_y = -f M_x$. The surface stress τ is given by $\tau = C_S \rho V_G$ where C_S is the skin-drag coefficient and V_G is the geostrophic surface wind (i.e., the wind at the top of the friction layer). Typical values of the form-drag are as follows: very high mountains, 0.5–0.9×10^{-2}; moderately high mountains, 0.2–0.5×10^{-2}; low relief with woods, 0.1–0.2×10^{-2}. For orography involving mountains of the order of 300 m high at 10-km intervals (e.g., the Appalachians), $C_f = 0.15 \times 10^{-2}$; for mountains 3 km high and 80 km apart (e.g., the Rocky Mountains), $C_f = 0.85 \times 10^{-2}$.

3.22. Following P. Graystone, *QJRMS*, 88 (1962), 256, the vertical velocity

$\left(\bar{w} \equiv \dfrac{dp}{dt} \right)$ induced by ground of height H is given by

$$\bar{w} = -g\rho \mathbf{V} \cdot \nabla H,$$

where ∇H is computed by finite-difference approximations.

3.23. See J. S. Malkus and M. Stern, *JM*, 10 (1953), 30, for the mathematics, also J. F. Black and B. L. Tarmy, *JAM*, 2 (1963), 557. In general, assuming (1) a steady state; (2) the atmosphere behaves as a perfect gas; (3) the heat supplied to the atmosphere by the island is dissipated within the airflow and not returned to the ground; (4) a unidirectional and uniform wind velocity upstream from the island; (5) a linear temperature gradient in the undisturbed air; (6) a geographically uniform temperature distribution at sea level; and (7) the island supplies an amount of energy to the atmosphere that is negligible in comparison with the total energy of the air-land-sea system, then the height M of the equivalent mountain is described as follows.

(a) Provided $\dfrac{K\sqrt{gs}}{U^2}, < \frac{1}{2}$ where U is the wind speed in the undisturbed air, s is the static stability, and K is the eddy diffusivity of the airstream, then M at distance x downwind from the windward edge of the island is given by:

$$M = 0, \text{ when } x < 0;$$

$$M = \frac{\tau}{sT_m} \left\{ 1 - \exp\left[-\left(\frac{K}{U}\right)\left(\frac{g^s}{U^2}\right)x \right] \right\},$$

when $0 \leqslant x \leqslant L$;

$$M = \frac{\tau}{sT_m} \left\{ \exp\left[-\left(\frac{K}{U}\right)\left(\frac{gs}{U^2}\right)x \right] \left[\left(\exp\left(\frac{K}{U}\right)\left(\frac{gs}{U^2}\right)L \right) - 1 \right] \right\},$$

where τ is the temperature increase due to the island, L is the length of the island parallel to the wind direction, and T_m is the average (absolute) air temperature.

(b) If $\dfrac{K\sqrt{gs}}{U^2} > \dfrac{1}{2}$, then $M = 0$ when $x < 0$,

$$M = \frac{\tau}{sT_m} \text{ when } 0 \leq x \leq L, \text{ and } M = 0 \text{ when } x \geq L.$$

(c) In general, $L = \dfrac{U^2}{\left(\dfrac{K}{U}\right)sg} \log_e \left(\dfrac{\tau}{\tau - MsT_m} \right).$

3.24. The model, devised by J. S. Malkus, *QJRMS*, 81 (1955), 558 shows that the sea breeze should reverse at a height h given by

$$h = \frac{\pi U}{2\sqrt{gs}}.$$

APPENDIX **4**

4.1. For details of the calculations, see: J. Charney, *HM*, pp. 292–97; H. Landsberg, *HM*, pp. 929–37; M. Milankovitch, "Mathematische Klimatlehre," in Köppen and Geiger, *Handbuch der Klimatologie* (Berlin, 1930), vol. IA. The basic principles are as follows.

The *flux* of radiation is defined as the total quantity of radiative energy traversing a surface from one side to the other per unit area per unit time. The total flux F through a unit area A (see Figure A4.1) whose normal is inclined at an angle θ to the direction L of the incident radiation is

$$F = \int I \cos \theta \, d\omega, \qquad [A4.1]$$

where $d\omega$ is a unit solid angle, and I is the intensity of the radiation. The radiation intensity in direction L is the flux per unit solid angle across a surface normal to direction L.

The flux of solar radiation reaching a point on the Earth's surface in the absence of an atmosphere is then : $\int I \cos d\theta \, \omega$ extended over the solid angle $\Delta\omega$ subtended at that point by the sun. Within this small angle, $\Delta\omega$, I can be regarded as constant. The flux is then equal to $I \cos \theta \, \dfrac{\pi R^2}{d^2}$, where R is the radius of the sun and d is its distance from the Earth. When θ equals $0°$ and $d = d_m$, the mean distance between Earth and sun, this expression gives the equation defining the solar constant S, defined as the intensity of solar radiation falling on a surface perpendicular to the sun's rays outside the atmosphere when the Earth is at its average distance from the sun:

$$S = \frac{\pi R^2 I}{d_m^2}. \qquad [A4.2]$$

Therefore the total flux F_T at a point on the Earth's surface is given by

$$F_T = \frac{S \cos \theta \, d_m^2}{d^2}. \qquad [A4.3]$$

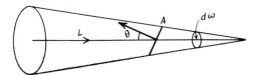

This expression, integrated over the period during which the sun is above the horizon, gives the total energy Q delivered per unit area on the Earth's surface during the course of a day:

$$Q = \int F_T \, dt = \frac{24 d_m^2}{\pi \, d^2} S \sin \phi \sin \delta (H - \tan H), \qquad \text{[A4.4]}$$

where $H = -\tan \phi \tan \delta$, ϕ represents latitude, δ is the solar declination (obtainable from a nautical almanac), and H is the hour-angle between sunrise and noon or noon and sunset. Thus the computation is purely astronomical. For a specific day of the year, d is equal to

$$d_m\left(1 - e \cos \frac{2\pi n}{y}\right),$$

where d_m is 92.9×10^6 miles, e is the eccentricity of the Earth's orbit (0.0167330), n is the number of days elapsed since January 1 (when Earth and sun are closest to each other), and Y is the length of the year in days (365.2563). Hence, knowing S, one can calculate the solar radiation received in any latitude on any day of the year in the absence of an atmosphere, from accurate astronomical tables.

The determination of the solar constant S by actual measurement is a difficult problem. The method adopted by C. G. Abbot and others at the Smithsonian Institute makes use of Beer's Law (see Appendix 4.2); the intensity of solar radiation at several wavelengths is measured several times during several hours as the angle between the zenith and the sun's actual height varies, i.e., as sec θ in Equation A4.9 varies. Assuming that k_λ and ρ in Equation A4.9 do not vary during the period of observation, extrapolating to zero air mass in the path of the solar beam gives the intensity of sunlight outside the atmosphere (I_0). The latest value determined for S is 2.00 cal per cm^2 per min (or 2.00 ly per min, one *langley* or ly equaling one cal per cm^2), which supersedes the value of 1.94 ly per min commonly used in previous climatological work.

All the preceding computations depend on knowing the solar constant, which is assumed not to vary. In fact, there seem to be both regular and irregular variations in the solar constant; see C. G. Abbot, *SMC*, vol. 101, no. 1 (1941), in which it is shown that variations in solar radiation affect the weather on Earth; see also C. G. Abbot, *SMC*, vol. 117, no. 110 (1952).

Empirically, the mean incoming radiation received at the Earth's surface (Q cal per cm^2 per min), after passing through an atmosphere of transparency k, is

$$Q = kh(1 - 0.0071C), \qquad \text{[A4.5]}$$

where h is the mean elevation of the sun above the horizon, C is the local value of mean cloudiness (as a percentage of the sky covered by cloud), and k varies from 0.023 at the equator to 0.027 in latitude 70°.

4.2. For a standard account of the principles of radiation physics, see J. G. Charney, *HM*, pp. 284–301. A convenient summary will also be found in *ITM*, pp. 114–27. The basic principles are derived from *Planck's law*, which states that the intensity of the radiation E_λ emitted by a black body depends only on its absolute temperature T and the wavelength λ of the radiation:

$$E_\lambda = \frac{\dfrac{2hc^2}{\lambda^5}}{e^{hc/\lambda kT} - 1}, \qquad [A4.6]$$

in which h is Planck's constant (6.55×10^{-27} erg sec), c is the velocity of light (3×10^{10} cm per sec), and k is Boltzmann's constant (1.37×10^{-16} erg per K°). If Equation A4.6 is integrated between the limits 0 and ∞, the resulting equation is

$$E = \frac{2\pi^4 k^4}{15c^2 h^3} T^4 = \frac{\sigma}{\pi} T^4, \qquad [A4.7]$$

where σ is the Stefan-Boltzmann constant (5.70×10^{-5} erg per cm^2 per sec per °K^4, or 8.312×10^{-11} cal per cm^2 per °K^4 per min). The radiative flux from a surface at a temperature of T°K is equal to πE, or to σT^4.

Other important principles are *Beer's law of absorption*, and the *equation of radiative transfer*. The former assumes that when a beam of monochromatic radiation is transmitted through an infinitesimal distance dl in an absorbing medium (see Figure A4.2), then the fraction of the radiative intensity absorbed, dI_λ, is independent of the original radiation intensity I_λ, but proportional to the density ρ of the absorbing substance and to the distance dl. Thus

$$\frac{dI_\lambda}{I_\lambda} = -k_\lambda \rho \, dl, \qquad [A4.8]$$

where k_λ is the *coefficient of absorption* of the medium. Beer's law states that

$$dI_\lambda = k_\lambda I_\lambda \rho \sec \theta \, dz, \qquad [A4.9]$$

or alternatively that

$$I_\lambda = (I_\lambda - dI_\lambda)e^{-k_\lambda m}, \qquad [A4.10]$$

where $m = \int_0^l \rho \, dl$, termed the *optical path length*, is defined as the mass of absorbing material in a column of unit cross section extending a distance l. The equation of radiative transfer is

$$\frac{dI_\lambda}{dm} = -k_\lambda(I_\lambda - E_\lambda), \qquad [A4.11]$$

where E_λ is the intensity of black-body radiation of wavelength λ, and I_λ is the actual intensity of radiation of wavelength λ. Provided that the distribution of absorbing substances (and their respective coefficients of absorption) are known, then Equation

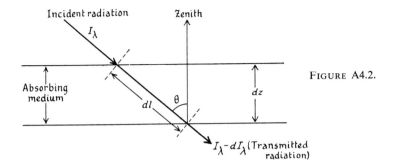

Incident radiation   Zenith

I_λ

Absorbing
medium

dl θ dz

$I_\lambda - dI_\lambda$ (Transmitted radiation)

FIGURE A4.2.

A4.10 enables the intensity of solar radiation at any point in the atmosphere to be calculated.

4.3. If the calculus rule for obtaining the maximum value of a function is applied to Planck's law, the result is

$$\lambda_m = \frac{a}{T},$$ [A4.12]

where a is equal to 0.288 cm °K. This is *Wien's displacement law*, which states that the maximum value of the wavelength of emitted radiation is inversely proportional to the absolute temperature of the emitting surface. Using the Stefan-Boltzmann law to compute the surface temperature of a black body whose emitted radiation is similar to that of the sun leads to a temperature of 5,750°K. In contrast, the surface temperature of the "black-body sun" according to Wien's law is 6,108°K. The sun therefore cannot be a perfect black body; the discrepancy between the two figures is attributed to the absorption of short waves in the outer layers of the sun.

4.4. The importance of scattering in the atmosphere is deduced from theory, not based on observation. The theory of scattering was largely developed by Lord Rayleigh; see *Phil. Mag.* 41 (1871), 107–20, 274–79, 447–54, and 47 (1899), 375–84 (also reprinted in Lord Rayleigh, *Scientific Papers*, Cambridge, vol. I, 1899, vol. IV, 1903).

Rayleigh proved from first principles that scattering is inversely proportional to the fourth power of the wavelength of the incident radiation. The scattering by very small particles is described by

$$I = I_{\pi/2}(1 + \cos^2 \theta),$$ [A4.13]

where I is the intensity of the scattered light in a direction at angle θ to the incident beam, and $I_{\pi/2}$ is the intensity at right angles to the beam. Therefore the greatest scattering is in the direction of the incident beam and in the opposite direction, and the least scattering in the direction at right angles to the beam.

Substitution of $k_\lambda + S_\lambda$ for k_λ in Beer's law (Equation A4.9 or A4.10) allows the depletion of the intensity of a beam of light by both absorption and scattering to be calculated. S_λ is the *coefficient of scattering*, defined by the expression

$$S_\lambda = \frac{32\pi^3}{3n\lambda^4}(\mu - 1)^2,$$ [A4.14]

in which μ is the refractive index of the air layer, and there are n scattering particles per cm^3 of air.

4.5. The word "actinometer" is sometimes used as a general term for all instruments that measure *solar* radiation, while "pyrheliometer" is restricted to actinometers that measure direct solar radiation. Pyranometers are sometimes termed *solarimeters*. The various types of radiation instrument are classified in F. Möller, *BWMO*, 6 (1957), 13; see also W. Mörikofer, *BWMO*, 6 (1957), 139.

4.6. The *efficiency* (as a percentage) is given by the heat represented by the distillate, divided by heat received, times 100. This equals

$$\frac{100(D + d)Ly}{ITA},$$

where D is the distillate during insolation (in cm^3); d is the distillate forming after insolation because of the lag of the instrument (in cm^3); L is the latent heat of vaporization of the liquid in the integrator (in cal per gm per °C); y is the density of this liquid (in gm per cm^3); I is the intensity of incident radiation (in cal per cm^2 per min); T is the duration of exposure to the radiation (in min); and A is the effective exposed area of the back receiver (in cm^2).

4.7. Above this threshold, the field response of the instrument is 84 cm^3 per kcal per cm^2. According to J. L. Monteith and G. Szeicz, *QJRMS*, 86 (1960), 91, the relation between the total distillate (D cm^3 per week) and the total radiation received on a horizontal surface (S kcal per cm^2 per week) is

$$D = 84.0S - 49. \qquad [A4.15]$$

4.8. K. Ångström, *Astrophys. J.*, 9 (1899), 332. The basic equation is

$$q = \frac{ri^2}{4.18ba}, \qquad [A4.16]$$

where q is the radiation received (in gm cal per sec per cm) on a strip b cm wide; a is the power of absorption of the strip, r its electrical resistance per unit length, and i the strength of the electric compensation current; r, b, and a need be determined only once, and i then observed in order to obtain radiation as an absolute measurement. That r changes with temperature very slightly must be allowed for in accurate work.

4.9. The heating of the strips is proportional to i^2; therefore

$$m_i' = m_o'(1 + \beta i^2), \qquad [A4.17]$$

where m_i' and m_o' represent the electrical resistances of the strips for current strengths i and o respectively. For platinum, $m_o' = 0.3092$ and $\beta = 0.303$. The temperature coefficient of resistance of the strips is found (using a Wheatstone bridge) by means of the expression

$$m_t'' = m_o''(1 + \alpha t). \qquad [A4.18]$$

For platinum, $m'' = 0.2970$ and $\alpha = 0.00216$. Then

$$m_{it} = m_o(1 + \alpha t)(1 + \beta i^2), \qquad [A4.19]$$

where m_{it} is the resistance of the strip to a current of strength i at a surrounding temperature of t. Thus the resistance may be calculated from t and i, since α and β are standard values for platinum.

Equation A4.19 is not quite exact, since the heating of the strips depends also on their rate of cooling, which is not constant. The error in the calculated resistance caused by current heating only amounts to 3 per cent at the most, and results in a maximum error of 0.3 per cent in the calculated radiation.

4.10. See C. G. Abbot, *SMC*, 56, no. 19 (1911); L. B. Aldrich, *SMC*, 111, no. 14 (1949). The disc changes its temperature according to the relation

$$C \frac{dT}{dt} = IA - k(T - T_c), \qquad [\text{A4.20}]$$

where

$$I = \frac{C}{At_1}(T_1 - T_o), \qquad [\text{A4.21}]$$

I being the radiation intensity, C the thermal capacity of the disc and A its area, T_0 the initial value of the temperature of the disc, T_1 its temperature after a (small) time interval t, and T_c the temperature of the inner case of the instrument. Corrections must be applied because C varies with temperature, and also if the ambient air temperature differs from the arbitrary standard value of 20°C chosen for the temperature of the exposed stem. The final corrected equation is

$$I = BR(1 - k[T - 30] - k'[T_a - 20]), \qquad [\text{A4.22}]$$

where T_a is the ambient air temperature, k' is a constant (0.00014), and k and B are constants that must be determined for individual instruments by comparison with an absolute standard instrument. Equation A4.22 gives the radiation intensity for a standard disc temperature of 30°C and a standard stem temperature of 20°C.

4.11. Following Robinson, *BAMS*, 36 (1956), 32, the angle β of the cone in Figure A4.11 made by the occulting strip is given by $\beta = \phi + h - 90$, where h is the solar altitude at noon and β is numerically equal to the sun's declination. The generating line R of the cone is given by

$$R = r_0 \, ctg\beta = r_0[tg(\phi + h)], \qquad [\text{A4.23}]$$

and the arc L is given by $L = h_{\max} - h_{\min}$, where $h_{\max}$ is the sun's altitude on the longest day and $h_{\min}$ is its altitude on the shortest day. L is therefore given by $L = \delta_{\max} - \delta_{\min}$, where δ is the sun's declination, and equals twice the angle of obliquity of the plane of the ecliptic, i.e., 46° 54'. The length of arc L that must be covered on the vertical circle is given by $L = \frac{\pi r_o \alpha}{180}$, where r_0 is the radius of the sphere. (All these angles are measured in degrees.)

4.12. According to Robinson and Stoch, *JAM*, 3 (1964), 179—see also A. J. Drummond, *JAM*, 4 (1965), 810—the error caused by the vertical member of the support is $\frac{2Db}{r}$, where b is the width of the vertical member, D is the sky radiation, and

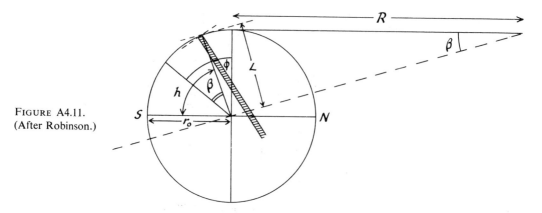

FIGURE A4.11.
(After Robinson.)

r is the radius of the sphere embodied by the shading frames. The energy absorbed by the shading frame is

$$2D\frac{w}{r}\cos\delta(t_0\sin\phi\sin\delta+\cos\phi\cos\delta\sin t_0), \qquad [\text{A4.24}]$$

where w is the width of the shading frame, δ is the sun's declination, t_0 is the hour angle of the sun at sunrise or sunset, and ϕ is latitude.

4.13. I. G. Galindo Estrada and E. M. Fournier D'Albe, *QJRMS*, 86 (1960), 270, found that $E = 6.13W + 59.5$ for Mexico City, where E is the daily total radiation (in g cal per cm^2) and W is the weight of paper burned (in mg).

4.14. Following F. H. W. Albrecht, *UCSCM*, p. 99, the global radiation is given by

$$(I+D)_0 = (I-I_w)\sin h\left(1-\frac{0.13}{\sqrt{\sin h}\sqrt{\frac{b}{760}}}\right)\left(1+0.153\alpha\sqrt{\frac{b}{760}}\right)\left(\frac{R_m}{R}\right)^2, \quad [\text{A4.25}]$$

where I is the intensity of direct solar, D of diffuse sky, radiation; I_w is the energy absorbed by the water vapor in the atmosphere; h is the height of the sun; b is the atmospheric pressure; α is the albedo of the ground; and R_m is the mean distance, and R is the actual distance, between the Earth and the sun.

4.15. See H. J. Albrecht, *GPA*, 30 (1955), 93, according to which the basic formula is

$$C_t Tk_t = C_a Tk_a + C_b Tk_b + C_c Tk_c + \cdots, \qquad [\text{A4.26}]$$

where C_t is the total capacitance in the tuned circuit and Tk_t is the resultant temperature coefficient of C_t; C_a, C_b, C_c, ... are single condensers in parallel forming C_t, and Tk_a, Tk_b, Tk_c, ... are the temperature coefficients of C_a, C_b, C_c, ...

4.16. R. H. Shaw, *BAMS*, 37 (1956), 205. For Ames (Iowa) the relation is $R_N = 0.75R_T - 21.4$ on cloudy days and $R_N = 0.87R_T - 82.0$ on clear days, where R_N represents the net radiation and R_T the total solar radiation.

4.17. Following V. E. Suomi, M. Franssila, and N. F. Islitzer, *JM*, 11 (1954), 276, the net radiation R_n is given by

$$R_n = (T_1 - T_2)[2\lambda + f(v)], \qquad [\text{A4.27}]$$

where λ is the constant given by the ratio of the thermal conductivity of the plate to its thickness; $f(v)$ is a function of windspeed, air temperature, and other factors that contribute to the cooling effect of the wind, and is the same on each side of the plate; and $(T_1 - T_2)$ is the temperature difference between the upper and lower sides of the plate. The equation used in practice is

$$a(R_1 + R_2) = 2a\sigma T_m^4 - 2f(v)(T_m - T),\qquad\text{[A4.28]}$$

where a is the absorptivity of the plate, T_m its mean temperature, and R_1 and R_2 the total hemispheric radiation incident on the upper and lower surfaces of the plate; $a(R_1 + R_2) = R_n$.

4.18. For details, see F. F. Abraham, *JM*, 17 (1960), 291. The Suomi radiometer gives values for downward flux of long-wave radiation that are 6 per cent higher than those given in the Beckman and Whitley total radiometer, and 1 per cent higher than those computed from the Elsasser chart.

4.19. L. J. Fritschen, *BAMS*, 41 (1960), 180, and *JAM*, 2 (1963), 165. Following the theory given in the latter paper, the net radiation, R_n, is given by

$$R_n = R_S - rR_S + R_A - R_E,\qquad\text{[A4.29]}$$

where R indicates radiation (in ly per min) and the subscripts S, A, and E refer to solar, atmospheric, and earth, respectively. If a thermal tranducer is used as the sensitive element instead of a blackened plate, then (see Figure A4.19) for the top surface of the transducer

$$R_n = (4\varepsilon\sigma T_t + 2k)(T_t - T_b),\qquad\text{[A4.30]}$$

and for the bottom surface

$$R_n = (4\varepsilon\sigma T_t + 2k)\frac{G}{BC} = gG.\qquad\text{[A4.31]}$$

Here, k is the thermal conductivity (in cal per cm per min per °C), G is the thermo-electric potential (in μV per °C), B is the thermoelectric power (in μV), C is the heat capacity of the plate at constant pressure (in cal per g per °C), and g is the calibration

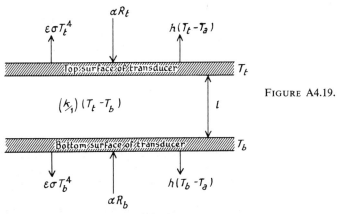

FIGURE A4.19.

α is the absorptivity of the transducer

constant, which depends on the temperature, emissivity, conductivity, and thickness of the transducer. The energy balance on the top surface is

$$aR_t = \varepsilon\sigma T_t^4 + h(T_t - T_a) + k(T_t - T_b),$$ [A4.32]

that on the bottom surface is

$$aR_b = \varepsilon\sigma T_b^4 + h(T_b - T_a) - k(T_t - T_b),$$ [A4.33]

and $R_n = a(R_t - R)$, where a is the absorptivity of the transducer and h is the thermal convection coefficient (in kcal per cm per min per °C).

4.20. For a general discussion, see: C. D. Walshaw, *SP*, 47 (1959), 67; W. M. Elsasser, *HMS*, no. 6 (1942); W. M. Elsasser and M. F. Culbertson, *AMM*, vol. 4, no. 23 (1960); and F. Möller, *CM*, p. 34. Following Walshaw, the fundamental description of the radiation field at any point P in the atmosphere is made in terms of the *specific intensity*, $I_\nu(\theta, \phi)$, which is defined by

$$I_\nu(\theta, \phi) = \lim_{\delta\nu, \delta A, \delta\omega \to 0} \left[\frac{E_\nu \, \delta\nu}{\delta\nu} \delta A \, \delta\omega \right],$$ [A4.34]

where $E_\nu \, \delta\nu$ is the energy flowing per unit time in the frequency range ν to $\nu + \delta\nu$ through an element of area δA at P which is normal to the direction (θ, ϕ) of the radiation impinging on the element of area. The radiation received on the latter comes from a cone of solid angle $\delta\omega$ as depicted in Figure A4.1 (see Appendix 4.1 and 4.2). The direction $\theta = 0$ is taken to point vertically upward, and is the azimuth angle. The subscript ν indicates that monochromatic radiation is being considered. The flux of radiation, F_ν, across a horizontal unit area in the atmosphere is given by

$$F_\nu = \int I_\nu(\theta, \phi) \cos\theta \, d\omega = \int_0^{2\pi} d\phi \int_0^{\pi/2} I_\nu(\theta, \phi) \cos\theta \sin\theta \, d\theta.$$ [A4.35]

The net flux, Φ_ν, is defined by $\Phi_\nu = \uparrow F_\nu - \downarrow F_\nu$, where $\uparrow F_\nu$ is the upward flux of radiation and $\downarrow F_\nu$ is the downward flux. The total net flux of radiation, Φ, is obtained by integrating the expression for Φ_ν over all frequencies. The total net flux determines the rate of heating caused by radiation. Thus the rate of temperature rise at a height with pressure p is given by

$$\left(\frac{g}{C_p}\right)\left(\frac{\partial\Phi}{\partial p}\right).$$ [A4.36]

According to Kirchoff's law,

$$\frac{E_\lambda}{a_\lambda} = f(\lambda, T),$$ [A4.37]

where E_λ is the (monochromatic) intensity of radiation of wavelength λ emitted in a specific direction in a unit time per unit area, and a_λ is the (monochromatic) fractional absorption, defined as the ratio of the incident radiation that is absorbed to the total incident radiation; the ratio $\frac{E_\lambda}{a_\lambda}$ is termed the *emissitivity* of the body. Kirchoff's law indicates that for all bodies, the emissivity is a function of wavelength and temperature only. Complete absorption (i.e., $a_\lambda = 1$ at all wavelengths) defines a *black body*, whereas $a_\lambda < 1$ and a_λ constant with wavelength defines a *gray body*, and a_λ varying with wavelength defines a *selective emitter*. The intensity of radiation emitted by an element δm of absorbing material (where the mass of absorbing material present per unit area is m) at X in the direction of P will be $B_v(T)k_v\,\delta m$, where k_v is the absorption coefficient of the material between X and P, and B_v is the black-body flux that emerges from a unit area of a black surface into all directions of the hemisphere. According to Lambert's law, that radiation transmission depends exponentially on the thickness of the absorbing layer, then the intensity of radiation emitted by δm at X in the direction of P will be attenuated by a factor τ_v on its way to P, where

$$\tau_v = e^{-k_v m}. \qquad [\text{A4.38}]$$

Therefore the radiation received at P is

$$\delta I_v = B_v(T)k_v \cdot e^{-k_v m}\,\delta m = -B_v(T)\,\delta\tau_v; \qquad [\text{A4.39}]$$

the last expression is negative because τ_v decreases as m increases. Consequently,

$$I_v(\theta, \phi) = \int B_v(T)\,d\tau_v. \qquad [\text{A4.40}]$$

In effect, the *Elsasser radiation chart* performs this integration graphically. Because k_v (and hence τ_v) is a rapidly varying function of v, owing to the rotational structure of the absorption bands, it is convenient to average τ_v over a frequency interval Δv centered on v, to give τ_I, the *transmission function*. τ_I is defined by

$$\tau_I = \int_{\Delta v} \tau_v \cdot \frac{dv}{\Delta v}; \qquad [\text{A4.41}]$$

Δv is chosen to contain enough rotation lines that rapid fluctuations of transmission are smoothed out. The transmission function does not, in general, obey Lambert's law.

4.21. For a simplified derivation, following J. Charney, *Handbook of Meteorology* (New York, 1945), pp. 291–92 and 301–3, consider an atmospheric layer, bounded by infinite horizontal planes, in which the absorbing substances are uniformly stratified in the horizontal (see Figure A4.21). The flux F_v of radiation of wavelength λ at a point P at the base of the layer, from an infinitesimal sheet at X of optical thickness du, is

$$F_v = \int_0^{2\pi} d\phi \int_0^{\pi/2} dI_v\, e^{-k_v u\, \sec\theta} \sin\theta \cos\theta\, d\theta. \qquad [\text{A4.42}]$$

The total flux in the downward direction (F) at P is given by

$$F = \int_0^\infty dv \int_0^u (F_v)_b\, \frac{d\tau_I}{du}\, du, \qquad [\text{A4.43}]$$

in which u is the *optical depth*, or *optical thickness*, defined as $\int_{z2}^{z1} \rho\, dz$, where this integral represents the number of grams of absorbing material in a vertical air column

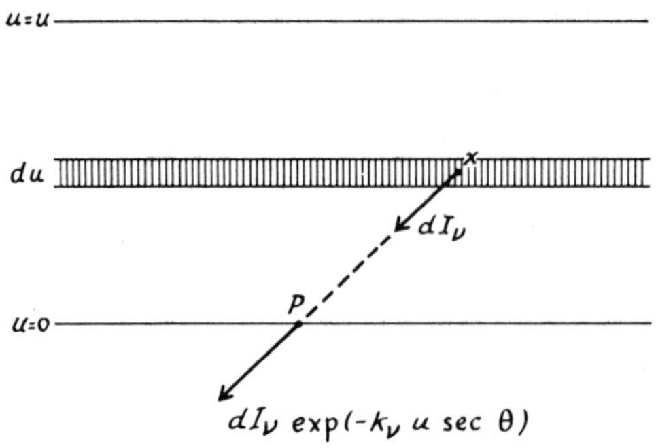

$u=u$ ————————————————————— ——————

du ▨▨▨▨▨▨▨▨▨▨▨▨▨▨▨▨▨▨▨▨▨▨▨▨▨▨

dI_ν

$u=0$ ———————————————— P ————————————

$dI_\nu \, exp(-k_\nu \, u \, sec \, \theta)$

FIGURE A4.21.
u is optical depth. du is a sheet of absorbing substances, of infinitesimal thickness.

one cm^2 in cross section extending from height z_0 to height z_1. Equation A4.43 cannot be integrated directly, and therefore a function Q is introduced, defined by

$$Q = \int_0^\infty \frac{d(F_v)_b}{dt} \tau_I \, dv. \qquad [A4.44]$$

The introduction of Q allows the integration to be performed by parts. In Equations A4.42–A4.44, v is the frequency of the radiation; i.e., $v = \frac{1}{\lambda}$, where λ is its wavelength.

4.22. According to D. L. Brooks, this is because the emissivity (ε) is not well-known for such short path lengths as are involved in microclimatology, and the slope of the ε-curve with respect to the water path must be known accurately if the method is to be used successfully.

4.23. For the procedure, see W. H. Klein, *JM*, 5 (1948), 119. The first step is to determine I_o, the direct solar radiation received on a horizontal surface at the top of the atmosphere, from

$$I_0 = \frac{S(\sin \phi \sin \delta + \cos \phi \cos \delta \cos H)}{r^2},$$

where S is the solar constant (in g cal per cm^2 per min), r is the radius-vector of the Earth, ϕ is latitude, δ is solar declination, and H is the hour-angle of the sun.

The atmospheric transmission coefficient, a, is defined by $a = \frac{I_c}{I_o}$, where I_c is the insolation received at the Earth's surface with a cloudless sky; a may be expressed in terms of precipitable water and air mass, where *air mass* is defined as the secant of the sun's zenith distance.

The transmission coefficient at altitude z, for the summer season, is given by

$$a_z = (0.123 + 0.051 \sec Z)\log z + 0.483 - 0.245 \sec Z,$$

where Z represents solar zenith distance. If d represents dust depletion, then $a = a'' - d$, where a'' is the transmission coefficient for dust-free humid air; d may be defined as the total fraction of incoming solar radiation that is scattered, diffusely reflected, or absorbed by all the solid and liquid particles suspended in a cloudless atmosphere.

The diffuse sky radiation, D_c, is given by $Q_c = I_c + D_c$, where Q_c is the total radiation, both D_c and Q_c referring to the radiation received on a horizontal surface with a cloudless sky. According to Klein, $D_c = 0.5sI_o$, where $s = 1 - a' + d$, a' being the transmission coefficient after scattering by the air and water-vapor molecules. Therefore,

$$Q_c = I_o(a + 0.5s) \quad \text{and} \quad R = \frac{0.5s}{(a + 0.5s)},$$

where $R = D_c/Q_c$. The radiation depletion due to clouds is described by $Q = Q_c[k + (1 - k)S_u]$, where S_u represents sunshine, Q is the total radiation received on a horizontal surface on a day with clouds, after terrain reflection, and k is the ratio of the total radiation with zero sunshine to total radiation with a cloudless sky.

Terrain reflection is defined as AQ_c, i.e., the amount of radiation reflected by terrain with albedo A. A fraction s of this reflected radiation will again be subjected to scattering and diffuse reflection in the atmosphere, so that the total radiation finally reaching the ground will exceed Q_c by the amount $0.5sAQ_c$. If no sunshine data are available, the expression $Q = Q_c[1 - (1 - k')C]$ should be used, where C is the cloudiness and k' the ratio of the total radiation received at the ground under a complete overcast ($C = 1$) to Q_c.

4.24. See *HBES*, pp. 28ff, for a detailed review. The Savino-Ångström formula is

$$Q + q = (Q + q)_o[1 - (1 - k)n],$$

where Q and q represent the direct and scattered radiation, respectively, the subscript o indicates clear-sky conditions, n is the mean cloudiness, and k is an empirical coefficient that varies both seasonally and diurnally. Ångström, *QJRMS*, 50 (1924), 121, gave the equation in the form

$$Q = Q_0\left[a' + \frac{(1.00 - a')^n}{N_o}\right];$$

for Stockholm, $a' = 0.25$. Later, Fritz and MacDonald found that, in general,

$$\frac{Q}{Q_0} = 0.35 + \frac{0.61n}{N_0}.$$

4.25. On the following equations, see *HBES*, pp. 33, 41, 42. The Efimov equation is

$$I = I_o[1 - (c_B n_B + c_C n_C + c_H n_H)]$$

where n_H, n_C, n_B represent the amounts of high, middle, and low cloud, respectively. The coefficients c_B, c_C, c_H have the values 0.20, 0.6 − 0.7, and 0.8 − 0.9, respectively. The Kuzmin equation used the expression $1 - 0.14(n_0 - n_H) - 0.67n_H$ for the ratio of actual to possible solar radiation, where n_0 is total cloud amount and n_H is the amount of lower clouds.

The Berliand equation is

$$I_0 = s\sigma\theta^4(0.39 - 0.058\sqrt{\bar{e}}),$$

where e is the vapor pressure in mm and I_0 is in cal per cm^2 per min; θ represents air temperature and I_0 is the effective outgoing radiation. The effect of clouds is taken into account by computing $I = I_0(1 - cn)$, where I is the effective outgoing radiation with the existing cloud cover, n is the cloud amount in tenths, and c is an empirical coefficient (0.75, on the average).

The Budyko equation is

$$I = I_0(1 - cn^2) + 4s\sigma\theta^3(\theta_w - \theta),$$

where the second term on the right is an estimation of the effect of the difference between the temperature of the underlying surface (θ_w) and that of the air on the effective outgoing radiation.

4.26. According to M. Weinstein, *MWR*, 89 (1961), 419, the energy balance for a mirror-backed hemisphere spinning in space, in the shadow of the Earth, is given by

$$\alpha\beta\,\frac{R\uparrow}{2} = 2\pi\varepsilon\sigma T_w^4 - H\,\frac{\partial T}{\partial t} - \kappa(T_w - T_m),$$

from which

$$R\uparrow = \frac{1}{\beta}\left[\frac{4\pi\varepsilon\sigma T_w^4}{\alpha} - \frac{2H}{\alpha}\frac{\partial T}{\partial t} - \frac{2\kappa}{\alpha}(T_w - T_m)\right]$$

and is the total radiation received by the bolometer; β is the solid angle to the Earth from the hemisphere; α and ε are the infrared absorptivity and emissivity, respectively, of the sensor; H is the time-constant of the sensor; and κ is the thermal conductivity between the sensor (at temperature T_w) and the mirror (at temperature T_m).

4.27. Following S. Fritz and J. S. Winston, *MWR*, 90 (1962), 1, the important equation is

$$\int_0^\infty \frac{\phi_\lambda B_\lambda(T_E)}{d\lambda} = \int_0^\infty \phi_\lambda I_\lambda\,d\lambda,$$

where $B_\lambda(T_E)$ is the black-body intensity (according to Planck's law) for a temperature T_E, and the right-hand side of the equation represents the actual intensity of radiation from Earth passing through the instrument. The left-hand side of the equation is the intensity of radiation that would be transmitted to the sensing element in the satellite from the black body at temperature T_E. Therefore T_E may be found from the satellite signals produced by I_λ.

4.28. According to W. Chiu and R. S. Greenfield, *JM*, 16 (1959), 271, the local temperature variation on a constant-pressure surface is given by

$$\left(\frac{\partial T}{\partial t}\right)_p = -\mathbf{V}\cdot(\nabla T)_p - w(\Gamma_d - \Gamma) + \frac{1}{C_p}\frac{dQ}{dt};$$

where $\mathbf{V}$ is the horizontal velocity vector, $(\nabla T)_p$ is the temperature gradient along an isobaric surface, and dQ is the heat added per unit mass to the air. The first term on the right-hand side of the equation gives the contribution to the total heating made by horizontal temperature advection, the second term gives the contribution by vertical convection of heat, and the third (negligible) term gives the contribution by non-adiabatic processes, such as radiation, heat conduction, and eddy transport of heat.

4.29. F. Möller, *JM*, 17 (1960), 566, showed that for Europe, North Africa, Greenland, and adjacent parts of the North Atlantic, the radiation balance, *B*, is given by

$$B = 159.8 - 0.119R_0 + 9.885N + 7.30\Delta T_{300},$$

where R_0 is the total daily insolation received on a horizontal surface, N represents cloud amount and type, and ΔT_{300} is the difference between the 300-mb temperature and the temperature at the Earth's surface.

4.30. Variation in reflectivity is defined by the expression $R = (1 - \alpha)S + L$, where R is the net radiation, L is the net incoming long-wave radiation, S is the incoming short-wave radiation, and α is the reflectivity.

4.31. J. L. Monteith, *QJRMS*, 88 (1962), 508, derived the following expression for rural areas:

$$\frac{T_2}{T_1} = \frac{1 - c(\rho + \phi)}{1 - c\alpha\rho}$$

where T_2 is the mean monthly incoming radiation, T_1 is the mean monthly total (direct plus diffuse) radiation, c is the fractional cloudiness, α is the surface reflection coefficient, and ρ is the cloud reflection coefficient.

4.32. According to S. Orvig, *JM*, 18 (1961), 199, the mean daytime net radiation flux, Y, over sub-Arctic surfaces is given by $Y = -2.84 + 0.01X$, where X is the total incoming short-wave radiation.

4.33. This equation is given by Brunt, *PDM*, pp. 130–33, as

$$\frac{\partial T}{\partial t} = K_R \frac{\partial^2 T}{\partial z^2}.$$

This is similar to the classical equation for conduction of heat in a solid, with the constant K_R replacing the thermometric conductivity κ, where $K_R = \dfrac{139}{60\rho c_p} \dfrac{T}{e} \dfrac{\partial E}{\partial T}$ where

E represents radiation in g cal per cm^2 per min.

4.34. J. R. Goss and F. A. Brooks, *JM*, 13 (1956), 482, give the following values for the constants:

	a	*b*
Algeria	0.48	0.058
Austria	0.47	0.063
California	0.05–0.66	0.032–0.039
England	0.55	0.056
France	0.60	0.042
India	0.47	0.061
Oklahoma	0.68	0.036
Sweden	0.43	0.082

The Robitzsch formula is given by F. Möller, *CM*, p. 38, as

$$R = \sigma T^3(0.135p + 6.0e),$$

where p is atmospheric pressure and e is vapor pressure, both at the time the surface air temperature, T, is measured.

4.35. According to W. E. Saunders, *QJRMS*, 75 (1949), 154, the appropriate formulae are

$$\text{with no inversion, } T_r = \tfrac{1}{2}(T_{\max} + T_d) - 0.6°\text{F, and}$$

$$\text{with an inversion, } T_r = \tfrac{1}{2}(T_{\max} + T_d) - 4.0°\text{F.}$$

4.36. For details of Bouguer's law, as developed by Lambert, see *PM*, pp. 77–78. The Bouguer-Lambert law forms a fundamental relationship in theoretical physics, and may be expressed in the form

$$\frac{dB}{B} = -\sigma \, dx,$$

where σ is the fractional change in brightness over distance Δx and

$$\sigma \, \Delta x \rightarrow \frac{dB}{B} \quad \text{when } \Delta x \rightarrow dx.$$

5.1 The fixed points are defined as follows. For the equilibrium point of CO_2,

$$T_p = -78.5°C + 0.0159(p - 760) - 0.0000111(p - 760)^2,$$

where p is the atmospheric pressure in inches of mercury. For the boiling point of water,

$$T_p = +100.00°C + 0.0367(p - 760) - 0.000023(p - 760)^2.$$

For the platinum resistance thermometer,

$$R_T = R_o(1 + AT + BT^2),$$

where R_T is the resistance at temperature T, R_o is the resistance at the melting point, and the constants A and B are determined by calibration at the fixed points. For interpolations at temperatures below 0°C,

$$R_T = R_o[1 + AT + BT^2 + C(T - 100)T^2],$$

where C is determined by calibration at the temperature of equilibrium between liquid oxygen and its vapor at atmospheric pressure (i.e., the " oxygen point ").

5.2. According to *MOHSI*, the maximum error over the range $-80°$ to $+40°C$ is 0.05°C at $-50°$ and $+40°C$. The equation employed is

$$R_T = R_0[1 + A(T - T_0) + B(T - T_0)^2]$$

For small intervals of temperature,

$$R_T = R_0[1 + A(T - T_0)],$$

where R_T is the specific resistance of the wire at temperature T. In both equations T_0 is the standard temperature.

5.3. For thermistors, $R = ae^{-\frac{b}{T}}$, where R and T are the resistance and temperature of the thermistor, and a and b are constants.

5.4. The equation describing the thermocouple principle is

$$E = A(T - T_0) + B(T - T_0)^2,$$

where E represents the emf, T_0 is the reference temperature, and A and B are constants. For most meteorological purposes, $E \propto (T - T_0)$.

5.5. If C is the speed of the wave-propagation of sound, then

$$C = \sqrt{\frac{\gamma R T}{m}},$$

where γ is the ratio of c_p and c_v, and m is the molecular weight of the air.

5.6. The *lag coefficient* (λ) is defined by the expression

$$\frac{dT}{dt} = -\frac{1}{\lambda}(T_T - T_m)$$

where T_m is the temperature of the medium and T_T is the temperature of the thermometer. If T_m is constant, then

$$T_T - T_m = (T_o - T_m)e^{-\frac{t}{\lambda}},$$

where T_0 is the value of T_T at time $t = 0$, and from which λ may be determined.

5.7. Following *HBES*, p. 7, a horizontal transfer of heat is described by

$$\frac{\partial Q}{\partial t} + u\frac{\partial \theta}{\partial x} = \frac{\partial}{\partial x}\left(K\frac{\partial \theta}{\partial z}\right),$$

where θ refers to the air temperature, x represents the horizontal coordinate along the wind direction in the lower layer of the atmosphere, u is the windspeed in this layer, and K is the coefficient of turbulent exchange. By integration of this equation, we obtain

$$\int_0^z \frac{\partial \theta}{\partial t}\,dz + \int_0^z u\frac{\partial \theta}{\partial x}\,dz - \frac{H}{\rho C_p} = K\frac{\partial \theta}{\partial z},$$

where H represents the heat flux between Earth and atmosphere. The second term on the left describes the direct effect of advection on the air layer under consideration. For heat balance at the Earth's surface, $z \to 0$, and consequently this term will equal zero. Heat balance at the surface is determined only by $K\frac{\partial \theta}{\partial z}$, the vertical heat flux.

5.8. According to *HBES*, since the average dryness of the soil increases with increasing energy gain from net radiation and with decreasing precipitation, we may write that (a) $\frac{f}{r} \to 0$ or $\frac{E}{r} \to 1$ as $\frac{R}{Lr} \to \infty$. The expenditure of heat in evaporation is compensated for only by radiation balance; therefore the upper limit of increase of LE is R. Hence for conditions of excessive moisture, we may assume that (b) $LE \to 0$ as $\frac{R}{Lr} \to 0$. The relationship between $\frac{E}{r}$ and $\frac{R}{Lr}$ is determined by (a) and (b); consequently,

$$\frac{E}{r} = \phi\left(\frac{R}{Lr}\right) \quad \text{for} \quad \frac{R}{Lr} \to 0 \quad \text{and} \quad \frac{R}{Lr} \to \infty,$$

where ϕ is some function to be determined. Budyko found empirically that

$$E = r\left(1 - e^{\frac{-R}{Lr}}\right) \quad \text{and} \quad E = \frac{R}{L}\tanh\left(\frac{Lr}{L}\right).$$

The geometric mean of these two expressions gives

$$E = \sqrt{\frac{Rr}{L}\tanh\left(\frac{Lr}{L}\right)\left[1 - \cosh\left(\frac{R}{Lr}\right) + \sinh\left(\frac{R}{Lr}\right)\right]}.$$

The relation between the components of heat and moisture balance is given by the following expressions for runoff (f) and runoff coefficient (f/r):

$$f = r - E; \frac{f}{r} = 1 - E.$$

5.9. The parameter H is given by

$$H = I_0 - RI_0 - B_a - S - LE,$$

where I_0 is the solar radiation reaching the ground surface, R is the ground surface albedo, B_a is the net long-wave radiation, S is the heat entering the ground, and E the water evaporated or transpired.

5.10. See N. E. Rider: W, 12 (1957), 241; *QJRMS*, 85 (1959, 51; *EAFM*, p. 58. According to classical heat-conduction theory,

$$\ln\left(\frac{R_1}{R_2}\right) = b(z_2 - z_1)$$

where R_1, R_2, the daily temperature ranges at depths z_1, z_2, below the surface of the ground;

$$b = \sqrt{\frac{\pi}{tK}},$$

where t represents time (in 24-hour periods usually) and K is the thermal diffusivity of the conducting medium (in cm^2 per sec). In the case of natural soils, a graph of $\ln\left(\frac{R_1}{R_2}\right)$ against $(z_2 - z_1)$ does not give a straight-line relationship. For details of the definition of K from fundamental heat-conduction theory, see M, pp. 106–8.

5.11. For this application, the Bowen ratio is written in the form

$$\beta = 0.64\frac{p}{1000}\left(\frac{t_w - t_a}{e_w - e_a}\right),$$

where t_w represents water temperature and t_a air temperature in °C, p denotes atmospheric pressure in mb, and e_w and e_a the vapor pressure in mb of the air immediately above the water surface and at screen level on the ship, respectively. Thus the Bowen ratio, β, is defined here as $\beta = Q_n/Q_e$. In computations for northern hemisphere oceans by W. C. Jacobs, *CM*, p. 1057,

$$Q_h = \beta L_v E$$

in cal per cm² per day, where L_v is the latent heat of vaporization of water at the average temperature of the sea surface, and E is the rate of evaporation. In computations for the southern hemisphere oceans by D. W. Privett, *MOGM*, vol. 13, no. 104 (1960),

$$Q_h = 2.29(t_w - t_a)_w,$$

where w is the windspeed in knots.

Following Jacobs,

$$E = k(e_w - e_a)w_a$$

in cm per day, where e_a and w_a, the vapor pressure and the windspeed, are both measured at height a above the sea surface, e_w is the vapor pressure at the sea surface, and k is a constant found to be 0.142 for the North Atlantic and the North Pacific Oceans. It should be noted that this equation for E makes use of climatological data, and gives the average, not the instantaneous, rate of evaporation. Following Privett,

$$E = k\rho(x_w - x_a)w,$$

where x represents the humidity mixing ratio, and the subscripts w and a denote measurements taken at the water surface and in the air, respectively. The term $k\rho$ equals 0.00587 for the southern hemisphere oceans. The latent energy, Q_e, equals $585E$ cal per cm² per day, where E is in cm per day. The heat supplied by condensing water vapor, Q_p, is defined by $Q_p = L_c P$, where L_c is the latent heat of condensation at temperature t and P is the rate of precipitation. This definition applies to the northern hemisphere oceans. For the southern hemisphere oceans, the relation $Q_p = \dfrac{1}{n} L_c P_s$ is found to work well, in which P_s is the seasonal precipitation rate (in cm per season), and the season contains n days. The availability of surplus energy is given by $Q_a - Q_{ph} = L_c(E - P)$; a positive value indicates that latent energy is locally surplus, a negative value that an excess of latent energy is being transformed into real heat in situ.

5.12. According to H. Wexler, *JM*, 1 (1944), 23, the equation employed was

$$\frac{\partial\theta}{\partial t} - w\frac{\partial\theta}{\partial z} = \frac{\theta}{T}\mathbf{V} \cdot \nabla_z T = \left(\frac{100}{9}\right)^{0.288}\mathbf{V} \cdot \nabla_z T,$$

where $\mathbf{V}$ is the mean geostrophic wind vector for the layer between mean sea level and 10,000 feet, $\nabla_z T$ is the horizontal ascendant of mean temperature in this layer, $\dfrac{\partial\theta}{\partial t}$ is the nonadiabatic heating, and $-w\dfrac{\partial\theta}{\partial z}$ is the adiabatic heating. See also E. J. Aubert and J. S. Winston, *JM*, 8 (1951), 111, for a study (including maps) of northern hemisphere atmospheric heat sources for monthly periods.

5.13. Following P. F. Clapp, *MWR*, 89 (1961), 147, from the definition of potential temperature and the first law of thermodynamics, it may be shown that

$$\frac{d\bar{q}}{dt} = c_p\frac{\partial\bar{T}}{\partial t} + c_p\bar{\mathbf{V}} \cdot \nabla\bar{T} + \frac{c_p T}{\theta}\bar{w}\frac{\partial\bar{\theta}}{\partial z} + c_p\overline{\mathbf{V}' \cdot \nabla T'} + \frac{c_p T}{\theta}\overline{w'\frac{\partial\theta'}{\partial z}},$$

where q is the amount of heat per unit mass of air, and a prime indicates the departure from a time average. The first term on the right-hand side of the equation describes the heat storage (i.e., the local heating); the second term gives the horizontal advection of heat (i.e., the horizontal heating); the third term gives the vertical advection of heat (i.e., the vertical heating); the fourth term gives the horizontal advection of heat due to eddies; and the fifth term gives the vertical heat advection due to eddies. The equation makes two assumptions: (a) that the local rate of change of pressure and its horizontal advection are small compared with the vertical advection of pressure; and (b) that $\dfrac{T}{\theta}$ is constant in time in the third term of the right-hand side of the equation.

This equation, by the thermodynamic energy equation method, in effect calculates the individual rate of heating of the air (per unit mass, averaged over a specific time interval), in terms of the rate of change of potential temperature with height, the horizontal gradient of temperature, and the vertical and horizontal windspeeds. The heat-balance method, described by the equation

$$\frac{d\bar{q}}{dt} = \frac{g}{\Delta p}\, (\overline{Lr} + \bar{R} + \bar{p}),$$

calculates the average heating per unit mass in an air column of unit cross section, in terms of the pressure difference between the top and the bottom of the column (Δp), the heating due to radiation to and from the Earth's surface (R), the turbulent heat exchange with the surface (p), and the net heat gain due to the condensation or evaporation of water vapor (Lr, where L is the latent heat of condensation or evaporation, and r is the amount of precipitation reaching the ground). This second equation assumes that: (a) net condensation and evaporation may be measured by the heat equivalent of precipitation reaching the ground; (b) the radiation effect is determined by the net radiative inflow at the top and at the bottom of the column; (c) transformations between kinetic energy and heat in the lower (friction) layer of the atmosphere, and any other sources of heating and cooling, may be neglected.

5.14. According to J. T. Tanner, *JAM*, 2 (1963), 473, the gradient (°F per 1,000 feet) of daily mean dry-bulb temperature in July varies with altitude from 2.5–3.2 in the Great Smoky Mountains, 3.0 in North Carolina, and 3.4 in the southern Appalachians, to 3.5 in central Europe.

5.15. Thom's method for the calculation of accumulated temperatures, in H. C. S. Thom, *MWR*, 82 (1954), 111, makes use of the expression

$$\bar{D} = N(b - \bar{\imath} + l\sqrt{N} \cdot \sigma_m),$$

where N is the number of days in the month, b is the base temperature, $\bar{\imath}$ is the average monthly mean temperature (°F), σ_m is the standard deviation of the monthly mean temperature, and l is a parameter determined empirically.

5.16. Following T. H. Kirk, *MOGM*, vol. 11, no. 90 (1953), the rate of change of sea-surface temperature, $\dfrac{\partial T_w}{\partial t}$, obtained by differentiation of the following harmonic representation of sea-surface temperature, T_w,

$$T_w = a_0 + a_1 \sin(t + A_1) + a_2 \sin(2t + A_2),$$

is given by

$$\frac{\partial T_w}{\partial t} = a_1 \cos(t + A_1) + 2a_2 \cos(2t + A_2),$$

where a_1 and a_2 are the amplitudes, and A_1 and A_2 the phase angles, of the annual and semiannual components, respectively.

6.1. Following WMO *International Cloud Atlas*, vol. I (Geneva, 1956), *meteors* are defined as phenomena other than cloud, observed in the atmosphere or on the solid surface of the Earth, which consist of a precipitation, suspension, or deposit of aqueous or nonaqueous liquid or solid particles, or of an optical or electrical manifestation. The following subdivisions are recognized.

Hydrometeors consist of an ensemble of liquid or solid water particles, falling through or suspended in the atmosphere. They include dew, drizzle, fog, rain, rime, and snow.

Lithometeors consist of an ensemble of mainly solid, nonaqueous particles, more or less suspended in the atmosphere or lifted by the wind from the ground. They include haze, dust haze, smoke, drifting or blowing dust or sand (including duststorms and sandstorms), and dust devils.

Photometeors are luminous phenomena produced by the reflection, refraction, diffraction, or interference of light from the Sun or Moon. They are found in clear air (mirages and shimmer), on or inside clouds (haloes, coronae, and glories), and on or inside certain hydrometeors or lithometeors (glories, rainbows, Bishop's rings).

Electrometeors are visible or audible manifestations of atmospheric electricity. They may be present as discontinuous electrical discharges (thunder and lightning), or as continuous phenomena (aurorae and St. Elmo's fire).

6.2. The nephohypsometer (on which see K. M. Laufer and L. W. Foskett, *J. Inst. Aeronaut. Sci.*, 8, 1941, 183) is really a photoelectric clinometer. With the ceilometer, a pulsating light beam is projected vertically from a mercury-arc source placed at a known distance from both photoelectric receiver and a recorder; the resulting light beam is nearly parallel. When it is aimed at a cloud, a diffusely reflected light from the lowest part of the cloud base is detected by the receiver, which scans the light beam once every six secs. A false reading can be caused by rain, but if the light spot on the (apparent) cloud base is viewed through a polarizing filter, one can determine whether it is due to rain or to clouds, since the light beam is partially polarized by reflection and refraction if it passes through raindrops. See D. L. Jorgénsen, *JM*, 10 (1953), 160.

The *rotating-beam ceilometer*, which was installed at the middle runway markers in almost 100 United States airports in 1959, is somewhat similar to the nephohypsometer. It uses an incandescent lamp, with shutter mounted at the focus of a 24-inch parabolic mirror rotating at five times a minute to project a beam obliquely on the cloud base. This beam is reflected vertically downward to a detector, which consists of a photocell at the focus of a second parabolic mirror. The detector signal is then amplified.

For a vertical beam, $h = l \tan \theta$, where h is the cloud height, l is the length of the baseline, and θ is the angle of elevation of the spot as measured from the other end of the baseline. If the beam is is not vertical,

$$h = \frac{l}{(\cot \theta + \cot \theta')},$$

where θ' is the angle of elevation of the beam as measured at the searchlight (see *MOHSI*, p. 375). Differentiating this equation, we get

$$\frac{\partial h}{\partial \theta} = \frac{l}{(\cot \theta + \cot \theta')^2 \sin^2 \theta} = \frac{h^2}{l^2 \sin \theta}.$$

Thus $\dfrac{\partial h}{\partial \theta}$ is minimum when $\theta = 90°$, i.e., when the spot is overhead. But for a vertical beam, $\theta' = 90°$, so θ cannot equal $90°$, and

$$\frac{\partial h}{\partial \theta} = \frac{h^2 + l^2}{l}.$$

Therefore the accuracy of measurement of h is limited by the accuracy of measurement of θ. The error Δh, caused by an error $\Delta \theta$ in measuring θ, is given by

$$\Delta h = l \sec^2 \theta \cdot \Delta \theta,$$

which is minimum when $\theta = 0$.

6.3. The velocity-height ratio is

$$v = \omega h.$$

where v is the speed of the cloud, h its height, and ω its angular velocity about a point vertically beneath the cloud, in radians per hour.

6.4. The velocity-height ratio is $\dfrac{a}{bt}$, where a is the distance between the spikes and b is the vertical distance from the top of the central spike to the observer's eye level. Usually, a is 6 inches and b is 60 inches; therefore the velocity-height ratio, multiplied by 3,600 (seconds in an hour) gives a speed of $360/t$ radians per hour, where t seconds is the time it takes a part of the cloud to move from one spike to the next.

6.5. The velocity-height ratio is $\dfrac{a}{bt}$, where a is the difference between the radii of the engraved circles on the instrument, b is the height of the pointer tip above the mirror, and t is the time required for the cloud image to travel distance a. The speed of the cloud equals $3,600 \dfrac{a}{bt}$ radians per hour.

6.6. A convenient and inexpensive cloud theodolite may be constructed by mounting an aircraft gunsight on an astrocompass; see P. B. MacCready, Jr., *BAMS*, 38 (1957), 460. The observer watches the movement of a small puff or irregularity in the cloud. The whole cloud is not observed, since it may be growing or even stationary, and therefore not moving with the wind. The observer notes (a) the time it takes for a point on the cloud to move from the center of the gunsight reticle to its outside edge; (b) the radical direction along which the observed point appears to move on the reticle; (c) the angle of elevation of the cloud; and (d) the theodolite azimuth. The cloud velocity may be found from

$$v = \frac{Rz \tan \phi}{rt \sin \theta}$$

where R is the radius of the reticle circle, r the length of the vector used in the direction calculation, z the cloud height above the observer, θ the angle of elevation of the cloud, t the time required for the cloud to move out of the reticle circle, and ϕ the angular radius of the reticle circle as seen by the observer ($2°$ in the original instrument).

6.7. For photogrammetry generally, see American Society of Photogrammetry, *Manual of Photogrammetry* (New York, 1952), and B. Hallert, *Photogrammetry* (New York, Toronto, and London, 1960). According to P. M. Saunders, *W*, 18 (1963), 8, sufficiently accurate measurements may be obtained from photographs taken with a conventional camera, without the use of special photogrammetric apparatus. At least two reference points must be established in the field of view, for each of which elevation and azimuth must be known. If enlarged photographs are employed, all dimensions measured from them must be multiplied by $\frac{1}{M}$, where M is the magnification. Elevation, θ, and azimuth, ϕ, of any point x, y, on the photograph may be found from

$$\tan \phi = \frac{x}{(f \sec \beta - y \sin \beta)}$$

and

$$\tan \theta = \frac{y \cos \phi \cos \beta}{f \sec \beta - y \sin \beta},$$

where f is the focal length of the camera lens, β is the angular elevation of the camera axis above the horizon plane, and the cloud is assumed to be at infinity. Coordinates x, y, are measured along OX and OY respectively, where O is the origin, OX equals $f \sec \beta \tan \phi$, and OY equals $f \tan \beta$.

6.8. In Figure A6.8, two cameras are directed parallel to a central azimuth, and

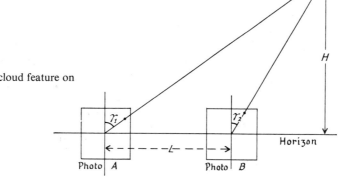

FIGURE A6.8.
The dot marks position of cloud feature on the photograph.

perpendicular to baseline L. The cloud height H is given by

$$H = \frac{L}{(\tan \gamma_1 - \tan \gamma_2)}.$$

The distance D of an object from the camera is given by

$$D = \frac{Hf}{(h \cos \theta)},$$

where h is the height of the cloud feature in photograph A above station A (in the same units as f), f is the focal length of the camera, and θ (equal to arc tan x/f, where x is the horizontal distance of this cloud feature from station A) is the horizontal angle of the cloud feature from station A.

6.9. To obtain a stereoscopic effect with movies taken from an aircraft at right angles to its line of flight, a rotating shutter or a polarizing chopper synchronized with the projector must be employed to ensure that one frame of the film is viewed with one eye and the succeeding frame with the other eye. If shots taken at 1 frame per sec are projected at more than 10 frames per sec, stereoscopic vision then results for clouds up to 5 miles distant if the aircraft is flying at 100 knots, or 15 miles at 300 knots. See W. S. von Arx, *JM*, 15 (1958), 230.

Movies taken at 10 frames per min from downward-pointing cameras suspended beneath Skyhook balloons have revealed that cumuli unfold and possess a radial outflow motion; see B. Vonnegut and B. M. Atkinson, *JM*, 15 (1958), 232. Neutral density filters were used to compensate for the greatly increased exposure times, and, since the balloons reached over 100,000 feet, precautions had to be taken to prevent the cameras from freezing during the night.

6.10. For example, the Japanese Widelux camera covers an angle of 140° horizontally and of 55° vertically. A 26-mm fixed-focus f2.8 lens is rotated by a ball-bearing mounting, so that a wide-angle photograph is produced by scanning the subject very rapidly. Such a camera must be oriented with respect to the horizon very carefully if photographs free of distortion are to be obtained. The basic problem with whole-sky photography is that of projection, both mathematical and optical. See W. S. von Arx, *Photogr. Eng.*, 4 (1953), 60.

6.11. Similar in principle to a planetarium, an *atmospherium* is a dome-shaped screen on which may be projected an image of the whole sky. The projection of time-lapse movie films enables a whole day's weather in any part of the world to be viewed in a few seconds. Climatic conditions requiring months or years to develop may be reviewed in minutes. For details of the atmospherium at the Desert Research Institute, University of Nevada, see *BAMS*, 42 (1961), 215.

6.12. See J. G. Sparrow *et al.*, *W*, 16 (1961), 29. As shown in Figure A6.12, provided $D > r$, where D is the distance from the lens to the center of the hemispheric mirror and r is the radius of the latter, then $\phi \approx 2\theta$, where $s = r \sin \theta$ and

$$\frac{2s^2}{r^2} \approx \frac{1 - h}{\sqrt{x^2 + h^2}},$$

where h is the height of the balloon and x is a distance on the Earth's surface (measured from a point vertically beneath the balloon) such that $x = h \tan \phi$, ϕ being the angle

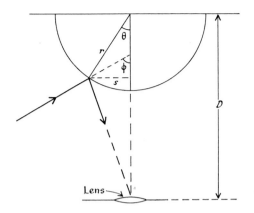

FIGURE A6.12.

subtended by x at the balloon. Looking horizontally, ϕ is 90° and s equals $\dfrac{r}{\sqrt{2}}$. A graph of x against $\dfrac{s}{r}$ then enables radial distances in the photo to be determined.

7.1. Average values of θ, in degrees, according to H. Neuberger, *CM*, p. 61, are: daytime, 21.0–29.0 in cloudy conditions, 22.5–34.0 in clear conditions; clear nights, 26.6–36.7 on moonlight nights, 30.0–34.0 in clear conditions; clear twilight, 32.2; 34.0 with no cloud; 31.5 with one- to three-tenths' cloud; 30.6 with four- to seven-tenths; 29.9 with ten-tenths' cloud.

7.2. According to H. G. Houghton in *HM*, the optical theory devised by H. Koschmieder, *Beitr. Phys. freien Atmosph.*, 12 (1924), 33 and 171, represents the brightness of a black object at distance D from an observer by the density of luminous flux produced on the retina of the observer's eye (see Figure A7.2). The object is viewed against the horizon sky through an atmosphere in which the scattering coefficient (i.e., the fraction of incident light scattered per unit distance along the beam) is k. The total apparent brightness, B_o, of the black object is given by

$$B_0 = cE_o \frac{\alpha}{d^2}(1 - e^{-kd}),$$

where E_0 is the density of luminous flux due to sunlight and skylight, assumed to be constant along the optical path (i.e., the sky is assumed to be perfectly clear or uniformly clouded), α is the area of the entrance aperture of the eye, and c, the fraction of the total light scattered toward the observer's eye per unit solid angle, is a function

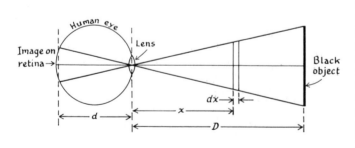

FIGURE A7.2.
Here dx represents the thickness of an elementary slice, through the cone, of cross sectional area $\sum$. The total quantity of light scattered by this elementary volume will be $kE_0 \sum dx$, where k is the scattering coefficient and E_0 is the incident radiation. Because light is lost by scattering between the elementary volume and the eye, only a fraction e^{-kx} of the light scattered by the elementary volume will reach the eye. The total scattered light flux from the elementary volume entering the eye is $cE_0 \sum \alpha(k/x^2)e^{-kx}dx$.

of the size of the scattering particles and the angular distribution of sunlight and skylight, and is assumed to be independent of x in Figure A7.2. The brightness of the background, B_b, is given by

$$B_b = cE_o \frac{\alpha}{d^2}.$$

In the absence of color contrast, the object will be visible as long as

$$\frac{(B_b - B_0)}{B_b} \geq \varepsilon,$$

where ε is the brightness contrast threshold of the eye.

Alternatively, if a black object is viewed at a distance l against an horizon sky of brightness B_h, then the apparent brightness of the object, B_a, is given by

$$B_a = B_h(1 - e^{-\sigma l}),$$

where σ is the sum of the scattering and absorbing coefficients of the atmosphere (i.e., the extinction coefficient). This assumes a uniform atmosphere, illumination by the sun, and a uniformly overcast or cloudless sky. For a non-black object

$$B_a = B_o e^{-\sigma l} + B_n(1 - e^{-\sigma l}),$$

where B_o is the intrinsic brightness of the object. In terms of contrast, if B_o and B_o' are the intrinsic luminances of two adjacent objects in the field of view, and B_a, B_a', their apparent luminances, then the contrast C between them is given by

$$C = C_0\left(\frac{B_0'}{B_a'}\right)e^{-\sigma \bar{L}},$$

where $\bar{L}$ is their mean distance away from the observer and C_0 is the intrinsic contrast (-1 for a black object). For objects seen against the horizon sky, this reduces to

$$C = C_o e^{-\sigma l}. \qquad [A7.1]$$

The visibility is equal to the distance D at which the brightness contrast equals the threshold value ε. D is equal to the standard visibility, defined by

$$V = \frac{1}{\sigma} \log_e \frac{1}{\varepsilon}, \qquad [A7.2]$$

in which ε is a constant value representing the threshold of contrast of the human eye; ε varies slightly from person to person, but equals from 0.01 to 0.04 for the average person's eye in daylight. Its value rapidly increases, however: (a) when the general illumination falls below a certain minimum value; (b) as the angle subtended at the eye by the object becomes less than $2°$ (due to the light diffracted by the edges of the object into the line of sight, and hence small objects such as chimneys and flagpoles are not good visibility standards); (c) if there is glare, i.e., a source of light such as the sun in the field of view. The value of ε doubles when visibility falls to 300 m, and quadruples when it falls to 50 m. A twofold increase in ε corresponds to a decrease in visibility of 15 per cent. The value of ε decreases if the object subtends an angle of more than $5°$ at the eye, and $\log_o \varepsilon$ decreases as the logarithm of the diameter (in minutes) of approximately circular objects increases, reaching a constant value of -2.5 when the log of the diameter is $+1$ or above. Hence ε is constant for large objects in full daylight, and it is usual to take it as equal to -0.02 for ordinary objects.

To derive an expression for the visual range, substitute $\varepsilon = 0.02$ and $c_o = -1$ in Equation A7.1. Equation A7.2 is valid for all weather conditions, but applies only to black objects. For details on the computation of the visual range in fog and low clouds, see D. R. Dickson and J. V. Hales, *JAM*, 2 (1963), 281.

The standard visibility, V, for a nonblack object is given by

$$V = \frac{1}{\sigma} \log_e \left[\frac{\left(1 - \frac{A}{2}\right)}{\varepsilon} \right],$$

where A is the *reflectivity* of the object. Thus the brightness of a nonblack object is regarded as being due to the scattered light in the optical path plus the brightness of the object itself. For a perfectly black object, $A = 0$; for a perfectly white object, $A = 1$. Thus the ratio of the visibility of a white object to the visibility of a black object (for uniformly clouded skies only), for the same ε, is 0.82. For a nonblack object with reflectivity A, and viewed against a reflecting background, the standard visibility is given by

$$V = \frac{1}{\sigma} \log_e \left[\frac{1}{\varepsilon} \left\{ \frac{(|A - A_b|)}{2} + \varepsilon - \frac{(\varepsilon A)}{2} \right\} \right],$$

where A_b is the reflectivity of the background.

According to D. Atlas, *JM*, 10 (1953), 486, the standard visibility may be obtained from records of precipitation intensity. The extinction coefficient is given by

$$\sigma = 2\pi \sum Nr^2,$$

where there are N spherical particles of radius r (large with respect to the wavelength of the light) per unit volume. For rain, $\sum Nr^2$ is quantitatively related to the rainfall intensity. The following relationships apply:

(a) for Bergeron-type precipitation,

$$\sigma \approx 0.25 R^{0.63} \quad \text{and} \quad V \approx 11.6 R^{-0.63};$$

(b) for warm orographic precipitation and drizzle,

$$\sigma \approx 1.2 R^{0.33} \quad \text{and} \quad V \approx 2.4 R^{-0.33}.$$

Here V is the visual range in km of objects seen against the horizon sky through an atmosphere with extinction coefficient σ, and R is the rainfall intensity in mm per hour. A value of 0.055 for the daylight threshold of contrast is assumed.

7.3. E. Gold, *QJRMS*, 65 (1939), 139. Suppose a small screen is interposed in the path of a light beam, so that the flux density of illumination is reduced from E to Ex, where $x < 1$. If n similar screens are used, the final flux density will be Ex^n. If x is chosen so that when there are 100 screens, the final flux density is one thousandth of its original value, a unit screen is said to have an *opacity* of *one nebule*, and 100 screens placed together have an opacity of 100 nebules; $x^{100} = (1{,}000)^{-1}$, therefore $x = 0.933$. Any length of the atmosphere acts in the same way as a number of screens, and hence may be characterized as having an opacity of a certain number of nebules.

7.4. According to Mie's theory, the coefficient b of attenuation of light by scattering in an atmosphere containing n water droplets each of radius r, per m^3, is given by

$$b = nk\pi r^2 \text{ per m,}$$

where k is the ratio of total amount of light scattered by a droplet to the total light incident on it. For large droplets, $k = 2$. For such droplets, absorption is negligible, hence

$$\sigma = nk\pi r^2 \text{ per m.}$$

7.5. According to Middleton, the contrast, C, is given by

$$C = \frac{(B - G)}{G},$$

where B is the brightness of the object and G is the brightness of the background. The main deficiency of the model is that for a black background, C is infinite, but the actual contrast is not. If B and G decrease simultaneously, as when a screen is interposed, the C remains constant but the actual contrast doer not. In fact, the threshold value of C (i.e., the value at which the object is just visible) is not constant, but increases as brightness decreases. For a small object, the threshold value may increase a thousand-fold as brightness diminishes.

Gold's definition of contrast is

$$C = \frac{(G - B)}{(G + g)},$$

where $g = 2 \times 10^{-2}$, and G is always the greater of the brightnesses of object or background.

7.6. According to E. Allard, *Mémoire sur l'intensité de la portée des phares* (Paris, 1876),

$$E_t = \frac{(IT^{R/b})}{R^2},$$

where E_t is the *visual illuminance threshold* (defined as the minimum density of luminous flux at the eye of the observer that will enable him to detect a light source); T is defined by $T = \tau^b$ or by $\tau^b = T^{R/b}$, where τ is the *atmospheric transmissivity* measured over a path of known length b; $\tau = e^{-\sigma}$, where σ is the extinction coefficient; R is the visual path; I is the intensity of the light source, which is assumed to vary inversely at the distance R (i.e., I is not a collimated beam).

Transmissometers measure the transmissivity of the atmosphere along a horizontal path of known length x. The extinction coefficient is then given by $\sigma = -\log_e \frac{\tau}{x}$.

A light source of known intensity is directed at a photoelectric cell at a known distance from the light source. The instrument is expensive and must be used in a fixed orientation. Climatological visibility observations are defined in directional terms, hence use of a transmissometer to provide them may result in unrepresentative measurements.

Runway visual range is the distance along the runway that a landing pilot can expect to see the high-intensity runway lights. *Approach-light contact height* is the height above ground at which a pilot making his final approach along the glide path can expect to establish visual contact with the approach lights. For details of a transmissometer used by British Meteorological Office stations, see G. J. W. Oddie, *W*, 23 (1968), 446. For details of runway visual range at British airports, see D. P. Smith, *MM*, 97 (1968), 51.

7.7. Allard's law may be rewritten

$$\tau_R = \frac{(E_t R^2)}{I},$$

where the atmospheric transmission along the slant-range R is given by

$$\tau_R = \left\{ \left(\frac{\Delta r}{\varepsilon} - r_1 \right) \frac{E_g}{L_h} + 1 \right\}^{-1}$$

in which Δr is the reflectance difference between two adjacent contrasting surfaces, r_1 is the reflectance of the brighter of the two surfaces, E_g is the general luminance of the terrain, L_h is the horizon luminance in the direction of vision, and R indicates slant visibility.

The effect of a complete cloud cover on R is given by

$$\log_e \tau_R = \sigma R_h(a - 1) + a\sigma R \log_e \left(1 - \frac{R_h}{R} \right),$$

in which R_h is the horizontal visibility and a is a local empirical constant. The equation assumes the applicability of a ceiling model, in which the cloud base is regarded as a sharply defined, impenetrable surface. This model is rather oversimplified, and a more useful result is obtained by regarding the ceiling as the upper limit of an atmosphere in which the attenuation of a light beam increases exponentially with height. The relation between τ_R and σ in other conditions is:

for smoke-haze, $\log_e \tau_R = -\sigma R$;

for snow, $\log_e \tau_R = -K\sigma R$,

where K is an empirical constant;

for radiation fog, $\log_e \tau_R = -\left(\frac{1}{\sigma R} + 0.123 \right)^{-1}$

7.8. For nighttime conditions, the standard visibility, V, is given by

$$V = \frac{1}{\sigma} \left(\log_e \left[\frac{I}{E_0} \right] - 2 \log_e V \right),$$

where I is the intensity of the light source (assumed to be a point source) in candlepower, and E_0 is the threshold value of the luminous flux density, depends on the background illumination, and is given empirically by

$$E_0 = 3.5 \times 10^{-9}\sqrt{B},$$

where B is the brightness of the background. For absolute darkness, $E_0 \approx 10^{-3}$ lumens per cm^2, and for full moonlight, $E_0 \approx 2 \times 10^{-11}$ lumens per cm^2, one *lumen* being defined as the luminous flux in one solid angle from a light source of one candlepower; E_0 varies more than does ε for different conditions.

If V_d is the standard visibility of a black object against the horizon sky in daylight, and V_n is the standard visibility of a light of candlepower I at night in an atmosphere with the same extinction coefficient, then

$$V_d = \frac{V_n \log \varepsilon}{(\log E_0 - \log I + \log V_n)},$$

where E_0 is defined as above. Climatological visibility at night (i.e., equivalent daytime visibility) is usually specified as that horizontal distance through an optically homogeneous atmosphere for which the transmittance of a light beam is 5 per cent.

7.9. For a point source of light, luminous intensity I, the diminution in illumination proceeds according to the inverse-square law, and

$$E = \frac{I}{d^2}\, e^{-\sigma d}.$$

For a parallel beam of light,

$$E = E_0\, e^{-\sigma d}.$$

In both cases, σ is assumed to be constant over the whole of the light path, E is the flux-density of the light, d is the optical path, and E_0 is the threshold value of the flux density at the eye.

7.10. The *visual threshold* is taken to be 0.15 km-candles, equivalent to 0.14×10^{-7} foot-candles. The sighting range of a lamp at night may undergo a 10-to-1 change in value, solely because of changes in the observer's visual adaptation. An instrumental system developed at the Scripps Institute of Oceanography at La Jolla, California, by J. H. Taylor and J. J. Rennilson, *JAM*, 1 (1962), 184, incorporates a calibrated viewbox to allow for this effect.

According to A. C. Best, *MM*, 88 (1959), 161, the relation between the candlepower C of a light, and the distance d at which it is just visible to the eye, is given by

$$\log_{10} C - 2 \log_{10} d = 0.03N + K,$$

where N nebules is the opacity of the imaginary optical screen between the light and the observer, and K depends on the visual acuity of the observer; K equals -6.9 for a "standard observer" who can just detect 0.15 km-candles. Assuming a value of 0.02 for ε, the standard visibility V in yards is given by $NV = 56d$. If two observers note the distance at which the light can be seen, obtaining estimates d_1 and d_2 $(d_1 > d_2)$ for distance d, and N_1 and N_2 represent the values of their respective optical screens, then

$$d_1 - d_2 = \frac{d_2}{160}(N_1 - N_2).$$

This is equivalent to $x = fy$, where f depends on d_2, and $y (= N_1 - N_2)$ has a normal frequency distribution with a standard deviation of 14 nebules. Therefore, for a fixed value of d_2, $(d_1 - d_2)$ has a normal frequency distribution with a standard deviation of $14f$.

7.11. Following *PM*, p. 84, the *telephotometer* principle is described by the relation

$$\varepsilon = \frac{(B_b - B_0)}{B_b},$$

where B_b is the brightness of the background, B_o is the brightness of the object under observation, and ε is the threshold of brightness contrast. The apparent brightness of the sky is compared with the apparent brightness of a dark target object; then if the background is the horizon sky, $B_b \to B_h$, and if the target is at distance $x_0 (x_0 < V_m$, where V_m is the visual range),

$$V_m = \frac{3.912 x_0}{\ln\left(\frac{1}{|C|}\right)},$$

where C is the apparent brightness contrast defined by

$$C = \frac{(B_0 - B_h)}{B_h};$$

C is measured for a target at a known distance x_0, and the instrument is calibrated in terms of the equation for V_m for a given value of V_m.

The *nephelometer* principle measures the coefficient of scattering, k_s, directly, by means of the relationship

$$V_m = \left(\frac{1}{k_s}\right) \ln\left(\frac{1}{\varepsilon}\right) = \frac{3.912}{k_s}.$$

The transmission coefficient of a light beam, τ, is measured over a known distance x_0 and substituted for k_s in the equation for V_w to give an equation of analogous form to

$$V_m = \frac{3.912 x_0}{\ln\left(\frac{1}{|C|}\right)}.$$

The *photometer* principle involves viewing a standard light of known intensity I_0 from a fixed distance x_0. The flux density, E_0, reaching the viewer is, in the absence of scattering, $E_0 = \frac{I_0}{x_0^2}$. A second standard light of known intensity, I_1, located distance x_1 from the viewer, is then viewed through the scattering atmosphere, giving flux density of illumination E_λ, defined by:

$$E_\lambda = \frac{I_1}{x_1^2 \, e^{k_s x_1}}.$$

A null method of comparison is used, so that $E_\lambda = E_0$, and therefore

$$k_s = \frac{1}{x_1} \ln\left(\frac{I_1}{I_0}\right) \left(\frac{x_0}{x_1}\right)^2.$$

This value for k_s is then substituted in the equation

$$D = \frac{1}{k_s} \ln \frac{C^*}{C_0} = \frac{V_m}{3.192} \ln \frac{C^*}{C_0},$$ [A7.3]

where C^*, the *intrinsic contrast*, is defined by

$$C^* = (B_0^* - B_b^*)B_b^*,$$

in which B_0^* is the intrinsic brightness of a target object at a specified wavelength. Intrinsic brightness is the total sky (or sun) light scattered toward the observer; i.e., intrinsic brightness is the light scattered from the sky by the particulate matter in the cone of air defined by the target and the observer's eye. C_0 is the threshold value of C, where

$$C = \frac{(B_0 - B_b)}{B_b}.$$

For the method of determination of D, the *visual range*, by means of Equation A7.3, see *PM*, p. 96.

7.12. According to *PM*, pp. 33–61, Rayleigh's theory shows that

$$k_s = \frac{32\pi^3}{3n\lambda^4}(m-1)^2$$

where m is the refractive index of the atmosphere, λ is the wavelength of the incident radiation, n is the number of molecules of air per cm^3 at sea level ($\approx 2.66 \times 10^{19}$ per cm^3), and k_s is the scattering coefficient. At any height in the atmosphere, therefore, $k_s \approx \lambda^{-4}$. Since blue light has a shorter wavelength than red light—$\lambda_{\text{blue}} \approx 425$ mμ and $\lambda_{\text{red}} \approx 650$ mμ—the ratio of $\frac{k_{s(\text{blue})}}{k_{s(\text{red})}}$ comes to 5.5. Therefore more blue light is scattered than red.

7.13. Following *PM* and K. Bullrich, *AGP*, 10 (1964), 99, Mie theory shows that Rayleigh scattering is merely the first term in the following expansion for all value of $\frac{a}{\lambda}$:

$$k_s = 24\pi_n^3 \left(\frac{m^2 - 1}{m^2 + 2}\right)^2 \frac{V^2}{\lambda^4} \left[1 + \frac{6\pi^4}{5}\left(\frac{m^2 - 2}{m^2 - 2}\right)\frac{a^4}{\lambda^4} + \cdots\right].$$

7.14. Following H. Neuberger, *CM*, p. 61, and *PM*, pp. 1–16, atmospheric *refractive index*, m, is defined by $m = cv$, where c is the speed of propagation of electromagnetic waves in a vacuum (2.998×10^8 m per sec), and v the speed of propagation in the atmosphere. For dry air, over a wide range of wavelengths of both visible light and microwave radiation,

$$(m - 1) \times 10^6 = \kappa\rho,$$

and for moist air,

$$(m - 1) \times 10^6 = A\frac{p}{T}\left(1 + \frac{Be}{Tp}\right),$$

where p and e are in mb, A, B, and κ are constants, and ρ is the density of dry air. For the visible spectrum, $\dfrac{B}{T} = -0.120$.

If the atmosphere is assumed to consist of a series of concentric, equidistant, isopycnic surfaces (see Figure A7.14,A), then a ray of extraterrestrial light penetrating the atmosphere will describe a curve given by

$$mr \sin Z = K,$$

where r is the distance of an isopycnic surface from the center of the Earth, Z is the angle of incidence of the ray, and K is constant. An observer at P would see an object O with an apparent zenith distance Z_A instead of true zenith distance Z. The *astronomical refraction* is then defined by

$$R_a = Z - Z_a.$$

For *terrestrial refraction*, defined as the angle α between the tangent PX and the straight line PO in Figure A7.14,B, which represents an observer at P viewing an object at O in the direction of incidence PX of the curved light ray (i.e., the object appears to the observer to lie along PX), there are two possibilities. Either the observer is located at P and views an object at O, giving the terrestrial refraction of α as stated; or the observer is located at O and views an object at P, in which case the terrestrial refraction is β. The *total terrestrial refraction*, t, is $\alpha + \beta$. When, as is normal, the light ray describes a circular arc, $\alpha = \beta$, and the path of the ray may be determined by its radius of curvature. Since the heights of P and O above the Earth's surface are very small compared with the radius of the Earth, r, and the angles t and ϕ are small,

$$\frac{t}{\phi} = \frac{r}{r_L}$$

and

$$\frac{t}{2} = \frac{\phi r}{2r_L}.$$

The *curvature* of the ray is defined as the reciprocal of r_L and is given by

$$\frac{1}{r_L} = (m-1)\frac{p}{T_v^2} \cdot \frac{273}{760}(\Gamma_c - \Gamma),$$

where p is in inches of mercury, T_v is the virtual temperature of the air through which the ray passes, and Γ_c is the *autoconvective lapse-rate* (i.e., the lapse-rate in an atmosphere in which density is constant with height). In general, therefore, the curvature of a light ray is proportional to atmospheric pressure and inversely proportional to the square of air temperature. If the actual vertical gradient of temperature, Γ, equals Γ_c, the ray will have no curvature. If $\Gamma > \Gamma_c$, the ray will curve upward. If $\Gamma \approx -114°C$ per km ($\equiv 6.3°F$ per 100 feet), the ray will have the same curvature as the surface of the Earth; and if there are larger inversions of temperature, the ray will bend back to the Earth's surface and may be reflected back into the atmosphere again.

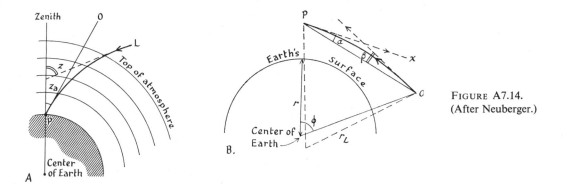

FIGURE A7.14.
(After Neuberger.)

7.15. In the Schlieren technique, a point source of light is placed near the center of curvature of a spherical mirror; all the rays from the source that reach the mirror will return to a point near the center of curvature of the mirror. If the refractive index of the medium is changed, for example, by using a small heater to generate a buoyant plume of convection in front of the mirror, then some of the light rays will be bent and will come to focus as a blurred point instead of a fine point on the mirror. A knife edge is mounted so as to intercept some of the rays at this blurred point, thus casting shadows on the plume and making it visible when observed from directly behind the knife edge.

7.16. The *Grashof number*, G, is defined by

$$G = \frac{g\beta \, \Delta T l^3}{v^2}$$

where β is the coefficient of expansion of the air $\left(\beta = \dfrac{1}{T}\right)$, ΔT is the temperature difference between the plume and its surrounding air in the lower part of the plume, l is a characteristic length parameter of the system, and v is the kinematic viscosity of the air.

7.17. According to *PM*, pp. 175–99, there are three rays at each interface in Figure A7.17, i.e., incident, reflected, and refracted rays, but to illustrate the principle involved, only the incident and refracted rays are needed, since the observer (at P) is able to detect only the refracted ray. If D is the angle of deviation, i.e., the total angle through which the ray is turned, then

$$D = 2(i - r) + 2n\left(\frac{\pi}{2} - r\right),$$

in which n is the number of internal reflections. Tracing the path of several rays will show that the refracted rays are concentrated about the ray with the least angle of deviation, usually termed the *Descartes ray*. From Snell's Law,

$$\sin i = m \sin r,$$

where m is the refractive index of the air. In addition,

$$\cos i = \sqrt{\frac{(m^2 - 1)}{(n^2 + 2n)}}$$

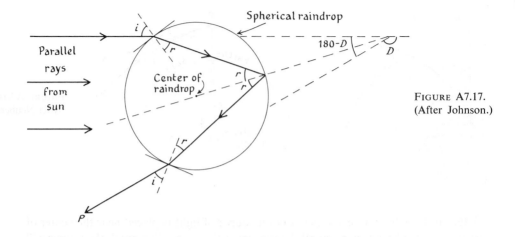

FIGURE A7.17.
(After Johnson.)

and

$$\cos r = \frac{(n + 1)}{m} \sqrt{\frac{(m^2 - 1)}{(n^2 + 2n)}}.$$

The equations for D, $\cos i$, and $\cos r$ enable the Descartes ray to be found if m and r are known. If $n = 1$ (as in Figure A7.17), the rainbow is a *primary rainbow*. For details of secondary $(n = 2)$ and tertiary $(n = 3)$ rainbows, see *PM*, p. 178. Since m is a function of λ, the wavelength of the light, drops at one point in the sky will register one color only at the eye of an observer at the ground. In general, relatively low drops in a rain sheet will produce blue, and relatively high drops will produce red.

7.18. The *primary bow* is the most intense rainbow visible at a given time, and consists of the colors of the spectrum, going from red on the outside of the bow, through yellow, to violet on the inside of the bow. *Supernumerary bows* are inside the primary rainbow. In general, the colors change from white (mist bows) for rainbows consisting of raindrops with a mean diameter of less than 0.05 mm, through various color combinations for drops of intermediate sizes (see *PM*, p. 184, for details), to almost pure red for drops 1 to 2 mm in diameter.

7.19. According to *PM*, pp. 185–89, the explanation of haloes assumes that the cloud producing them consists of ice needles of hexagonal cross section, every linear dimension of which is large compared with the wavelength of light. The crystals are assumed to orient themselves, when falling through the air, with their long axes parallel to the Earth's surface, and to spin on this axis; they would then have optical properties similar to those of a 60° prism (see Figure A7.19). The deviation D of a narrow pencil of light rays OP passing through a prism of angle P is given by

$$D = i_1 + i_2 - A.$$

For D to be minimum,

$$i_1 = \frac{1}{2}(D + P) \quad \text{and} \quad r = \frac{P}{2}.$$

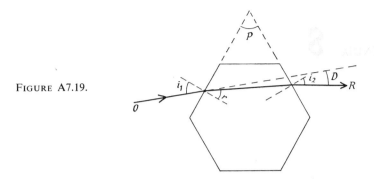

FIGURE A7.19.

Snell's Law is that

$$\sin \frac{1}{2}(D + P) = m \sin \left(\frac{P}{2}\right),$$

so that the Descartes ray may be found from P and m. For $P > 99° 28'$, no Descartes ray exists. Therefore only two haloes are commonly seen: the 22° halo produced by ice needles with a 60° prism angle, and the 46° halo produced by needles with a 90° prism angle. In general, the ice crystals must fall with their long axes vertical for the latter situation to obtain, and so 46° haloes are not reported as frequently as 22° haloes.

7.20. In general, the sine of the half-angle of a corona is inversely proportional to the diameter of the water drops or ice crystals producing it. For the detailed theory, see *PM*, pp. 192–97.

7.21. According to *PM*, p. 97, a *glory* is a complete, circular rainbow. If the mean diameter of the raindrops producing the bow is below 0.10 mm, the rainbow becomes a glory.

7.22. The purple light is currently thought to be caused by volcanic dust particles so located as to scatter light at a small angle. The sunlight beam is made progressively more red as it passes through the dust layer in the atmosphere. This red light is reinforced by blue light, which has been scattered downward from clearer air above the dust, and the eye interprets the combination of blue and red as purple.

APPENDIX **8**

8.1. The map is by H. P. Bailey, *GA*, 40 (1958), 1966. Dalton's law states that the rate of evaporation is proportional to $e_s - e$. Temperature is transformed to a term proportional to $e_s - e$ by assuming a humidity environment given by setting wet-bulb depression equal to 1.025^t where t is in °F. The moisture index (*ep*) is defined as $\frac{p}{1.025t}$, where p is the precipitation in inches, using monthly values. Summing the twelve monthly values of moisture index gives the annual index (*EP*). The *EP* values corresponding to moisture provinces are: $\geqslant 16.2$ for perhumid (*A*); 8.7–16.1 for humid (*B*); 4.7–8.6 for subhumid (*C*); 2.5–4.6 for semiarid (*D*); and <2.5 for arid (*E*). The Bailey method can be applied to negative temperatures (the Thornthwaite system cannot) so long as *EP* is less than the annual precipitation, and, unlike the Köppen and Thornthwaite systems, is suitable for analyzing polar and subpolar climates.

8.2. B. Rodhe, *GA*, 34 (1952), 175, gives the complete theory. A function of air temperature, the τ-function, which is a particular integral of the expression $\frac{d\tau}{dt} = k(T - \tau)$, is used, where T represents air temperature, τ is sea-surface temperature, t is time, and k represents a constant with dimensionality of $\frac{1}{\text{time}}$, which is calculated from the equation

$$\tau_n = \tau_{n-1} + (1 - e^{-k\,\Delta t})(T_n - \tau_{n-1}).$$

A graph may be constructed giving normal air temperatures in different months with the respective τ-curves for different values of k^{-1}. Ice forms first in small, sheltered bays, then in archipelagoes, and finally in the open sea. Thus small bays, archipelagoes, and open seas represent water systems whose heat content lags behind air temperature by differing amounts. The lag of τ behind T depends on k^{-1}, so that ice formation may be expected to be correlated with a specific value of τ in a particular time period. It may be shown that a relation exists between the date of sea-ice formation and the value of k^{-1} that corresponds to a τ-series that becomes 0°C at the same time. The theory may be simplified by employing a parameter z instead of k^{-1}, and a close relation exists between the value of z and the time of ice formation; z is obtained from the expression

$$k^{-1} = z + \Delta z, \; \Delta z \approx \frac{-\phi n}{z\,\dfrac{\partial\tau}{\partial z}}.$$

Here ϕ is the weighted mean of air temperature during the period t_0 to t_n, where

$$\phi = \frac{\sum\limits_{v=1}^{v=n} T_v \, e^{-k(t_n - t_v)}}{\sum\limits_{v=1}^{v=n} e^{-k(t_n - t_v)}}$$

in which n is the number of days it is required that the air temperature should remain below 0°C, v refers to any specific observation, t is time in days, and T_v is the mean air temperature for the interval from t_{v-1} to t_v.

8.3. According to B. E. Eriksson, *Geographica*, no. 34 (1958), 65, the glacier wind may be regarded as an air stratum placed between two planes represented by the glacier surface and the air layer immediately overlying the glacier wind. These two planes drag the air stratum with them, although at each plane it has zero velocity. Taking the x-axis as coincident with the lower plane and the z-axis vertically downward, i.e., perpendicular to the glacier surface, then the gravity component in the direction of the wind (see Figure A8.3) is given by $g_z \sin \beta$, where g_z is the downward force per unit volume of air. If ρ_g is the density of the cooled air over the glacier and ρ is the density of the air surrounding the glacier at the same level, then by Archimedes' principle, $g_z = g\rho_g - g\rho$ and, if α represents the coefficient of expansion of the air, then $g_z = g\rho_g \alpha \, \Delta T$, where ΔT is the depression in air temperature caused by the presence of the glacier. Considering a stratum of air of thickness dz parallel to the bounding planes, then the difference between the stresses on the two faces of this stratum is $K_m \rho_g \dfrac{d^2 u}{dz^2} \, \delta z$ per unit area. This force is balanced by the gravity component, so that:

$$-\rho_g K_m \frac{d^2 u}{dz^2} \, \delta z = g\rho_g \alpha \, \Delta T \sin \beta \, \delta z$$

or

$$\frac{d^2 u}{dz^2} = -k,$$

where k is a constant. Integration then gives

$$u = \frac{g\alpha \, \Delta T \sin \beta}{2K_m} \, z(h - z),$$

where h is the distance between the two planes and $u = 0$ at $z = 0$ and at $z = h$. Thus the velocity of the glacier wind may be estimated.

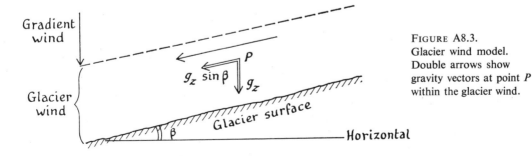

FIGURE A8.3.
Glacier wind model.
Double arrows show
gravity vectors at point P
within the glacier wind.

8.4. T. Kawamura, *TJC*, 2, no. 2 (1965), 38, says that the parameter is given by $\dfrac{1}{c_p\sqrt{\kappa}}$, where κ is the specific conductivity of heat for the ground. The value of the parameter varies from 34.2 for asphalt, 33.3 for concrete, 28.4 for roofing tiles, to 50.2 for soil, and 50.5 for sand, the units being cm^2 °C per cal per $\sqrt{sec}$.

8.5. According to R. L. Ives, *AAAG*, 39 (1949), 143, an analysis of the phase relations between temperature and rainfall in mathematical form enables, for example, different types of precipitation to be isolated from the curve of mean annual precipitation. Assuming a harmonic form for both temperature and precipitation, there are three possible cases: (a) if temperature and precipitation prove (after analysis) to be in phase, or nearly so, then a dominance of convectional rainfall is indicated; (b) if the temperature and rainfall curves are four or more months out of phase, frontal rainfall is likely to dominate; and (c) if the rainfall curve indicates a definite surge or pulse near the time of the temperature maximum, a monsoon tendency exists. Mixed regimes may also be studied in this manner. A simplified form of harmonic analysis may be employed in the preliminary analysis of the temperature and precipitation curves, of the form

$$P = A + B \sin \theta,$$

where P is the precipitation at any given time (i.e., during a short interval, such as one week), A is the precipitation for the year, B is the maximum rainfall experienced during the year, and θ represents the time elapsed since the precipitation curve crossed the line for A, expressed as an angle, i.e., days elapsed divided by 365.244, then multiplied by 360. Either mean or actual values may be analyzed in this fashion.

8.6. C. L. Godske, *GA*, 41 (1959), 85, advocates the following procedure. Assume the existence of b weather types in a given area, each of which contains a weather situations. The variability of any climatic element (minimum temperature, for example) represents a "total" variability that is measured by the sum of the squares of the deviations of the values from their arithmetic "total" mean. There are $ab - 1$ degrees of freedom. The "total" variability, u_1, is the sum of the variability between types u_2 and the variability within types u_3, provided that the conditions are such that u_2 and u_3 are statistically independent. The statistical pattern is given in Table A8.6.

TABLE A8.6
The Statistical Pattern

Variability	Sum of Squares	Degrees of Freedom	Mean Square
Between types	u_2	$b - 1$	$\dfrac{u_2}{b-1} = s_2$
Within types	u_3	$b(a - 1)$	$\dfrac{u_3}{b(a-1)} = s_3$
"Total"	u_1	$ab - 1$	F^*

$$* F = \frac{\left[\dfrac{u_2}{b-1}\right]}{\left[\dfrac{u_3}{b(a-1)}\right]}$$

If F is great, then we know, with a certain probability, that the weather types we have defined really do produce significantly different values of the variable in question. Thus a systematic search may be carried out for that classification of types which gives a maximum variability between, and a minimum variability within, types. The types should obviously prove to be the same, whatever the variable is, if they are real features of nature.

The Godske procedure is based on a mathematical model of the climatic variables which may be expressed as

$$V_v = f_v(g; s, a, d),$$

where V_v is the value of a climatic variable v that varies with geographical position g (including latitude, longitude, and height above mean sea level) and time, the latter having three components, s (the secular component, with a time unit of one year), a (the annual variation, with a time unit of one day), and d (the diurnal component, with a time unit of one minute or one second). The form of the function f_v will be different for each climatic variable, and must be found, by considering a fixed locality (i.e., so that $g = g_0$), and a specific date and time of day (i.e., $a = a_0$ and $d = d_0$). Therefore

$$V_v = f_v(g_0; s, a_0, d_0) = \phi_v(g_0; s, a_0 d_0) + \varepsilon_v(g_0; s, a, d_0),$$

where ϕ_v represents a regular secular variation and ε_v an irregular, erratic variation. The second term on the right may be replaced by the expression $f_i(g_0; s, a, d)$, which in turn may be represented by the sum of a regular annual-variation term, a regular diurnal-variation term, and an erratic term that forms a stochastic variable whose statistical parameters may depend on g_0, s, a, and d. Part of the variation exhibited by this last term will be a consequence of the existence of different weather types w, whose influence may differ for different values of y, s, a, and d. Hence we may write:

$$V_v = \phi_v(g_0; s, a_0, d_0; w_0) + \psi_v(g_0; s, a, d_0; w_0) + \chi_v(g_0; s, a, d; w_0)$$
$$+ \theta_v(g_0; s, a, d; w) + \varepsilon_v(g_0; s, a, d; w),$$

in which the term containing ψ_v represents the regular annual variation, that containing χ_v represents the regular diurnal variation, and the terms containing θ_v and ε_v represent the stochastic variable. In practice, the secular variation is ignored, so that the equation represents a stationary time-series with independent terms. Analysis of variance may then be applied in order to derive the statistical properties of the random terms for each locality, day, and hour, provided that a long enough set of observational data is available.

8.7. See R. A. Bryson, *JTG*, 17 (1963), 46. The starting point of Bryson's investigation was the equation of continuity expressed in the form

$$\frac{\partial D}{\partial t} + \frac{\partial (C + D)}{\partial r} = - \frac{C_r D}{r},$$

where C_r is the radial velocity of air in the subtropical anticyclone, r is the distance from the center of the anticyclone at which C_r is measured, and a circular typhoon is assumed, extending through a depth D of homogeneous atmosphere. From the equation of continuity he deduced

$$\frac{\partial^2 C_r}{\partial t^2} + f^2 C_r + \frac{\partial}{\partial t}\left(\frac{C^2}{r}\right) = g\left[\frac{\partial^2 (C_r D)}{\partial r^2} + \frac{1}{r}\frac{\partial (C_r D)}{\partial r} - \frac{C_r D}{r^2}\right],$$

yielding the solutions

$$C_r = A J_1\left(\frac{k}{\sqrt{gD_0}}\, r\right)\cos\sqrt{f^2 + k^2}\, t,$$

$$C_t = C_0 - \frac{fA}{\sqrt{f^2 + k^2}} \cdot J_1\left(\frac{k}{\sqrt{gD_0}}\, r\right)\sin\sqrt{f^2 + k^2}\, t,$$

in which k is a constant of integration; A is an empirical constant, J_1 is a Bessel function of the first kind; C_0 is a steady tangential velocity on which the perturbation is superimposed; C_t is the total tangential velocity; D_0 is the depth (constant) of the anticyclone in the local area of consideration, and f is the Coriolis parameter. The equations describe perturbations that must arise on the flank of a subtropical anti-cyclone, sufficiently distant from its center for the centripetal acceleration to be small. They indicate that the perturbation must be elliptical, the ellipticity depending on k, and the amplitude of the perturbation must be nil at the center of the anticyclone, increase to one main maximum on its flank, then decrease, ultimately becoming reversible at great distances from the anticyclone's center. The theory suggests that the most likely latitude for an incipient typhoon to form in an easterly wave is the latitude at which the local inertial period (i.e., the local period of free oscillation about geostrophic balance) is a multiple of the period of the easterly wave. The basic assumption is that easterly waves are internal waves that can only exist in a stratified medium, hence some of their energy will be transformed when they enter a region of latent instability. Empirical data indicate that easterly waves are periodic, with a period closely related to the half-pendulum day at the latitude of the subtropical anticyclone in which they originated. The theory indicates that when an internal (i.e., easterly) wave causes the latent instability of the air to be manifested, vortices should develop in the latitude where the local inertial period is a multiple of the easterly wave period. Synoptic data for the Pacific and Atlantic, and climatological data for the North Atlantic, appear to confirm the theory.

Abbreviations and Symbols

The following abbreviations are used in the bibliographical references in the notes at the end of each chapter and in the Appendixes.

AAAG.	*Annals of the Association of American Geographers.*
AFGP.	*Air Force Geophysics Research Papers.* (United States Air Force, Geophysics Research Directorate, Cambridge Center).
AFSG.	*Air Force Surveys in Geophysics.*
AG.	*Australian Geographer.*
AGP.	*Advances in Geophysics* (American).
AGS.	*Australian Geographical Studies.*
AM.	*Antarctic Meteorology* (New York, 1960).
AMGB.	*Archiv für Meteorologie, Geophysik, und Bioklimatologie.*
AMM	Meteorological Monographs of the American Meteorological Society.
ANYAS.	*Annals of the New York Academy of Science.*
AS.	*The Advancement of Science* (Journal of the British Association for the Advancement of Science).
ASM.	B. Bolin, ed., *The Atmosphere and the Sea in Motion* (New York, 1959).
BAMS.	*Bulletin of the American Meteorological Society.*
BWMO.	*Bulletin of the World Meteorological Organization.*
CD.	C. E. Anderson, ed., *Cumulus Dynamics* (New York, 1960).
CG.	*The Canadian Geographer.*
CM.	T. F. Malone, ed., *Compendium of Meteorology* (Boston, 1951).
CPRMS.	*Centenary Proceedings of the Royal Meteorological Society* (London, 1950).
CTP.	Technical Papers, Division of Meteorological Physics, Commonwealth Scientific and Industrial Research Organization (Australia).
DC.	R. L. Pfeffer, ed., *Dynamics of Climate* (New York, 1960).
DMWF.	C. L. Godske, T. Bergeron, J. Bjerknes, and R. C. Bundgaard, *Dynamic Meteorology and Weather Forecasting* (Washington, D.C., 1957).
EAFM.	H. H. Lettau and B. Davidson, eds., *Exploring the Atmosphere's First Mile*, vol. I (New York, 1953).
EDM.	A. H. Gordon, *Elements of Dynamic Meteorology* (London, 1962).
G.	*Geography* (Journal of the Geographical Association of the United Kingdom).

GA.	*Geografiska Annaler* (Swedish).
GC.	*Glaciers and Climate*, special volume of *Geografiska Annaler*, vol. 31 (1949).
GJ.	*Geographical Journal* (Royal Geographical Society, London).
GPA.	*Geofisica Pura y Applicata* (Italian).
GR.	*Geographical Review* (American Geographical Society).
GRTMU.	*Geographical Reports* (Tokyo Metropolitan University).
GS.	*Geographical Studies* (British).
HBES.	M. I. Budyko, *The Heat Balance of the Earth's Surface*, trans. N. Stepanova (Washington, D.C., 1958).
HM.	F. A. Berry, E. Bollay, and N. R. Beers, eds., *Handbook of Meteorology* (New York, 1944).
HMS.	*Harvard Meteorological Studies.*
HSM.	C. E. P. Brooks and N. Carruthers, *Handbook of Statistical Methods in Meteorology* (London, 1953).
ITM.	S. L. Hess, *Introduction to Theoretical Meteorology* (New York, 1959).
JAM.	*Journal of Applied Meteorology* (American Meteorological Society).
JAS.	*Journal of the Atmospheric Sciences* (American Meteorological Society).
JGR.	*Journal of Geophysical Research.*
JM.	*Journal of Meteorology* (American Meteorological Society).
JPC.	*Japanese Progress in Climatology.*
JTG.	*Journal of Tropical Geography* (Singapore).
LSGA.	Lund Studies in Geography, Series A (Swedish).
M.	O. G. Sutton, *Micrometeorology* (New York, 1953).
MAB.	*Meteorological Abstracts and Bibliography* (American Meteorological Society).
MAE.	*Meteorology and Atomic Energy* (Washington, D.C., 1955).
MC.	V. Conrad and L. W. Pollak, *Methods in Climatology* (Cambridge, Mass., 1950).
MGAB.	*Meteorological and Geoastrophysical Abstracts and Bibliography* (American Meteorological Society).
MJTG.	*Malayan Journal of Tropical Geography.*
MM.	*Meteorological Magazine* (Meteorological Office of the United Kingdom).
MO.	Meteorological Office of the United Kingdom.
MOGM.	*Meteorological Office Geophysical Memoirs.*
MOHSI.	*Meteorological Office Handbook of Meteorological Instruments, Part I: Surface Instruments* (1956).
MOHUI.	*Meteorological Office Handbook of Meteorological Instruments, Part II: Upper-Air Instruments* (1961).
MOOH.	*Meteorological Office Observer's Handbook* (1952).
MOPN.	*Meteorological Office Professional Notes.*
MOSP.	*Meteorological Office Scientific Papers.*
MWR.	*Monthly Weather Review* (United States Weather Bureau).
N.	*Nature* (British).
NZG.	*New Zealand Geographer.*
PAS.	*Polar Atmosphere Symposium, part I* (New York, 1958).

PC.	*Publications in Climatology* (Laboratory of Climatology, C. W. Thornthwaite Associates, New Jersey).
PDM.	D. Brunt, *Physical and Dynamical Meteorology* (Cambridge, Eng., 1941).
PIGU.	*Proceedings of the International Geographical Union.*
PISRSM.	H. Wexler and J. E. Caskey, Jr., eds., *Proceedings of the First International Symposium on Rocket and Satellite Meteorology* (Amsterdam, 1963).
PM.	J. C. Johnson, *Physical Meteorology* (New York, 1954).
PPOM.	Papers in Physical Oceanography and Meteorology (M.I.T. and Woods Hole Oceanographic Institute).
PSTM.	W. Hutchings, ed., *Proceedings of the Symposium on Tropical Meteorology, Rotorua* (Wellington, N.Z., 1964).
PTMC.	*Proceedings of the Toronto Meteorological Conference* (London, 1953).
QJRMS.	*Quarterly Journal of the Royal Meteorological Society.*
RSI.	*Review of Scientific Instruments* (British).
SAGJ.	*South African Geographical Journal.*
SG.	*Soviet Geography.*
SGM.	*Scottish Geographical Magazine.*
SMC.	*Smithsonian Miscellaneous Collections* (Smithsonian Institution, Washington, D.C.).
SP.	*Science Progress* (British).
SPIAM.	*Scientific Proceedings of the International Association of Meteorology* (Rome, 1954).
T.	*Tellus* (Swedish).
TAGU.	*Transactions of the American Geophysical Union.*
TIBG.	*Transactions of the Institute of British Geographers.*
TJC.	*Tokyo Journal of Climatology.*
TMA.	D. J. Bargman, ed., *Tropical Meteorology in Africa* (Nairobi, 1960).
UCPG.	University of California Publications in Geography.
UCSCM.	*UNESCO Canberra Symposium on Climatology and Meteorology* (Paris, 1958).
USCC.	*UNESCO Symposium on Changes in Climate* (Paris, 1963).
W.	*Weather* (Royal Meteorological Society).
WAF.	S. Petterssen, *Weather Analysis and Forecasting* (New York, 2nd ed., 1956), vols. I and II.
WW.	*Weatherwise* (American Meteorological Society).

In the Appendixes, symbols are employed as defined here, unless otherwise stated.

Quantities and parameters		*Basic units*
c.	Horizontal windspeed	cm sec^{-1}
C_d.	Drag coefficient	dimensionless
C_p.	Specific heat of air at constant pressure	cal gm^{-1} deg^{-1}
C_v.	Specific heat of air at constant volume	cal gm^{-1} deg^{-1}
e.	Vapor pressure	dyne cm^{-2}
e_s.	Saturation vapor pressure	dyne cm^{-2}

Quantities and parameters   *Basic units*

h.	Height above mean sea level	cm
H.	Eddy heat flux	erg cm^{-2} sec^{-1}
K_E.	Eddy viscosity	cm^2 sec^{-1}
K_H.	Eddy conductivity	cm^2 sec^{-1}
K_W.	Eddy diffusivity	cm^2 sec^{-1}
L_v.	Latent heat of vaporization of water	cal gm^{-1}
p.	Actual atmospheric pressure	dyne cm^{-2}
p_0.	Pressure reduced to mean sea level	dyne cm^{-2}
p_h.	Pressure at height h above mean sea level	dyne cm^{-2}
q.	Specific humidity	dimensionless
r.	Radius of the Earth	cm
R_n.	Net radiation	erg cm^{-2} sec^{-1}
Re.	Reynolds number	dimensionless
Ri.	Richardson number	dimensionless
Ro.	Rossby number	dimensionless
u.	Horizontal windspeed component in the west-east direction	cm sec^{-1}
u_*.	Friction velocity	cm sec^{-1}
v.	Horizontal windspeed component in the south-north direction	cm sec^{-1}
V_g.	Geostrophic windspeed	cm sec^{-1}
V_{gs}.	Surface geostrophic windspeed	cm sec^{-1}
w.	Windspeed component in the vertical direction	cm sec^{-1}
x.	Distance measured in the west-east direction	cm
y.	Distance measured in the south-north direction	cm
z.	Distance measured in the vertical direction.	cm
z_0.	Roughness length	cm
Γ.	Lapse-rate	deg cm^{-1}
Γ_a.	Adiabatic lapse-rate	deg cm^{-1}
Γ_d.	Dry-adiabatic lapse-rate	deg cm^{-1}
Γ_e.	Environmental lapse-rate	deg cm^{-1}
ω.	Angular velocity	radians sec^{-1}
Ω.	Angular velocity of the Earth at the equator	radians sec^{-1}
ρ.	Density	gm cm^{-3}
ε.	Emissivity	cal cm^{-1} min^{-1} $micron^{-1}$
v.	Kinematic viscosity, degrees of freedom	dimensionless
σ.	Stéfan-Boltzmann constant, standard deviation	cal cm^{-2} sec^{-1} deg^{-4}
κ.	Thermal conductivity	cal cm^{-1} sec^{-1} deg^{-1}
τ_0.	Surface tangential (shearing) stress	dynes cm^{-2}
θ.	Potential temperature	deg K
ϕ.	Latitude	degrees
λ.	Coriolis parameter	radians sec^{-1}
ψ.	Stream function	dimensionless

Further Reading

The references in the notes at the end of each chapter of *Techniques of Climatology* are intended (1) to indicate the sources of the information given in the text, (2) to guide the student to useful reviews of the literature, and (3) to indicate classic or basic papers on specific topics, many of which may interest the climatologist more than the meteorologist, because the climatologist may be concerned with analysis of data from what the meteorologist may consider an outdated era of atmospheric science. Most of the topics discussed in this book are matters of current interest, and for up-to-the-minute information (in the English language) about the state of knowledge, the reader should see the following journals and periodicals.

1. For reviews of progress in specific fields: *AGP, AMM, BAMS, BWMO, Reviews in Geophysics* (Washington, D.C.), *SP, TAGU, WMO Technical Notes* (Geneva, Switz.).

2. For reviews and original research: *Agricultural Meteorology* (Amsterdam), *Atmospheric Environment* (Oxford), *International Journal of Biometeorology* (Leiden, Holland), *MM, N, Science* (Washington, D.C.), *The Marine Observer* (London), *W, WW*.

3. For original research: *AMGB, Atmospheric and Oceanic Physics* (Izvestiya, Academy of Sciences, U.S.S.R.; English translation by the American Geophysical Union, Washington, D.C.), *Australian Meteorological Magazine* (Melbourne), *GA, Geophysical Magazine* (Tokyo), *Indian Journal of Meteorology and Geophysics* (Dehli), *Japanese Journal of Geophysics* (Tokyo), *JAM, JAS, Journal of the Meteorological Society of Japan* (Tokyo), *MM, MWR, QJRMS, T*.

5. For reviews of progress and occasional original research papers in geographical climatology: *AAAG, AGS, CG, G, GA, GJ, GR, JTG, LSGA, NZG, SAGJ, SG, TIBG*.

The reader will find additional information concerning various topics covered in this book in the following books.

ON SOLAR AND ATMOSPHERIC RADIATION

R. M. Goody, *Atmospheric Radiation*, vol. I (Oxford, 1964).
N. Robinson, ed., *Solar Radiation* (New York, 1966).

ON INSTRUMENTS

W. E. K. Middleton, *A History of the Barometer* (Baltimore, 1964).
W. E. K. Middleton, *A History of the Thermometer and Its Use in Meteorology* (Baltimore, 1966).

ON SATELLITE METEOROLOGY

R. K. Anderson, E. W. Ferguson, and V. J. Oliver, *The Use of Satellite Pictures in Weather Analysis and Forecasting*, *WMO Technical Notes*, no. 75 (1966).
W. K. Widger, Jr., *Meteorological Satellites* (New York, 1966).

ON CLOUDS AND PRECIPITATION

H. R. Byers, *Elements of Cloud Physics* (Chicago, 1965).
N. H. Fletcher, *The Physics of Rainclouds* (Cambridge, Eng., 1962).
W. E. K. Middleton, *A History of the Theories of Rain* (London, 1965).

ON NUMERICAL METHODS

P. D. Thompson, *Numerical Weather Analysis and Prediction* (New York, 1961).

ON GEOGRAPHICAL CLIMATOLOGY

J. Blüthgen, *Allgemeine Klimageographie* (Berlin, 2nd ed., 1966).
The only comprehensive account of world climates, the vastly out-of-date Köppen-Geiger *Handbuch der Klimatologie* (Berlin, 1930), is now being replaced by H. E. Landsberg, ed., *World Survey of Climatology* (Amsterdam, 12 volumes). The first two volumes of this work to be published are D. F. Rex, ed., *Climate of the Free Atmosphere* (vol. IV, 1969), and H. Arakawa, ed., *Climates of Northern and Eastern Asia* (vol. VIII, 1969).
See also I. Tsuchiya, *The Climate of Asia* (Tokyo, 1964), the first of a six-volume series on world climatography (in Japanese, but with guide in English).

ON APPLIED CLIMATOLOGY

Jen-hu Chang, *Climate and Agriculture* (Chicago, 1968); R. G. Fleagle, ed., *Weather Modification* (Seattle, 1969); J. L. Griffiths, *Applied Climatology* (Oxford, 1966); W. P. Lowry, *Weather and Life* (New York, 1969); W. J. Maunder, *The Value of the Weather* (New York, 1970); G. Y. Narovlyanskii, *Aviation Climatology* (Jerusalem,

1970); R. S. Scorer, *Air Pollution* (Oxford, 1968); L. P. Smith, *Seasonable Weather* (London, 1968), on phenology; J. A. Taylor, ed., *Weather Economics* (Oxford, 1970), especially pp. 84ff on the cost/loss utility ratio; UNESCO, *Agroclimatological Methods* (Geneva, 1968). *W.M.O. Technical Notes*, nos. 59 (windbreaks and shelter-belts), 69 (locusts), 106 and 114 (pollution), 108 (urban climates), and 109 (building climatology).

For data sources and data problems, see *Monthly Climatic Data for the World* and *Climatological Data, U.S.A. National Summary,* both published by the U.S. Department of Commerce (Ashville, N.C.); also W.M.O. *World Weather Watch* Planning Reports.

RECENT PUBLICATIONS

Significant topics covered in publications appearing while this book was in proofs include the following:

Books. B. R. Bean and E. J. Dutton, *Radio Meteorology* (London, 1968); R. Claiborne, *Climate, Man, and History* (New York, 1970); D. M. Ludlam, *The History of American Weather* (Boston, 1969); W. E. K. Middleton, *Invention of the Meteorological Instruments* (New York, 1969); R. K. Pilsbury, *Clouds and Weather* (London, 1969); C. S. Ramage, *Monsoon Meteorology* (New York, 1971); S. Teweles and J. Giraytys, *Meteorological Observations and Instruments* (Boston, 1970); R. A. R. Tricker, *Introduction to Meteorological Optics* (London, 1970); *World Survey of Climatology* (New York), vol. 8 on Asia (Arakawa, 1969), vol. 5 on Europe (Wallén, 1970), vol. 13 on Australia and New Zealand, (Gentilli, 1971). *W.M.O. Technical Notes*, nos 87 (polar meteorology), 100 (data processing), 103 (sea-surface temperatures), 111 (station networks), and 114 (radiation).

Journals. For satellite cloud climatology, see *JAM*, 8 (1969), 687 (Kornfield and Hasler); *MM*, 9 (1970), 86 (Holdsworth); and *T*, 23 (1971), 183 (Saha). For satellite radiation climatology, see *JAM*, 9 (1970), 215 (Raschke and Bandeen), and *JAS*, 28 (1971), Vonder Haar and Suomi). For satellite dynamic climatology, see *JAM* 9 (1970), 508 (Martin and Salomonson), *MM*, 100 (1971), 117 (Lloyd), and *TIBG*, no. 50 (1970) (Barrett). For satellite precipitation estimations, see *MWR*, 98 (1970), 321 (Barrett), and *W*, 26 (1971), 279 (Woodley and Sancho). On topoclimates, see *World Survey of Climatology*, vol. II, New York, 1970 (Geiger). On the use of Clerk Maxwell's physical interpretation of the Laplacian in mapping, see *MM*, 99 (1970), 151 and 294 (Kirk and Dixon). For a revival of Huntington's ideas concerning climate and history, see *GR*, 60 (1970), 347 (Chappell). On electric climate, see *W*, 25 (1970), 350 (Markson and Nelson), on the Andes glow, and *JAM*, 9 (1970), 194 (Pierce), on latitudinal variation of lightning parameters. On climatic change, see: *JAM*, 9 (1970), 960 (Sellers); *JGR*, 76 (1971), 4195 (Saltzman and Vernekar); *Science*, 171 (1971) (Rasool and Schneider); *SG*, 11 (1970), 849 (Borisov); and *T*, 21 (1969), 610 (Budyko).

See also: *AMGB*, 17, series B (1969), 325, for radiation balance of the southern hemisphere (Bridgeman); *AMGB*, 18, series B (1970), 253, for precipitation/temperature periodicities (Dehsara and Cehak); *Erdkunde*, 23 (1969), 127, for deficiencies of monthly means in studying droughts, in equatorial lowlands (Brunig); *GA*, series A, 51 (1969), 160, for radiation topoclimatology (Rouse and Wilson); *loc. cit.*, p. 176, for diurnal variation of precipitation in Sweden (Anderson); *GA*, series A, 52 (1970),

160, for nighttime energy/moisture balance in Death Valley (Terjung *et al.*); *JAM*, 8 (1969), 908, for a numerical model of the urban heat island (Myrup); *GR*, 60 (1970), 31, on energy balance of the city-man system (Terjung); *JAM*, 9 (1970), 373, for the floating drag-plate lysimeter (Goddard); *loc. cit.*, p. 379, on the pressure-sphere anemometer for winds within one meter of the ground (Thurtell); *JAM*, 10 (1971), 186, for the relation of variance spectra of temperature to the large-scale atmospheric circulation (Dickson); *MM*, 98 (1969), 201, for atmospheric-circulation/sea-temperature patterns (Murray and Racliffe); *MM*, 99 (1970), 93, on the tethered radiosonde (Painter); *MM*, 99 (1970), 261, on the Fast Fourier Transform (Rayment); *MOSP*, nos. 30 (1969) and 31 (1970), on polynomials (Dixon and Spackman); *MWR*, 98 (1970), 335, for multilevel radar precipitation maps (Marshall and Holtz); *loc. cit.*, p. 70, for a power-density spectrum of surface winds (Hwang); *QJRMS*, 95 (1969), 576, for eigenvectors of 500-mb height (Craddock and Flood); *QJRMS*, 96 (1970), 226 on long-range forecasting by sea-surface-temperature/air-pressure lag associations (Ratcliffe and Murray); *loc. cit.*, p. 14, for the nighttime microclimate temperature profile (Oke); *SG*, 11 (1970), 38, for climatic classification in relation to cold resistance of machines (Kolyago); *SG*, 12 (1971), 66, on photosynthetically active radiation (Yefimova); *W*, 25 (1970), 520, for the use of ships' log books in studying past climates (Oliver and Kington); *W*, 26 (1971), 55, for high-level stratocumulus (Bigg and Meade); *loc. cit.*, p. 150, on photometeors (Ripley and Saugier); *loc cit.*, p. 157, on multiple reflection in Arctic regions (Catchpole and Moodie); *loc. cit.*, p. 164, on black-bulb thermometer data (Lawrence).

Indexes

Name Index

This index includes (a) names mentioned in the main body of the text, and (b) names given in the notes (at the end of each chapter) as authorities for statements made in the book.

Geographical Index

References to figures are in boldface type. Except for large, famous cities, references to towns and cities are given under the name of the country or, for the United States and Canada, under the name of the state or province.

Subject Index

Only the most important discussions of each subject are indexed. Boldface numbers indicate Figures.